EDITORIAL RESEARCH REPORTS
ON THE

SCIENTIFIC
SOCIETY

Published by Congressional Quarterly, Inc.
1735 K Street, N.W.
Washington, D.C. 20006

About the Cover

This color photograph of the Orion nebula, an immense body of highly rarified gas in interstellar space, was provided by courtesy of the U.S. Naval Observatory in Washington, D.C. Editorial Research Reports art director Howard Chapman is responsible for the overall design and execution of the cover.

Published August 1971

Library of Congress Catalogue Card Number 71-168709
International Standard Book No. 0-87187-021-5

Editorial Research Reports
Editor Emeritus, Richard M. Boeckel
Editor, William B. Dickinson, Jr.
Associate Editor for Reports, Hoyt Gimlin

Contents

FOREWORD

WHEN the history of the 20th Century is written, it surely will contain several chapters on the scientific revolution. This has been a century of epochal technological and scientific achievements. The airplane, the atom, the computer and radio-television are only a few of those that come to mind. Biological and medical science achieved massive breakthroughs, and before 2000 A.D. we may find a way to prevent that most feared and unyielding of human afflictions, cancer.

Those who live in industrially advanced societies find their entire lives reshaped by the new technology that science inspires. Not all these changes are wholly benign. The airplane speeds movement but also contributes to noise and air pollution. The atom was tamed first for a weapon of war, not for peaceful purposes. The computer accepts orders to assemble dossiers on private citizens just as readily as it handles programs for the solution of complex mathematical problems. The television set may enlighten the viewer or it may be a device for dulling his sensibilities. And the prolongation of life provided by modern medical techniques can destroy the dignity of the individual kept alive long after life has lost its meaning.

People have a right to be concerned over the future role of science in a society which can land men on the moon but seemingly lacks answers to pressing human problems on earth. But science is not anti-life. If its accomplishments bring disaster, it will be because of the disparity between the rapid progress achieved in the natural sciences and the slower progress in the sphere of morals. Science cannot assure that the new world of options that it opens for man will be used wisely. What is required is the best social intelligence that man can muster. That is the challenge laid down by science to 20th century society. The reader of the Reports in this volume can decide for himself whether the challenge has been accepted.

William B. Dickinson, Jr.
Editor

Washington
August 1971

Human Engineering

by

Helen B. Shaffer

BIOLOGIC MANIPULATION OF HUMAN LIFE
Man's Rising Ability to Intervene in His Makeup
Ethical Problems Raised by Biologic Advances
Experiments Leading Toward 'Test Tube' Babies
Possibility of Genetic Selection Among Humans
Penetrating the Mystery of Aging; Body Repair

RESEARCH INTO MANAGEMENT OF THE MIND
Communication Between the Brain and Computer
Use of Electricity to Turn Emotions On and Off
Drugs for School Children to Help Them Learn
Transfer of Knowledge Through Chemical Action

FUTURE CONTROLS AND VISIONS OF NIRVANA
Apprehension About New Powers Over Inner Man
Statutes on Sterilization and Donation of Body
Limits on Man's Ultimate Control of His Destiny

**1971
May 19**

HUMAN ENGINEERING

MOST VISIONS of the future emphasize the tremendous changes to be wrought by advancing technology on the physical and social environment of the human species. Much speculation has centered on how man as we know him today will adjust to that push-button era of the future. Often overlooked, except in the fantasies of fiction writers, is the likelihood that science-based technology will be capable of changing not only the environment of man, but man himself.

The human animal has never been reluctant to doctor himself with a variety of potions and unguents to improve on nature and he has been clever about modifying his immediate environment to meet his needs and desires. But the extraordinary breakthroughs of recent years in the biologic sciences point to a totally new dimension of the power of human self-manipulation. They promise a capability of scientific intervention in the life of man that extends beyond cradle-to-grave. For the intervention can begin before the potential human being has even developed into an egg mature enough for fertilization. As for the grave, today's promise of future human-modification would prolong life by some 40 years—and those extra years, happily, in the very prime of life. Even "the ultimate in biologic engineering [the addition or exchange of genetic material in a potential fetus]...is not as far-fetched as it seemed a scant decade ago," according to a new massive survey of the biologic sciences prepared by panels of some 175 leading scientists for the National Academy of Sciences.[1]

In the years between birth and death, new possibilities are continually emerging for remodeling the individual physically and psychologically. Scientists are cautious and many warn not to exaggerate the potential application of current discoveries. The brave new world, for good or ill, will not arrive after breakfast. But it is impossible not to sense the undercurrent of excitement as new findings are reported, each adding another

[1]National Academy of Sciences, *Biology and the Future of Man* (1970, edited by Philip Handler), p. 239.

3

small piece of knowledge of the biological mechanisms that make man what he is today.

"A scientific revolution, led largely by biologists, promises to bring the very essence of nature, as well as alteration of human nature, under man's control," wrote Harvey Wheeler, Senior Fellow of the Center for the Study of Democratic Institutions.[2] "Biology is in a period of explosive growth that promises to revolutionize our concept of ourselves and to alter profoundly the conduct of our lives," Walter Eckhart of the Salk Institute for Biological Science in San Diego observed in a report on genetic modification of cells.[3]

Dr. Jose M.R. Delgado of the Yale University School of Medicine, a pioneer in studying the effects of electrical stimulation of the brain, believes the human race is at an evolutionary turning point. "We're very close to having the power to construct our own mental functions, through a knowledge of genetics...and through a knowledge of the cerebral mechanisms which underlie our behavior. The question is what kind of humans would we like, ideally, to construct?"[4]

The new view of man's potential power over his own biologic system suggests a new definition of human engineering to mean not merely the designing of objects to fit the human frame but action taken to modify the biologic condition of man. This new power will come not only from further discoveries but from the development of incredibly delicate and complex instrumentation. The computer figures importantly in opening new routes to modification of the human organism as does the more venerable surgeon's scalpel.

With new knowledge, new skills and new instruments, scientists are ready to penetrate what the National Academy of Sciences called the "only two truly major questions [that] remain enshrouded in a cloak of not quite fathomable mystery: (1) the origin of life...and (2) ...the physical basis for self-awareness and personality." The ultimate explanations may still be "only dimly perceived," it added, but "great strides have been made in the approaches to both of these problems."

---

[2] Harvey Wheeler, "The Rise of the Elders," *Saturday Review,* Dec. 5, 1970, p. 14.

[3] Walter Eckhart, "Genetic Modification of Cells by Viruses," *BioScience,* Feb. 15, 1971, p. 173.

[4] Quoted by Maggie Scarf in "Brain Researcher Jose Delgado Asks—'What Kind of Humans Would We Like to Construct'?" *The New York Times Magazine,* Nov. 15, 1970, p. 46.

Human Engineering

Though current explorations still represent only a few steps into what is still largely unknown ground, leading scientists believe the time has come to think about establishing public policies pertaining to the social consequences of possessing these extraordinary new powers. A warning to that effect recently was given before a congressional science group by Dr. James D. Watson, who won the Nobel prize in biochemistry in 1962 as co-discoverer of the structure of the heredity molecule, DNA (deoxyribonucleic acid).[5]

Ethical Problems Raised by Biologic Advances

In his remarks to the lawmakers, Watson predicted that within a year—other scientists say five years—it will be possible for a woman to give birth to a normal baby whose embryonic growth had been started by fertilization outside her body. "When that happens all hell will break loose," Watson said, meaning ethical issues will arise that will be profoundly disturbing to many persons. He urged the United States to take the lead to form an international commission that would explore such questions as: "Do we really want to do this?" Watson feared that "if we do not think about that matter now, the possibility of our having a free choice will soon be gone."

Sen. Walter F. Mondale (D Minn.) introduced on March 24, 1971, a resolution co-sponsored by 19 other senators calling for the establishment of a national commission to study legal, ethical and social issues posed by rapid advances in the biomedical field. Hearings were held in 1966 on a similar resolution he had sponsored then in the wake of the excitement over the transplantation of human organs.

Not long after Dr. Watson warned about the prospect of a test-tube baby, Dr. Landrum B. Shettles, an obstetrician at Columbia Presbyterian Hospital in New York, reported that he had fertilized an ovum—the female egg cell—in a laboratory and successfully implanted it in the woman from whom it had been removed.[6] The egg had been taken in an early (oöcyte) stage of development from its follicle in the woman's ovary *before* ovulation had occurred; that is, before it leaves the follicle and starts down the fallopian tube, a journey that results in fertilization if the egg unites with male sperm.

---

[5] Watson is author of *The Double Helix* (1968), in which he gives his account of the discovery he shared with the Englishman Francis Crick. Watson's remarks cited above were made Jan. 28, 1971, before the House Science and Astronautics Committee at the committee's annual meeting with its advisory panel of scientists. For further information on DNA and the Watson-Crick discovery, see "Genetics and the Life Process," *E.R.R.*, 1967 Vol. II, pp. 905-907.

[6] See *Science News*, March 6, 1971, p. 166.

Dr. Shettles kept the early-stage egg in its follicular fluid at body temperature. After a certain number of hours he placed it in a covered glass dish (petri dish) with cervical mucous from the woman's body; semen from her husband was added to the dish and the dish was placed in an incubator. Cellular division began, as in the case of normal conception. After five days, the ovum reached the blastocyst stage—the stage at which a fertilized egg in its natural environment will attach itself to the wall of the uterus and continue embryonic growth. The blastocyst was then placed in her uterus.

The experiment could not be carried out fully because, two days after the implantation, the woman had to undergo a hysterectomy, an operation for removal of the uterus. But it could be detected that in those two days the blastocyst had grown several hundred cells and appeared to be as normal as if the fertilization had occurred within her body. Dr. R.C. Edwards of Cambridge University in England, a leading figure in the field of reproductive biology, recently expressed his conviction that "eggs fertilized in the laboratory and cultivated to the blastocyst stage could be transferred back to the mother with an excellent chance of completing development normally."[7]

Experiments Leading Toward 'Test Tube' Baby

As early as 1934, Dr. Gregory Pincus, who later became one of the developers of the birth-control pill, published findings suggesting that he might have succeeded in fertilizing rabbit eggs. Six years later, Dr. John Rock, another developer of the pill, reported that he had fertilized the human egg and had witnessed initial cell division—the growth process. Their conclusions were challenged within the scientific community, however. One of their challengers, Dr. M.C. Chang of the Worcester Foundation for Experimental Biology in Shrewsbury, Mass., conducted extensive research and reported, in 1959, he had been able to fertilize a rabbit egg. Since then several other scientists have made similar reports on their work with small mammals.

As for humans, a physiologist at the University of Bologna in Italy, Dr. Daniele Petrucci, said in 1961 he had succeeded in fertilizing the human egg. Moreover, he said a heartbeat became discernible in the embryo but that after 29 days he

---

[7] R. C. Edwards and Ruth E. Fowler, "Human Embryos in the Laboratory," *Scientific American*, December 1970, p. 45. Edwards and Fowler are husband and wife. He is a physiologist and she is a geneticist.

had to destroy the embryo because it was deformed. His work was reported in the Italian press—and denounced by the Vatican—but he offered no photograph to support his findings. Nor did he publish them in a scientific journal.

Dr. Edwards and his research group have fertilized human eggs[8] since 1969, according to accounts published in the English journal *Nature,* but so far he has declined to implant the eggs in the womb for fear that the laboratory manipulation of eggs and sperm might have damaged the chromosomes and thus would cause birth defects. The benefit that Dr. Edwards has uppermost in mind at present is that of providing a means of motherhood for women who are infertile because the fallopian tube is blocked or non-existent. This is the reason that about one-fourth of all sterile women cannot conceive; the sperm cannot reach the egg.[9] The fallopian tube is bypassed when the egg is removed from the ovary, fertilized in the laboratory and then implanted in the womb.

There are efforts now being directed at devising an artificial womb which would, in theory at least, allow the embryo to develop into a child entirely in the laboratory. The main interest of science in this regard, however, is not in displacing the mother's role in gestation but in saving babies born prematurely. Building an environment that duplicates the natural environment of the fetus requires far more than is provided by today's incubators. These can sustain life in most infants born after seven months and some as young as six months but none expelled from the womb in an early stage of fetal development.[10]

Possibility of Genetic Selection Among Humans

Scientists in their tinkering with the genetic material of animal embryos have suggested what lies ahead for man. As long as a decade ago, the possibility arose in frog experiments conducted by Dr. John Gurdon at Oxford University that man

---

[8] The eggs are obtained in sizable numbers by a new surgical technique. As practiced by Dr. Patrick Steptoe, a surgeon in the Edwards research group, women volunteers are injected with a hormone, gonadotrophin, which causes them to "superovulate" several eggs. (Usually only one egg matures in each menstrual cycle.) The eggs are then removed by a slender hollow tube with which the surgeon has pierced the adomen and inserted in an ovary.

[9] Conversely, if the male is sterile but the woman's reproductive organs function normally, she can conceive by means of artificial insemination. It is estimated that 10,000 to 20,000 children are born annually in the United States of such conception. In many cases the sperm is provided by anonymous donors who have been screened by medical researchers for hereditary defects.

[10] For a description of the medical research involved in the new branch of medicine called "fetology," see Edward Grossman's article "The Obsolescent Mother" in *Atlantic,* May 1971, pp. 45-46.

might someday reproduce himself asexually ("clone") in his *identical* image. The English zoologist based his work on the understanding that every cell of an animal theoretically carries a load of genetic material unique to that individual. Taking an unfertilized egg cell from a frog, Gurdon destroyed its nucleus with radiation and replaced it with the nucleus of an intestinal cell taken from a tadpole of the same species. The egg, now with a full set of chromosomes instead of the half set found in unfertilized eggs, responded by beginning to divide as if it had been fertilized normally. There developed a tadpole that was genetically identical, not just similar, to the tadpole that had provided the transplanted nucleus.

Time magazine, recounting the event, said the implications raised such prospects as "a police force cloned from the cells of J. Edgar Hoover, [or] an invincible basketball team cloned from Lew Alcindor."[11] The notion of human cloning seemed far-fetched at first, James D. Watson has recalled. The human egg was still relatively inaccessible for laboratory manipulation and the technique of implantation was as yet untried. But now, he added, "if the matter proceeds in its current nondirected fashion, a human being born of clonal reproduction most likely will appear within the next twenty to fifty years, and even sooner if some nation should actively promote the venture."[12]

Other advances are in the offing, whether in the near or distant future. The National Academy of Sciences said in its survey on biology[13] that "surgical techniques, already in use in some primates, will enable the removal of the fetus from the uterus for the correction of defects and its return to complete full-term development." *Nature* recently stated that it is now "possible in principle" to remove genetically defective cells from a human patient, treat them by introducing new genes, and then replace the cured cells in the patient. While genetic engineering of this type was not exactly "imminent," the journal said, "these experiments offer the exciting prospect, in theory at any rate, of evolving a therapy" for defective human genes. Geneticists have determined that every person carries between five and ten flawed genes, which if not directly harmful to him hold potential harm for his children or their offspring.

---

[11] *Time*, April 19, 1971, p. 38. *Time* devoted a 15-page special section of that issue to "Man Into Superman." It was subtitled "The Promise and Peril of the New Genetics."

[12] James D. Watson, "Moving Toward the Clonal Man," *Atlantic*, May 1971, p. 52.

[13] See footnote 1.

Looking to the future, it may also be possible for potential parents to exercise a degree of selection in the genetic make-up of their children—including the choice of whether they will be boys or girls. The Edwards research team has noted that "choosing male or female blastocysts is one possibility that has already been achieved with rabbits."[14] If such selection becomes possible, then presumably genetic material could be examined in advance and rejected if it were found to be defective.

Penetrating the Mystery of Aging; Body Repair

Equally remarkable prospects for human engineering lie at the other end of the life span. The new probings of biologic science are beginning to penetrate the mysteries of why people grow old and die. These studies go beyond research on the killer diseases of post-middle age. The conquest of these alone might raise the average life expectancy of adults by five to 10 years. Interest here is directed toward the possibility of extending the normal life span by several decades as a result of measures taken to slow down the aging process itself.

Not everyone may agree with the writer who said "Death is an imposition on the human race, and no longer acceptable."[15] But the prospect of "at least the partial conquest of death" is not preposterous, writes Harvey Wheeler. "Until quite recently, the hard scientists—the molecular biologists—were of two minds about the possibility of elongating the human life span.... Today, however, even the 'conservatives' are persuaded that within 10 to 25 years man's internal clocks will be alterable. It will be possible to make them run longer, to reset them, even to turn them back."[16]

Several theories on the cause of aging are being explored. Much interest centers on what happens to the human cell as age advances. Dr. Helger Hyden of Sweden suggested at the Third International Conference on The Future of the Brain Sciences, held in New York in 1968, that DNA with a "high intrinsic orderliness" might be added to aging brains to overcome mental symptoms of aging. Other medical researchers are experimenting with the injection of enzymes on the theory that the lack of a certain enzyme may cause cell death.

---

[14] Edwards and Fowler, *op. cit.*, p. 53.
[15] Alan Harrington, *The Immortalist* (1969), p. 3. See also "Life Prolongation," *E.R.R.*, 1966 Vol. II, p. 502.
[16] Harvey Wheeler, "The Rise of the Elders," *Saturday Review*, Dec. 5, 1970. Wheeler cites the work of Alex Comfort, professor of biology at the University of London whose studies on aging have earned him the nickname "Mr. Gerontology."

Other advances in the life sciences have already given great powers, and promise still more extraordinary powers, for remodeling the physical structure of man and influencing his mental and emotional makeup. Surgery can now perform what would only a few years ago have been considered miracles of body repair.[17] Nature does in fact provide tissue replacement growth, as in the healing of wounds. But some scientists dare suggest the possibility of growing new limbs or organs as certain lower forms of life do. Failing such powers as yet, medical technology is moving ahead to create more and better artificial substitutes for vital parts or better devices to take over eroded vital functions. The heart pacemaker, which requires implantation of electrodes directly in the heart, for example, has become an acceptable device for control of heart blockage. Efforts to improve such devices—to make them smaller, safer, with fewer undesirable side effects—is continuous.

Other forms of surgery, sometimes known as "psychosurgery," have been directed specifically to obtaining psychic results. This is true, for example, of surgery to transform or give more specific definition to the sex of an individual. Surgery has been the recourse of the transsexual, described as someone who "has the feeling of 'being in reality a woman' whom nature by some cruel mistake has burdened and embarrassed with male genitalia." In some cases the psychic distress has been so great and the prognosis by psychotherapy so poor that surgery was employed to make the victim's body conform to his or her self-concept.[18]

Research into Management of the Mind

PREFRONTAL LOBOTOMY, a surgical operation separating the frontal brain lobes from the thalamus to relieve extreme anxiety, was enthusiastically praised when it came into use in the 1930s but then fell out of favor, partly because the opera-

---

[17] Transplantation of body parts, beginning with grafting of skin, then bones, teeth, corneas and now vital organs, represents a major advance of the 20th century in surgery. The first kidney transplant took place at Peter Bent Brigham Hospital, Boston, in 1954; the first heart transplant in South Africa on Dec. 3, 1967. See "Heart Surgery and Transplants," *E.R.R.*, 1968 Vol. I, pp. 385-404.

[18] "If such conversion surgery is not readily available...the transsexual may well attempt self-castration..., suicide, or fall into...psychosis....Normally...[surgery is] advised only after psychiatric and hormonal treatment of the patient has failed."—David W. Meyers, *The Human Body and the Law* (1970), pp. 48-49.

tion caused mental deterioration and partly because the introduction of tranquilizer drugs[19] in the early 1950s provided a less-drastic method of calming mental patients. While brain surgery as a form of psychological therapy is not extinct, far more interest in recent years has centered on applying electric current to the brain. Experiments in this area differ from the earlier electric-shock treatment, which has been used as an aid to psychotherapy. The newer forms of electrical brain stimulation provide a means of behavior and mood control that revive the classic nightmare of subjecting the mind of man to the machinations of the mad scientist.

The discovery that electricity applied to an animal's body could produce a muscular response dates back nearly 200 years to the experiments of the Italians Galvani and Volta.[20] That the brain could be electrically stimulated to cause body movements was demonstrated in 1870 on an anesthetized dog. In 1932 a Swiss neurophysiologist, the Nobel laureate W.R. Hess, hooked up an electric circuit to the brain of a cat and found that by releasing current to the diencephalon area, the cat reacted as if it were expecting to be attacked; its fur rose, its ears flattened, it growled and spat and raised its paw to swat.

Hess devised a procedure for implanting very fine steel wires in the brain of an anesthetized animal, leaving terminals available externally so he could send an electric charge at will to the points of contact in the brain. This permitted investigation of brain activity when the animal was free to move around at will. Before that the interior workings of the brain could be studied only during surgery while the subject was under anesthesia. Since the Hess experiments, thousands of animals in laboratories all over the world—and a number of human patients as well—have had their brains wired. As technology improved, it became possible to increase the points of contact. Some laboratory monkeys have had as many as 100 electrodes implanted in their brains for as long as four years.

There is no need to open the skull to do this. A tiny hole is drilled and micromanipulators guide the electrodes, which are teflon-coated wires of stainless steel, to their designated points in the brain. This procedure is said to be neither dan-

---

[19] See "Drugs and Mental Health," *E.R.R.*, 1957 Vol. II, pp. 859-874.
[20] Luigi Galvani (1737-98), a physician and anatomist, and Allessandro Volta (1745-1827), a physicist, whose names are today applied to such terms as galvanize and volt.

gerous nor damaging to brain function and the presence of the wires causes no apparent pain or discomfort. Some microelectrodes are only 25/10,000ths of an inch in diameter. They lead to an outlet box placed on the top of the skull.

Communication Between the Brain and Computer

A tiny instrument devised by Dr. Delgado at Yale has made possible two-way radio communication to and from the brain. The instrument, which he calls a stimoceiver, is a combined radio transmitter, receiver and electrical stimulator. Delgado's team of scientists succeeded in 1970 in establishing direct two-way radio communication between a chimpanzee's brain and a computer. In tests repeated a number of times, the following sequence occurred: (1) electrodes implanted in the animal's brain picked up electrical waves from one part of the brain; (2) the brain waves were then transmitted by the stimoceiver to a computer; (3) the computer on recognizing a certain brain-wave pattern produced a signal that was converted into an FM radio wave; (4) this radio signal was transmitted back to the animal where it was picked up by the stimoceiver which sent an impulse to another part of the brain; (5) that part of the brain responded by turning off the brain waves originally registered though the setup.

This was said to constitute a form of communication from one part of a brain to another, via the stimoceiver and the computer, with no sensory response to external stimuli having been involved at all. The specific effect on the behavior of the chimpanzee that played this historic role in brain-behavior research was a reduction of aggressiveness and excitability and a falling off of interest in food. This effect lasted for several weeks after the computer was disconnected.

Important implications for medical application are foreseen, especially in treating such conditions as multiple sclerosis, Parkinson's disease, anxiety, obsessions and conditions giving rise to uncontrollable violence. The possibility of helping the blind and the deaf by providing direct stimulation of specific areas of the brain may have been advanced. Even before he completed the chimpanzee-brain experiment, Dr. Delgado believed that "the miracle of giving light to the blind and sound to the deaf has been made possible by implantation of electrodes, demonstrating the technical possibility of circumventing damaged sensory receptors by direct electrical stimulation of the nervous system."[21] ESB (elec-

---

[21] Jose M.R. Delgado, *Physical Control of the Mind* (1969), p. 201.

trical stimulation of the brain) has already been widely used in treating epilepsy, involuntary movements, intractable pain and several mental disturbances. "Some patients have been allowed to stimulate their own brains repeatedly by means of portable stimulators," Delgado wrote. Where it has been found medically necessary to destroy brain tissue, a sufficient amount of current has been passed to the electrode contact; this is a safer and more controllable substitute for open-brain surgery.

Use of Electricity to Turn Emotions On and Off

More sensational even than medical uses of ESB are reports of instantaneous control of mood and behavior of test animals according to the will of the experimenter. It has long been known that electrical stimulation of the brain can induce physical responses. But the physical responses may be without emotional content. The hissing and growling produced by stimulating the anterior hypothalamus of a cat's brain is considered "false rage" while the cat's display of aggression produced by stimulating the lateral hypothalamus, is identified as "true rage."

Emotions can be turned on and turned off by ESB. Stimulation of the temporal lobe may induce fear. Violent behavior may follow stimulation of the amygdaloid. There are pleasure centers in the brain too. Test animals have learned to press levers that activate stimulation to these centers. Dr. Delgado wrote:

> Watching a rat or monkey stimulate its own brain is a fascinating spectacle. Usually each lever pressing triggers a brief 0.5-to-1.0-second brain stimulation which can be more rewarding than food. In a choice situation, hungry rats ran faster to reach the self-stimulation lever than to obtain pellets, and they persistently pressed this lever, ignoring food within easy reach. Rats have removed obstacles, run mazes and even crossed electrified floors to reach the lever that provided cerebral stimulation.

Delgado cited the cases of several epileptic patients who became pleasurably excited, more communicative and emotionally responsive to the therapist on receiving electrical stimulation at specific points of their temporal lobes.

ESB has been used to stop a normal activity instantly. It has immobilized a milk-lapping cat with its tongue out. It has turned a starving animal away from food and compelled a satiated animal to eat more. A doting mother monkey instantly lost all interest in her offspring, despite the baby's piteous cries. The "boss" of a monkey colony, who normally maintains his supreme position by fierce gestures and sounds, was so changed in manner by ESB that normally cowed subjects took over his turf without fear. "The old dream of an individual overpowering the strength of a dictator by remote control has been fulfilled, at least in our monkey colonies, by a combination of neurosurgery and electronics," Delgado reported. In this case, subject monkeys learned to press a lever that triggered the inhibition of the boss monkey's aggressive posture.

Perhaps the most theatrical of all experiments with ESB was the stopping of a "brave bull" in full charge. A brave bull is one bred for ferocity; normally he will attack anyone. But when Delgado pressed the right button to deliver stimulation by radio to the appropriate point in the bull's brain, the animal stopped short in the middle of a charge and walked away. Several stimulations caused lasting inhibition of his aggressiveness. There are limits to the game, however, as Delgado pointed out: "With the present state of the art, it is very unlikely that we could electrically direct an animal to carry out predetermined activities such as opening a gate....We can induce pleasure or punishment and therefore the motivation to press a lever, but we cannot control the sequence of movements necessary for this act in the absence of the animal's own desire to do so....We can evoke emotional states which may motivate an animal to attack another or to escape, but we cannot electrically synthetize [sic] the complex motor performance of these acts."[22]

Drugs for School Children to Help Them Learn

The use of drugs to influence states of mind goes back thousands of years. A number of drugs favored today were long available in natural form to primitive peoples. Biologic research has clearly demonstrated the important role played by various chemical substances in governing physiological processes, mental activity and emotional states. The body itself manufactures such substances and modern science can reproduce many of these in the laboratory.

Though excessive drug use was a familiar story to most Americans, nevertheless a shockwave swept the country when it was reported in the summer of 1970 that some doctors had been routinely dosing overactive children with amphetamines so they would behave better in school. Not only was this practice abhorrent to many persons, but amphetamines were

---

[22] *Ibid*, pp. 113-114.

widely known as pep pills—"speed" to the drug culture—and it seemed an extraordinary thing to give them to children who were already hyperactive. However, it has been learned that (1) the pills have apparently been serving a medically valid purpose for children suffering a specific neural handicap and (2) amphetamines that stimulate normal adults have the opposite effect on children with this handicap.

Use of the drug for this purpose grew out of studies of children with learning difficulties. For some children the trouble was found to lie in a malfunctioning of the nervous system. It sent so many messages that the brain was unable to sort them out. As a result the children became compulsively active, often obstreperous and unable to concentrate, a condition known as hyperkinesis. Amphetamines helped, apparently by blocking out some of the messages that cause the confusion. In the normal adult they block out the neural messages of fatigue and hunger; hence they are frequently used by night workers, truck drivers, students on the eve of exams and weight-watchers.[23]

A government-appointed panel of child psychiatrists and drug experts[24] issued a report on March 10, 1971, approving the use of the drug in true cases of hyperkinesis, but warned against its indiscriminate use on restless children with learning problems. It was necessary to make sure that the problem was not of some other origin; the panel mentioned specifically "hunger, poor teaching, overcrowded classrooms or lack of understanding by teachers and parents." The report provided a set of guidelines to prevent misuse of the drug.

Transfer of Knowledge Through Chemical Action

Another possibility of mind-management, the dream of every student, is the transmission of learning without the bother of studying. Student hopes of absorbing knowledge in their sleep from a recording are as yet unfulfilled. But experiments in a number of laboratories have indicated that certain kinds of learning may be transferred by chemical means. "Here is a facet of the biochemical revolution that is perhaps as important to man's future as the discovery of the DNA molecule of the gene," said science writer D.S. Halacy Jr. "Because it involves RNA and is in some ways similar to

---

[23] See Edward T. Ladd, "Pills for Classroom Peace," *Saturday Review,* Nov. 21, 1970, p. 66, and Mark A. Stewart, "Hyperactive Children," *Scientific American,* April 1970, p. 94.

[24] Appointed by the Office of Child Development in the Department of Health, Education and Welfare, and headed by Dr. Daniel X. Freedman, chairman of the Department of Psychiatry, University of Chicago.

genetic coding, memory transfer...is an exciting and frightening possibility for the future."[25]

A research team headed by Dr. Georges Ungar at Texas Medical Center in Houston apparently proved that a learned fear of the dark could be transferred from one rat to another by injection. First the scientists taught a rat to be afraid to enter a dark hole. They did this by the classic mode of animal conditioning: they gave the rat an electric shock whenever, true to his nature, he sought refuge in the hole. When his training was complete the rat was killed and chemicals were taken from his brain. The scientists isolated what they believed was the rat's "memory molecule for fear of the dark." They then decoded it and built their own molecule in a test tube. This man-made "memory molecule" was injected into the brain of an untrained rat who was then put into a compartment with a dark hole. The rat refused to enter it.[26] Ungar explained the phenomenon as resulting from a "mechanism of chemical recognition between neurons."

> Ungar [Halacy wrote] assumes that training synthesizes great amounts of chemical or molecular "connectors" and that this is the extract that binds itself to the proper sites in brains injected with extract....Because the brain probably manufactures the transfer factors only during the early period of intensive training, only this learning and not...long-established memory can...be transferred.

Excitement over the transfer of learning dates from the early 1960s when James V. McConnell at the University of Michigan fed cut-up pieces of "educated" flatworms to "uneducated" flatworms. The "education" consisted of conditioning the creatures to respond to light by contracting their bodies. Again this was done through the use of that favored conditioning instrument, the electric shock. The untrained flatworms were tested after their cannibalistic meal and found to have acquired the learned response by mere ingestion, or at least to have acquired a capacity to learn the lesson with uncommon speed.[27]

---

[25] D.S. Halacy Jr., *Man and Memory* (1970), p. 220. RNA (ribonucleic acid) is the substance that carries heredity messages from DNA (deoxyribonucleic acid) in the cell.

[26] Dr. Ungar told a *New York Times* interviewer on Feb. 13, 1971, that he had just completed a synthesis of a chemical that he believed was the rat brain's code message for fear of the dark and was seeking another chemical that tells the animal not to worry about loud bell ringing.

[27] The experiment gave rise to the joking suggestion that old professors be ground up and fed to college students. It also brought to mind the description in Jonathan Swift's *Gulliver's Travels* of a 'cephalic tincture" that, fed to schoolboys, would carry knowledge up from the gastric system to the brain.

Experimenters later injected the untrained flatworms with only the RNA from the trained one. The results were similar, suggesting that RNA and possibly a form of genetic coding was the source of the transfer factor. Others, however, question whether the process is true learning-transfer. Some suggest that the injected or ingested substance merely provides a temporary stimulant to learning ability.

Many persons would be happy, in the absence of a knowledge pill, to settle for an effective learning stimulant. The galloping pace of research on the brain offers some leads in that direction. In one line of animal experimentation, scientists are grafting one brain on top of another to produce a super-brain, giving the bearer superior learning powers. In another, rats are fed minute quantities of strychnine to improve learning ability. Other researchers are following up the discovery of Dr. Holger Hydèn of Sweden that an increase in RNA and consequent formation of certain proteins are involved in the process of storing and retrieving information in the brain. Research of this kind is moving ahead despite the pessimistic view that man can never fully comprehend his brain because his brain is the only mechanism he has with which to study it. The 10-billion-cell human brain with its associated nervous system has been described as "probably the most complicated piece of matter in the universe."[28]

Future Controls and Visions of Nirvana

ADVANCING CAPABILITIES in human engineering inspire visions of a golden future and nightmares of dehumanized horror. Despite all the benefits expected from the biologic revolution, a distinct uneasiness pervades discussions of what man might do with the extraordinary powers he is gaining over his own biology. Part of the uneasiness stems from fear of the misuse of biologic controls by tyrants or even by well-meaning bureaucrats. Delgado reassuringly advises that no mad scientist or evil dictator is likely ever to be able to force people to commit atrocities by means of remote radio control of their brain

---

[28] Lee Edson, "Science Probes a Last Frontier: the Brain," *Think* (publication of International Business Machines Corp.), November-December 1970, pp. 2-8.

waves. "The inherent limitations of ESB make realization of this fantasy very remote," he said.

Moreover, geneticists have sought to cool "exuberant, Promethean predictions of unlimited [genetic] control... [that] have led the public to expect the blueprinting of human personalities."[29] But scientists themselves are responsible for laying out the course of future man-modification and none can promise that the new knowledge will not be misused. Man's potential for changing himself "is doubly frightening," said Harvard zoologist Ernst Mayr, "when one thinks of the Nazi horrors perpetrated under the guise of improving man."[30]

An even deeper source of psychic discomfort on contemplating the great victories of biologic science is due to the threat they may seem to present to the individual's sense of his own human identity. Where is the real man when his body, his mind and his emotions can be changed by drugs, by surgery, by electricity and now by genetic manipulation? Perhaps most unsettling of all is the encouragement modern science has given to the concept of man as a mere biologic machine: turn a button here or there and he can be switched from happy to sad, from passionate, brilliant, loving to phlegmatic, dull, desexed. The new powers from biology add one more worry to those emanating from progress in other sciences. As Bernard Davis of the Harvard Medical School has asked: "With nuclear energy threatening global catastrophe and with so many other technological advances visibly damaging the quality of life, who would wish to have scientists tampering with man's inner nature?"

Such fears are not likely to stem advances in human engineering. Despite psychological deterrents, "there seems to be a general readiness for behavior technology," especially in its application to social problems.[31] Similarly there can be no turning away from genetics research for the reason that it promises so many medical benefits. The only hope for preventing misuse is to lay down guiding principles, possibly even legislation, in advance.

---

[29] Bernard Davis (Harvard Medical School professor), "Prospects for Genetic Intervention in Man," *Science*, Dec. 18, 1970, p. 1270.

[30] Ernst Mayr, "Biological Man and the Year 2000," *Daedalus* (journal of the American Academy of Arts and Sciences), Summer 1967, p. 833.

[31] Perry London, *Behavior Control* (1969), p. 8. Dr. London, a psychologist, referred to problems of overcoming anti-social attitudes related to crime, delinquency and school dropouts.

Some forms of human engineering that have been available to man for many years have evoked a body of laws and a set of social attitudes. Civilized nations, for example, no longer condone castration, which was practiced for thousands of years as punishment, as a means of controlling captives and slaves, and as a religious practice. David W. Meyers counted 26 states with eugenic sterilization laws. In 23 they are of a compulsory nature.[32] Most of the provisions apply to the severely mentally retarded, although some apply to the mentally ill, epileptics, and criminals. Two states, California and Washington, provide sterilization as punishment for carnal abuse of a child.

Statutes on Sterilization and Donation of Body

Sterilization laws date from early in this century when it was believed the tendency to commit anti-social acts was directly inheritable from parent to child. A total of 63,678 Americans, including hundreds of boys in reform schools, had been sterilized under laws by 1964, Meyers reported. The practice has fallen off considerably in recent years. A test case on a state sterilization law was dropped by the Supreme Court in 1970 because the state, Nebraska, repealed the law while the appeal was pending.[33] Only three states—Connecticut, Kansas and Utah—have laws forbidding voluntary sterilization except as a medical necessity.

Another area of human engineering where the state has intervened as regulator is in the use of the human body and its parts for medical and scientific purposes. Meyers counted 42 states that have statutes dealing with those uses of unclaimed bodies. The laws vary in their limits and in assignment of veto powers to relatives, a legal situation that has been characterized as one of "inadequacy, diversity and confusion."

In the years ahead there will doubtless be pressures for legislation to govern new forms of human engineering. Perhaps the most difficult fields of human engineering to regulate will be those affecting eugenic and population controls. The current debate on abortion laws[34] gives a foretaste of the inten-

---

[32] David W. Meyers, *The Human Body and the Law* (1970), p. 28. All statistics herein cited on sterilization matters have been compiled by Meyers.

[33] The law had given the state authority to sterilize "mental defectives" under certain conditions. It was challenged by the guardian of a woman who had been offered release from a mental institution if she would consent to sterilization. The woman was 37, unmarried, had an I.Q. of 71, and was the mother of eight children.

[34] See "Abortion Law Reform," *E.R.R.*, 1970, Vol. II, p. 543.

sity of the conflict to be expected over the ethical, moral, religious and social aspects of these questions. The extent of experiments on human beings is another area likely to come up for further regulation; advances in human engineering, while they often originate in animal research, inevitably require human guinea pigs.

Limits on Man's Ultimate Control of His Destiny

How far can man carry his quest for controls over his own biologic destiny? Faith that he will move ahead on it indefinitely is by no means universal. "The uniformity of nature and the general applicability of natural law set limits to knowledge," said biologist Bentley Glass in his presidential address before the American Association for the Advancement of Science in Chicago on Dec. 28, 1970. "...The laws of life, based on similarities, are finite in number and comprehensible to us in the main....There are still innumerable details to fill in, but the endless horizons no longer exist."

Gunther Stent, a geneticist and molecular biologist, believes the main goal of genetics, the understanding of the mechanism of transmitting hereditary information, has been "all but reached" while two major goals of biology, the origin of life and the mechanism of cellular differentiation, are certain to fall soon before "the host of biologists now standing ready to do battle and the armory of experimental hardware at its disposal." As for a third great mystery of biology—the mechanisms of consciousness—that is "beyond the realm of scientific research."

Stent envisions a future "golden age" of no-change, when the arts and sciences will no longer progress and life will attain a universally pleasurable but static quality. In place of "Faustian man"—man seeking power, as he has all through history—there will be "a golden race of mortal men who, thanks to technology, will live like gods, without sorrow of heart, remote and free from toil and grief, but with legs and arms never failing, beyond the reach of all evil." Today's "Faustian man" may not desire such a nirvana on earth, but the scientist warns that this is where we are heading if we continue our "frantic efforts" to improve on nature.[35]

---

[35] Gunther S. Stent, *The Coming of the Golden Age* (1969), pp. 111, 123-124. Stent describes his view of the future as "an impressionistic vision rather than an objective forecast."

Approaches to Death

by

Helen B. Shaffer

SOCIETY'S AWAKENING INTEREST IN DEATH
Disappearance of Taboos on Discussion of Death
Psychic Distress From Denial of Ultimate Fate
New Institutes, Seminars on Meaning of Dying
College Courses and Stirring of Popular Interest

CHANGED WAYS OF COPING WITH DEATH
Removal of the Dying From Home to Institution
Technology and Growth of the Funeral Business
Obsession With Death Theme in Earlier Times
Modern Child's Meager Experience With Death

ETHICAL PROBLEMS AND MAN'S ANXIETY
Moral Issues Involving Life-Sustaining Devices
Question of Defining Death in Medicine and Law
Demand for Psychological Support of the Dying
Man's Hope for Overcoming His Fear of Death

1 9 7 1
Apr. 21

APPROACHES TO DEATH

IN THIS OUTSPOKEN age, no aspect of human life seems barred from flagrant public exposure. Yet one subject of profound interest to every man—death—has remained in the shadows, shunted away from popular discourse, ignored until recently by social research. Modern man has tended to avoid pondering his ultimate fate. Only in its formal aspects has death come forward for public consideration, as in the rites for a departed public figure or in abstract terms such as casualty statistics.

Death is omnipresent—more than 5,000 Americans die every day—and death by violence is commonplace in the news and in popular entertainment. But the actual experience of dying and the experience of bereavement are private matters. There is little sharing of them, for the dying are hustled out of sight into institutions and the bereaved are expected to hold public expression of grief in check, then to put their grieving quickly behind them.

Everyone owes God a death, as Shakespeare put it,[1] but "why dwell on it until you have to?" appears to be the prevailing view. Good form requires that the curtain be drawn on the dying, the dead, and the bereaved to shield the rest of the community from these reminders of man's mortality. "Our society views dying as being in questionable taste," a man who knew he had a mortal illness wrote shortly before his death.[2] Even doctors and clergymen, whose professions have the most intimate and continuing concern with death and dying, have tended to avoid discussing the subject beyond what is required for carrying out their duties.

The taboos are beginning to fall away, however. A growing movement, led chiefly by sociologists and members of the health professions, seeks to bring the subject into the open for

---

[1] "A man can die but once; we owe God a death."—*Henry IV*, Part II, Act 3, Scene 2.
[2] Howard Luck Gossage, " 'Tell me, doctor, will I be active right up to the last?' " *The Atlantic*, September 1969, p. 55.

study and discussion. The object is not ghoulish. Promoters of this movement are convinced that life would be more fulfilling, dying less painful, and bereavement more endurable if people, individually and as a society, were more accepting of the reality of personal death. The movement's leaders are particularly critical of their own professions for having failed to give sufficient understanding and psychological support to the dying and the grief-stricken.

Psychic Distress From Denial of Death

That everyone must die is not an easy "fact of life" for egocentric man to accept. But facing up to man's ultimate fate is preferable to persisting in various subterfuges of denial. Much of the psychic distress of modern man is attributed to his unresolved fear of death, his inability to come to terms with his own mortality. Philosophers, poets, theologians, and scientists in the past have often asserted the unity of life and death. Modern psychiatry shows how repression of awareness of death impairs full expression of the life instinct. "The horror of death is the horror of dying with what Rilke called unlived lines in our bodies," writes psychoanalyst Norman O. Brown. "...The construction of a human consciousness strong enough to accept death is a task in which philosophy and psychoanalysis can join hands."[3] Or as philosopher-historian Arnold Toynbee put it:

> Man alone...has foreknowledge of his coming death...and, possessing this foreknowledge, has a chance, if he chooses to take it, of pondering over the strangeness of his destiny.... [He] has at least a possibility of coping with it, since he is endowed with the capacity to think about it in advance and...to face it and to deal with it in some way that is worthy of human dignity.[4]

The movement to bring discussion of death into the open is giving prominence to an unfamiliar word: thanatology—the study of death—from the Greek word for death, *thanatos.* Thanatology has been building up slowly over the past two decades, with momentum picking up in the past few years.[5] Its continuing growth may well result in the modification of

---

[3] Norman O. Brown, *Life Against Death: The Psychoanalytical Meaning of History* (First Wesleyan University paperback edition, 1970), pp. 108-9.

[4] Arnold Toynbee, *Man's Concern With Death* (1969), p. 63.

[5] "Death as an area of study of social scientists has only recently broken loose from its taboo status. Although fear of death has been recognized as a researchable topic since 1896 [an article on "Old Age and Death" having appeared that year in the *American Journal of Psychology*] relatively little study of this important issue has been conducted prior to the past two decades."—Andie L. Knutson (professor of behavioral sciences, University of California at Berkeley), "Cultural Beliefs on Life and Death," *The Dying Patient* (1970), p. 43.

institutionalized practices that have relegated the dying and the bereaved to a limbo of social neglect. Such changes will come when the interest now rising in certain professions— medicine, psychiatry, hospital management, sociology, nursing, the ministry—reaches out widely to the general public. This is likely to happen because members of these professions are usually present in authoritative roles at critical times in the lives of most individuals—during serious illness, when shocked by sudden death, or when undergoing the sorrow of bereavement.

New Institutes, Seminars on Meaning of Dying

A signpost of the trend was the scheduling of a symposium on "Problems in the Meaning of Death" as part of the program of the 1970 annual convention of the influential American Association for the Advancement of Science (AAAS). The session, held in Chicago, was sponsored by the Institute of Society, Ethics, and the Life Sciences, a foundation-supported study center established in 1969 at Hastings-on-Hudson, N.Y. The symposium constituted an open "work-in-progress" meeting of participants in a three-year research program of the institute on "Death and Dying." The scope of the study includes "the adequacy of present definitions of death, current medical practices in the care of the dying patient, professional and legislative codes pertaining to death, and present philosophical and theological understandings of the meaning of death."

The Equinox Institute in Boston also exemplifies a new type of organization created to encourage greater attention to human problems pertaining to death. It sponsored a series of six seminars, March 1-April 6, 1971, at Lemuel Shattuck Hospital in Boston for professional personnel and trainees in medicine, nursing, religion, and social work, on "Death, Dying and Bereavement." An unusual feature of this seminar, but one which is beginning to take hold in thanatology despite some professional resistance, is the conduct of interviews with patients and their families, a procedure by which those experiencing the imminence of death and bereavement, in effect, teach those whose professional function is to serve them.

Other seminars are planned for the fall of 1971 to train teachers and ministers in dealing with children and parents on family deaths. Still others, for non-professionals, will be "dedicated to open exploration of the meaning of aging and dying as an intellectual as well as a deeply personal emotional experience." A series of educational television programs on aging, dying and bereavement is in the planning stage. Ultimately the Equinox Institute hopes to become a center for research, consultation service, and information exchange in this growing field.

The Foundation of Thanatology is another new organization with similar objectives, founded by four members of the faculty of the College of Physicians and Surgeons of Columbia University. Its moving spirit is Dr. Austin H. Kutscher, professor of dentistry, who was led to thanatology by a searing experience of anxiety and grief after learning that his wife was fatally stricken with cancer. Two aspects of the experience were significant in leading to establishment of the foundation: (1) Dr. Kutscher's near-total lack of preparation for facing the tragedy that struck in the midst of a busy and satisfying professional and family life, and (2) the inadequacy of the help available to him during his wife's terminal illness and after her death in 1967 from the so-called "helping professions."

Among the many difficult questions Kutscher faced—as they are faced by many others in similar circumstances—and for which the foundation will seek answers are: Should he (or the physician or anyone else) have told his wife that her illness would be fatal? If so, when and how should she be told? When and how should he have told their children? How can one tell if a patient knows that his illness will be fatal but is afraid to let on to loved ones?

After the death, problems ranged from practical ones—arranging for the funeral, finding a caretaker for the children, deciding how to dispose of the deceased's clothing—to more difficult questions pertaining to the physical and mental state of the bereaved. What symptoms of grief may be considered normal? Which abnormal? What symptoms require medical care? How soon after the death should persistence of grief symptoms—depression, sense of alienation, loss of appetite—be considered unhealthy? How to deal with the inevitable sense of guilt in the grieving?[6]

The foundation's major production to date is a book, *Loss and Grief: Psychological Management in Medical Practice*

---

[6] Dr. Kutscher has described his experience before and after his wife's death in terms of a case study in a chapter, "Practical Aspects of Bereavement," in *Loss and Grief: Psychological Management in Medical Practice* (1970), pp. 283-290.

(1970), written by 24 contributors, which has been described as the first textbook on thanatology. The foundation has laid out an ambitious program for 1971 and 1972 to publish more books, hold seminars at medical schools and elsewhere, make surveys on education in medicine and other fields, and provide special services to the distressed through the establishment of such facilities as a "Bereavement Crisis Center" and a "Widow-to-Widow" program. It is also reaching out in various ways to other organizations likely to be sympathetic to its goals: the Child Welfare League, Parents Without Partners, the National Association of Funeral Directors, and the like. The Foundation also is publishing *Thanatology Archives*, a record of its activities, and the new *Journal of Thanatology.*

College Courses and Stirring of Popular Interest

Thanatologists complain that medical and other professional schools in the appropriate fields are deficient in preparing future members of the professions to deal realistically and sympathetically with the problems of the dying and the bereaved. Filling this deficiency is one of the movement's major objectives. Some response from the institutions is beginning to take place. Stanford, Chicago, Oregon and Pennsylvania State are among universities that now schedule seminars or courses on death. Western Kentucky University was an earlier recruit to thanatological interest. The entire first issue of its publication *Sociological Symposium* (fall 1968) was on the subject of "The Sociology of Death."

New York University added a course on "The Meaning of Death" to the spring 1971 schedule. Dr. James P. Carse, chairman of the history and literature of religion department, who will teach it, said "pressure to get into this course was tremendous." The total enrollment of 220 could have gone to 300 or 400 if a limit had not been set. Students will study the various approaches to death that have been taken by people in different cultures, then prepare an original personal statement on the meaning of death. This may take the form of a term paper, or a painting, a work of sculpture, a dance, a film, musical composition or any other form of expression.

Literature on contemporary problems in dealing with death and dying is accumulating with some rapidity. The major flow dates from the publication in 1959 of a symposium, *The Meaning of Death*, edited by a clinical psychologist, H. Feifel. Six years later another spur to concern on this subject appeared with publication of *Death and Identity*, a collection of reprints with comments by the editor, Robert Fulton, a sociologist who has written extensively on the way death is dealt with in contemporary society. *Death and Identity* contains a bibliography of 400 books and articles, nearly all from professional journals. The fields of sociology and psychiatry evinced the greatest interest in the subject.

Several recent books reflect a growing demand for more humane treatment of the patient in terminal illness. Among these are *The Patient As Person* (1970) by Paul Ramsey, professor of religion at Princeton University; *The Dying Patient* (1970), a symposium published by the Russell Sage Foundation; and *On Death and Dying* (1969), an account by Elisabeth Kubler-Ross, a psychiatrist, of her precedent-breaking program at the University of Chicago's Billings Hospital of holding seminars in which the center of interest was interviewing patients in terminal illness.

Most of the authors are, aside from their professional interests, primarily knowledge-seekers, reformists, and philosophers. They express a belief that the times—or the state of man's psyche at this particular moment of history—require a change in the way men and society approach the crisis of inevitable death. No consensus on the specifics of what is needed has developed yet, but most want the subject opened for exploration. Their biggest hurdle has been the reluctance of the average person—which includes many members of their professions—to deal with a subject he would rather forget. However, there are stirrings of popular interest. *Newsweek*, for example, devoted its cover story of April 6, 1970, to "How America Lives With Death"; *Psychology Today* featured articles on death in its issue of August 1970; and the *Honolulu Star-Bulletin* ran a series of nine articles on death between May 12 and May 22, 1970, which were then compiled and reprinted by the newspaper.

Changed Ways of Coping With Death

MAN HAS NOT CHANGED much over the millenia of recorded history in his aversion to death. "When we look back in time and study old cultures and people, we are impressed that death has always been distasteful to man and will probably always be," writes Elisabeth Kubler-Ross. "What has changed is our way of coping...with [it]."[7] Or, as sociologists Robert Fulton and Gilbert Geis put it, "the cultural context within which death is experienced in the United States and the institutionally sanctioned response to it" have undergone marked change.[8]

The most evident change has been the removal of the dying and the dead from the home to the institution, with relegation of attendant duties to paid personnel. Three-fourths of all deaths in the United States now take place in institutions, mostly in hospitals. The dead are then removed with dispatch to funeral homes. Reliance on the specialist, so marked in the life of modern urban man, thus extends to his death.

To point up the contrast with the past, Dr. Kubler-Ross has described from a memory of her childhood in Europe the death of a farmer who had been injured in a fall:

> He asked simply to die at home, a wish that was granted without questioning. He called his daughters into the bedroom and spoke with each one...alone for a few minutes. He arranged his affairs quietly, though he was in great pain.... He asked his friends to visit him once more, to bid goodbye.... [The children] were allowed to share in the preparations of the family...[and] to grieve with [it].... When he did die, he was left at home...among his friends and neighbors who went to take a last look.... In that country today there is still no make-believe slumber room, no embalming, no false makeup to pretend sleep.

Technology stands behind the shift from this traditional picture of a death in the family. The vast array of life-saving equipment in the modern hospital makes it a foregone conclusion that the seriously ill or injured will be taken there as soon as possible, often by siren-screaming ambulance to a busy emergency ward. Here the sophistication of the facilities is almost an exact measure of the impersonality of the care.

While few would forego hospitalization in a medical crisis, the patient and his family pay a psychological penalty. The patient must submit at a critical time to impersonal authority and the application of mechanical devices. He becomes an object of professional attention, bound over to the rules and procedures of the institution. His family must stand by helplessly, their services unneeded, their presence a nuisance to the professionals. Neither the patient nor family can make any of the pertinent decisions.

---

[7] Elisabeth Kubler-Ross, *On Death and Dying* (1969), pp. 2, 4.
[8] Robert Fulton and Gilbert Geis, "Death and Social Values," *Death and Identity*, (1965, edited by Robert Fulton), p. 67.

Death often comes while the patient is connected to monitoring machines or other gadgets but disconnected from his
human associations. The family waits numbly in a dingy waiting room. Or, if the waiting period becomes unbearably long,
family members may detach themselves emotionally from the
drama of dying. Cases are reported of even parents turning
away from a dying child as a self-defensive reaction. The
dying person thus becomes more isolated than ever. "In the
modern hospitals of today, beautiful and factory-like in efficiency," a psychiatrist wrote several years ago, "I have rarely
seen a patient die in peace."[9]

Noting the "increasing subjection of the process of dying
and death to rational, technical, specialized control," Dr.
William F. May of Indiana University has expressed doubt
that "these technological feats have been accompanied by a
corresponding spiritual confidence in our relationship to
dying." On the contrary, he thought they reflected an "inner
sense of bankruptcy" in the face of death. Use of modern
medical devices and reliance on professional experts were
obviously used to promote recovery of the sick, "but a latent
social function of this specialization is the avoidance of an
event with which we cannot cope."[10]

Technology and Growth of the Funeral Business

The growth of medical technology and the institutionaliza-
tion of the dying have coincided with the emergence of the
funeral business as a near-indispensable social institution.
Like medicine, the funeral business moved ahead on a technological track. The perfection of arterial embalming in the
early days of the 20th century was the key achievement,
according to William H. Porter Jr., a sociologist who prepared
a special study for the National Foundation for Funeral Services. By delaying decomposition of the body, embalming encouraged the elaboration of pre-burial services. In the early
years of the century, embalmers did their work on a "house
call" basis. As more deaths began to take place in hospitals,
families consented to the removal of the body to the funeral
home. Thus "a certain amount of control tend[ed] to pass from
the family of the deceased to the funeral director."[11] The role

---

[9] James A. Knight, *A Psychiatrist Looks at Religion and Health* (1964), p. 181.
[10] Paper presented at the annual meeting of the American Association for the Advancement of Science, Dec. 29, 1970, in Chicago.
[11] William H. Porter Jr., "Some Sociological Notes on a Century of Change in the
Funeral Business," *Sociological Symposium* (of Western Kentucky University), fall
1968, p. 37.

of the church diminished too with installation of the funeral chapel.

Other refinements of the mortician's art were soon added. Advances in plastic surgery and cosmetology permitted improvement of the appearance of the corpse. The ravages of age, illness and injury were softened and the death pallor was covered over. An effort was made to dissipate some of the lugubrious connotations of the undertaker's trade. The furnishings of the funeral parlor became less somber and cemeteries became landscaped parks. Some of the changes were due to the findings of motivational research and market analysis to which the funeral business turned. The primary purpose, according to Porter, was "preservation of sentiment, without which there would be no funeral business in the modern sense of the word."

The funeral business flourished, however, against a counterpoint of criticism and a vein of hostility, rooted in an aversion to the concept of making money out of death. Never a popular business, undertaking came under particularly sharp fire in such works as Jessica Mitford's *The American Way of Death* (1963). The business defended itself on the ground that it provided a needed service and aided in "grief psychology." Porter stated: "Advocates of grief psychology contend that the modern funeral, with its well-prepared body, beautiful casket and well-directed service can provide appropriate emotional outlets for the mourners."

Every culture has its peculiar death ceremonials. There is evidence of burial of the dead as far back as the Stone Age. "Burial rites in primitive society, and to some extent in higher types of culture too," wrote anthropologist Franz Borkenau, "are directed towards a twofold goal: to keep the dead alive and to keep them away. These incompatible aims reflect the basic contradiction in the human attitude toward mortality."[12] The question is often raised whether the modern funeral serves as well as death rites did in the past to satisfy basic psychological and spiritual needs. Some persons object that the funeral is a needless ordeal for the bereaved. Others complain that secular influences have de-spiritualized the funeral. And still others say it is a way of denying or glossing over the reality of death.

---

[12] "The Concept of Death," *Death and Identity* (1965, edited by Robert Fulton), p. 42.

A sociologist has described the dilemma of the funeral director caught between conflicting demands: "On the one hand he is encouraged to mitigate the reality of death for the survivors who may no longer receive the emotional support once provided by theology.... On the other hand, he...must focus attention on the body. In so doing...he invites the anger and hostility of a society that is experiencing a growing need to repress death."[13] The continuing services of the funeral director indicate that the American family still feels dependent on him when death occurs. The few studies of attitudes toward the funeral business that have been made indicate that most of the critics are of the white, upper middle class.

Obsession With Death Theme in Earlier Times

Suppression of thoughts of death in the daily life of modern man contrasts sharply with most previous ages, when the hazards of premature death were far greater than today. No time contrasts so sharply with the present as the late Middle Ages. Mendicant friars and woodcut artists, through their sermons and pictures, reduced the meditations on death of cloistered monks to vivid and simplistic images, then cast the brutal message out to the uneducated multitudes. Death was portrayed in its most grotesque and lurid forms, in terms of decaying flesh, feasting worms, the corruption of youth and beauty, and agonized suffering.

"Pious exhortations to think of death" vied with "profane exhortations to make the most of youth" in the popular art and literature of the period. The *danse macabre* theme emerged at this time—the dead (then Death itself) dancing with the living who were soon to join them. The only consolation in these presentations was the message that death was the equalizer of men. The poorest beggar could gloat that the mightiest of men would fare no better than he as food for worms. Apparently few fled from the gruesome reminders. "The medieval soul," historian Johan Huizinga wrote, was "fond of a religious shudder."[14] One reason was that medieval man lived with death at such close quarters at all times that he was familiar with the reality of physical decay. He may even have enjoyed a satisfying sense of temporary triumph over the eternal enemy.

---

[13] Robert Fulton, "The Sacred and the Secular: Attitudes of the American Public Toward Death, Funerals, and Funeral Directors," *Death and Identity, op. cit.*, p. 101.

[14] Johan Huizinga, *The Waning of the Middle Ages* (first published in 1924, Doubleday Anchor reprint), pp. 138-151.

Nor did the early settlers of the New World appear squeamish about the topic of death. The Puritan jurist Samuel Sewall (1652-1730) made endless entries in his three-volume *Diary* about deaths in his family, and among friends and acquaintances, and he often commented matter-of-factly on the state of body decomposition and similar aspects of death. American literature well into the present century is replete with the theme of death in the family, whether on the homestead or in the city.

Modern Child's Meager Experience With Death

Many people today grow up without ever having experienced the loss by death of a single person of significance in their lives. The sick tend to be put in institutions and the aged rarely live under the same roof with grandchildren. Diseases which once took off multitudes of children have largely been brought under control. A child's death was once a common experience. "Of all children who are born," Jean Jacques Rousseau wrote in 1762, "scarcely one-half reach adolescence." In America, as late as the end of the 19th century, one-half of the children born did not survive to age 40; today nine-tenths can expect to.[15] Many others were orphaned early. Parents as well as children were taken off by appendicitis, tuberculosis, smallpox, typhoid fever or any of a dozen other dread maladies; mothers often died in childbirth.

> Until very recently death struck almost at random.... More often than not its coming seemed both senseless and tragic, and most men lived under a constant shadow of fear for themselves and their children....Many men spent much of their time trying to understand, explain and render acceptable the fact.[16]

Children today do not grow up without fear of death, for the awakening of the realization of personal mortality is a universal experience that comes at an early age—usually well before the fifth birthday. But the infrequency of death and death rites at home have limited the child's experience. In addition, parents tend to shield children from reminders of death.[17] Parents who speak openly on sex, deploring the

---

[15] See "Infant Health," *E.R.R.*, 1970 Vol. II, pp. 901-918. The Rousseau quotation is from *Emile* (Everyman Library edition, 1963), p. 42.

[16] Robert S. Morison, preface to *The Dying Patient* (Russell Sage Foundation, 1970), pp. x-xi.

[17] The most influential child-raising manual for mothers in modern times advises: "Try to make the first explanation [of death] casual, not too scary. You might say, 'Everybody has to die someday. Most people die when they get very old and tired and weak and they don't want to stay alive any more'....[Or one could say] 'God took him to heaven to take care of him.' " Parents were urged not to convey a feeling of dread. "Remember to hug him and smile at him and remind him that you're going to be together for years and years."—Dr. Benjamin Spock, *Baby and Child Care* (1957), p. 365.

Victorian parents' reticence, avoid the mention of death or try to disguise its reality when the subject does come up.

"The Victorians may well have terrified children with their realistic descriptions of dying and death and with their details of God and of hell and heaven," a London pediatrician observed in commenting on a similar shielding of British children. But "we allow them to be terrified by our secrecy and by our private and often furtive misery."[18] The British anthropologist Geoffrey Gorer once observed that repression often masks an underlying preoccupation—it was so about sex in the Victorian era and about death in the 20th century.[19] But the reticence on death may now be going the way of reticence on sex. Some see in today's youth culture a longing for more connection with life's fundamentals, including an acceptance of life as mortal. Others see "a morbid preoccupation with death implicit in the youth cult...violent, youthful death as opposed to death from a natural old age."[20]

"Young people are kind of obsessed with death and execution," writer Kurt Vonnegut Jr. has observed. "They get hooked in a sort of a pornographic way, read all they can about electrocution, about the rack."[21] Movies cater to their obsession, film reviewer Richard Schickel has noted, with close-ups of gray-faced corpses and lingering camera studies of the way a body sprawls in its final moments of life.

Ethical Problems and Man's Anxiety

MAN'S NEW EFFORTS to shed the taboo on talking about death can be attributed to a number of factors. Some say his pervasive anxiety is leading him to seek solace for his underlying fear of personal death. Some think the interest in death portends a religious revival as he recognizes the emptiness of living on a purely corporeal plane. The threat of annihilation by atomic war or by some other global catastrophe has given the issue of human mortality an unprecedented and attention-

---

[18] Simon Yudkin, "Death and the Young," in *Death and Dying* (1968, by Arnold Toynbee and others), p. 52.
[19] Geoffrey Gorer, "The Pornography of Death," *Modern Writing* (1956, edited by W. Phillips and P. Rahv), pp. 56-62.
[20] Suzanne Lessard, "America's Time Traps: the Youth Cult, the Work Prison, the Emptiness of Age," *The Washington Monthly*, February 1971, p. 34.
[21] Quoted by Richard Schickel in *Life*, Feb. 26, 1971, p. 6.

demanding dimension. The persistence of the war in Southeast Asia, the growing number of Americans affected directly by its casualties,[22] and the vivid portrayal of war's atrocities in the media, including television, have all played a part in breaking through the psychological barriers to facing the reality of death. Moreover, the failure of medical science to stem the number of prime-of-life deaths from heart attacks and cancer may have shaken popular faith in modern medicine as the victorious warrior against the eternal enemy.

Medicine's successes rather than its failures have made one aspect of death a major topic of debate at this time. Various devices and procedures to sustain life, while a great boon to many, raise questions when used on terminally ill patients whose death at best can be deferred for only a short time and often at the expense of great pain or a vegetable-like existence. The questions are important in a society that views each human life as sacred and especially to a profession dedicated to saving lives. Thus moral and ethical considerations are involved in a clinical appraisal of when death actually occurs.

Few issues in medical ethics have stirred as much soul-searching within concerned professions as this one. The traditional view was that life ends when the heart stops beating. But that definition seemed inadequate when it became possible to keep the heart beating artificially after other vital organs had ceased to function. The view has been put forth that death is not necessarily a specific happening at a moment of time. Dr. Henry K. Beecher of Harvard Medical School has described dying as "a continuous process," adding that: "While death may occur at a discrete time, we are not able to pinpoint it (except in certain traumatic situations). We can only describe its presence when it occurs.... At whatever level we choose to call death, it is an arbitrary decision."[23]

Question of Defining Death in Medicine and Law

Support has been growing for equating death with cessation of brain activity. This has become a hot issue in medicine because of the development of organ transplantation and the need to keep the organ to be transplanted in a viable condi-

---

[22] A professor of psychiatry estimated that more than one million Americans have had a close family member killed or seriously wounded in Viet Nam—Dr. E. James Liberman, "Americans No Longer Know How to Mourn," *Potomac* (Sunday magazine of *The Washington Post*), Dec. 20, 1970, p. 7. According to official figures, 44,876 American servicemen in Southeast Asia had been killed in hostile action as of April 3, 1971, while 9,408 others had died in noncombat situations.

[23] Paper presented at the annual meeting of the American Association for the Advancement of Science, Dec. 29, 1970, in Chicago.

tion. The question is whether the potential donor of an organ is still alive when brain function has ceased but other vital functions are artificially maintained. In practice, this may become the question of when a doctor is ethically—and legally—correct in turning off a heart-lung machine, and how soon thereafter he may remove an organ for transplantation. "In those first frantic months [of human heart transplantation] in late 1967 and early 1968," a medical journal reported, heart-lung machines were turned off on such patients. Then "the attending physician and transplant teams waited while the last sign slowly ebbed away in the next 10 minutes and more...[and] irreparable damage was being done in these minutes to transplantable organs."[24]

An Ad Hoc Committee to Examine the Definition of Brain Death, at Harvard, proposed in 1968 a new definition of death based on irreversible coma. Dr. Beecher, who was committee chairman, told the AAAS meeting in Chicago that the new definition had been widely accepted by physicians but "often not by philosophers and theologians" who tended to fear it might lead to a disregard of the rights of the donor. And lawyers were said to be leery of its legality. However, the *New England Journal of Medicine* reported Feb. 4, 1971, that the committee recommendation was being followed world-wide.

The new definition of death, as explained by Dr. Beecher in Chicago, stipulates the following: "Deep unconsciousness with no response to external stimuli...no movements; no breathing (except artificially maintained); no reflexes (excepting occasional spinal reflexes), and a flat (isoelectric) electroencephalogram." The EEG was said not to be essential in diagnosing brain death but provided "useful confirmatory evidence."

Kansas in 1970 became the first state anywhere in the world to enact a statutory definition of death. It established alternative criteria for determining death: (1) "absence of spontaneous respiration and cardiac function" and failure of attempts to resuscitate, and (2) "absence of spontaneous brain function" if it appears that attempts at resuscitation or "supportive maintenance" will not succeed. The law follows the Ad Hoc Committee recommendation that "death is to be pronounced before artificial means of supporting respiratory and circulatory function are terminated and before any vital organ is removed for purposes of transplantation."

---

[24] William J. Curran, "Law-Medicine Notes," *The New England Journal of Medicine*, Feb. 4, 1971, pp. 260-261.

Meanwhile, the Euthanasia Educational Fund is distributing copies of a so-called "living will" in which the signer requests that no measures be taken to prolong his life if he should be stricken with an incurable and painful illness. The will does not ask for "mercy killing." But it requests that "drugs be mercifully administered to me for terminal suffering even if they hasten the moment of death." According to the Fund, which is based in New York, 20,000 persons had requested copies of the will as of March 1, 1970. A "right to die in dignity" bill, before the Florida legislature, would provide that a patient in terminal illness would have the right to ask that his life not be prolonged by artificial means.

Demand for Psychological Support for the Dying

One positive gain likely to result from cracking the taboo is greater consideration for the psychological needs of the dying patient. Studies indicate that fear of death is often more a fear of dying alone and in pain than of death itself. Modern medicine has done much to overcome physical pain, little to assuage the aloneness. "The sting of death is solitude," theologian Paul Ramsey said. "...Desertion is more choking than death, and more feared."[25] But many people do not know how to approach the dying. "The terminally ill patient has very special needs which can be fulfilled if we take the time to sit and listen and find out what they are," wrote Dr. Kubler-Ross. The most important thing is "to let him know that we are ready and willing to share some of his concerns."[26]

The growing literature on the dying patient cites many cases. Some terminally ill patients are cantankerous, some uncomplaining, some giving no outward signs that they know they are approaching death, but all suffering from an inner turmoil begging for release. The psychological neglect of the dying described in these reports is a compound of many things: the doctor's sense of failure, the hospital staff's tendency to favor patients likely to survive, the professional accent on saving lives and restoring physical health rather than comforting the dying, the fear of death and the guilt of the well toward the dying, and the anticipatory grief.

All of these things add up to a premature isolation of the patient from life and increase his psychic distress. The false cheeriness of visitors, the evasive replies of doctors and nurses,

---

[25] Paul Ramsey, *The Patient as Person* (1970), p. 134.
[26] Kubler-Ross, *op. cit.*, p. 240.

and the application of uncomfortable mechanical devices to his body only accentuate the afflicted person's sense of removal from the living before he is ready to accept that removal. "The dying patient is often defined as 'irresponsible,' " report two public health professors at Johns Hopkins University. "It is tragic that, with so terribly little time left, the very meaning of life—consciousness, self-control, decision-making —is taken away."[27]

A few psychiatrists have sought to break this pattern by offering themselves as therapists, counselors or simply as listeners to terminally ill patients. The doors to this kind of treatment have not been wide open. Many physicians object that their patients would be upset by such a visitation; some consider it cruel. Hospital personnel, trained to be brisk and matter-of-fact in the face of tragedy and usually averse to emotional scenes on the premises, are often resistant. Nevertheless there have been some breakthroughs.

The new concern for what is called "management of the dying" renews interest in the old question of whether a patient should be told he has a fatal illness. The few limited studies available indicate the great majority of doctors prefer not to tell most patients. Psychiatrists often do not recommend blurting out the truth either. But their experience indicates that most patients in such straits come to know how seriously ill they are without having to be told.[28]

Man's Hope for Overcoming His Fear of Death

How an individual accepts the news that his own death is imminent depends greatly on how he has handled his primitive fear of death, which exists in everyone.[29] Modern urban man is said to have a particularly hard time handling this fear; hence his efforts to repress it. Robert Jay Lifton, the author and psychiatrist, attributes this condition in part to the imprint that nuclear weapons make on the human mind. Lifton lists five concepts or "modes" of immortality that have served man: (1) religious belief in some form of afterlife; (2) living

---

[27] Sol Levine and Norman A. Scotch, "Dying as an Emerging Social Problem," *The Dying Patient* (Russell Sage Foundation, 1970), p. 218.

[28] Dr. Kubler-Ross found that among the 500 terminally ill patients of her experience, two-thirds of them had not been informed of the seriousness of their illness. "All of them were quite aware of the final stage of their life and most of them attempted on several occasions to convey this awareness to either their family or their physician, who often either denied the seriousness of their illness or bluntly minimized it, or changed the topic of conversation."—Paper presented at AAAS meeting, Dec. 29, 1970.

[29] "Fear, as Lucretius said, was the first mother of the gods. Fear, above all, of death."—Will Durant, *Our Oriental Heritage* (1935), p. 1.

on through one's descendants; (3) survival of man's creative works; (4) the continuing world of nature; and (5) "experiential transcendence [which] includes various forms of ecstasy and rapture...encountered in religious and secular mysticism...sometimes with the aid of drugs, starvation and other ordeals."

None of these modes of immortality can withstand the threat of the destruction of all life, not by an act of God, but by man himself. "The young have grown up with an inner knowledge that their world has always been capable of exterminating itself—and thus with an anxiety related not so much to death itself as to an underlying terror of premature death and unfulfilled life."[30]

The best deaths would seem to be the deaths of those who feel fulfilled in life. Literature and history present a number of inspiring models. Socrates spent his last days in contented discourse with his pupils. "There is nothing...more illustrious in the life of Socrates," wrote Montaigne 20 centuries after the Greek sage drank the hemlock, "than having had 30 whole days to ruminate his death sentence, having digested it all... without emotion...and with a tenor of actions and words rather lowered and relaxed than strained and exalted."[31] Catherine Drinker Bowen, the biographer, recalled the last words of old John Quincy Adams with admiration. "This is the last of earth!" he said. "I am content."[32]

Life and death—*eros* and *thanatos* as Freud called them— are interlocked. Philosophers say life would be less precious without knowledge of death. Thus, awareness of mortality may be necessary to zest in life. It may give drive to creativity. "Those who have lived with the knowledge of their finitude... have lived each moment as though it was their last.... A corollary of the *Todesangst* [death anxiety] of 'great men' has been their intense application of life. The knowledge that life is ephemeral has been a spur to their work and their humanity."[33]

---

[30] Lifton's views on death are expounded in *Boundaries* (1969), pp. 21-34, and in "The Politics of Immortality," *Psychology Today*, November 1970, pp. 70-73, 108-110.

[31] "Judging the Death of Others," *The Complete Essays of Montaigne* (Doubleday Anchor edition, 1960), Vol. III, p. 312. For himself, Montaigne preferred a "collected, calm and solitary" death where he would not be "wrung with pity to hear the laments" of friends nor wrung with "anger to hear...laments that are feigned."—"On Vanity," p. 213.

[32] "A biography is disappointing if it does not include the death scene.... Today's public still shows avid interest in a man's final words."—Catherine Drinker Bowen, *Biography: The Craft and the Calling* (1968), pp. 3, 34.

[33] David Cole Gordon, *Overcoming the Fear of Death* (1970), pp. 17, 19-20.

Hans Zinsser, the physician-author, wrote shortly before his death of the heightened appreciation of life that came to him after being informed that he had a fatal illness. Referring to himself in the third person, he wrote: "He found—to his own delighted astonishment—that his sensitiveness to the simplest experiences...was infinitely enhanced,....[his mind] more alive and vivid than ever before, [his] affections stronger."[34]

A similar statement was made by Abe Maslow, the psychologist-author, shortly before he died in 1970. Recalling his feelings after an earlier heart attack, he wrote of the satisfaction of having just completed what he considered his most important work. "I had really spent myself. This was the best I could do and here was a good time to die.... It was like a good ending [in a play]." Maslow keenly enjoyed what he called his "post-mortem life"—the time remaining to him after the heart attack. "If you're reconciled with death or even if you are pretty well assured that you will have a good death, a dignified one, then every single day is transformed because the pervasive undercurrent—the fear of death—is removed."[35]

Such equanimity toward the approach of death may be possible only to those who have lived—and lived fully—most of their allotted span. The really bad deaths are the deaths of the young, cut down by accident, by war,[36] by suicide, including the subconsciously motivated suicides of those who sank into self-destructive modes of living. These remain, as they have always been, the deaths that only an abiding faith in a divine plan can make acceptable.

---

[34] Hans Zinsser, *As I Remember Him* (1940), pp. 438-9.

[35] "Editorial," *Psychology Today*, August 1970, p. 16.

[36] A notable exception, perhaps, was the death of Army Maj. John Alexander Hottell, who at age 27 was killed July 7, 1970, in Viet Nam. About a year before his death he wrote his own obituary, which was posthumously published in the West Point alumni quarterly, *The Assembly*, and reprinted in *The New York Times*, March 3, 1971, where it attracted a flurry of letters to the editor. "The Army is my life, and is such a part of what I was that what happened is the logical outcome of the life I loved.... I have lived a full life in the Army, and it has exacted the price. It is only just."

VIRUS RESEARCH

by

Ralph C. Deans

1 9 7 0
Sept. 16

VIRUS RESEARCH

A S FAR AS SCIENCE has been able to determine, only snakes, yeasts, fungi, mollusks and cone-bearing evergreens are immune to viruses. All other forms of life, from germs to blue whales, are subject to viral attack. Man may be most vulnerable of all. It has been estimated that these unimaginably tiny entities account for 60 per cent of all human illness.[1] They can kill or maim on a massive scale—cause blindness, deafness, paralysis, heart trouble and mental defects. Moreover, medical researchers are coming to believe that viruses cause some forms of cancer in people and that virus research will sooner or later reveal a cure.

Virologists are now in the forefront of the fight against disease—a position once held by bacteriologists. The sulfonamides discovered in the 1930s and the antibiotics developed in the 1940s have vastly reduced the danger of bacterial infections.[2] It is viruses, which are largely resistant to chemical treatment, that now occupy center stage. Virology has become one of the "big" sciences of the 20th century. No other field of research holds more promise for the immediate benefit of mankind. The continuing study of viruses is sure to cast new light on other areas of scientific inquiry, including genetics, chemistry, evolution and the life process. Some scientists believe that viruses may have been the first form of life, created from lifeless chemicals through the action of radiation or electrical impulses in the primitive atmosphere.[3] Other scientists describe viruses as an "evolutionary slip-up"—a backward step. They argue that viruses could not have reproduced into higher forms because there were no cells for them to replicate within.

---

[1] Frank L. Horsfall Jr., *Viral and Rickettsial Infections of Man* (fourth edition, 1965), p. 1.

[2] Bacterial pneumonia, the world's greatest killer of man in 1900, has dropped to sixth place; tuberculosis, once the second worst killer, is now 13th; and inflammation of the stomach and bowels, once third, is now 19th.

[3] Simple amino acids—basic building blocks of life—have been created by bombarding an artificial atmosphere of water, methane, ammonia and carbon dioxide with electricity.

A new understanding of viruses is coming perhaps none too soon. Joshua Lederberg, the Nobel laureate biologist, writes that viral infections are "a global time bomb against which we have few defenses."[4] Lederberg is especially concerned that viral infections might be let loose upon the earth, accidentally or otherwise, by nations that maintain arsenals of chemical and biological weapons.[5] Even harmless viruses can mutate rapidly into deadly strains, as did the influenza virus which claimed 20 million lives throughout the world in 1918-19. Mutations occur faster than vaccines can be perfected to deal with them. With rare exception, the vaccine developed for one virus is ineffective against another.

New Understanding of Viruses; Their Definition

Exactly what constitutes a virus is a matter on which not all scientists agree. Viruses are classified as to their nucleic acid type, their size, structure, serological nature, behavior in cell structure, their susceptibility to various chemical agents, the hosts they will infect, and the mode of transmission. Size is one standard of identification on which there is basic agreement. To be a true virus, one of its sides must measure not more than 200 millimicrons[6] in length. Viruses are thus the smallest organisms known. They lie, in most cases, just beyond the resolving range of the most powerful optical microscopes. Under optimum conditions, the unaided human eye can perceive a speck eight-thousandths of an inch in diameter. Such a dot is 650 times bigger than the largest viruses and more than 13,000 times the size of the smallest viruses.

An electron microscope (see page 701) is almost always needed to see viruses. They display an amazing assortment of shapes. Viruses may appear rod-shaped, like the tobacco mosaic virus; sphere-shaped, as in the polio virus; icosahedron-shaped, a solid with 20 triangular faces, such as the insect virus *Tipula iridescent*; or tadpole-shaped, in the manner of many varieties of bacterial viruses or bacteriophages.

Viruses are also the simplest form of life, or borderland life, known to science. They are composed essentially of a nucleic acid core enclosed in a protein coat called a capsid. The core consists of either RNA (ribonucleic acid) or DNA

---

[4] In a newspaper column appearing in *The Washington Post*, March 7, 1970. Lederberg, a professor at the Stanford University School of Medicine, won the Nobel Prize in physiology and medicine in 1958 for his research in genetics.

[5] See "Chemical-Biological Weaponry," *E.R.R.*, 1969 Vol. I, pp. 453-470.

[6] A millimicron is one-thousandth of a micron; a micron is one-millionth of a meter (39.37 inches).

(deoxyribonucleic acid) but not both. DNA is the genetic material found in the nucleus of living cells; RNA is a closely related substance found generally in the cellular structure. These nucleic acids have been described as the master chemicals of life because they contain the genetic information that induces cells to reproduce exact duplicates of themselves.

Whereas bacteria actively attack cells and consume them, viruses act as subtle parasites, invading cells and subverting their normal processes. Textbooks describe viruses as "obligatory cellular parasites" because they must infect cells to survive. It is only within living cells that viruses can reproduce. Outside a cell they are apparently lifeless and inert, dependent upon molecular bombardment for movement. The so-called T2 phage,[7] which infects the coli bacterium found in the human intestine, provides an example of parasitic behavior.

Without any power of movement, the tadpole-shaped T2 comes into contact with the bacterial wall and "locks" into it chemically. "Tail" fibers of the T2 force an opening in the wall and squirt viral genetic material into the cellular cytoplasm. The virus then begins to control many cellular functions—substituting its own genetic pattern on the organism. Instead of manufacturing new cellular components, the cell's chemicals begin to replicate the viral DNA.[8] The virus also induces the construction of protein units which eventually combine with the DNA to form new viruses. About 30 minutes after infection, the cell bursts open, freeing up to 300 mature virus particles to infect the next bacterial cells they encounter.

Phages do not always kill their hosts immediately. The viral DNA may remain dormant for 40 or 50 generations of bacteria and only then destroy the cell. And not all viruses replicate in exactly the way the T2 does. Influenza and mumps viruses, for example, emerge slowly from the host cell and leave it intact. Viruses that cause tumors stimulate cells to grow abnormally, instead of killing them. Smallpox kills

---

[7] Phages—short for bacteriophages—are probably the most intensively studied of all viruses. They were discovered by an Englishman, F. W. Twort, in 1915 and by a Frenchman, Felix D. d'Herelle, independently, two years later. The discoveries created hope that bacteria could be killed with viruses. This line of research was unfruitful but it encouraged a great deal of virus research.

[8] Once inside the bacterial cell, the viral DNA induces the production of various enzymes which break up the genetic material in the bacterium's chromosomes. An enzyme is a form of protein which either breaks down or builds up other chemical substances. Enzymes act as catalysts, speeding up reactions between other chemical compounds without undergoing change themselves.

human cells in a small area, leaving pustules which may heal but cause scars.

Viruses display a bewildering range of characteristics. They not only have various shapes and sizes but different methods of infection. Usually a virus has an affinity for only one or two kinds of cells. The rabies virus is one of the few viruses that will produce the same fatal disease in a wide variety of animals, including man. "The riddles and the mysterious paradoxes that make the virus so difficult to define are the very things that make it invaluable as a source of answers to the fundamental questions of biology. Its intimate involvement with the complex machinery of living cells, plus its own relative simplicity, make the virus a research tool of unprecedented importance....[It is] a scientific Rosetta stone...helping us to relate the structure and chemical behavior of inert molecules to the structure and function of living cells."[9]

Discovery of Lethal Viruses; Lassa Fever Scare

New viruses are frequently being discovered, some of them lethal. The deadly Machupo virus which causes Bolivian hemorrhagic fever, was isolated only in 1962. The Marburg virus, identified in West Germany in 1966, brought death to seven of 31 persons stricken with the virus. A more recent discovery of a dangerous virus was that of Lassa fever in Nigeria. It killed three of five Americans who contracted it in 1969 and was suspected of taking 10 other lives during an outbreak of illness in Nigeria early in 1970.

Laura Wine, an American missionary nurse, was the first known victim of Lassa fever. She fell ill in January 1969 while working at the Church of the Brethren's mission station in Lassa, a northern Nigerian village. She was flown to Jos where she died 24 hours after admission to the Sudan Interior Mission's hospital. Charlotte Shaw, one of Miss Wine's nurses, came down with the disease a week later and died after a 10-day illness. Lily Pinneo, a missionary nurse who assisted at Miss Shaw's autopsy and who had cared for both patients in Jos, began to show symptoms of the disease—a fever as high as 107 degrees, mouth ulcers, a skin rash with tiny hemorrhages and severe muscle aches[10]—but she survived.

---

[9] Wendell M. Stanley and Evans G. Valens, *Viruses and the Nature of Life* (1961), p. 9.

[10] It is now suspected that, unlike other viruses, Lassa fever virus invades many of the body's organs, producing—among other things—heart infection and kidney damage.

Miss Pinneo was removed to Lagos where she spent four days in a hospital under the care of a U.S. Public Health Service physician, and then to Columbia Presbyterian Hospital in New York. Miss Pinneo eventually recovered, although her doctors could prescribe no treatment except excellent nursing care. The woman lost 28 pounds and most of her hair during nine weeks of hospitalization. Her throat became so swollen that she could not eat; she had to be fed intravenously for five weeks.

From samples of blood taken from the victims, a research team at the Yale (University) Arbovirus Research Unit isolated the virus particle and, following a well-established tradition, named it for the place where it first appeared. It was established that the virus was an entirely new entity and not the mutant strain of an existing species; that the virus had an incubation period of six to ten days; and that it contained RNA genetic material. The researchers noted similarities to a virus that causes cerebral meningitis and to viruses that cause Argentinian and Bolivian hemorrhagic fever—the last being the new-found Machupo virus.

One of the Yale research scientists, Dr. Jordi Casals, fell ill in June 1969 with what was later diagnosed as a laboratory-acquired infection of Lassa fever. In a therapy common in pre-vaccine days but rarely used today, doctors injected plasma separated from Miss Pinneo's blood into Dr. Casals' veins. Antibodies[11] in the plasma neutralized the virus almost immediately. Casals returned to his laboratory but research into Lassa fever was terminated there when laboratory technician Juan Roman, who had no known contact with the virus, mysteriously contracted it and died on Dec. 8, 1969.

The research project was moved in March 1970 to a new "maximum security" laboratory at the National Communicable Disease Center in Atlanta. Unlike the "limited access" lab at Yale, the Atlanta facility has safeguards similar to those at the lunar receiving laboratory in Houston. Air flowing from the laboratory is burned before being vented through filters; waste water is boiled and steam-filtered before being released into the sewage system. Dr. Brian E. Henderson, chief of the Virology Section's Arbovirology Unit at Atlanta, told Editorial Research Reports that scientists were trying to confirm that the later outbreak in Jos, during January and February 1970, was Lassa fever.

---

[11] Antibodies are large protein molecules which combine with viruses, blocking their normal processes.

Dr. U. Pentti Kokko, director of the Laboratory Division at Atlanta, said that the new virus may not be as dangerous as it was thought at first. "The early history of Lassa fever is similar to that of some other rather common diseases such as *histoplasmosis* and *coccidioidomycosis*,[12] which are endemic in the United States," he said. "There is already some indication that illness of less severity, with specific antibody rise, has occurred among American missionaries working in the Lassa area." Kokko added that fatal infections may constitute only a small percentage of the total cases and, if so, "the situation would be similar to that in polio."—"Paralytic cases constitute a very small percentage of all infections with polio virus."[13]

Laboratory Search for Cancer-Causing Viruses

Nowhere is scientific research more intense than in the hunt for a direct link between human cancer and viruses. Cancer is expected to kill 330,000 Americans during 1970, about one-third of the number of persons who will be treated in this country for the disease. A probable cause-and-effect relationship has been established between smoking and cancer. Various things are believed to induce cancer—among them soots, tars, chemicals, and radiation. It is often suggested that cancer might be produced by a combination of factors in a way that is not yet understood. One theory is that chemicals, genetic factors and possibly a "harmless" virus could trigger a cancer-causing virus that would otherwise remain dormant. If a viral invasion can be proved to cause cancer in humans, the way might be opened for development of a vaccine.

Dr. Peyton Rous of the Rockefeller Institute demonstrated a virus-cancer link as long ago as 1911 when he isolated the virus responsible for a tumor growing in the breast muscle of a Plymouth Rock hen. Surprisingly, this lead was not followed up until decades later and it was only in 1966 that Rous received the Nobel Prize for his discovery.[14] Some 80 viruses have now been proven to cause cancer in animals. Evidence pointing to a viral factor in human cancer, especially leukemia,

---

[12] *Histoplasmosis* is characterized by irregular fever, emaciation, lesions of the respiratory passages and other symptoms; it is caused by an infection apparently spread by airborne contamination in dust. One form of *coccidioidomycosis* is known as valley fever or desert fever, a disease common in the San Joaquin Valley of California and certain additional areas of the Southwest. It is caused by the inhalation of a fungus and its symptoms resemble pneumonia.

[13] Quoted in *The Word* (publication of the National Communicable Disease Center in Atlanta), January-February 1970, pp. 1-2.

[14] See "Cancer Research Progress," *E.R.R.*, 1967 Vol. I, pp. 235-236.

became so strong that Congress authorized the National Cancer Institute in 1964 to undertake a special virus-leukemia study program. Since then the search has become even more intense. The 1971 federal budget almost doubled the previous year's outlay for the program, up from $22 million to $41 million.

As yet there is no proof that viruses cause human cancer, despite the strongly suggestive results of many experiments. While viral particles have been isolated from cancerous tissue, the only way to prove that they cause cancer is by injecting them into healthy living tissue. It is possible that viruses are only harmless passengers infecting the tumorous cells and are not responsible for their uncontrolled growth. But the danger is too great to expose human beings to such viruses for test purposes. Despite the difficulties, Surgeon General Jesse Steinfeld has said, "the probability is very high that viruses responsible for at least some human cancers will be found."[15] An even more hopeful note came from Dr. Frank J. Rauscher Jr., director of virology at the National Cancer Institute. Rauscher predicted early in 1970 that "we are going to crack the cause of a couple of major human cancers and take giant steps toward prevention in the next couple of years."

A cancer virus known as "c type" has been isolated by scientists at the Sloan-Kettering Institute for Cancer Research in New York City from the milk of women with breast cancer. Another virus, known as EB, has been found in African children stricken with a form of cancer called Burkitt's Lymphoma. Still another virus, Herpes simplex 2, similar to a virus that causes cold sores, has been isolated from the tumors of women with cervical cancer and from the sperm of their sex partners. The Baylor College of Medicine is now studying this virus in relation to the sexual habits of various women, including nuns, who have the lowest incidence of cervical cancer, and prostitutes, who have the highest.

A great deal of work has been done recently on an RNA "c type" virus that causes leukemia in cats. Drs. Timothy E. O'Connor and Peter J. Fischinger of the National Cancer Institute showed in 1969 that the virus, known as FeLV, crossed the species barrier and replicated in human embryonic cells cultivated in the laboratory. Dr. Oswald Jarrett and his colleagues at the University of Glasgow have also infected human cells with FeLV. O'Connor emphasized that these find-

---

[15] Quoted in *The Wall Street Journal*, April 27, 1970.

ings did not necessarily mean that the virus would cause leukemia in humans. Noting that less than one-fourth of all leukemia patients have cats, he told a reporter: "There is no evidence that this cat virus is a cause of disease in man."[16]

A five-man team of scientists[17] working with newly isolated strains of another "c type" virus known as FSV—which causes *feline fibrosarcoma* (hard tumors)—found in 1970 that human cells are "extremely susceptible" to it. Fifteen days after human cells were infected with FSV, microscopic cancer-like changes could be seen. These cells eventually grew until they could be seen with the naked eye as clusters of cancerous tissue. The scientists discovered that human cells were less susceptible to the virus than were cat or dog cells. They reported in the May 29, 1970, issue of *Science* magazine:

> It is therefore conceivable that human cells that are quite susceptible to infection in vitro [in a test tube] may have some degree of susceptibility in vivo [in life] such as when the virus is introduced...through a bite or scratch by a viremic cat with or without clinical symptoms of leukemia. Although there is no evidence to implicate feline leukemia and sarcoma viruses in human cancer, further studies are necessary....

Even if viruses are found to be the causative agent of some of the 150 forms of human cancer, a safe vaccine for general use might be years away. An improperly prepared product containing live viruses might spread the disease rather than prevent it. A spokesman at the National Cancer Institute has been quoted as saying that at least two years are needed to develop a suitable experimental vaccine and perhaps five additional years to test it.

The virologist is not alone in hunting for a cancer-causing agent. The whole range of biologic sciences is involved. Dr. Robert Huebner, who took part in the FSV studies and is chief of the institute's viral carcinogenesis branch, said genetic factors may be significant in the susceptibility—or immunity—of humans to cancer. Antigens, which produce antibodies that ward off cancer viruses, have already been detected as hereditary factors in animal fetuses. Huebner told the National Academy of Sciences in Washington on April 27, 1970, that the institute was trying to find a similar antigen that can affect cancer in human beings.

---

[16] Barbara J. Culliton, *Science News*, Jan. 3, 1970, p. 23.

[17] Drs. Padman S. Sarma and Robert J. Huebner of the National Cancer Institute; Drs. John F. Basker and Lee Vernon of Microbiological Associates, Inc.; and Dr. Raymond V. Gilden of Flow Laboratories, Inc.

Growth of Knowledge About Viruses

THE WORD VIRUS comes from the Latin, meaning poison. But it has taken on different additional meanings since it first entered the medical vocabulary in the Middle Ages. At first the word was a synonym for venom and applied also to any liquid poison, especially the effluvia from a wound. By the time the French chemist Louis Pasteur (1822-95) was developing his theory of immunization, "virus" was used to describe any infectious agent. When it was discovered late in the 19th century that this infectious agent could pass through a very fine filter—fine enough to hold back all bacteria—medical experimenters began to refer to "filterable viruses." Present-day scientists generally speak of them simply as viruses.

There is no reason to doubt that viral diseases have attacked man from the earliest times. Egyptian hieroglyphs show men with withered legs—probably the result of poliomyelitis. Ancient laws often compelled citizens to dispose of mad dogs—evidence that rabies existed thousands of years ago. The plagues that killed millions during medieval times were usually bacterial, such as the bubonic plague which was spread by rats and the fleas they carried. But one of the worst was smallpox, a viral disease. Many who survived it were often left in poor health and disfigured for life by scars. Thomas Macaulay (1800-59), the British historian, wrote that smallpox turned a baby "into a changeling at which the mother shuddered, and making the eyes and cheeks of a betrothed maiden objects of horror to the lover." Greer Williams, recalling Macauley's grim portrayal, added:

> In Europe in the eighteenth century, perhaps as many as one person in ten died of smallpox, more than half of them children. Different strains of smallpox virus moved through the big cities. The death toll among those infected ranged from as low as one or two per cent to as high as thirty-three per cent. About ninety-five per cent of the Europeans who survived into adulthood had had smallpox.[18]

Even in ancient times, people recognized that a person who survived a smallpox infection would thereafter be immune. Intentional infection was popular among the Chinese, who induced their children to use the powdered scabs of a small-

---

[18] Greer Williams, *Virus Hunters* (1960), p. 3.

pox victim as snuff. Arabs gathered pus from mild cases of smallpox and injected it into healthy persons on the point of a needle—a remarkable precursor to modern vaccination methods. Greeks, too, used this form of inoculation and the patients usually survived.

Jenner's Development of Smallpox Immunization

The practice of inoculation was introduced into Britain in the 18th century; by 1746 there was a hospital in London for smallpox inoculation. While the death rate was low—reportedly about one person in 300—these inoculations often became a source of fresh epidemics. Dr. Edward Jenner (1749-1823), a British physician, was the first man to devise a safe smallpox vaccine—a remarkable feat since he had no idea of what viruses were. Jenner incorrectly believed that cowpox and horsepox—often transferred to milkmaids and blacksmiths—were weaker forms of smallpox. In 1796, he performed the first vaccination by transferring lymph matter from beneath the skin of an infected milkmaid to the arm of an eight-year-old boy. When Jenner subsequently inoculated the boy with smallpox taken from a patient, the disease did not develop.

While Jenner was not the first to use cowpox as a protection against smallpox, he was the first to conceive of transferring the vaccine from person to person, thus ensuring a constant supply. It is known today that cowpox *(vaccina)* and smallpox *(variola)* are different but related viruses, and that the weaker one gives immunization against its virulent relation. Such immunization ("cross immunization") rarely works with other viruses. Smallpox vaccine is now cultivated in laboratories, instead of being taken from infected persons or cows.

Jenner became famous in his own time and was known as "vaccine clerk to the world" and "the father of vaccination." His method soon spread to continental Europe—Dr. Luigi Sacco of Milan and his assistants vaccinated 1.3 million persons in eight years; Germany and Sweden made the practice compulsory. Dr. Benjamin Waterhouse championed Jenner in America. After vaccinating his own family with cowpox shipped to him on a thread from London, Waterhouse informed Thomas Jefferson of the deed. President Jefferson wrote Jenner in 1806 that "one evil more is withdrawn from the condition of man."

It was later discovered that smallpox vaccination did not always confer lifelong immunity, as Jenner thought. While the same method of vaccination is used today, the Public Health Service requires a person entering the United States to have had a smallpox vaccination within three years. It is often recommended that a person living in a smallpox-epidemic area, such as India, be vaccinated once a year.

Beijerinck as Father of New Science of Virology

Louis Pasteur, already famous for his process of pasteurization and his proof of the germ theory of disease, found the antidote to a second viral infection—rabies or hydrophobia. Although rabies has always been a rare disease it was, then as now,[19] a fearful one. The virus attacks nerve tissue and produces convulsions, paralysis and eventual death in its victims. Although he was never able to see or cultivate the rabies microbe, Pasteur described it as a virus and suspected it was an ultra-microscopic entity.

With his assistants Charles Chamberland and Emile Roux, Pasteur infected healthy animals by injecting them with saliva from rabid ones. He later discovered that the unseen virus could be weakened by aging it in the spinal cords of rabbits—a process called attenuation. Pasteur took bits of these rabbit spinal cords and injected them into a dog. New injections became progressively stronger day after day until finally the test animal was subjected to the bites of a rabid dog. Nothing happened—Pasteur's treatment had made the test dog completely immune to the disease. The first human use of rabies vaccine came in 1885, to save a nine-year-old French boy, Joseph Meister, who had been bitten by a rabid dog.[20] Pasteur's process has been modified several times but the technique remains essentially unchanged.

Pasteur's many experiments marked the dawn of modern medicine. "Now the ancient fallacies, the turtle-breeding mud, the miasmas, humors, and demons that had ordered the course of medicine for so many centuries began to vanish, never to be seen again. Before the close of the century, most of the major bacterial agents of disease were discovered and

---

[19] Although few humans contract the disease, authorities are concerned about its spread among wild and domestic animals. About 30,000 Americans underwent treatment for suspected rabies infection during 1969, according to Dr. John M. Hejl, director of the Veterinary Biologics Division, Agricultural Research Service, U.S. Department of Agriculture.

[20] See Rene J. Dubos, *Louis Pasteur: Free Lance of Science* (1950).

described and many could be prevented. Virology was born in this setting as an unwanted child of bacteriology."[21]

Dr. John Brown Buist (1846-1915) became the first man to see a virus, in 1887. The Scots surgeon noted tiny granules in a microscopic slide prepared from the fluid of a smallpox vaccination and assumed they were spoors. It is now known that he had discovered the large cowpox virus. Amédée Borrel (1867-1936), a French bacteriologist, glimpsed fowlpox and cowpox virus particles as early as 1904. He and Buist believed the bodies they saw were infectious agents but could not prove it. Many scientists subsequently caught sight of what they called "Borrel bodies" or "inclusion bodies" of various pox diseases.

The birth of virology as a distinct science is usually attributed to the Dutch botanist Martinus Willem Beijerinck (1851-1931), who found a virus as the cause of the tobacco mosaic disease in plants. Some historians gave a share of this distinction to a contemporary, Russian bacteriologist Dmitri Iwanowski. Both discovered that the disease could be induced in healthy plants with a fluid that had been processed from diseased tobacco leaves and passed through an unglazed porcelain filter. The filter had been invented by Pasteur's assistant Chamberland. Iwanowski believed the infectivity resulted from either a defect in the filter or a toxin released by a bacterium. Beijerinck was closer to the truth in believing the filtrate—the substance that passed through the filter—was the essential causative factor of the disease. Later scientists used filters to discover many other viruses—including the yellow fever virus, the first detected in humans. It was isolated by the Americans Walter Reed and James Carroll in 1901.

Technological Advances in the Study of Virology

Proof that the infective agent was a particle of a definite size, however, awaited the development of collodion membrane filters. Dr. William J. Elford of the National Institute for Medical Research in London performed a series of brilliant experiments in 1931, proving that membranes made from cellulose nitrate had pores of different sizes. He developed a set of graded filters with openings between 10 and 3,000 millimicrons across.

Collodion filters were one of several new tools and techniques that had to be developed before research into viruses

---

[21] Helene Curtis, *The Viruses* (1965), p. 8.

could advance. One of the most important devices used in virology today is the electron microscope, developed to the stage of practical application by Ernst Ruska and Bodo von Borries in Germany. Viruses are so small that they will not block a wave of light and are therefore invisible under an optical microscope. A beam of electrons, with a much smaller wave length than light, can actually pick out a single atom. In 1939, German physicists G. A. Kausche, E. Pfrankuch and Helmut Ruska (a brother of Ernst Ruska) took the first electron micrograph of a virus—the same type virus that Beijerinck and Iwanowski worked with in determining the cause of the tobacco mosaic disease.

Five years later, physicist-astronomer Robley C. Williams and biophysicist Ralph Wyckoff developed the technique of "metal shadowing"[22] a virus particle so that it stood out in an electron micrograph and provided scientists with a second means of measuring it. Scientists were given the means to collect concentrations of viruses quickly and cheaply by use of the centrifuge, invented by Swedish chemist The Svedberg and further developed by American physicists Jesse W. Beams and E. G. Pickels. The centrifuge and its refinement, the ultracentifuge, are apparatuses for separating solid particles in fluid by revolving the fluid at great speed.

Virologist Gilbert Dalldorf, in his book *Introduction to Virology* (1955), said the rapid development of virology was due to three technological advances in the 1930s: the ultracentrifuge, the electron microscope and, particularly, the method of growing viruses in hatching eggs—a method developed by Dr. Ernest W. Goodpasture and his associates at the Pathology Laboratory at the Vanderbilt University School of Medicine in Nashville, Tenn. "The fertile egg is our least expensive experimental host, is free of confusing latent infections, and provides a large harvest of many viruses," Dalldorf wrote.

Dr. Wendell M. Stanley, an American chemist, crystallized a virus in 1935 and thereby raised both scientific understanding and controversy. His experiment revived an old argument over whether viruses are truly living. Before then, only chemicals had been crystallized. Stanley himself at first sided with those who believed that viruses were non-living infectious agents. Others in the scientific community argued that since

---

[22] The virus matter is fixed on a small disk and placed in a vacuum chamber. A metal such as uranium or palladium is heated inside the chamber until it vaporizes and throws off tiny droplets which coat the virus particle.

viruses can replicate they must be living. In fact, as it is now known, viruses have properties of both life and non-life. They are essentially non-living until they infect a cell. Thus they are referred to as borderland or twilight forms of life.

Stanley's experiment was a long, complicated process of removing, one-by-one, the various chemical constituents of diseased tobacco leaves. "What this sane, solid, thirty-one-year-old, already bald Indiana chemist had done was to reduce about one ton of infected Turkish tobacco plants to less than a tablespoonful of white, sugary, crystalline powder that was almost all protein and almost all tobacco-mosaic virus."[23] It took two weeks to perform Stanley's original experiment. With the help of an ultra-centrifuge, it can be done today in about two hours.

Successful Quest for Effective Polio Vaccines

With the aid of techniques devised in the 1930s, Dr. Max Theiler, an American microbiologist, discovered a safe and effective yellow fever vaccine. It was only the third disease, after smallpox (1798) and rabies (1885), and the first in this century, to be brought under control through vaccination. But since then the barriers have fallen faster. Today vaccines are available for some 20 diseases. One of the latest conquests has been that of poliomyelitis.

Although it is an ancient disease, poliomyelitis is first known to have reached epidemic stage in 1887, with 44 cases reported in Stockholm. The first outbreak of noticeable size in this country came seven years later when 119 cases were counted in Otter Creek, Vt. In the summer of 1916—the worst year—6,000 Americans died of the virus and 27,000 were left paralyzed. The scientific research for a vaccine was moving slowly when in 1938 President Franklin D. Roosevelt, the world's most famous polio victim, established the National Foundation for Infantile Paralysis. Pasteur's method of attenuation yielded a vaccine of sorts in tests on monkeys but it provided immunity only on a hit-and-miss basis.

Polio is now known to infect tissues in the alimentary canal, where it does little damage. Early researchers did not know this but recognized polio as a viral infection of the spinal cord and the brain—where it usually caused terrible damage. Dr. John Kolmer of Temple University in Philadelphia and Dr. Maurice Brodie at New York University had developed polio

---

[23] Williams, *op. cit.*, p. 76.

vaccines. Brodie believed he had killed the polio virus by washing it in formaldehyde; Kolmer tried to weaken it by repeated passage through monkeys. Both the Brodie and Kolmer vaccines were discredited in 1935 when the *Journal of the American Medical Association* listed 12 cases of poliomyelitis associated with their use.

Four basic discoveries led to the eventual development of a successful vaccine. They were that: (1) polio viruses could be grown in various human cells cultivated in vitro; (2) there were three distinct types of polio; (3) polio probably entered the body through the alimentary canal; and (4) preliminary vaccine tests showed that formaldehyde-killed viruses produced specific antibodies and no ill effects in monkeys. This work led to the vaccine developed by Dr. Jonas E. Salk in Pittsburgh during the early 1950s.

Salk found the precise "mix" of formaldehyde, temperature and time to kill the most virulent strain of polio virus. He vaccinated more than 12,000 children and adults with his "killed virus" prior to a large field trial of 1954. In that year, the old Kolmer-Brodie argument reappeared with new champions. Dr. Albert Bruce Sabin at the Children's Hospital Research Foundation in Cincinnati objected that Salk was moving with "undue haste." Sabin, like Kolmer, believed that an attenuated live-virus vaccine would offer longer protection. After exhaustive trials examined by a blue-ribbon medical committee that included Sabin, it was announced on April 12, 1955, that the Salk vaccine was safe and effective. Sabin later developed effective oral vaccines against polio.[24]

New Frontiers of Virus Research

SCIENTISTS HAVE IDENTIFIED more than 500 viruses, of which some 300 affect humans. The precise number of viruses in existence is not known but is believed to run to several thousand. More viruses have been discovered than scientists have found the time to categorize. Some appear to be "harmless passengers," infecting human cells but causing no discomfort. Others remain dormant in the body, like the cold sore virus, until they are activated. There is some speculation that

[24] See "Degenerative Diseases," *E.R.R.*, 1955 Vol. I, pp. 379-380.

the many viruses that cause colds, grippes and flu-like ailments may act in this way.

The common cold has been a research target since German and American scientists proved more than half a century ago that it resulted from virus infections. The proof was established by taking the mucous secretion of patients with colds and forcing it through a filter that blocks out all except viruses. The filtrate will infect healthy persons with regularity. British virologist Christopher H. Andrewes worked unsuccessfully for four years, 1946-50, on a project to cultivate cold viruses and attenuate them into a vaccine. Since then it has been learned that any of a great number of viruses can cause cold-like symptoms. In addition to this obstacle to developing a vaccine, there is evidence that cold viruses mutate almost constantly—even in moving from person to person. Many medical researchers doubt that a quick, simple and effective remedy for the common cold will ever be found. Those who maintain hope believe any vaccine developed will have to include perhaps 25 or 30 different viruses. The common cold represents one of the most tangled puzzles faced by virologists today.

Impact of Recent Advances in Biologic Sciences

Improvements in the electron microscope and other laboratory tools will undoubtedly advance the knowledge of viruses. A microbiologist has written, "We may expect the number of effective vaccines to increase rapidly as new agents are being found and studied in their relation to disease."[25] Seemingly unrelated advances in other biologic sciences often help to increase knowledge about the nature of viruses. The discovery of the four nucleotides, or bonding agents in nucleic acid molecules, has been essential to the study of chemistry, biology, genetics and virology.

Discoveries within the past two years are expected to have a profound, if as yet undefined, impact on virology. Scientists at two American laboratories announced in January 1969 they had artificially created the enzyme ribonuclease which breaks down RNA. The DNA molecule was soon photographed, and then even a single atom. New fruits of virus research continued to pour forth through 1969 and into 1970. Scientists deciphered the complete structure of the complicated antibody gamma globulin. Dr. Solomon Spiegelman, a microbi-

---

[25] Isabel Morgan Mountain, "Virus," *The Encyclopedia Americana* (1968), Vol. 28, p. 171.

ologist at the University of Illinois, received a patent for the synthesis of RNA, work that is potentially important to the development of anti-cancer substances; the New York University School of Medicine announced the artificial creation of a lysosome, one of the building blocks of a living cell. One of the most recent advances was announced on June 2, 1970, by Dr. Gobind Khorana and his research team at the University of Wisconsin. They reported the first complete synthesis of a gene, a discovery particularly of benefit to virology.

Viruses, cancer, genes, and questions concerning the nature and origin of life are tied together by a whole series of interrelationships. Viruses can act as genes and genes can act as viruses under certain circumstances. Viruses can cause cancer and, within limits, can destroy cancerous tissue. Furthermore, viruses offer us a unique key to understanding the function of nucleic acid, and perhaps, therefore, to understanding the nature of life itself.[26]

The creation of artificial life has emerged as a seldom-stated but obvious goal of many related lines of research. It was at first thought in 1955 that biochemist Heinz Fraenkel-Conrat had "created life" in his research into the tobacco mosaic virus at the Virus Laboratory at the University of California. But Fraenkel-Conrat never accepted this line of reasoning, and it later proved to be false.

Future Studies; Search for Virus-Inhibiting Agent

A further understanding of the cellular processes by viral infection is important for progress in several other areas of biological and medical science. A report prepared by a number of scientists for the National Academy of Sciences said research is necessary to answer the following basic questions: (1) What is the mechanism of virus absorption into the host cell? (2) Where in the cell does viral nucleic acid reproduce itself? Is it in the nucleus, as most evidence would suggest, or in the surrounding cytoplasm, as is the case with non-viral protein? (3) What is the mechanism of growth control exerted by DNA tumor-producing viruses? (4) What is the significance of the similarity between some tumor-virus DNA and the DNA found naturally within cells? (5) Do all viruses become uncoated as they are injected into the cell?

Generally, the report was optimistic that further research would eventually crack the mysterious nature of virus infection. In concluding a section on viruses, the report said, "It will be evident that the huge intellectual triumph of the past decade will, in all likelihood, be surpassed tomorrow—and to

---

[26] Stanley and Valens, *op. cit.,* pp. 206-207. See "Genetics and the Life Process," *E.R.R.*, 1967 Vol. II, pp. 905-922.

the everlasting benefit of mankind."[27] It called for a continuing search for chemical agents which could inhibit virus attack, even though past research has proved fruitless.

Science News reported Aug. 22, 1970, that the chemical compound tilorone hydrochloride, derived from coal tar, had excited virus researchers. The magazine explained the background of its development in this way: "During the last decade, interferon, a native protein that forms the first line of defense against viruses of all types, has been the object of continually expanding research. Many investigators believe that eventually interferon will be to viral diseases what antibiotics are to bacterial infections. Their challenge is to find an effective and relatively simple way of artificially inducing interferon either to protect individuals against viral infections or to ensure that those that do take hold are rapidly knocked out." Tilorone hydrochloride was described by its researchers as an inducer of interferon, the first to be administered orally.

"Should tilorone become clinically available within the next few years," *Science News* commented, "it could be put to work in a variety of circumstances. A pregnant woman exposed to rubella, for example, could be protected from infection. And in situations in which large numbers of persons were in danger of being exposed to a virus during an epidemic, the [tilorone] pill could be downed as a prophylactic." A leading authority on virus infections, Dr. John F. Enders of the Harvard University Medical School, expressed doubts recently in the *New England Journal of Medicine* about the long-range effectiveness of a vaccine licensed in 1969 for general use against rubella—German measles. The disease is especially dangerous to an unborn child if the mother contracts it.

The future of viral research is, of course, impossible to predict. Human cancer viruses may be found and following that, a vaccine. The common cold may eventually be beaten. Mental diseases, multiple sclerosis, and even the aging process may be found to have a viral cause and come within reach of vaccine benefits. On the other hand, it is just as possible that a common and relatively harmless virus may mutate into a killer and bring on fresh plagues. Some immunologists fear that if science manages to protect humanity from too many forms of viral illness, the body will lose its ability to produce antibodies to fight infections when they do develop.

---

[27] *Biology and the Future of Man* (edited by Philip Handler, president of the National Academy of Sciences, 1970), pp. 150-162.

POLLUTION TECHNOLOGY

by

Helen B. Shaffer

1 9 7 1
Jan. 6

POLLUTION TECHNOLOGY

THE EMERGENCE of environmental pollution as everybody's issue may be a victory for those foresighted scientists and conservationists whose warnings of impending crisis went unheeded for so many years. But winning popular support for an all-out attack on pollution is but the bare beginning of a solution. A big question is how to go about doing the job. And here the experts often do not agree. An even bigger question is how much people are willing to pay to restore their damaged earth, air, and waterways, and to prevent further depredations. Pay means not only money, which will run into many billions of dollars, but possibly the sacrifice of conveniences and luxuries so familiar as to seem necessities to most Americans.

To date, pleas for a cutback of consumption in the interests of environmental protection have won few converts. It is true that some housewives take their grocery bags and egg boxes back to the supermarket for re-use, and here and there conservation zealots ride bicycles instead of drive cars. But these symbolic gestures do not make a perceptible dent in the pollution problem. The load, in fact, has been growing. Air pollution alone has increased over the past four years from an annual outpouring of 142 million tons of pollutants to more than 200 million. Solid wastes cast off by U.S. municipalities and industries now add up to 360 million tons a year; the total is estimated at 3.5 billion if wastes from agriculture, mining and fossil fuel production are included.[1]

Obviously the solution, at this stage at least, is not being sought in sacrifice. When it is cold, householders will turn up the heat regardless of noxious emissions from power plants; and few couples today deny themselves a baby for the sake of population control. Elaborate packaging and throwaway cans have become luxuries not easily given up. Americans typically

---

[1] Richard D. Vaughan (director of the Bureau of Solid Waste Management, Department of Health, Education and Welfare), "Reuse of Solid Wastes: A Major Solution to a Major National Problem," *Waste Age*, April 1970, p. 10.

look to experts rather than abstinence. Technology made the mess, the prevailing view proclaims, so let technology clean it up. The catch is that clean-up technology must be at an acceptable price. The engineering problems, therefore, are closely tied to the economics of pollution control.

Criticism of Slow Pace of Technical Advances

Is sufficient technology available at the present time to maintain a reasonably clean environment without lowering present levels of consumption? The question hardly lends itself to a direct answer. Considerable technology for limiting the contamination of air, water, and soil has been available for many years and more was added after the government moved decisively into the anti-pollution field.[2] The government spurs technological advance in pollution control in two ways: (1) by providing funds for research and development and (2) by regulatory action that compels polluters to develop more effective environment-protection systems.

One hindrance to an accurate assessment of the state of pollution-control technology is that the pollution problem does not stand still. Not only does the population grow, industry adds to the burden of wastes with new ingredients and combinations that result from changes in processes, development of new products and new consumption patterns of the people. What worked to keep air and water clean yesterday may not be sufficient today.

Though pollution-control technology has made appreciable advances in recent years, it is not hard to find experts who deplore its present state. "Primitive" would suit better than "sophisticated" to describe pollution-control technology now, according to the editor of the American Chemical Society's monthly journal. "It's no secret that pollution control has traditionally drawn upon old concepts, old equipment, and the least skilled work force available," he wrote. "Pollution-control technology has evolved at a snail's pace."[3]

"Cleaning up this country's air and water will be a much tougher, slower, and costlier job than politicians and environmentalists sometimes make it sound," Gene Bylinsky of

---

[2] Major federal programs in environmental protection date from the Water Pollution Control Act of 1948 and the Air Pollution Control Act of 1955. These were greatly expanded by new congressional mandates during the 1960s. Legislation in 1970 extended government activity still further. See Congressional Quarterly *Weekly Report* of March 27, 1970, p. 852; Oct. 16, 1970, pp. 2546-2547; Dec. 4, 1970, p. 2897; Jan. 1, 1971, pp. 14-16.
[3] D. H. Michael Bowen, "Build a Better Mousetrap" (editorial), *Environmental Science and Technology*, November 1970, p. 877.

Fortune magazine wrote. "Without advances in technology the big cleanup can only plod along at best. And there are serious lags in new pollution control technology, as well as in the readiness of business, government and the public to encourage, apply and pay for technological improvements."[4]

Not only is technology laggard but basic scientific knowledge is limited. Scientists need to know more about the character of existing and potential contamination and of specific effects of pollutants in their multitude of forms and combinations. This knowledge lays the foundation on which technology builds. "Existing systems for measuring and monitoring environmental conditions and trends and for developing indicators of environmental quality are still inadequate," President Nixon said Aug. 10, 1970, in a message to Congress accompanying the first report of the Council on Environmental Quality.[5] These systems are needed to provide data for determining what measures are needed to reduce pollution and to assess the effectiveness of those used. "We need to know far more about the effects of specific pollutants, about ecological relationships, and about human behavior in relation to environmental factors," the President said.

Existing data on levels of particulate matter in the urban atmosphere, a scientist told his colleagues at a recent seminar, indicated that they knew the exact chemical composition of "less than 40 per cent of the dirt in the air of our cities."[6] There were so many factors to be considered—size of the particles, climate, presence of different bacterial and viral organisms, interactions of the various pollutants. The damage to health or well-being of a local environment from gross contamination may be readily assessed. But the long-term effects of lesser levels of contamination are still something of a mystery. "Determining the adverse effects of exposure to very low levels of pollutants over long periods remains an urgent but difficult task," the National Air Pollution Control Administration reported. "Gaps in the arsenal of biological and medical research tools pose still other difficulties."[7]

---

[4] Gene Bylinsky, "The Long, Littered Path to Clean Air and Water," *Fortune*, October 1970, p. 112.

[5] The Council on Environmental Quality was established under the National Environmental Policy Act of 1969 to advise the President. Russell E. Train, chairman, Robert Cahn and Gordon J. MacDonald are the council members.

[6] Glenn L. Paulson (secretary of New York Scientists' Committee for Public Information), "A Piece of the Action," *Air Pollution* (A Scientists' Institute for Public Information Workbook, 1970), p. 19.

[7] *Progress in the Prevention and Control of Air Pollution* (third report of the Secretary of Health, Education and Welfare to Congress), March 1970, p. 3.

With knowledge so limited it is no wonder that scientists differ in their over-all outlook on the problem of environmental protection. "Harmful effects on the environment or on other organisms are often assumed, without evidence, to imply biological damage in man," the director of the International Agency for Research in Cancer observed. "In fact there is a surprising dearth of factual data on these relationships."[8] He said limited data did not support the "simplistic" view that any chemical modification of the environment was necessarily bad for man. There are still scientists who protest the phasing out of DDT as an unrealistic rejection of a useful pesticide. Differences among experts as to tolerable levels of radioactivity are well-known.

Main Problems in Reducing Water Contamination

During the past decade, the water pollution problem has become increasingly severe. Since 1952, about $15 billion has been spent in the United States in building 7,500 municipal sewage treatment plants and other water-treatment facilities. But some 1,400 U.S. communities and hundreds of industrial plants still drop untreated waste into the waterways. Only 140 million of the country's more than 200 million people are served by any kind of sewer system. "Over 1,000 communities outgrow their treatment systems every year," the Federal Water Quality Administration, an agency in the Department of the Interior, reported in *Clean Water for the 1970s: A Status Report.*

"Fortunately, there is the technological knowledge to deal effectively with municipal wastes," the agency added. "This technology has not been applied to the extent needed to prevent pollution." Stubborn technical problems remain nevertheless. Industrial wastes challenge clean-water technology because of the variety and novelty of alien substances. "Many of the new chemicals are a challenge to detect, much less control," the agency reported. Agricultural pollution of the waterways taxes existing measures for water protection because it combines animal and vegetation wastes in a variety of combinations. Pollutants in the return flow of irrigation waters are particularly difficult and expensive to control.

The Council on Environmental Quality contends that abatement technology is generally available for reducing pollution

---

[8] John Higginson, "International Research: Its Role in Environmental Biology," *Science*, Nov. 27, 1970, p. 935. The cancer research agency which Higginson directs is based in Lyon, France.

from industrial sources. Nevertheless, it has pointed to many gaps in basic knowledge pertaining to water-quality control. A full understanding of the connection between pollutants and eutrophication is lacking. Eutrophication is the process by which an excess of nutrients in a body of water produces an undue growth of algae. Algae consume oxygen, resulting in fish-kill. Eutrophication takes place naturally in lakes but over a very long time. The nutrient load discharged into the water by man greatly speeds up the process.

Phosphates in fertilizer and household detergents have been held responsible for most of this despoliation. But recently some experts have begun to say that while phosphates are involved, other elements—nitrogen, heat, carbon—also contribute to eutrophication.[9] Eutrophication has taken place in lakes that had minimal amounts of phosphates.[10]

Sulfur Dioxide Increase and Air Pollution Control

Techniques for dealing with many air pollution problems arising from stationary sources—electric generating plants, space heating systems, industrial operations, and incinerators —are generally satisfactory, according to the National Air Pollution Control Administration. But adequate technology is lacking for "the most significant of these problems."

> There is a widening gap between the rising trend of sulfur dioxide emissions and the nation's technological capability for bringing the problem under control, partly because the total national investment in research and development on the problem has not been sufficient to support all the potentially fruitful work that could have been undertaken in the past few years. Of particular importance is the need for practical techniques applicable to electric generating plants....

> Even with rapid application of the control techniques now under development...it is unlikely that sulfur dioxide emissions in 1980 will be reduced even to the 1968 level. The rapid growth of the electric utility industry (from 300,000 megawatts in 1969 to an anticipated 600,000 megawatts by 1980) and slower-than-predicted growth of nuclear electric generating capacity are compounding the problem.[11]

Businessmen engaged in anti-pollution work tend to be optimistic about the eventual conquest of pollution. Karel A.

---

[9] Two scientists with the Agricultural Research Service in Fort Collins, Colo., reported still another and more subtle source of eutrophication—ammonia from cattle urine in feedlots, carried in the air to nearby lakes where it is absorbed by the water.—*Science News*, Nov. 14, 1970, p. 384.

[10] See Philip H. Abelson, "Excessive Emotion About Detergents," *Science*, Sept. 11, 1970.

[11] *Progress in the Prevention and Control of Air Pollution* (third report of the Secretary of Health, Education and Welfare), March 1970, p. 26. See also "Electric Power Problems," *E.R.R.*, 1969 Vol. II, pp. 939-956.

Weits, president of the Industrial Gas Cleaning Institute, told the Muskie Subcommittee[12] in March 1970: "We believe our industry has the necessary skills and facilities to develop equipment to meet many unsolved problems once they are defined as problems." Equipment was available to reduce particulate emissions from industrial sources to acceptable levels, he said, though industry would have to speed up production of equipment to handle gaseous pollutants. Where profit is the motivation, he added, industry can move ahead fast—faster than government—to find the technical solutions.

If legislation provided sufficient economic incentive, Aaron J. Teller, head of Teller Environmental Systems, Inc., told the subcommittee, all emissions of sulfur dioxide—the major pollutant gas discharged by electric power plants—could be eliminated from this source within five years. "The technology is here," he said. "We are now in our second and third generation systems." From his experience in working with both government and industry, he found that "where processes are to be developed to be exploited and sold, you will find the major advances will occur in industry."

Protest of Auto Industry Over Emission Criteria

Makers of products that cause pollution, either as industrial waste during production or as waste cast off after consumer use, tend to be less than optimistic about the swift attainment of goals demanded by environmentalists. Automobile manufacturers, for example, fought against emission standards provided in the Clean Air Act of 1970 on the ground that they lacked the technology to meet the standards. This measure, which the President signed into law on Dec. 31, requires a 90 per cent reduction in major automobile pollutants—by Jan. 1, 1975, for hydrocarbons and carbon monoxide and by Jan. 1, 1976, for nitrogen oxides. Automobile manufacturers will be allowed a one-year extension of each deadline if they can demonstrate that they tried hard but failed to meet the standards in the allotted time. They could apply for extensions by Jan. 1, 1972, for hydrocarbons-carbon monoxide, and by Jan. 1, 1973, for nitrogen oxides.

Automobile officials insisted they could not meet the deadlines. "Unless the science and technology of emission control move ahead much faster than we believe is possible," L.A.

---

[12] Formally, the Subcommittee on Air and Water Pollution, a unit of the Senate Committee on Public Works. It is known by the name of its chairman, Sen. Edmund S. Muskie (D Maine).

Iacocca, then executive vice-president (later president) of Ford Motor Co., said on Sept. 9, 1970, "we will not be able to meet the standards prescribed by the bill." Even if they were technically feasible, he said, it would take at least two years after establishing an emission standard to perfect the technology and make necessary changes in vehicle design and in manufacturing plants.

Officials of General Motors and Chrysler took similar positions. They cited the difficulty of finding materials capable of withstanding the high temperatures required of exhaust-control devices. Another problem was the incompatability of new systems with many gasoline compounds currently in use. The oil industry has been under pressure to hasten changes in fuel composition to help the automobile industry meet rising pollution-control standards. President Nixon early in 1970 requested Congress to levy a special tax on leaded gasoline to discourage pollution from this source; on Oct. 26 he ordered that federal vehicles use low-lead fuel whenever possible and urged governors to take similar action with state-owned vehicles. The House Ways and Means Committee decided on Nov. 23 to postpone consideration of the tax proposal until 1971.

Industry takes the position that any regulation that discourages or forbids the use of a particular ingredient in a product inhibits the ability to develop more effective pollution abatement and stifles innovative approaches. Environmentalists contend, however, that without specific governmental restrictions industry will take its time developing the necessary techniques.

Industries that contribute to pollution are beginning to complain of confusion in the government's rush to regulate them. The *Wall Street Journal* on Dec. 23, 1970, described the situation as "a Pandora's box of changing and conflicting law, bureaucratic snarls and technical impossibilities." Industry officials say there are too many regulating agencies, standards shift too quickly, and there is "technical chaos" due to disagreement on the nature of pollution, the degree of danger it presents, how it should be measured, and differences as to goals. Some confusion may subside now that various federal anti-pollution activities have been consolidated in the new Environmental Protection Agency. The new agency, the product of an Executive Branch reorganization plan approved by Congress in 1970, formally came into being Dec. 2 with William D. Ruckelshaus, a former Assistant Attorney General, as its director.

Advances in Environmental Protection

TECHNOLOGY, THE APPLICATION of science to practical tasks, is a two-edged sword. "It concentrates on immediate effects and often ignores long-range environmental impact. Yet, at the same time, its inventions promise to reverse the trend toward environmental degradation."[13] The development of technology to diminish pollution has moved ahead with government regulation. Historically, the tendency has been to permit pollution until it becomes intolerable. At that point, government action is taken to restrain polluters, who then hasten to develop a more efficient method of compliance than is already available to them.

Government Action as Spur to Pollution Control

Outbreaks of communicable diseases in the 19th century led to government action that resulted in early improvements of municipal sewage disposal systems. Filtering devices developed at the Lawrence (Mass.) Experiment Station, established in the wake of regulatory action taken by Massachusetts in the 1880s, are still useful in treating municipal wastes.[14] Obstruction of shipping led to enactment of the first federal clean-water law, adopted in 1886, which compelled polluters to find other means of disposing of refuse than dumping it in New York harbor. The Rivers and Harbors Act of 1899 extended the prohibition on dumping to other navigable waters.[15]

The 1899 law, rarely enforced for 71 years, became the basis for an executive order President Nixon issued Dec. 12, 1970, requiring industries to obtain federal permits before they discharged waste materials into the nation's waterways. Permits would henceforth be issued by the U.S. Corps of Engineers only after state and interstate agencies certified that industrial discharges met existing water quality standards. Existing industrial plants were given until July 1, 1971, to obtain permits. New plants must have them when they begin operations. The Corps of Engineers has estimated that 40,000 factories

---

[13] Harvey Lieber, "Water Pollution," *Current History*, July 1970, p. 29.

[14] See James Ridgeway, *The Politics of Ecology* (1970), pp. 31-32.

[15] One section of the 1899 law declares that "it shall be unlawful to throw, discharge or deposit...any refuse matter of any kind or description whatever" other than municipal sewage "into navigable water of the United States" or place it nearby or in tributaries where it could wash into navigable waters. Another section of the law does sanction discharges into waterways, but only if they are specifically permitted by the U.S. Corps of Engineers.

discharge fluid wastes into navigable waters and thus would be affected by the executive order. Russell E. Train, chairman of the Council on Environmental Quality, called the order "the single most important step to improvement of water quality that this country has taken."

Fear aroused by illness and death from severe local episodes of smog[16] led to enactment in 1955 of the first comprehensive federal law for air pollution control, which was followed by improvements in abatement technology. Crisis, followed by government regulation, still puts spur to anti-pollution advance. The government's new concern over the harmful presence in commercial fish of mercury, a known poison that has been discharged into waterways since the beginning of the industrial revolution, has sparked a race to find a way of neutralizing mercury residues in lake and river bottoms. Among suggested techniques, none of them perfect, are covering the sediment with inert clay, dredging to remove tainted sediment, or adding other products to dissipate the mercury.[17]

Testing of Microbes to Remove Oil, Phosphates

New or long-neglected problems in environmental pollution control continually seem to rise to critical levels. Oil spills from sea-going tankers and the dumping of ship bilge are not new occurrences but the rising tide of coastal pollution has fostered a demand for quick remedies. Accidents in offshore oil drilling have dramatized the need for better systems of prevention. The open seas, viewed from time immemorial as a safe dumping ground, now are being seen in a new light. Oceanic pollution is emerging as a new problem for the experts to solve.[18]

The rush is on to find effective ways of dealing with oil slicks. Past efforts to cleanse concentrations of oil spillage with chemicals have backfired, augmenting the damage.[19] Much hope rests on the use of concentrations of micro-organ-

---

[16] The London "killer smog" of 1952 contributed to the death of 4,000 persons. Perhaps the worst occurrence in the United States was at Donora, Pa., in December 1948. In the period of a few days, 14,000 residents of Donora and nearby communities were taken sick and 18 deaths were attributed to the smog.

[17] See "Mercury in the Environment," *Environmental Science and Technology,* November 1970, p. 890.

[18] "About 48 million tons of wastes were dumped at sea in 1968...dredge spoils, industrial wastes, sewage sludge, construction and demolition debris, solid wastes, explosives, chemical munitions, radioactive wastes, and miscellaneous materials," many of them toxic to human and marine life.—Council on Environmental Quality, *Ocean Dumping: A National Policy,* October 1970, pp. v. 1.

[19] See "Coastal Conservation," *E.R.R.* 1970, Vol. I, pp. 141-146.

isms, which do the job in nature but at a much slower pace than suits man. Bioteknika International Inc., of Alexandria, Va., claims to have found the answer in a mixture of 20 different micro-organisms. They break down the oil, changing its molecular structure and rendering it harmless. The company said that tests it conducted on a Potomac estuary showed that the microbe packet can clean 100 square feet of thick black oil in four days. The microbes will be marketed in dried, frozen form.

Microbes may be put to work on the phosphate-eutrophication problem too. Biospherics Inc., a Washington concern, announced it had devised a process that induces bacteria to consume not only organic wastes in sewage—a normal process—but to extract phosphate as well. After the bacteria settle to the bottom of the sewage (the sludge), they are removed to a separate basin where they are induced, by denying them oxygen, to emit the phosphate. The concentrated solution of phosphate is then treated to separate it from the flow of sewage and the bacteria are returned to repeat the process on incoming sewage. This process is said to be readily adaptable to existing treatment plants.

Major detergent makers have been speeding up a search for a phosphate substitute. The best found to date from a cleansing standpoint is NTA (nitrilo triacetic acid). But questions of human safety have been raised and manufacturers are reluctant to continue its use until the government completes studies on its effect on environment and human health. Dr. Samuel S. Epstein, chief of the Laboratories of Environmental Toxicology and Carcinogenesis at the Children's Cancer Research Foundation Inc., in Boston, has warned that laboratory studies have shown that NTA is taken up cumulatively by the bones of test animals, that high doses of NTA apparently caused chromosome breakage in cultured human cells, and that a breakdown product of NTA might combine under certain conditions with nitrite to form nitrosamines, which are highly carcinogenic at even very low doses.[20]

Old and New Ways of Purifying Water Supplies

Though basic methods of treating sewage have changed little over the past quarter-century, the growing burden of municipal wastes has fostered refinements and additions. The basic methods are (1) screening sewage waters to remove

---

[20] Samuel S. Epstein, "NTA" (based on Dr. Epstein's testimony before the Senate Subcommittee on Air and Water Pollution) *Environment,* September 1970, pp. 3, 7, 9.

large solids and the settling of smaller solids in a sedimentation tank (primary treatment) and (2) bacterial action to remove organic matter, filtering, and chlorination (secondary treatment). Tertiary treatment removes more of the undesirable chemicals in municipal sewage, but it is very costly, involving construction of additional plants at a time when many communities have insufficient primary and secondary treatment facilities. Among the more advanced processes are:

Use of lime or alum to force suspended solids to clump, speeding up separation of solids and liquids and thus increasing the capacity of existing systems;

Ridding water of persistent organic matter by passing the effluent over a bed of activated carbon to which the organic matter adheres;

Forcing unwanted salts from water by passing the effluent through an electrodialysis cell.

"Properly designed and applied...[these methods] will be able to supply any quantity of water for any re-use. But none of these processes will stand alone. They must be used in a series or a parallel plan."[21] Because of the expense, much of the inventive energy in this field is directed toward developing devices or procedures that can be grafted at relatively small expense on existing water treatment systems. "Ultimately, entirely new systems will no doubt replace the modern facilities of today," the Federal Water Pollution Control Administration has predicted. In the future may be such revolutionary techniques as the use of reverse osmosis to take pure water out of the waste, rather than take pollutants out of the water.

Sensitive Devices for Measuring Air Pollution

Establishment of air quality criteria has set the pace for technology. Technical documents setting forth criteria for carbon monoxide, particulate matter, sulfur oxides, hydrocarbons, and photochemical oxidants have already been issued by the National Air Pollution Control Administration. Criteria for other substances will follow.[22] To conduct the studies needed for establishing criteria and for monitoring the

---

[21] Federal Water Pollution Control Administration, *A Primer on Waste Water Treatment* (1969), p. 14.

[22] Air quality criteria for nitrogen oxides, lead, fluorides, and polynuclear organic compounds will be published in 1971; odors, asbestos, hydrogen chlorides, beryllium, and chlorine gas in 1972; arsenic, nickel and vanadium in 1973; barium, boron, chromium, mercury, and selenium in 1974, and pesticides and radioactive substances in 1975.

atmosphere for compliance has required the development of sophisticated instruments. These devices check both on the general condition of the atmosphere in an area and on particular emissions.

A prototype instrument capable of identifying and measuring malodorous sulfur compounds in the parts-per-billion range has been constructed. An electro-chemical sensor for sampling sulfur dioxide in exhaust stacks is being tested. Research promises to provide for the first time a satisfactory device for measuring nitric oxide and nitrogen dioxide. "Diffusion models," which simulate the movement of pollutants in the air, are being used to guide predictions of pollutant concentrations likely to occur under various weather conditions and emission rates. A "climatological model" may enable planners to forecast the extent to which the future growth of a community, or the application of control measures, could be expected to alter average pollution concentrations. A national air data bank and a system for storage and retrieval of air data obtained from the nationwide federal-state-local air sampling network have been established. The National Air Pollution Control Administration is working on standardization of analytical techniques so that data from one community can be readily compared with that of another.

Two approaches are being taken to handle the troublesome problem of sulfur oxide pollution from fuel combustion: (1) to apply devices that remove sulfur from stack gas before it escapes into the air and (2) to develop low-sulfur fuels. Fuels may be naturally low in sulfur or they may be treated to reduce sulfur content. Techniques have been developed to remove sulfur from a low-grade imported fuel oil and a number of companies have been building desulfurization plants. The supply of low-sulfur oil, however, remains low. Efforts to improve coal-cleaning processes for more efficient removal of sulfur are also under way.

The most practical measures developed for application to electric generating plants are new processes utilizing limestone. In one process, limestone is injected into the boiler where it causes sulfurous particles to form. These particles can then be removed from stack gas by electrostatic precipitators. An even more effective limestone process would use scrubbers to remove sulfurous material before it reached the stack. Testing of the first (dry limestone) process is to be completed by mid-1971, when testing of the second (wet limestone) will be initiated.

Long-term gains may come from studies to achieve more efficient combustion so that less waste is thrown off. The United States and Britain have agreed to share technical information on "fluidized bed combustion," a process which holds considerable hope for stack gas reduction. In addition, the government has reviewed the testing of a number of inexpensive, commercially manufactured devices to improve combustion in oil-fired furnaces and it has singled out a flame-retention device. It is said to reduce pollutant emission without impairing the furnace's operating efficiency.

Standards to Reduce Motor Vehicle Emissions

The Public Health Service estimates that up to two-thirds of all pollutants released into the air in America come from the gasoline engines that propel the country's 105 million cars, trucks, and buses. Pollution-control technology for motor vehicles, as first applied in the early 1960s to meet California state standards, dealt largely with reducing hydrocarbon emissions from the crankcase by means of a ventilating system that recycled gases to the engine intake. Crankcase "blow-by" gases are believed to account for up to 25 per cent of all hydrocarbons emitted from automobile engines. Anti-pollution controls were later applied to the exhaust system to reduce its outpouring of both hydrocarbons and carbon monoxide.

To meet federal standards for hydrocarbons and carbon monoxide, beginning with 1968 models, auto makers made changes in the fuel system, redesigning combustion chambers and adjusting fuel-air ratios to achieve maximum burning of pollutants and minimize evaporation losses. One new system involves recirculation and reburning of exhaust gases to reduce wastes. A more advanced approach under development involves use of a catalytic device through which exhaust gases would pass. The catalyst would convert pollutants to harmless derivatives—carbon monoxide and hydrocarbons to carbon dioxide and water, nitrogen oxides to nitrogen and other byproducts.

The automobile industry claims that a car 65 to 80 per cent "cleaner" than pre-regulation models has already been attained.[23] The following standards were reported to Congress in March 1970 by the Department of Health, Education and Welfare:

---

[23] L. A. Iacocca of the Ford Motor Co. said on Sept. 7, 1970: "The 1971 standards and technology to meet them will produce an 80 per cent reduction in hydrocarbons....A 70 per cent reduction in carbon monoxide has already been achieved."

AUTOMOBILE EMISSION STANDARDS

	1968 national standards	1970 national standards	1971 national standards
	% reduction since 1963	% reduction since 1963	% reduction since 1963
Hydrocarbon emissions	66.9	73.5	85.6
Carbon monoxide emissions	55.6	70.9	70.9

However, the Federal Air Pollution Control Administration, which has authority to establish national standards, expressed concern that "air pollution control systems installed in mass-produced vehicles often lose some of their effectiveness more rapidly than prototype systems do." Another concern was that increases in motor car use would offset gains from limited emission controls. Foreign cars brought into the United States by private persons would be immune to agency standards, but not those brought in for sale.[24]

The air might be cleaner if the government required older cars to be equipped with devices now available for $10 to $20. "If the engine is in decent shape," Iacocca said, "[Ford's] system will reduce emissions by 30 to 50 per cent." The air-cleaning gadgets have not been big sellers and there is much skepticism about their effectiveness, especially in the absence of good engine maintenance. Hope rests more with the sophisticated systems to be built into future models.

Some experts believe auto pollution will linger as long as the internal combustion engine remains, no matter how many devices and refinements are added. Auto manufacturers have not shown much enthusiasm for such alternatives as electric cars, steam engines and gas turbines.[25] However, General Motors startled the automotive world in November 1970 by announcing that it had agreed to pay $50 million over the next five years for a worldwide license to manufacture a new type of German-made engine known as the Wankel. The purpose of the purchase, General Motors said, was to conduct "intensive research and development studies of the Wankel rotary combustion engine to determine whether it is suitable

[24] The Federal Air Pollution Control Administration has established auto emission standards for hydrocarbons and carbon monoxide, but not yet for nitrogen oxides and particulates. The 1975-76 standards set by the 1970 Clean Air Act will be superimposed on agency standards.

[25] See "Steam and Electric Autos," E.R.R., 1968 Vol. II, pp. 583-601.

for GM...applications." The Wankel is not only much smaller and lighter but is expected to be much more adaptable to emission-reduction equipment than the standard piston engine.

Economics of Pollution Technology

THE RISING DEMAND for more effective measures to protect the environment has stimulated considerable growth in the pollution control business. "Almost overnight the market for anti-pollution equipment has exploded into a five-billion-dollar-a-year business," *U.S. News & World Report* stated Aug. 31, 1970. "That includes new devices, bricks and mortar, and engineering and contracting services. Plant-operating and labor costs add hundreds of millions more."

The 1970 annual directory of air pollution firms, published in the November issue of *Environmental Science & Technology,* showed a 70 per cent growth in the number of new companies dealing in pollution control. Many of them were said to offer "novel, sophisticated and frequently exciting concepts in technology." Sales were low, however, and some companies have confided that if they do not receive orders soon, they will be out of business. The magazine attributed the unexpectedly slow sales to high prices. Expectations of a good future for this industry, however, are indicated by the decision of Dun & Bradstreet Inc. to undertake a survey of U.S. plants, beginning in January 1971, to determine the best markets for pollution-control devices in the years ahead.

Estimates of what it will cost to put and keep the environment in good order are imprecise and variable, but all agree that it will be enormously high. When President Nixon proposed in his 1970 State of the Union message that the country spend $10 billion through 1974 for municipal waste-treatment plants, leading Democrats described his plan as inadequate. Sen. Edmund S. Muskie (D Maine) said $25 billion was required "if we were to catch up on the backlog of untreated municipal wastes alone" and perhaps $50 billion if industrial wastes were added. Stewart L. Udall, former Secretary of the Interior, said a comprehensive program to control water pollution would cost $30 billion in the four years ahead.

EXPECTED AIR QUALITY CONTROL COSTS, 1971-75*

(in thousands of dollars)

1	Chicago	801,300	19 Youngstown/Warren, Ohio	46,000
2	New York	338,300	20 Boston	45,800
3	Pittsburgh	287,900	21 Toledo	45,500
4	Detroit	263,700	22 Hartford	43,700
5	St. Louis	257,500	23 Allentown/Bethlehem/	
6	Cleveland	209,600	Easton, Pa.	42,800
7	Philadelphia	199,300	24 Saginaw/Bay City, Mich.	42,400
8	Steubenville, Ohio/		25 Dayton	42,200
	Weirton/Wheeling, W.Va.	166,600	26 Los Angeles	42,200
9	Cincinnati	162,800	27 Grand Rapids/Muskegon/	
10	Buffalo	129,300	Muskegon Heights, Mich.	41,000
11	Louisville	114,800	28 Houston/Galveston/	
12	Milwaukee/ Kenosha/		Texas City	38,500
	Racine, Wis.	109,600	29 Atlanta	35,300
13	Washington	98,900	30 Scranton/Wilkes-Barre, Pa.	30,900
14	Birmingham	94,800	31 Kansas City, Mo.	29,200
15	Baltimore	84,200	32 Tampa	27,800
16	Denver	67,100	33 Charleston, W.Va.	26,200
17	Minneapolis/St. Paul	57,800	34 Peoria, Ill.	25,000
18	Indianapolis	46,000	35 Portland, Ore.	23,100

Figures include expected cost of controlling emissions from industry, government facilities, and private households, according to conditions prevailing in 1967.

SOURCE: Department of Health, Education and Welfare report to Congress, *The Cost of Clean Air*, March 1970, pp. 58-79.

Marshall I. Goldman, a professor of economics at Wellesley College, has estimated that $130-180 billion will be required to construct needed facilities to control both air and water pollution and $12-17 billion will be required yearly to operate them. However, he notes that "technological breakthroughs could reduce costs significantly."[26] *U.S. News & World Report* foresaw $71 billion of new spending over the next five years to clean up the nation's air, land, and water.[27]

Alan K. Browne, senior vice-president of the Bank of America in San Francisco, found a composite of various estimates indicated that a total of $80-85 billion needed to be spent over the next five years for pollution control, including rapid transit developments, or $300 billion by the year 2000. Whatever the actual figure, Browne wrote in *The New York Times,* Oct. 11, 1970: "We do know for certain...that the cost will be enormous and that control cannot be met alone by nickel-and-dime taxes or nonreturnable bottles...[or] imposts on industrial polluters."

Who will pay the piper? The man in the street knows that he will, whether in the form of higher taxes or higher prices.

---

[26] Marshall I. Goldman, "The Costs of Fighting Pollution," *Current History,* August 1970, p. 79.

[27] "Pollution Price Tag: 71 Billion Dollars," *U.S. News & World Report,* Aug. 17, 1970.

But he now appears more willing to pay for pollution cleanup than he did in the past. A nationwide poll conducted by Louis Harris and Associates in April 1970 indicated that 54 per cent of the people were willing to pay $15 a year more in taxes to finance federal programs of pollution control. The same polling organization reported in November 1970 that Americans felt "the most serious problem" facing their communities was pollution. Of those interviewed, 34 per cent cited pollution as the foremost problem, whereas 25 per cent cited crime and 14 per cent cited drugs.

Environmentalists point out that ultimately the cost of controlling pollution will be offset by savings that accrue from cleaner surroundings. No one is quite sure how to measure the pollution toll. It is costing Americans $35 billion in ill health alone, by the reckoning of Dr. Paul Kotin, director of the National Institute of Environmental Science. *U.S. News & World Report* calculated that clean air could save the American economy $11 billion a year—the difference between an estimated $13.5 billion damage from air pollution and $2.6 billion spent to control it.

Adding of Anti-Pollution Costs to Pricing System

Economists say pollution-control technology lagged because the price system was not structured to take into account the cost of pollution. There was no incentive for manufacturers to spend money on developing and installing pollution-control devices; to do so would merely add to the cost of production without providing for an increase in return. The editor of *Environmental Science and Technology,* D. H. Michael Bowen, has pointed out an inevitable consequence of this situation: the striking contrast between the magnificent technology of new industrial plants and the technologically backward sewage system that serves to dispose of their waste products.

Interest is rising in finding some way of putting a price tag on pollution that could be applied somewhere in the marketing system to reflect its actual cost. And actual cost is coming to mean more than specific monetary losses from pollution, such as medical expenses, cleaning bills, repair or replacement of damaged materials, and so on. It also means "social costs" on which no monetary valuation can be put. Social costs are the sundry discomforts and offenses to the senses that individuals suffer as well as the disruptions of social order that are associated with a debased environment. President Nixon reflected the current interest in applying this concept to the economics of environmental protection when he mentioned, in his Aug. 10 message to Congress, "the failure of our economy to provide full accounting for the social costs of environmental pollution."

The costs, both social and monetary, could be offset by taxing or penalizing polluters. This practice would induce private individuals to refrain from practices that pollute, such as burning leaves and driving cars with poorly maintained engines. It would also motivate industry to find a way to cut down on pollution at minimal cost. Interest in cost-cutting is apparent in trade-journal advertisements of manufacturers of pollution-control devices. The best-selling devices, according to a recent trade survey, were those that promised low-cost operation as well as pollution control. General Motors' interest in the Wankel engine is attributed, in part, to the hope that, because of simpler structure, it will be less expensive to produce than the piston engine, thus offsetting the cost of pollution control.

Under the Tax Reform Act of 1969, an individual or corporation taxpayer is allowed to amortize the cost of a certified pollution-control facility over a period of five years. The facility must be certified by both state and federal agencies. Several states also provide tax relief for industrial companies to help them absorb the initial costs of pollution abatement. (See list, opposite page.) This relief is generally in three forms: (1) allowing the purchaser to accelerate the depreciation write-off value over a period of one to five years for purposes of income or franchise taxes, (2) exempting the purchase of pollution-abatement equipment from sales and use taxes, and (3) exempting pollution-abatement installations from property taxes.

Ultimately, experts say, entirely new systems for maintaining a clean water supply may be necessary. Starting afresh would be feasible in new planned communities but not in older cities where "new technologies cannot be tried...because they are incompatible with existing systems and obsolete legal, labor and taxation codes."[28] Much of the new technology is directed toward compatibility with existing systems, obsolete or not. Often the selling point is that the new device will make unnecessary the construction of an additional major facility. Controlling pollution at its source is often

---

[28] Athelstan Spilhaus, "The Experimental City," *America's Changing Environment* (1970, Roger Revelle and Hans H. Landsberg, editors), p. 222. See also "New Towns," *E.R.R.*, 1968 Vol. II, pp. 818-819.

STATES WITH TAX LAW INCENTIVES
FOR INDUSTRY POLLUTION ABATEMENT

ALABAMA: Exemptions from property, sales, use, income and franchise taxes granted for pollution abatement facilities.*

ARIZONA: A deduction for income and franchise taxes is granted with respect to the amortization of the facilities.

ARKANSAS: Certain exemptions from sales and use taxes.

CALIFORNIA: For purposes of personal or corporation income taxes, air and water facilities may be amortized for a period of one or five years.

CONNECTICUT: Exemptions from property, sales and use taxes; corporations are allowed a tax credit on income and franchise taxes.

FLORIDA: In computing property taxes, air and water equipment is assessed at no more than salvage value.

GEORGIA: Exemptions from property, sales and use taxes.

HAWAII: Air facilities exempt from real property assessment and from 4 per cent excise tax; for income and franchise taxes, five-year amortization permitted for air and water facilities.

IDAHO: Exemptions from property taxes.

ILLINOIS: Exemptions from sales, service occupation and use taxes.

INDIANA: Exemptions from property taxes, and from sales and use taxes unless facilities are constructed so as to be part of realty.

KENTUCKY: Exemptions from sales and use taxes.

LOUISIANA: Ten-year exemption from state or local property taxes.

MAINE: Delineated facilities exempt from sales and use taxes; property tax exemptions for all disposal systems that produce no salable product.

MARYLAND: Partial exemptions from sales and use taxes.

MASSACHUSETTS: Exemptions from property and sales and use taxes; deduction granted under income or franchise tax for water pollution equipment only.

MICHIGAN: Exemptions from property and sales and use taxes.

MINNESOTA: Property tax exemptions; income or franchise tax credit.

MISSOURI: Exemptions from sales and use taxes.

MONTANA: Exemptions from property taxes for air pollution control facilities only.

NEW HAMPSHIRE: Exemptions from property taxes.

NEW JERSEY: Exemptions from property, sales and use taxes.

NEW YORK: Exemptions from property, sales and use taxes; a deduction from income or franchise taxes allowed for pollution control expenditures. (Exemptions apply also to local sales taxes upstate but not in New York City.)

NORTH CAROLINA: Partial exemptions from property taxes; persons and corporations may claim state income tax deductions.

OHIO: Exemptions from personal property, sales and use, and franchise taxes.

OKLAHOMA: Exemptions from sales and use taxes; income tax credit granted.

OREGON: Taxpayer may choose either property tax exemption or income tax credit.

PENNSYLVANIA: Exemptions from sales and use taxes.

RHODE ISLAND: Exemptions from property, sales and use taxes; an income or franchise tax deduction may be claimed.

SOUTH CAROLINA: Exemptions from property, sales and use taxes.

TENNESSEE: Exemptions from property taxes.

VERMONT: Property tax exemptions.

VIRGINIA: An income tax deduction may be claimed.

WASHINGTON: Taxpayer may choose exemption from sales-use tax or take a credit against sales-use tax, business and occupation taxes, and public utilities taxes.

WEST VIRGINIA: Exemptions from sales and use taxes

WISCONSIN: Property tax exemptions; income tax deduction allowed.

WYOMING: Property tax exemptions.

All facilities mentioned refer to both air and water pollution control unless only one or the other is specified.

SOURCE: National Association of Manufacturers.

recommended too; this means simply not throwing so much contaminating stuff into the air or water. A major change in the basic industrial process itself may be necessary to achieve this goal.

Low Economic Return on Salvaged Waste Matter

Where technology of the future can be most helpful, both from the environment-protection and the cost-saving standpoint, is by developing feasible means of capturing valued elements in wastes and processing them for re-use. To some extent this is already done but not nearly enough to meet its potential. It has been estimated that 45 per cent of the iron and steel, 42 per cent of the copper, 25 per cent of zinc, 21 per cent of paper, and 12 per cent of rubber is recovered and re-used in American production.[29]

The reason there has not been more re-use of castoff materials is that it has not been profitable for garnerers of waste. Industry has usually preferred to get fresh materials rather than go to the trouble and expense of reclamation. The costs of collection, separation, cleansing, transportation, and reclamation will probably continue to be a barrier to re-use until an economic incentive can be applied somewhere in the marketing system.

Explaining some of the hurdles that lie between the city dump and the industrial stockpile, Robert R. Grinstead wrote: "Two changes in our official attitudes seem called for....First, we need to treat waste material industries on at least an equal basis with virgin material industries. In fact, until the technology of recycling matures, we may have to go further and favor it for a time, using such devices as subsidies...and a reduction or elimination of existing depletion allowances, favored tax positions, and lower freight rates...." Grinstead added that if the material producer can somehow be given at least part of the responsibility for the disposal problem, a powerful brake on extravagance would exist.[30]

Many possibilities for greater re-use are being explored. Sanitary landfill, probably the best-known practical use for solid wastes, is being extended to create recreational areas, modest hills for landscaping and ski slopes, and is even being considered as a foundation for airports on what is now shallow water close to shore. Another possibility is use of

[29] Walter O. Spofford Jr. (of Resources for the Future, Inc., Washington, D.C.), "Closing the Gap in Waste Management," *Environmental Science and Technology,* December 1970, p. 1109.

[30] Robert R. Grinstead, "The New Resource," *Environment,* December 1970, p. 17.

heat from incineration of community wastes as energy for the generation of electric power. The diversion of sewage sludge, with its many nutrients, to fertilize farmland is another usage that could be greatly extended, though it presents such problems as harmful chemicals and transportation costs.[31] The Federal Water Quality Administration has committed $2 million for a project to demonstrate that sewage and factory effluent now going into Lake Michigan can be diverted to fertilize barren land in Michigan.

Recycling: Hope for Turning Wastes into Assets

Recycling is the key word in environmental protection now. Recycling involves not merely the piecemeal salvage of this or that waste product, but a totally new approach to the use of resources, which conceives of them as passing through a closed system of use-reuse, so that nothing is essentially wasted though it goes through various stages in its passage through the closed cycle. As expressed by Kenneth E. Boulding, professor of economics at the University of Colorado:

> The great unsolved problem of technology is that of creating what is being called a "looped economy" in which man finds a comfortable life in the middle of the process which is essentially circular, that is, in which the waste products of human activity are all used as raw materials for the next cycle of production.
>
> We are still a very long way from this kind of technology, although there are the beginnings of it....Ultimately it is clear we will have to use the atmosphere, the oceans and the soil as inexhaustible material resources in the sense that what we take from them we will also put back into them.[32]

The "closed loop" principle is gaining in practical application. Dow Chemical Co., for example, is building 28 new cooling towers at its Midland, Mich., plant with the aim of reusing 50 per cent of its cooling water. A new Eastalco Co. aluminum plant at Frederick, Md., will contain a facility for treating contaminated water so that it can be returned to the plant for re-use rather than being dumped into the waterway. The current drive for returning soft drink bottles[33] represents a simple form of recycling. Other reuses of glass in crushed form, for paving materials for example, are being explored.

---

[31] See "Waste Disposal: Coming Crisis," *E.R.R.*, 1969 Vol. I, pp. 200-201.

[32] Kenneth E. Boulding, "The Scientific Revelation," *Bulletin of the Atomic Scientists,* September 1970, p. 16.

[33] "The increased use of non-returnable bottles for soft drinks has been phenomenal in the past decade. Shipments...advanced at an annual rate of 30 per cent."—"Glass Industry: Auspicious Future," *The Magazine of Wall Street*, Nov. 7, 1970.

To make recycling work on a grand scale that would truly safeguard "spaceship earth," it is necessary to view all of nature and all of man's activities affecting nature as parts of a single system. This would put technology into harmony with nature rather than at cross-purposes with it. As Dr. Lee A. DuBridge, former science adviser to President Nixon, recently pointed out, no one really wants to "go back to nature" in a true sense and few would forcibly stop the technology if they could. But technology is not, as some imply, an uncontrollable force; it is up to man to use it for his own good.

REAPPRAISAL OF COMPUTERS

by

Richard L. Worsnop

COMPLAINTS AND FEARS ABOUT COMPUTERS
Rising Opposition to Computerized Data Banks
Problem of Correcting Erroneous Computer Data
Disenchantment of Business With the Machines
Computer Vulnerability to Piracy and Sabotage

RAPID GROWTH OF THE COMPUTER INDUSTRY
Computation Aids From the Abacus to Computer
Basic Types of Computers: Analogue and Digital
Fundamental Principles of Computer Operation
Tasks That Computers Perform Most Efficiently
Rising Value of Computers to Medical Profession

SPECULATION ABOUT COMPUTERIZED FUTURE
Debate Over Prospect of Artificial Intelligence
Use of Computers for Translation, Composition
Possibility for Usage of Computers in the Home

1 9 7 1
May 12

REAPPRAISAL OF COMPUTERS

T HE ELECTRONIC COMPUTER, the president of Burroughs Corp. asserted several years ago, "has a more beneficial potential for the human race than any other invention in history."[1] At the time, few knowledgeable persons would have quarrelled with that appraisal. Computers, it was confidently predicted in the early 1960s, would make business more creative and efficient and thus help to raise per-capita income. In addition, the machines would eliminate drudgery and increase leisure time. The almost universal view of the computer was that of a benevolent servant of man.

Today, many former enthusiasts are having second thoughts. Businessmen have found that computers, though indispensable in many ways, lack the intuition needed to arrive at sound executive decisions. They have found also that the machines are highly vulnerable to sabotage and that electronically stored information can be pirated almost as easily as can files committed to paper.

Almost every consumer is indirectly acquainted with computers. Numbering about 60,000, the machines handle the vast majority of the nation's banking transactions as well as the customer accounts of utilities, credit-card companies, and large department stores. Rare is the person dealing with such establishments who has not encountered an error in his computerized bill. In most instances, the process of getting the error acknowledged—much less rectified—is lengthy and frustrating.

The most worrisome aspect of computers at present, however, derives not from weakness but from strength—their enormous capacity to store and retrieve information of all kinds. Civil libertarians charge that the United States is moving inexorably toward a "dossier society" in which all important details of an individual's life—his credit and police records, tax returns,

---

[1] Ray R. Eppert, president of Burroughs, 1958-66, quoted by Gilbert Burck, *The Computer Age* (1965), p. 1.

scholastic scores, and the like—can and will be stored in computer "memories." If this should happen, it is said, personal freedom would suffer a grievous, perhaps fatal blow.

Some degree of disenchantment with computers was inevitable. Every new invention has its drawbacks, which come to light only through experience. The computer age still is in its infancy, so no doubt additional shortcomings will come to light. Nevertheless, the computer is here to stay; if it were done away with, chaos would surely result. To cite just one example, the National Science Teachers Association stated that "by 1980, if all records were handwritten and hand-processed, an army of workers equal in number to the present population of the United States would be needed to process the records produced by the federal government alone."[2]

Problems and fears associated with computers are not the fault of the machines but of the people who use them. It is true that computers, capable of solving complex problems at fantastic speed, are in many ways a match for the human brain. At the same time, the computer is totally dependent on the operating skill of humans. For it to solve a problem such as shown on the opposite page, the problem must be broken down into sequential steps involving yes-no answers.

Rising Opposition to Computerized Data Banks

Concern about computers acting as giant dossier libraries began to surface six years ago. A study in 1965 sponsored by the Social Science Research Council proposed that a national data center be established to (1) centralize widely scattered statistics gathered by the federal government; (2) preserve such data more efficiently than was currently being done; (3) make it more accessible to users. Edgar S. Dunn Jr., former deputy Assistant Secretary of Commerce for Economic Affairs, reviewed the proposal for the Bureau of the Budget and recommended its adoption.

Opponents were soon heard. The House Subcommittee on the Invasion of Privacy, headed by Rep. Cornelius E. Gallagher (D N.J.), held hearings in July 1966 on the suggested data bank; the purpose, Gallagher said, was "to create a climate of concern, in the hope that guidelines can be set up which will protect the confidentiality of reports and prevent invasion of

---

[2] National Science Teachers Association, *Computers—Theory and Uses* (1964), p. 79. The publication was prepared by the association's Project on Information Processing under a grant from International Business Machine Corp.

Typical Computer Flow Chart

Problem: Program a computer for getting a reluctant boy to school

Solution: Divide his activity into a sequence of steps

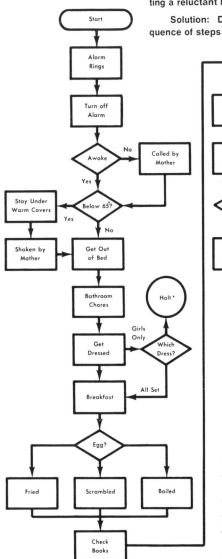
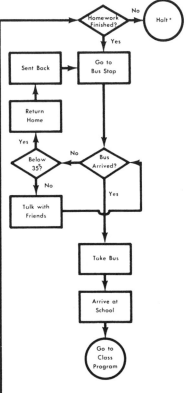

FLOW-CHART SYMBOLS

An oval box represents the starting of an operation.

A rectangular box indicates any processing operation except a decision.

A diamond indicates a decision.

A square box indicates the stopping of an operation.

A circle indicates a conditional halt pending some further operation.

Arrows indicate the direction of flow through the diagram.

SOURCE: National Science Teachers Association, *Computers-Theory and Uses* (1964).

individual privacy."[3] Similar hearings were held in March 1967 by the Senate Subcommittee on Administrative Practice and Procedure. The net effect of both sets of hearings was to arouse considerable opposition to the data bank among the public and the news media.

The data bank proposal, however, remains very much alive. It was examined again in hearings conducted in February and March 1971 by the Senate Subcommittee on Constitutional Rights. Opening the hearings, Chairman Sam J. Ervin Jr. (D N.C.) said: "Led on by the systems analysts, state and local governments are pondering ways of hooking their data banks and computers onto their federal counterparts, while federal officials attempt to 'capture' or incorporate state and local data in their own systems." Sen. Charles McC. Mathias (R Md.) said March 9 that the Justice Department alone had seven computerized data banks;[4] he added that they "are bringing the ammunition for persecution, harassment and idle gossip within the reach of every prosecutor and part-time deputy sheriff in the land."

Justice Department officials vigorously denied that the information contained in its data banks was abused. "I think it quite likely," Assistant Attorney General William H. Rehnquist told the Ervin subcommittee March 9, "that self-discipline on the part of the executive branch will provide an answer to virtually all of the legitimate complaints against excesses of information gathering." In a statement delivered to the subcommittee on March 17, FBI Director J. Edgar Hoover described the scope and operation of the bureau's National Crime Information Center.

He said the NCIC contained around 2.5 million active records on wanted persons, stolen vehicles, vehicles wanted in felonies, and other identifiable stolen property such as firearms and securities. It handled 60,000 transactions a day —inquiries, new entries, clearances, etc.—and was linked to 104 law-enforcement control terminals in the 50 states and Canada. While the majority of terminals employed manual teletype-like devices, the system also contained 35 computer-to-computer—or "interface"—links. Hoover asserted: "A person having a record has no absolute right to see it."

---

[3] See "Protection of Privacy," *E.R.R.*, 1966 Vol. I, p. 281.

[4] The FBI's National Crime Information Center; the Bureau of Narcotics and Dangerous Drugs' file on narcotics users; the FBI's Known Professional Check Passers file; the Organized Crime Intelligence System; the Civil Disturbance System; a file on offenders based on federal prison records; and the records of the Immigration and Naturalization Service.

A national data bank could prove beneficial in many ways. Arthur R. Miller wrote in *The Assault on Privacy:* "Some information specialists have suggested providing everyone with a birth number to identify him for tax, banking, education, social security, military and various other purposes....The goal is to eliminate much of the existing multiplicity in record-keeping while...expediting the business of society...For example, if a person falls ill while away from home, a local doctor could use the patient's birth number to retrieve his medical history and drug reactions from a central medical data bank."[5]

On the other hand, Miller added, "The identification number given us at birth might become a leash around our necks and make us the object of constant monitoring through a womb-to-tomb dossier." Setting up a national data bank would be easier than most people imagine. The Univac Division of Sperry-Rand Corp. has advertised a "nonfatiguing photochromic material...that can be used as a reservoir for computer information." According to Univac, use of this material may someday make it possible "to store the medical records of every American in the space of a cold capsule. Or the tax records of the nation may fit in one file cabinet."[6]

Problem of Correcting Erroneous Computer Data

Concern about computerized data banks is by no means confined to this country. Britain, Denmark and Sweden have set up committees to consider ways of protecting the individual from computers. The Dutch government has gone further: It is preparing a bill in the States-General (parliament) which would, if passed, enable citizens to learn what personal data are on file in government and business computers and to correct errors. A similar bill has been introduced in Congress by Rep. Edward I. Koch (D N.Y.). Under the Koch bill, every government agency maintaining records on individuals would have to (1) notify the person that such a record exists; (2) notify him of all transfers of such information; (3) disclose information from such records only with the consent of the individual or when legally required; (4) keep a list of all persons examining the record; and (5) permit the individual to inspect his records, make copies of them and supplement them. The bill provides exceptions in cases of national security, investigations for criminal prosecutions and safeguards for informers.

---

[5] Arthur R. Miller, *The Assault on Privacy* (1971), p. 4.
[6] Advertisement in *Time,* Sept. 27, 1968, p. 51.

Computer memory is superior to human memory in that a fact stored in the computer is never "forgotten" unless it is erased by an outside operator. The trouble is that the computer, lacking judgment or compassion, is unable to distinguish between facts and information that may be half-true or totally erroneous. Consumer advocate Ralph Nader contends that the files maintained by credit bureaus contain a good deal of inaccurate or incomplete information because "most of the material is obtained from others (merchants, employers) and not verified by them."

> The introduction of computers [Nader states] can create its own set of problems. Although mechanical errors in the handling of information by people may be reduced, the probability of machine error is increased. In addition, credit data are taken directly from a creditor's computer to a credit bureau's computer without discretion. Your payments may have been excused for two months, due to illness, but the computer does not know this, and it will only report that you missed two payments. Storage problems alone will prevent the explanation from being made. Your rating with that creditor may not be affected, but with all others it will be.[7]

The Fair Credit Reporting Act, which took effect April 25, 1971, is designed to cope with the type of problem Nader described. The new law requires all agencies that report credit information to follow reasonable procedures to assure the accuracy of their data. If a person is refused credit, insurance or employment on the basis of adverse information, he must be informed of the reason. Furthermore, the law grants a consumer the statutory right to sue in court to learn where the adverse information came from; if the information were proved to be erroneous, he could then sue for damages.

Disenchantment of Business With the Machines

Public disenchantment with computers is shared by numerous businesses that have invested heavily in the machines. In a special survey of computers in business, *The Economist* of London came to the startling conclusion that "many senior executives now regard their computer systems as the worst investment their companies have ever made."[8] Computers have enabled companies to speed up operations and thereby provide better service, but they rarely have reduced the cost of operations, even in routine clerical work.

Part of the trouble lies with the companies. Many of them rushed to buy computers without knowing if the machines

---

[7] Ralph Nader, "The Dossier Invades the Home," *Saturday Review*, April 17, 1971, p. 19.

[8] "Behind the Glamour," *The Economist*, Feb. 27, 1971, p. vii.

were suited for the task at hand; in some cases, the purchase of computers appeared motivated by a "keeping-up-with-the-Joneses" attitude. This attitude has penetrated the federal government also. The General Accounting Office reported to Congress in May 1971 that the various federal agencies were using 5,277 computers—but not using them to full advantage.[9] That number had increased almost tenfold in the past decade.

Now that the limitations of the machines are known, disillusionment has set in among businessmen. A leading cause for complaint is that the many variables that enter into an executive decision are difficult to translate into computer language. An attempt to force a decision from a computer in these circumstances could be disastrous. Mayford L. Roark, director of the Ford Motor Co. computer-systems office, said two years ago that "the academician, fortified with his technique, is likely to warp the problem so that he can find a solution. He may omit variables that don't fit in and get an elegant solution to something that is not the problem."[10]

Even when computers produce the desired result, the vagaries of human nature can negate it. "A simple example of this," Tom Alexander recounted in *Fortune,* "occurred when Johnson & Johnson instituted a new computerized inventory-control system based on typical patterns of orders for its products. But it turned out that the new system was so successful and dependable that customers learned they never had to worry about J. & J. being out of stock—so they changed their own patterns of buying and stocking inventory."[11]

Computers sometimes are faulted by businessmen for being *too* efficient; they complete complex tasks in a few minutes or hours and then stand idle and unproductive. Meanwhile, the computer industry continues to develop machines with greater speed and capacity and to urge them on companies that have yet to make full use of the computers they already have. Some companies are now converting from second-generation (transistor-based) to third-generation (microcircuit) computers, not because they need them, but because they fear the older machines may shortly become obsolete.

---

[9] The GAO did not recommend a specified number but suggested more sharing of computers among agencies. It noted that in 1969 the government spent $560 million to buy and rent computers.

[10] Quoted by Tom Alexander, "Computers Can't Solve Everything," *Fortune,* October 1969, p. 128.

[11] *Ibid.,* p. 168.

Computer users have learned to their dismay that it is relatively easy for an outsider to destroy or pilfer magnetically stored data. As Arthur R. Miller put it: "A simple magnet can be used to erase the information stored on a reel of magnetic tape. In one incident, a disgruntled employee virtually wiped out the records of a business enterprise 'in no time at all.' In another, a small band of anti-war protesters erased over a thousand Dow Chemical Company computer tapes that apparently contained data on napalm, poison gases and chemical weaponry. If technical expertise is lacking, a match, or a hammer in the case of a disc or a data cell, will do the same job more crudely in a minute or two."[12]

"Shared-time" systems, in which several companies make use of the same computer, offer a particularly tempting target for the data thief. One of the soft spots in such systems occurs when computerized information moves from the central processor through the communications links to customers. In addition to the relatively simple task of bugging the transmission line and recording the electronic communications passing through it, the wiretapper might attach his own computer terminal to the line and join the group sharing the system's services.

He could do so by intercepting a user's communication and substituting his own; by invading the system while a remote-access user has his channel open but is not transmitting, or by intercepting and canceling a user's sign-off signal in order to continue operating the system under that user's name. "Depending upon the sophistication level of the system's internal security procedures," Miller wrote, "the piggybacking tapper might be able to gain entry to all of the time-share customers' files."

The computer-security problem has given rise to a peripheral industry. Data Processing Security Inc., one of about a dozen companies in the field, will prepare a report pointing out security gaps, estimating the cost of reconstructing lost or destroyed data and recommending procedures and equipment to remove weaknesses. The fee for this service is between $3,000 and $5,000. The concern also offers, for $25,000, a security system that includes a double door with electric locks, magnet sensors and closed-circuit television to control access to the computer. Backup power systems, essential

---

[12] Arthur R. Miller, *op. cit.*, p. 28.

where computers cannot be permitted to shut down during a power failure, can cost anywhere from $25,000 to $1 million. Data Processing expects its 1971 revenue to reach $4 million —more than 10 times that of the previous year.

Rapid Growth of the Computer Industry

MAN HAS DEVISED machines to lighten his physical and mental labors from the earliest days of civilization. The abacus, a rudimentary counting device developed centuries ago in China and Egypt and still employed there today, was the forerunner of a gear-driven adding and subtracting machine built in 1642 by Blaise Pascal. Thirty years later, the German philosopher Wilhelm Leibnitz exhibited before the Academy of Paris a device that could multiply, divide and extract roots as well as add and subtract.

Charles Babbage, a mathematics professor at Cambridge University, is regarded as the father of the computer. Subsidized by the British government, he undertook in 1823 to construct a "difference engine" to calculate and print such things as tables of logarithms. The machine was almost completed when, in 1833, Babbage abandoned it. He devoted the remainder of his life to development of an "analytical engine" that would automatically evaluate any mathematical formula. But this scheme proved too ambitious; the analytical engine was unfinished when Babbage died in 1871.

The next important step in evolution of the computer came in connection with the United States census in 1890. The 1880 census, which enumerated 50 million persons, had taken seven years to compile. As a result, there was concern that full results of the 1890 count of 63 million persons would not be available before the end of the decade. Such a delay would have prevented reapportionment of seats in the House of Representatives as required by the Constitution. It was to avert a crisis of this kind that the U.S. Census Bureau made use of an invention of Herman Hollerith, a statistician from Buffalo, N.Y. In 1889, Hollerith had constructed an automatic data-handling machine that employed perforated cards. Thanks largely to this device, the census taken in 1890 was compiled in little more than two years.

Research initiated during World War II led to development of ENIAC,[13] the first electronic computer. Two and one-half years in the making, ENIAC was introduced at the University of Pennsylvania in 1946. It solved its first problem, a question involving atomic physics, in two weeks. The machine contained 18,000 vacuum tubes and could carry out 5,000 additions a second. In contrast, today's computers contain transistors or microcircuits instead of tubes and can make as many as several billion additions a second.

Basic Types of Computers: Analogue and Digital

Computers usually are divided into two broad groups: analogue and digital. Digital computers work by using specific information which is commonly in the form of numbers. Analogue computers, on the other hand, usually process continuous information. They are often likened to speedometers, which translate rotation of the wheels into an approximation of speed, shown on a dial in terms of miles per hour. In much the same way, an analogue computer uses constantly changing measurements to solve thousands of interrelated equations. Because of their ability to handle simultaneously such variables as wind drift, aircraft speed, ship speed, range and direction, analogue computers now form the backbone of military tracking systems.

Digital computers are descendants of the machines invented by Pascal, Leibnitz, Babbage and Hollerith. Technically, the mathematics done by any computer consists only of addition, although problems involving subtraction, multiplication, division and more complicated calculations are continually run through the machines. The apparent contradiction between the technical limitations and the problems actually processed by the machines can be explained in part by the speed at which computers work. For instance, multiplication is done by many rapid additions, subtraction by the addition of complementary numbers, and division by many rapid subtractions.

The ideal, all-purpose computer has yet to be developed. If it were, it probably would be a hybrid combining the best features of digital and analogue models. The nature of the job at hand determines what kind of computer is best suited to handle it. "Analogue computers are best when direct physical measurements are to be used as data for the calculations, when each problem has only one sequence of operations needed

---

[13] A name, pronounced EENY-ack, is formed from the initial letters of the words Electronic Numerical Integrator and Calculator.

COMPUTER LOGIC AND BOOLEAN ALGEBRA

Computer "logic circuits" are based on some very simple concepts. They utilize some of the same rules of logic used to design home lighting circuits—the rules of Boolean algebra, invented in the 19th century by an English mathematician named George Boole.

A simple example is a light that can be turned on or off at either end of a room. Every time you flip that light on or off you are operating a logic machine, working in Boolean algebra and making the light circuit operate in basically the same way as the adding circuit in a digital computer.

A Boolean proposition is either true or false and these two conditions can be represented in any two-state (binary) device, such as a simple on-off electrical switch, relay, vacuum tube or transistor. Boole devised a set of algebra-like rules of manipulating true-false propositions, using the three fundamental connectives: *Or, And, Not.*

In its simplest form, an *Or* connective means that a condition, C, is true if either A or B or both are true. The simple electrical circuit shown below illustrates the idea. With two on-off switches in parallel, current flows and the light goes on when either switch A or switch B or both are closed. The numeral I represents a switch closed and current flowing and O represents a switch open and current not flowing.

A	B	C
O	O	O
O	I	I
I	O	I
I	I	I

A slight revision of the circuit, with the two switches in a series, illustrates in the following table a Boolean *And* function. In this case, condition C is true (current flows and the light goes on) only when A and B are true (when both switches are closed).

A	B	C
O	O	O
O	I	O
I	O	O
I	I	I

The Boolean *Not* operation is used in digital computer circuits to invert the "logic" of electrical signals—for example, to convert an impulse input to a no-pulse output, a one to a zero, or vice versa.

Combinations of *And, Or* and *Not* functions, carried out electronically, enable the computer to handle functions of arithmetic and logic.

SOURCE: Adapted from International Business Machines Corp., *More About Computers* (1969), pp. 18-20.

for its solution, and when accuracy of high degree is not needed. Digital computers are most useful when the rules for solving the problem can be stated in a yes-no fashion, when the problem to be solved requires a flexible program, and when a high degree of accuracy and precision is desired."[14]

Fundamental Principles of Computer Operation

Understanding computers is not easy, even for experts in the field. *The Economist* pointed out: "On the hardware side, it requires considerable study just to grasp how the electronic circuitry shuttles electrons around to accomplish its tasks. It would take a good understanding of physics and chemistry to see how these circuits can be designed into small chips of solid material. Magnetic tape, on which much data is stored, is a separate technology. The delicate recording heads that read and write the data on the tape make up yet another specialization. On the programming side, the master programs in today's computers are themselves beyond all but a few experts."

Putting complexities aside, however, it can be said that computers have five principal segments:

1. *Input*—the portion of the machinery which allows the coded information to get from the outside into the computer. Input also is defined as the information fed into the machine.

2. *Storage*—the place where the information to be processed and the rules to be used in processing it—the program—are stored until needed.

3. *Control*—the portion of the computer which directs the flow of information in and out and the manipulation of the information inside the computer.

4. *Arithmetic or logic*—the portion in which the processing of information actually takes place.

5. *Output*—a means of getting the results of the information processing out of the computer to the people who want it. The same term ordinarily is applied to the processed information itself.

Computers cannot "think" for themselves—at least not yet; they are unable to learn from experience and cannot "change their minds." Every step or instruction must be provided for the computer in great detail by a human programmer. In preparing a set of instructions, programmers usually go through four steps: (1) they analyze the problem completely and break it down into parts; (2) they prepare a plan, called a flow chart, for solving the problem one step at a time; (3) they translate the flow chart instructions into the particular machine lan-

---

[14] National Science Teachers Association, *op. cit.*, p. 70.

SYSTEM OF BINARY NUMBERING

In both the binary and decimal systems the first two digits have identical values when standing alone.

Binary   Decimal

0   equals   0
1   equals   1

Beyond that the differences begin. Two is written in the binary system as 1 followed by 0, thus 10. Three is written as 11 (10 plus 1). But four becomes 100. Why? That is because of a correlation between the binary system and the decimal system's method of raising a number to some "power." A power is shown by an exponent, like 2^2 (2 x 2) or 2^3 (2 x 2 x 2).

Since

Binary number		Decimal number
10	equals	2

Then

100 (10 x 10)	equals	4 (2 x 2)

And since the decimal number 3 is expressed in binary terms as 11 (10 plus 1), then the decimal number 5 is expressed in binary terms as 101 (100 plus 1). Six is 110, seven is 111 and so on.

guage that the computer is designed to handle; (4) and finally, they test the completed program on sample data to see if it works properly.

The language of the computer is based on a code of high-frequency electrical pulses, or "bits,"[15] generated and timed by an electronic clock in the machine. For this reason, computers employ the binary number system (see above), which consists of the digits one and zero; the one represents a pulse, or bit, and the zero represents no pulse, or no bit. Combinations of ones and zeros, arranged in proper sequence, can be made to represent any quantity, alphabetical character, punctuation mark or special symbol.

This basic binary language of computers has spawned a number of dialects to serve specific programming purposes. One of the first to be developed was FORTRAN (an acronym for Formula Translation), which is based on the language of algebra plus a few rules of grammar and syntax imposed by the computer. FORTRAN was devised primarily for engineers and scientists; it was of little use to the businessman who wanted to solve prob-

---

[15] Short for binary digit.

lems of inventory control, payroll, or billing. To plug this gap, COBOL (Common Business Oriented Language) was developed in 1960. COBOL permits business problems to be expressed in an approximation of business English; it consists of a basic vocabulary of 250 key words, such as "compute," which can be used in combination with other words and symbols.

Most early computers were designed separately for business or scientific applications because the requirements in the two areas were quite different. But the differences have blurred in recent years. Many business problems now require sophisticated mathematical techniques for such jobs as forecasting and sales analysis, while scientists find themselves increasingly involved in the analysis of business statistics. Accordingly, a new all-purpose computer language called PL/I was devised to serve both classes of users. Said to combine the best features of other high-level computer languages, PL/I permits inputs of various kinds and thus affords greater flexibility.

Another development in the manufacture of computers holds the promise of faster access to stored data. A tiny chip of silicon, about one-eighth of an inch square, is being used on some models to replace an established memory device called "core." An individual core is capable of storing one "bit"—one yes or one no of the computer code. In contrast, Gene Bylinsky of *Fortune* wrote, "a single silicon chip only a little bigger than an average core can hold hundreds, even thousands of such coded instructions in its labyrinthine circuits."[16]

Tasks That Computers Perform Most Efficiently

Computers perform thousands of different tasks that affect the daily lives of everyone. In some large cities, the machines regulate traffic flow by measuring the number of motor vehicles on the road and then altering the sequence of traffic signals to speed movement. Engineers and architects can design a machine or a building in much less time with the aid of a computer than by using a desk calculator and a slide rule. And the entire space program would have been virtually impossible without computers to monitor such things as fuel conditions and calculate flight paths.

To date, the computer has proved most valuable as a kind of super-clerk, capable of handling repetitive operations millions or trillions of times without ever becoming bored or tired and therefore error-prone. The fact that computers are able

---

[16] Gene Bylinsky, "Little Chips Invade the Memory Market," *Fortune*, April 1971, p. 101.

to perform such chores in a tiny fraction of the time required by human beings is an added advantage. The machines thus have transformed the field of information storage and retrieval. They are invaluable in keeping track of payrolls, bills, banking transactions and inventories; in helping law-enforcement officers to determine in seconds whether an automobile is stolen; in facilitating reservations on trains and airplanes and in hotels.

The storage and retrieval capacity of computers has found wide application in publishing also. As *Publisher's Weekly* pointed out: "The applications that are most suitable to the versatility of electronic technology are in manipulation and rearrangement of listing, such as indexes, directories, parts catalogues, compendia—any listing in which the tape used for one purpose can be reused for another."[17] The many computer-produced publications in existence today include the whole array of catalogues and indexes issued by the Government Printing Office; telephone directories; and *The New York Times Index*. Computers used in printing can be programmed to justify lines of type, but they do not always produce the correct hyphenation of individual words.[18]

Both military and industrial planners use computers to simulate complex processes that would be too costly or dangerous to test in actual situations. At the Pentagon, nuclear war games are played by machine; at aerospace companies, design problems are worked out by computer before the first prototype is built. In industries with automated processes, computers are more reliable than workers. For example, "A computer can run a turret lathe, a milling machine, and many other machine tools more rapidly and accurately than a man can operate a machine tool."[19]

Rising Value of Computers to Medical Profession

Computers are becoming increasingly valuable in the undermanned, overworked medical profession. In hospitals, the electronic processing of records and accounts is commonplace. Moreover, computers can perform such routine tests as urinalysis and blood typing as easily as they can run a machine shop. And they can record answers to yes-or-no questions about past illnesses or symptoms of new patients.

[17] "Using the Computer for the Appropriate Job," *Publisher's Weekly*, Dec. 4, 1967, p. 64.
[18] See "Computers in Publishing," *E.R.R.*, 1968 Vol. II, p. 505.
[19] Robert G. Middleton, *Computers and Artificial Intelligence* (1969), p. 12.

In time, the computer may handle much of the diagnostic work now done by physicians. Some doctors in rural areas already rely on computers for this type of assistance. One of them is Dr. Billy Jack Bass of Salem, Mo., a town of about 4,000 people situated 80 miles southeast of St. Louis. His office in Salem is connected by special telephone line to the University of Missouri medical school at Columbia. Electrocardiograms taken by Bass are transmitted to Columbia and fed into a computer; the computer's analysis of the EKG is returned immediately to Bass.

Bass believes that such computer hookups will encourage more physicians to set up practice in rural areas, which are severely short of doctors at present. He also sees a more substantial benefit: "Having all this material recorded [on computers] will produce a different kind of physician, a more thoughtful man."[20]

Some medical authorities foresee the day when computers will enable every American, rich or poor, to have a comprehensive health examination once a year. The cost per person is expected to be well under $100 a person, or much less than is now charged for such a checkup performed by a physician working alone. The availability of such examinations could bring dramatic improvement in the state of the nation's health, for illness is most amenable to treatment when diagnosed early.

Speculation About Computerized Future

THE QUESTION about computers that arouses the most in-interest and apprehension in laymen is whether the machines ever will become intelligent in the human sense of the word. Arthur C. Clarke, author of numerous books of science fiction and nonfiction, believes the answer is yes. "It should be realized that as soon as the borders of electronic intelligence are passed, there will be kind of a chain reaction, because the machines will rapidly improve themselves," Clarke wrote in 1968. "In a few generations—*computer* generations, which

[20] Quoted in *Los Angeles Times,* April 9, 1971.

by this time may last only a few months—there will be a mental explosion; the merely intelligent machine will swiftly give way to the ultra-intelligent machine."[21] On the other hand, Jeremy Bernstein asserted in *The Analytical Engine* (1966) that "the distinction between a schoolboy doing a multiplication a minute and a computer doing a hundred thousand multiplications a second is only one of degree."

The theme of a man creating a creature or a machine more powerful or intelligent than himself is a familiar one in fiction. It appeared in the last century in Mary Shelley's *Frankenstein* and since then in any number of science-fiction stories and films. One of the most vivid expositions of the man-versus-machine motif was contained in the film *2001: A Space Odyssey*, written by Clarke and director Stanley Kubrick. To most viewers the dominant character of the picture was not any of the astronauts or scientists but HAL,[22] the talking computer. The climax of the movie occurred when the sole surviving astronaut on the spaceship to Jupiter in effect murdered HAL by disconnecting all his memory cells.

To a large extent, the debate on the possibility of artificial intelligence begs the question of what exactly intelligence is. It is generally agreed that the computer's superiority to man in the area of processing data at great speed does not constitute intellectual superiority. Herbert Simon, a social scientist, and his associate, Allen Newell, suggested that what the computer lacks is the ability to produce "effective surprise."

> Suppose [they wrote] a computer contains a very large program introduced into it over a long period by different programmers working independently. Suppose that the computer had access to a rich environment of inputs that it has been able to perceive and select from. Suppose—and this is critical—that it is able to make its next step conditional on what it has seen and found, and that it is even able to modify its own program on the basis of past experience, so that it can behave more effectively in the future.
>
> At some point, as the complexity of its behavior increases, it may reach a level of richness that produces effective surprise. At that point we shall have to acknowledge that it is creative, or we shall have to change our definition of creativity.[23]

Ulric Neisser of the Massachusetts Institute of Technology made the additional point that computer intelligence "does

---

[21] Arthur C. Clarke, "The Mind of the Machine," *Playboy*, December 1968, p. 116.

[22] An acronym for Heuristically programmed ALgorithmic computer.

[23] Quoted by Gilbert Burck, *op. cit.*, pp. 136-137.

not grow, has no emotional basis, and is shallowly motivated." He added: "These defects do not matter in technical applications, where the criteria for successful problem-solving are relatively simple. They become extremely important if the computer is used to make social, business, economic, military, moral and government decisions, for there our criteria of adequacy are as subtle and as multiply motivated as human thinking itself."

Use of Computers for Translation, Composition

Computers already have been assigned tasks that would seem to call for human intelligence, but the results only serve to point up the machines' limitations. An example is that of translation. In the first language-translation computers a small bilingual dictionary was employed, with the corresponding words in the two languages recorded on magnetic tape. It was soon apparent that word-for-word translation was virtually incomprehensible. The addition of a dictionary of phrases brought only marginal improvement. An attempt to translate the maxim "out of sight, out of mind" into Russian and then back to English illustrates the difficulties involved. The computer's English retranslation read: "The person is blind, and is insane."[24]

Computers also have been programmed to compose music and poetry, and the results bring to mind Dr. Samuel Johnson's remark about the dog that could walk on its hind legs —"It is not done well, but you are surprised to find it done at all." Almost everyone who has heard computer music describes it as disembodied, lifeless and superficial. Computer verse, on the other hand, often contains striking or amusing images, as in this example.

> To belch yet not to boast, that is the hug,/The high lullaby's bay discreetly crushes the bug./Your science was so minute and hilly,/Yes, I am not the jade organ's leather programmer's recipe./As she is squealing above the cheroot, these obscure toilets shall squat./Moreover, on account of hunger, the room was hot."

Actually, the production of computerized "poems" is a relatively minor achievement of programming: The machine is fed a vocabulary and instructions on rhyming and syntax. It then is able to compose a stream of verses at random.

Some authorities, nevertheless, believe that computers eventually will display at least a low level of creativity. Two

---

[24] Robert G. Middleton, *op. cit.*, p. 3.

physicists associated with the Lawrence Radiation Laboratory at Livermore, Calif., recently stated that "in a decade or two, research laboratories may well have programmed computers to write acceptable potboiler detective novels."[25] It has been suggested also that simple, comprehensible verse and popular songs may someday be within the capability of computers.

An unanswerable question at present is whether computers ever will possess the ability to consistently turn out works of poetry or music, or whatever, that genuinely engage human emotions. It is often said that computers do not duplicate human behavior; instead, they simulate it. A "creative" computer, therefore, probably would be one that is programmed to anticipate human responses to patterns of notes or words. Needless to say, such a program would be enormously difficult to write. There is the added difficulty that any emotional reaction to a computer-made sonnet might be lessened by knowledge that the poem came from a machine, not a human brain.

Possibility for Usage of Computers in the Home

A new appliance, the "home computer," is expected to come into widespread use before the end of the present decade. It probably will be no larger than a suitcase and will come equipped with a display screen, typewriter console, and numerous command buttons. The home computer will perform at the domestic level many of the tasks that larger machines perform in business and industry; it will act as switchboard, information storehouse and calculator.

In taking over these and other routine chores, the home computer may well alter the pattern of daily living. Engineers, scientists, clerks, bookkeepers and many salesmen would be able to work almost entirely at home, using computers, teleprinters and face-to-face television. Occupations that supply ideas and information could become decentralized and thereby relieve some of the pressure of urbanization.

In combination with cable and cassette television, the home computer holds promise of transforming any living room into a complete audio-visual entertainment center. The machine could serve as access point to home libraries of recorded music or television programs—or to much larger libraries of such

---

[25] Gregory Benford and David Book, "Promise Child in the Land of the Humans," *Smithsonian* (magazine of the Smithsonian Institution), April 1971, p. 62.

material controlled by a central computer. And it would handle the bills not only for pay-television offerings but also for utilities, charge accounts and other household expenses.[26]

It is conceivable that the home computer will take on some of the more tedious housekeeping jobs, such as carpet-cleaning. As envisioned by Gregory Benford and David Book, it would work like this:

> Suppose that underneath the rug there is a grid of wires through which a tiny current flows. Suppose the chairs and tables and other furniture have metal plates in their legs and bases, each of a different size or shape. Lying above the grid, these plates respond to the current in the floor by developing induced currents.
>
> They in turn reach back on the grid. The resulting disturbances in the grid current can be analyzed to show exactly where each object is standing. Now, imagine a self-propelling steerable vacuum cleaner controlled by the central computer. If the computer reads the location of each object in the room from the grid current, it can guide the cleaner around the room, avoiding furniture without missing part of the rug.

Home computers, like others, would operate by electric current. It is conceivable, then, that the machines would place intolerable demands on the nation's limited supply of electric power.[27] There is the added danger that, in the event of a power brownout or blackout, home computers would malfunction or cease to operate altogether. In a highly computerized society, such an occurrence could cause much more disruption than power failures do at present. On the other hand, computers in the home and at central locations might well be programmed to monitor electricity use and thus head off potential power crises.

The multi-purpose home computer might do more than increase leisure time. By acting quite literally as a servant of man, it could ease the apprehensions of people who regard computers primarily as job rivals or repositories of derogatory information. Any such criticism of computers *per se* is misdirected; the shortcomings of the machines only reflect the shortcomings of the people who make and use them. In the final analysis it is up to computer manufacturers, government, and private citizens to insure that computers are used in ways that enhance the quality of human life.

---

[26] See "Video Revolution: Cassettes and Recorders," *E.R.R.*, 1971 Vol. I, p. 227; also "Cable Television: The Coming Medium," *E.R.R.*, 1970 Vol. II, p. 667.
[27] See "Electric Power Problems," *E.R.R.*, 1969 Vol. II, p. 937.

ENCOUNTER GROUPS

by

Helen B. Shaffer

1 9 7 1
Mar. 3

ENCOUNTER GROUPS

THE RAPID GROWTH of the encounter group movement has raised serious questions about the efficacy and even the hazards of a wholesale plunge by millions of Americans into consciousness raising and sensitivity stimulation, as it is now practiced in many places all over the country. The questioning, however, is seldom critical of the basic principles of the encounter group process, which are grounded in the findings of the behavioral sciences.[1] It is the rare critic who does not find something to praise in the group encounter procedure, provided it is properly conducted; that is, if it is supervised by a qualified leader, if the participants are screened to eliminate those who might be hurt by the experience, and if the process is directed toward specific and reasonable goals.

Many see in the encounter group a great potential for improving human relations and thus for reducing social tensions. Much hope rests with its yet-unproved usefulness for creating a bridge of mutual sympathy between hostile groups: between blacks and whites, police and community, elders and youth. The most enthusiastic of its adherents believe the encounter group holds a key to the creation of a better society composed of more joyous and lovable human beings. Others make more modest claims for it: that it teaches individuals how to work and live more satisfactorily with others—on the job, in the home, and in society. Still others dismiss the encounter group movement as a mere cult. And there are some who condemn it as immoral and anti-American.

Concern over the development of encounter groups applies largely to the lack of legal or professional controls over their operations. The door, unfortunately, is wide open for charlatans and misguided amateurs. Their appeal is all the more insidious in view of the vulnerability of the troubled, the emotionally unstable, and other persons with problems that might

---

[1] The behavioral sciences are psychology, sociology, anthropology, and areas of other disciplines that deal with human behavior.

more suitably be brought to the attention of a psychiatrist or a family counselor.

Responsible leaders try to make it clear that the encounter group experience is not intended as a form of psychotherapy, although some of the techniques are similar to those used by licensed practitioners in group therapy. The encounter group is intended for the mentally and emotionally well-balanced, the satisfactorily functioning individual who wishes to improve his performance, extend his perceptions, enlarge his outlook, make his life better or more productive in some way.

Though it has many aspects of a fad, and may even already have passed its peak of popularity, the encounter group movement appears to be settling in for a long stay on the American scene. Like listening to rock 'n' roll or watching television, participating in an encounter group is becoming one of the many things Americans can do with their time, money and energy. It may be viewed as a newly available route to self-knowledge or self-improvement, more enticing to many than taking courses, attending lectures, reading uplift books, consulting counselors or participating in all-night rap (talk) sessions with good friends. In some ways, the encounter group experience embraces all of these, plus an extra ingredient of untrammeled self-expression.

Even if the attractions of signing up for an encounter group weekend or series should pall as a popular activity, the movement is likely to have a lasting effect on standard practices in education, personnel management, the conduct of voluntary organizations, counseling, and many other activities that require interpersonal relationships. Even the professional practice of psychiatry is gaining and applying new knowledge from non-psychiatric encounter groups.

Diversity of Group Practices and Participants

It may not be generally known that one form of the encounter group has been around for a quarter of a century. Known earlier as sensitivity training or laboratory training, it has long been used as part of a practical training procedure by big business, school systems, voluntary organizations, churches, law enforcement agencies, and other pillars of the American establishment. The original purpose was to improve small-group performance in the organizations. Then it was found that improving group performance involved an exciting process of individual acquisition of self-knowledge. This discovery led to the growth of more fanciful programs that center on conscious-ness-raising of the individual. But all types of encounter group activity seek to knock down communication barriers between individuals, particularly the barriers of social decorum. And all forms of encounter group activity seek to awaken in the individual a deeper concern for others in his group and thus, it is hoped, a deeper concern for the "other" in the world at large.

It has been estimated that at least 20 million Americans have been exposed to the encounter group movement in one form or another.[2] Warren Bennis, a psychologist and leading authority on group techniques, recently ventured his belief that some six million Americans were then participating in encounter groups. Citing the Bennis estimate, Bruce L. Maliver added: "So many people are participants...that organizing and leading encounter groups has become fantastically profitable...[and] could even be regarded as a growth industry of the 1970s."[3]

"Encounter groupers" may be businessmen, top executives or middle managers of big corporations, teachers, policemen, students, doctors, lawyers, psychologists, writers, secretaries, housewives—virtually anyone who hankers for and can afford the experience. Some encounter groups may be part of an in-training program of a business or a public agency. Such sessions are usually directed toward a particular goal of the sponsoring organization. In other cases an individual seeks to join or organize a group for various personal reasons and the group engages a leader. By and large, the encounter group is middle-class and white. However, the encounter group experience has reached the inner city through its use by police departments and other urban agencies.[4]

One sign of the rising popularity of encounter groups is the rapid accumulation of "literature" on the movement. Articles on sensitivity training or encounter groups have appeared not only in newspapers and magazines of general readership, but also in journals addressed to educators, businessmen, church groups, law enforcement officials, manpower specialists, psychologists and medical practitioners. Among recent books are Jane Howard's *Please Touch* (1970), a reporter's account of

---

[2] Robert Fulmer (professor of management, Georgia State University), "Making Sense of Sensitivity Training," *Association Management*, May 1970, p. 49.
[3] Bruce L. Maliver (New York psychologist-psychoanalyst), "Encounter Groupers Up Against the Wall," *The New York Times Magazine*, Jan. 3, 1971, p. 4.
[4] See Robert Schrank and Susan Stein, "Sensitivity Training: Uses and Abuses," *Manpower*, July 1970, p. 3.

her personal experiences in a dozen different types of encounter groups; William C. Schutz's *Joy* (1967), an exposition of the theory and techniques used at Esalen at Big Sur, Calif., one of the largest and best known encounter group establishments; *Encounter* (1969), a symposium edited by Arthur Burton; John Mann's *Encounter: A Weekend With Intimate Strangers*, a semi-fictionalized account of a two-day group experience; Martin Shepard and Marjorie Lee's *Marathon 16* (1970), a report on a 16-hour encounter session; and *Carl Rogers on Encounter Groups* (1971), a view that the movement is a by-product of affluence.

Numerous textbooks and other instructional material, such as brochures, pamphlets, reprints of journal articles, tapes and films, have been issued by establishments that provide leaders and settings for encounter group sessions. Leading sources of such material are the National Training Laboratories Institute for Applied Behavioral Science (known as NTL), an independent affiliate of the National Education Association in Washington, D.C., and the Esalen Institute.

Proliferation of Centers for Personal Growth

No one knows for sure how far the movement has spread. No central agency exists to gather data of this sort, nor is there a national organization to speak for the entire movement. There is not even a clear definition of the limits of this kind of activity. At one end of the spectrum the encounter group is hardly more than a new form of in-service training or adult education; at the other end it is a half-mystical experience of psychic rebirth. In fact, the encounter group is only one phase of a larger development known as the human potential movement, which embraces a variety of efforts to tap more of the individual's innate capacity to work, to produce, to love, to express creativity, to get along with others.

According to an informal survey published in April 1970, there were at least 200 encounter groups in the immediate vicinity of Palo Alto, Calif.[5] Another account speaks of more than 100 "growth centers" that have sprung up recently across the nation.[6] Growth centers provide a variety of group experiences either on their premises or elsewhere—perhaps at a

---

[5] American Psychiatric Association, *Task Force Report: Encounter Groups and Psychiatry*, April 1970, p. 4.
[6] Bruce L. Maliver, *op. cit.*, p. 4.

school, a client's conference room or a resort hotel.[7] An informal directory compiled by NTL staff members lists 150 organizations that offer training in "personal growth." Much of the activity does not have a fixed location or an institutional character. There are free-wheeling encounter group leaders, sometimes referred to as "circuit riders," who travel about, setting up encounter sessions wherever there is a demand for their services. Some encounter grouping is *ad hoc*—simply a group that comes together with or without a trained leader to try out encounter group techniques.

That an appreciable growth has taken place is undeniable. Jane Howard, who spent a year investigating various manifestations of the movement across the country, drew up a list— admittedly incomplete—of 77 growth centers and kindred establishments in the United States, half of them in California. Her tally also included 66 colleges and universities that have been involved, officially or otherwise, in encounter group activity, 44 corporations that have used various forms of "laboratory training," and 12 organizations that sell or offer sensitivity training materials and services. Encounter group agencies exist in Atlanta, Austin, Boston, Cleveland, Denver, Kansas City, Los Angeles, Portland, Ore., and New York, among many other places. They are found not only in big cities but also in "retreats" far from the urban hurly-burly.

The movement is not limited to the United States, though it has had its greatest growth here. Miss Howard tells of meeting an emissary from the Mitsubishi Bank in Tokyo who was spending four months visiting sensitivity training centers in this country. He told her the movement had affected 3,000 Japanese to whom it was known as "Western Zen," because "it is a way to pursue truth." Among other nations she cited as having establishments similar to NTL and Esalen, or groups that have experimented with encounter group techniques, were Australia, Bolivia, Canada, Chile, Colombia, Denmark, France, Holland, India, Norway and Paraguay. Esalen has reported that institutes patterned after it have been set up in Canada, Mexico, and England, as well as approximately 100 in the United States. According to an Esalen flyer: "Visits by reporters from the BBC, German TV, French TV, Italian TV, [Paris]

---

[7] The Concord Hotel in New York's Catskill Mountains added an optional encounter group session to its popular "singles weekend" series for unmarried guests.—See Judy Klemesrud's "Having Wonderful Encounter," *The New York Times Magazine*, Dec. 20, 1970.

L'Express and other foreign media indicate increasing interest abroad."

Common Characteristics Among Diverse Groups

Growth of the movement was so rapid that the American Psychiatric Association, which has a natural interest in a development outside its control that impinges on its professional bailiwick, appointed a task force to investigate what was going on. The Task Force, which was headed by Dr. Irvin D. Yalom, reported in April 1970 that the field was so new, its sources so diffuse, and its implications so far-ranging that any report at this stage must remain tentative. The investigative team regarded the development as "a radical change and rapid growth of the small group field." The change was considered so great that "it is difficult to define the boundaries of the field and the related problematic issues."

Even the psychiatrists were confused by the great diversity of small group activity. "No doubt we court semantic confusion by attempting to cluster a wide array of group approaches under a single rubric, for there has been such a spate of new techniques that no one term can characterize the field." It cited the following names now in use to indicate the diversity of approaches: T-groups, sensory awareness groups, marathon groups, truth labs, psychological karate groups, human relations groups, personal growth groups, psychodrama groups, and human potential groups.

Whether each of these terms indicates a distinct technique could not be determined. Certain features, however, were nearly universal: All attempt to provide participants with an intensive group experience. The groups are small—usually with six to 20 members—permitting considerable face-to-face interaction. They focus on the "here-and-now," that is on what is actually happening in the group encounter and what the participants feel and think at the moment. This is in contrast to therapy, which is much concerned with the past in a search for the origins of present conditions.

Sensitivity groups strive to increase "inner awareness." They encourage expression of strong emotion. A prevailing theme is that of change. Individuals in the group are seeking some kind of change—a change in their own behavior, in their outlook on life, on their attitude toward other people, or they are trying to learn how to effect beneficial change in the functioning of a group. Whether such change comes about and, if so, whether it is lasting and beneficial remains in debate. Answers to these questions may determine the future of the encounter group.

Quarter-Century of Group Development

THE GENESIS of the encounter group movement can be traced to a workshop for community leaders held on the campus of the Central Connecticut State College (then State Teachers College) at New Britain in June 1946 under the auspices of the Connecticut Interracial Commission. The object of the workshop was to train local leaders in ways of winning wholehearted support in their communities for the state's Fair Employment Practices Act. The time was ripe for such an undertaking. World War II had heightened professional and popular interest in the sources of racial and religious prejudice, and efforts were being made to knock down barriers to good relations between people.

The interdisciplinary field of behavioral sciences had already explored some of the psychological and social factors influencing the behavior of individuals in groups. Findings from these prewar studies had attracted the interest of business executives, some of whom began to engage the services of behavioral scientists to guide them in making decisions affecting personnel relations.[8]

At war's end, a wave of optimism arose from a belief that techniques could be developed on the basis of this new knowledge to improve relations between people. One-worldism was in the air, racial and religious prejudice was becoming less tolerable, and all over the country official and non-official commissions were being formed to seek ways of helping people of diverse backgrounds work together harmoniously. Such a meeting was the 1946 workshop in New Britain. The Connecticut Interracial Commission had asked the Research Center for Group Dynamics at Massachusetts Institute of Technology to help train leaders to deal with intergroup tensions in their

---

[8] A discovery that was to influence the future of practical application of behavioral science theory was made in the late 1930s from a study of the effects of lighting on worker efficiency in a Western Electric plant. Worker efficiency improved, it was discovered, not so much because of a change in lighting but because the workers responded to the attention they received during the course of the investigation.

home communities. The guiding spirit of the Research Center was Kurt Lewin, a psychologist and refugee from Nazi Germany.

Lewin's studies had convinced him that few persons change their ingrained attitudes as the result of receiving instruction from others. He believed that an individual is more likely to change his true feelings when he has had an emotional experience that gives him new insight into himself and his relations with others. This thinking survived Lewin's death in 1947. Applied to leadership training, it meant providing trainees "with opportunities for feeling and knowing what their customary way of acting does to themselves and others by practicing it before their own group."[9]

The New Britain workshop brought about a fortuitous association of training leaders and researchers. Besides Lewin, they included Kenneth D. Benne of Columbia University who had been investigating "the social-psychological processes of building a community out of conflicting orientation";[10] Ronald Lippit, an educator who had worked with Lewin on studies of autocracy and democracy in youth groups; and Leland P. Bradford, an adult education specialist who had experimented with techniques of informal group discussion as a more effective procedure for learning than that of the standard classroom. These three men shared Lewin's enthusiasm "for making scientific theory and research relevant to social action."[11]

Chance Discovery of Impact Upon Participants

The workshop planners assigned observers to report on the behavioral interaction of workshop participants. Originally the observers were to report only to the workshop leaders. But the participants themselves asked if they could listen in on the observers' reports. The request was granted and thus was born the first of what has developed into today's encounter groups. The planners were totally unprepared for the impact that the "feedback" had on the participants.

The open discussion [Benne recounted] of their own behavior and its observed consequences had an electric effect both on the

---

[9] Alfred J. Marrow, "Events Leading to the Establishment of the National Training Laboratories," *The Journal of Applied Behavioral Science*, April-June 1967, p. 147.
[10] Leland P. Bradford, "Biography of an Institution," *The Journal of Applied Behavioral Science*, April-June 1967, p. 129.
[11] As described by the NTL Institute for Applied Behavioral Science in its *Midyear Catalogue of Programs and Services* (1970), p. 4.

ENCOUNTER GROUPS

Sensory awakening

Meditation exercises

Sensitivity training

Personal-growth laboratories

Gestalt therapy

chapman

participants and on the training leaders. What had been a conversation between research observers and group leaders...was inexorably widened to include participants who had been part of the events being discussed.... Participants reported that they were deriving important understandings of their own behavior and of the behavior of their groups.

To the training staff it seemed that a potentially powerful medium and process of re-education had been, somewhat inadvertently, hit upon.[12]

This first experience led to other meetings in which various theories and techniques of learning from the experience within the group itself were tested. Support came from universities, government agencies, foundations, private business, voluntary organizations. So many disparate groups were interested because it was felt the lessons learned in one group-learning session could be applied to almost any other session. The common ground, a pioneer of the movement wrote in 1948, was found in the role of the "change agent." Each of the interested organizations wanted to know how its own "change

[12] Kenneth D. Benne, "History of the T-Group in the Laboratory Setting," *T-Group Theory and Laboratory Method: Innovation in Re-education* (1964), Leland P. Bradford, Jack R. Gibb, and Kenneth D. Benne, eds., pp. 82-83.

agent" could function better "to produce changes in the under-standings, attitudes, and skills of the persons and groups with whom they work."[13]

Enthusiasts for this form of group learning sought to extend its activities nationwide. NTL was formed in 1947 to continue experimentation in "laboratory education," as it was called. A grant from the Carnegie Corporation in 1950 made possible establishment of a year-round office and program in the National Education Association, which had supported previous efforts. NTL became a separate division of NEA in 1962, then was reformed in 1967 as an independent, non-profit affiliate of NEA under the name NTL Institute for Applied Behavioral Science.

Today NTL is a flourishing enterprise constituting a national, and to some degree international, network of several hundred behavioral scientists and 600 NTL-trained group leaders. Its activities, greatly expanded from its shoestring beginning, are reported to bring in an annual income of $4 million. In addi-tion to the central establishment in Washington, NTL has seven branch institutes or laboratory sites.[14] It not only issues "experiential learning" materials and provides consultant ser-vices but regularly schedules dozens of conferences, training courses, workshops, "organization development" programs and the like throughout the year at many places besides its own sites. Nearly all of these activities utilize the principles of sensitivity training or laboratory learning of the human poten-tial movement.

Many of its programs are designed for managers and execu-tives. A Key Executive Conference, for example, was held on three separate weeks in 1970 for presidents, vice presidents and general managers only. (Cost: $500 plus room and meals, approximately $300.) Some programs are set up for members of a single profession—teachers, social workers, manpower trainers, and so forth. There are "cousin labs" for workers in related fields and "stranger labs" for participants from dif-ferent walks of life. NTL programs include sessions for purely personal as well as organization improvement. A program held at Cedar City, Utah, Aug. 2-8, 1970, was for members of

---

[13] Kenneth D. Benne, "Principles of Training Method," *The Group,* January 1948, p. 17.

[14] At Bethel, Maine; Cedar City and Salt Lake City, Utah; Lake Arrowhead and Los Angeles, Calif.; Kansas City, Mo.; and Portland, Ore.

families who wanted to learn "better ways to express their needs, their anger, their affection" and to try out "new patterns of relating to one another." (Cost: $350 for tuition and registration, plus room and meals, approximately $70 per adult, $50 for child under 16.)

Esalen Institute Emphasis on Inner Awareness

NTL's growth coincided with the flowering of the human potential movement and NTL became the prototype for other establishments. What later became known as the encounter group thus emerged from a larger innovative development in adult education. The term "encounter group" originated on the West Coast where the movement placed greater emphasis on awakening of the individual than on improved functioning of the group. Of all elements of the human potential movement, the encounter group has obviously exerted the greatest human appeal.

The founding of Esalen Institute in 1961 did much to transform sensitivity training developments into a kind of secular revivalist movement. Like NTL, Esalen rests on the underpinnings of behavioral science theory and its approach is that of humanist psychology, which stresses the "affective"—the feeling—side of learning. But Esalen added other ingredients, some borrowed from the Orient. One of its co-founders, Michael Murphy, had spent nine years in ascetic study of Eastern disciplines.[15] He and his partner, Richard Price, both of whom had majored in psychology at Stanford University, believed it might be possible "to apply certain Eastern modes of thought and action to Western problems."[16] They decided to use Murphy's Big Sur property, which contained a motel and mineral baths on a height above the Pacific Coast, for seminars and lectures on this subject. As the Esalen programs grew, influences were exerted by the participation or the writings of psychologists, humanists, mystics, theologians, philosophers, and human relations specialists, some of whom became advisers to the Institute.[17] In time, "experiential learning" sessions were added and from them developed the encounter group.

---

[15] George B. Leonard, *Education and Ecstasy* (1968), p. 210. Murphy was said to have practiced meditation six to eight hours a day.

[16] Esalen Institute, *What Is Esalen?* p. 1.

[17] Among influential figures in Esalen development were Aldous Huxley, Abraham Maslow, Teilhard de Chardin, the Indian mystic Sri Aurobindo, Arnold Toynbee, Paul Tillich, Rollo May, and the late Bishop James A. Pike.

Esalen programs today are so extensive and variegated that it is almo,st impossible to categorize them. The 1971 Esalen catalogue cites the encounter group as one of four basic approaches being taken in its various experiential programs. But it adds that "various disciplines being practiced at Esalen have influenced one another" so that the encounter group partakes of some of the qualities of the other three: gestalt therapy, sensory awakening and sensory awareness, and meditation.

Gestalt therapy is intended "to increase a person's awareness of the instant moment." "Emphasis is placed on having a participant explore his dreams, fantasies, expectations, gestures, voice, and other aspects of personal functioning, by enacting them before the group." In this way he "becomes aware of the unity of his...mind, body, emotions and feelings."

Sensory awakening, also used in encounter groups, involves "a series of exercises designed to...deepen the participants' awareness of his sensory existence."

Meditation exercises are usually mental and involve a concentrated awareness. "The aim...is to break through ordinary consciousness to various levels of being, ranging from relaxed but alert awareness of self to the deepest levels of mystical experience in which the identity of self and cosmos become apparent."

The catalogue described the encounter group approach as follows: "Participants are encouraged and helped to express their feelings about themselves and other members....There is often an emphasis on eliciting emotions which lead to positive or negative confrontation between participants....The focus of encounter groups is on exploring interpersonal relations, and the expression and exploration of self through interpersonal relations." Participants at Esalen sign up for single evening events, for a weekend or longer—up to three weeks. These programs are not inexpensive. Two-week workshops, including accommodations on the premises, cost from $500 to $600.

The success of the Big Sur programs led to the founding of other so-called "growth centers" that now offer variants of the encounter group experience. Since the over-all spirit of the movement was experimental from the beginning, there has been a tendency to trot out new devices, exercises, games and procedures to heighten the experience and response of the participants.

Body and Verbal Activities for Emotional Release

No two encounter groups are alike because it is essential to the learning experience that each group work out its own fate. The small classic group meets with no plan, no agenda, no general discussion subject, no directional structure. The learning comes from the way individuals in the group behave in this formless, rudderless situation. The leader is not a true leader in the sense of giving orders.[18] His job is to explain the freewheeling situation to participants in the beginning, watch for signs the session may be getting out of hand, interject occasionally with an interpretation of a development, or suggest at strategic moments an "exercise" for increasing sensory awareness. Leaders vary, however, in the degree to which they refrain from masterminding the sessions. Leaders of some of the newer groups are virtual gurus.

In the typical encounter group, according to psychologist Carl Rogers, an enthusiast, there is "a period of initial confusion, awkward silence, polite interaction, mild to extreme frustration, and great lack of continuity." This situation leads to irritability and a member of the group may strike out verbally against another. This is the beginning of the stripping away of the facade of polite behavior that covers up real feelings. Ultimately, in this milieu of free expression of real feeling, a sense of group solidarity develops.

> As the sessions continue, different threads interweave and overlap. One...is the increasing impatience with defenses. In time, the group finds it unbearable that any member should live behind a mask or front....The expression of self by some members...has made it very clear that a deeper and more basic encounter is possible and the group seems to strive intuitively and unconsciously toward this goal....
>
> In this freely expressive interaction, the individual quickly gets insights into how he appears to others....All this can be decidedly upsetting, but as long as it comes in a context of group caring it seems highly constructive.[19]

Participation in an encounter group can be an emotionally exhausting experience. Long-repressed resentments boil to the surface. Group members get into fiercely angry disputes. Free-flowing tears follow self-revelation. But one of the great things about the encounter group, its advocates say, is that it arouses a warm feeling of caring for others, especially for those who are temporarily feeling miserable during the encounter.

In some groups, leaders suggest certain "exercises" or "games" to facilitate the release of repressed feelings. Some of

---

[18] In some groups, the leader is called a "facilitator" to remove the impression of leadership.

[19] Carl Rogers, "Community: The Group Comes of Age," *Psychology Today*, December 1969, p. 27.

these, like role-playing or psychodrama, are borrowed from psychotherapy. Participants may act out family situations by taking the roles of father, mother or child. Masks are sometimes used as props for role-playing. Or participants are asked to pair off and stare eyeball to eyeball, then tell what they felt during that time.

Some of the exercises are non-verbal. In some groups, participants are encouraged to yell, scream out whatever sounds they want to make, to jump, dance or do anything that frees them from normal inhibitions. Some leaders bring pillows for participants to pummel as a symbol of a person in their lives they would like to thrash.

Other exercises are designed to encourage trust in others, like falling backwards into the arms of another or being led blindfolded. Bodily contact, but not sexual relations, are encouraged. Participants may be asked to touch each other or huddle together with arms interlocked. Embracing, especially to comfort the distressed, is a frequent spontaneous response during the encounter. So is wrestling, often in anger. The influence of Eastern mysticism shows in the introduction of meditation as part of the experience in some centers. Esalen gained a certain fame for having participants of both sexes wade nude in large tubs fed by natural hot springs. The experience is said to suggest a blissful return to the womb.

Problems and Promise in Encounter Work

REASONS FOR THE POPULARITY of encounter groups and other forms of sensitivity training are not hard to find. "During a time when Americans are torn with conflict and beset by fear, loneliness, and alienation, many are searching for something of meaning," Ted J. Rakstis said in an article on sensitivity training in an American Medical Association publication.[20] He quoted Thomas Bennett, an NTL fellow and official of George Williams College: "In our culture it's extremely difficult to find experiences with other people [that] provide a degree of freedom and intimacy and a real opportunity to deal with persons at a fairly intense level. A lot of that has led to the growth in sensitivity training."

[20] Ted J. Rakstis, "Sensitivity Training: Fad, Fraud, or New Frontier?" *Today's Health*, January 1970.

For some individuals the motivation may be longing for more meaning in life or desire to overcome a personality handicap, or it could be curiosity, a desire for kicks, bravado, or simply a tendency to go along with the "in" thing. One view is that it is another entertainment of an affluent society: "Leisured Americans have always had a weakness for costly regimens that promise to make them better than they are."[21] For some the appeal is akin to the religious impulse.

The confessional aspect of the experience is doubtless a potent factor in its appeal and contributes greatly to its emotional impact. In this it is related to other movements in which confession is encouraged as a route to self-betterment: Alcoholics Anonymous, Synanon, and Weight Watchers, for example. The Moral Rearmament movement of the 1920s and 1930s[22] similarly encouraged participants to confess their deepest feelings and their most moving experiences at group meetings known as "house parties." The current women's liberation movement has spawned untold numbers of group encounters of a special sort, in which women sit in the characteristic circle of the encounter group and "let down their hair" on the sexist issue.[23]

All of these manifestations of voluntary communing in a small group reflect the growing acceptance of the belief that help for the troubled comes best from those who have suffered similar pangs. That this concept has influenced psychiatry is evident in the growing use of group therapy by practicing psychiatrists. Such groups are far more controlled by the leader—the psychiatrist or psychotherapist—than encounter groups are, but the same process of blunt outspokeness and emotional catharsis goes on in therapy sessions.[24] The Task Force on Encounter Groups of the American Psychiatric Association observed that "a number of the technical

---

[21] Henry S. Resnik, "Techniques for Psychic Survival," *Saturday Review*, July 25, 1970, p. 21.

[22] The Moral Rearmament movement had a religious but non-denominational base. It originated among upper middle-class Britons and was known also as the Oxford Movement. Its adherents were called Buchmanites for its founder Frank Buchman.

[23] See Vivian Gornick's, "Consciousness ♀," *The New York Times Magazine*, Jan. 10, 1971, p. 22.

[24] Ronald D. Laing, the British psychiatrist, carried the self-help approach still further in the treatment of schizophrenics. Under his plan, the mentally ill lived together in communal style, free of the hierarchal controls of medical authorities or the rigid rules of the ordinary institution. A favorable account of the Laing experiment at Kingsley Hall in London is given by James S. Gordon, an American psychiatrist, in *Atlantic*, January 1971, pp. 50-65.

innovations employed by various encounter group leaders may have applicability in traditional therapy groups" and that "group psychotherapy has, in fact, already profited considerably from innovations arising from sensitivity training groups in the human relations field."

The Task Force also sought to pinpoint the meanings behind the surge of popularity of encounter groups. "We suspect that the groups have arisen in response to a pressing need in our culture," it stated. The mobility of the people, their loss of roots, the instability of the family, the diminishing connection with the immediate neighborhood, the "ever changing competitive culture"—all these were cited as conditions favorable to the growth of encounter groups.[25]

Uses and Misuses in Police Work and Schools

For organizations, the motivation is more obvious. They are simply looking for ways to improve their operations. Or they are facing new problems as a result of social change and find the old ways of operating won't do. Prevailing attitudes of their personnel may need changing. This is true, for example, of schools with racial friction or student rebellions and public service organizations having difficulty reaching out to people beyond the traditional middle-class orbit.

The applicability of sensitivity training to police work, especially where relations are bad between the policeman and the neighborhood, is expounded in an article by Donald Bimstein, a former member of the New York City Police Department who is now associate professor in the police science program of Northern Virginia Community College. Most police departments today have some sort of program for improving police-community relations. Some police departments—in Grand Rapids and Houston, for example—have participated in sensitivity training sessions. Bimstein urges more to do so.

"It is fine to conduct departmental programs covering human relations, but merely telling or explaining can never carry the impact exerted by experiencing," he wrote. "It is not until the officer knows at first hand what it is like to be on the receiving end that the program will accomplish meaningful results....It will be a rare officer who can emerge from these [sensitivity training] sessions unchanged in any way. At

---

[25] American Psychiatric Association Task Force Report, *Encounter Groups and Psychiatry*, April 1970, pp. 9-11.

the very least he will become conscious of the fact that his personal behavior can stand improvement. At its very best the end product will be a more efficient officer who will be constantly garnering goodwill for the department and enhancing its prestige."[26]

The Young Men's Christian Association has reported enthusiastically on the results of sensitivity training with teenagers. The YMCA's interest dates from 1948 when a top official of the organization attended an NTL human relations laboratory at Bethel, Maine. Since then some 7,000 teen-agers have participated in sensitivity training under YMCA auspices. Investigations of results, backed up by parents' comments show positive change for the better. The major gains were in healthier acceptance of self and better understanding of others. "Teen-age participants felt that they had become more open, more affectionate, more wanted, and more willing to try to express themselves. They said that they understood others, trusted other more, and accepted others' differences."[27]

Max Birnbaum, associate professor of human relations and director of Boston University Human Relations Laboratory, believes sensitivity training holds great promise for improvement of education, especially because of its emphasis on "affective" learning. But he warns that participation in the groups by school personnel "too often remains a memorable experience, but not one that produces change." In the school system of one large city, he said, the result of an encounter group including teachers, mostly white, and students, mostly black, "was the opposite of what was sought—increased physical and verbal hostility of students to the teachers in the school." He said the fault lay in the failure to manage the encounter properly.

Birnbaum expressed fear that "this promising innovation may be killed before its unique properties have a fair chance to demonstrate their worth." The most serious threat is not from the critics but from "a host of newly hatched trainers, long on enthusiasm or entrepreneurial expertise, but short on professional experience, skill, and wisdom." Too often sensitivity training as used in school situations is no more than a

---

[26] Donald Bimstein, "Sensitivity Training and the Police," *Police,* May-June 1970, p. 79.
[27] Charlotte Himber (associate director, Community Affairs Department, National Council of YMCA of U.S.A.), "Evaluating Sensitivity Training for Teen-Agers," *The Journal of Applied Behavioral Science,* July-September 1970, pp. 309, 314.

"routine application of what are basically gimmicks to an involved and highly charged area."[28]

Business Objections; Issue of Privacy Invasion

Like the schools, big business has learned that sensitivity training or similar forms of learning are not cure-alls for institutional problems. *The Wall Street Journal* reported in 1969 that some companies were becoming disenchanted with sensitivity training. One of the troubles was that executives who participated returned to their posts so sensitized they became dysfunctional on the job or else they tried but failed to apply the lessons of the sensitivity experience and then slid back to the old way of doing things.[29] More recently, *Dun's* magazine reported that "companies are becoming increasingly aware of the limitations of the behavioral science approach in solving their problems." Some successful executives were beginning to wonder why they should expose themselves "to the rash accusations of complete strangers, painful self-analysis, and at times, a brutal battering of defenses built up over a lifetime."[30]

Similar complaints have arisen in federal agencies, according to Sen. Sam J. Ervin (D N.C.) and the Senate Judiciary Subcommittee on Constitutional Rights, which he heads. Supervisors and some of their subordinates have objected to the senator and the subcommittee that they were forced to take part in soul-baring "psychological confrontation sessions" that delved into their attitudes on racial matters—in the interest of reducing job discrimination against Negroes. Ervin in a speech at Carlisle, Pa., Feb. 9, 1971, called this practice "tyranny over the mind of the grossest sort because it subjects employees to a probe of their psyches, to provoke and indeed to require disclosure of their intimate attitudes and beliefs during emotionally charged situations...."

The Montgomery County (Md.) Federation of Teachers charged that teachers at Montgomery Blair High School were required to attend discussion workshops which "were in reality sessions in sensitivity training." The federation's executive committee said in a news release, Jan. 4, 1971, that the teachers were subjected involuntarily "to abusive treatment

[28] Max Birnbaum, "Sense About Sensitivity Training," *Saturday Review*, Nov. 15, 1969, p. 82.

[29] *The Wall Street Journal*, July 14, 1969.

[30] George Berkwitt, "Behavioral Science: Is the Cure Worth It?" *Dun's*, May 1970, pp. 39-41.

under the guise of procedures which are commonly regarded to be private/medical psychological functions."

While the movement is thus raising questions among civil libertarians over the issue of privacy, it is being buffeted from another direction. Some quarters of the political right tend to see it as a threat to the American way of life. Rep. John R. Rarick (D La.) has charged that "the accelerated use of 'sensitivity training' as a tool to indoctrinate the masses for a 'planned change' in the United States has resulted in confusion, frustration, and wholesale disorientation among our unsuspecting people." He placed in the *Congressional Record* of June 10, 1969, 13 pages of documents equating sensitivity training with Communist brainwashing and sex education.

Psychiatric Casualties; Questions of Certification

There is serious concern that the experience may be harmful to individuals who cannot take the psychological buffeting. A report to the Archives of General Psychiatry cites two case histories of individuals whose participation in T-groups precipitated acute psychotic reactions. "The T-group is a social setting that encourages openness, intimacy and closeness," the authors observed, but it "has little structure for handling the anxiety that is provoked."[31]

Such cases are rare and occur only among participants with a history of or predilection toward neurotic or psychotic disturbance. The problem is how to screen applicants in order to eliminate the vulnerable. Good leaders supposed to watch for signs that a participant is not strong enough for the experience, but the danger may not be apparent during or even at the conclusion of an encounter. One casualty of an encounter group was a woman subject to depression whose spirits were greatly lifted by the experience. A few months later, however, she committed suicide, unable to endure a disappointment in love.[32]

Tragedies of this kind—as well as physical injuries that sometimes occur during group encounters—might be avoided if controls of some sort could be instituted to make sure that only qualified leaders conduct encounter groups. NTL has developed a set of standards for group training and is working

---

[31] Steven L. Jaffe and Donald J. Scherl, "Acute Psychosis Precipitated by T-Group Experiences," *Archives of General Psychiatry*, October 1969, p. 446.

[32] Bruce L. Maliver, *op. cit.*, p. 4. Many participants in encounter group sessions are euphoric about the experience at the conclusion. Psychiatrists warn that such emotional highs following an intense experience are often followed by lows.

toward developing some form of an accreditation system.[33] The American Medical Association's Council on Mental Health issued a statement in September 1969 cautioning prospective participants about sessions conducted by leaders who are not professionally trained and qualified. "In view of reports of psychotic and neurotic sequelae [aftereffects]," the council said, "physicians have a responsibility of pointing out these possible dangers to persons who are interested in becoming involved in these programs." There are no laws governing the practice of encounter group leadership as there are for psychotherapists and others in the healing professions.

Another question often raised is whether the benefits are lasting. Reports on this vary. Some say the effects, like the effects of seeing a stirring play, eventually dissipate. One enthusiast for the movement claims that even so the encounter group experience is useful. "Personal growth laboratories give an experience of a better order. It does not last beyond the end of the laboratory because the laboratory is a temporary little world that is insulated from the everyday world [but] experiments are useful to test ideas, to create visions that may be realizable."[34]

Despite the warnings, and the skepticism over its benefits, most informed observers believe the encounter group or a similar form of activity is likely to be around for a long time.[35] Its influence on education, on business practices, on psychiatry, and elsewhere in society seems assured. Whether it will ultimately improve man and society in the benign way its humanist founders hope it will—this remains for the future to decide.

---

[33] The first known damage suit resulting from sensitivity training was filed in Washington, D.C., December 1970. A woman employee of the Department of Health, Education, and Welfare sued the government and NTL for half a million dollars because of injuries she allegedly received during a T-Group session at Plymouth, N.H., in the summer of 1968. She claims she was injured when thrown to the floor during an episode when the leader urged participants "to physically demonstrate aggression and hostility." HEW was named in the lawsuit because it had recommended the session to employees.

[34] Herbert A. Shepard (visiting professor of philosophy, Yale University School of Medicine), "Personal Growth Laboratories: Toward an Alternative Culture," *The Journal of Applied Behavioral Science*, July/August/September 1970, p. 259.

[35] "Though the bizarre aspects may fade, the encounter group is based on a solid foundation and appears destined to survive for some time to come."—American Psychiatric Association Task Force Report, *Encounter Groups and Psychiatry*, April 1970, p. 9.

Blood Banking

by

Helen B. Shaffer

1 9 7 1
M a y 5

BLOOD BANKING

THE NATION'S LEADING blood bankers—the American Red Cross and the 1,300 institutions that belong to the American Association of Blood Banks—are pushing hard to put blood donations on a totally volunteer basis. The primary purpose is not to save money but to assure a sufficient supply of safe, good-quality blood to meet present and expected needs of the population. The drive for more volunteers has been accompanied by rising criticism of the long-standing practice of buying and selling blood for cash. So-called "commercial blood" is said to present more hazards than "volunteer" blood to patients receiving transfusions.

This attack on commercial blood gained new ammunition with the publication early in 1971 of *The Gift Relationship,* a book by Richard M. Titmuss, a British sociology professor. He compares the American blood-banking system unfavorably with the British on a number of counts: on safety, on sufficiency of supply, on costs, on orderly distribution, on prevention of waste, and on the quality of mercy involved in the giving of blood. Titmuss attributes most of the faults he found in the American system to "the commercialization of blood and donor relationships."[1]

Attacks on the quality of commercial blood have caused concern in some blood-banking circles because of the possible effect on the total supply, particularly in view of the rising demand for blood in medical practice. Hard data on blood sources for the nation are lacking but a panel of experts from the National Research Council (NRC) has estimated that approximately 15 per cent of the 6.6 million units[2] of blood drawn each year to meet U.S. needs comes from individuals who receive pay for it in cash. According to Titmuss, at least one-third of all blood donations in the United States in 1965-

---

[1] Titmuss, however, applies the term "commercial" to all blood for which the donor receives any kind of benefit, whether cash, credit against future blood need, or only a day off from work.

[2] About 480 milliliters per unit, roughly a pint.

67 were cash deals; by his reckoning, and standards, only 9 per cent was given outright, that is without some inducement or benefit to the giver.

Obviously it would be disastrous to eliminate blood purchases without assurance of an adequate additional supply of volunteer blood. And getting a steady and reliable roster of volunteer givers has been a continuing problem since blood banking began about 30 years ago. Blood Services, a nonprofit corporation that operates 27 blood banks in 11 states, acknowledges that it relies largely on a list of paid donors; it defends the practice as both safe and necessary.

The problems of supply and safety are closely related. The 6.6 million pints that flow annually through the complex channels of acquisition, processing, distribution, and utilization are roughly sufficient to meet day-by-day needs. But there is little margin of safety. The NRC panel has described the supply-demand situation as one of "critical balance."[3] The thin line of adequacy is evidenced by the frequent appeals for instant donors to meet emergencies requiring rare blood or uncommonly large quantities for a single patient. Operations are often postponed because of an insufficient blood supply. Whole blood is perishable, even though refrigerated. It remains usable no longer than three weeks, hence it cannot be stockpiled indefinitely against future needs.

Donations usually fall off during holiday and vacation seasons, and these are the times when demand rises because of accidents. New York City, for example, underwent a shortage crisis in the late summer of 1970, which was eased by public response to urgent appeals from Mayor John V. Lindsay and others. While doctors usually prefer volunteer blood, commercial blood will be used with little hesitation if the other is not available. The risk of side effects carries little weight against a patient's urgent need for blood.

Higher Infection Rate From Blood of Paid Givers

Like most helpful medications, the transfusion of blood entails certain risks. "A unit of blood delivered to the bedside has been exposed to a great many opportunities for human error," the NRC reported. "The acceptability of the donor must be judged, asepsis must be ensured throughout the pro-

---

[3] National Academy of Sciences-National Research Council, *An Evaluation of the Utilization of Human Blood Resources in the United States,* October 1970, p. 1. The academy is the council's parent organization.

BLOOD COMPONENTS AND DERIVATIVES			
Cellular elements are 45% of blood	Red cells	Can serve for 30-50% of all blood transfusions. For treatment of anemia.	
	White cells	For combating infections. Clinical use experimental at present.	
	Platelets	Essential for blood clotting. For treatment of leukemia, cancer, and other diseases associated with platelet deficiency.	
Plasma is 55% of blood	Plasma	Fresh-frozen plasma used to control bleeding. Single-donor liquid plasma used to restore plasma volume.	
	Cryoprecipitate		Prepared from fresh-frozen plasma. Contains AHF (antihemophilic factor). Therapeutic for hemophiliacs.
	Plasma proteins	Serum Albumin	For treatment of shock and low blood protein.
		Gamma globulin	Contains antibodies against diseases. For prevention of hepatitis and measles. For treatment of gamma globulin deficiency.
		Specific immune globulins	A special variety of gamma globulin prepared from plasmas with specific antibodies.
		Clotting factor concentrates; e.g., fibrinogen, AHF, and II, VII, IX, X	Essential for blood clotting. For controlling bleeding due to deficiencies of specific coagulation factors.
SOURCE: American Association of Blood Banks.			

133

cessing program, blood-typing tests must be interpreted accurately, the products must be labeled appropriately, and the labels must be read and understood correctly. It is not just a handful of men at two or three manufacturing establishments who might make mistakes, but rather the blood collectors, laboratory personnel and persons administering transfusions at well over 5,000 collecting stations, blood banks, and hospitals."

Supreme Courts in two states—Illinois and Pennsylvania —recently held hospitals in their jurisdictions liable for damages in cases where patients suffered ill effects from transfusions, even though there was no evidence of specific negligence.[4] In the Illinois case, a patient had sued the hospital for $50,000 damages for having allegedly contracted hepatitis after a transfusion. The court ruled that blood was a product, not a service, and carries an "implied warranty" against defects. The AMA *Journal* commented in an editorial that decisions of this kind "will lead to rules which make it difficult for the physician to exercise his medical discretion in prescribing blood." It added that if the rules are too drastic, "the blood money saved in minimizing hospital exposure to blood transfusion claims may be at the expense of human lives."[5] Medical and blood-bank leaders in a number of states are seeking legislation to remove liability in furnishing blood unless negligence is proven.

The chief hazard at the blood-banking level is that of infection. Of all infections that might be transmitted the greatest concern in recent years has been over the risk of hepatitis, a liver disease, especially when fresh whole blood is used. Some 30,000 cases of transfusion-connected hepatitis occur yearly in the United States, according to the NRC report, and between 1,500 and 3,000 of them are fatal. Because reporting on the disease is known to be incomplete, the incidence and mortality actually "may be much higher." It is possible that there are as many as five "subclinical" cases—cases without manifest symptoms—for every case that is identified.

Hepatitis has been hard to control because blood bankers have lacked a means of weeding out carriers of the hepatitis virus. Since clinical tests were unavailable, it was necessary to depend on the donor's own account of his medical history.

---

[4] Cunningham *v.* MacNeal Memorial Hospital, Ill. Sup. Ct., Sept. 29, 1970, and Hoffman *v.* Misericordia Hospital of Philadelphia, 267 A 2d 867, 1970.

[5] "Blood Money," *Journal of the American Medical Association,* Jan. 4, 1971, pp. 109-110.

Suspicion fell on the paid donor because it was believed that he had a stronger incentive to conceal a disqualifying medical condition than did those who received no money.

Suspicion of the link between commercial blood and hepatitis received impressive confirmation from a study of 110 patients who underwent open-heart surgery, a procedure requiring 15 to 25 pints of blood in each case. Volunteer blood was available for only 28 of the patients so an effort was made to give it to the 28 who were most susceptible to hepatitis. The other 82 patients received blood from paid donors. None of the patients receiving volunteer blood contracted hepatitis, but the disease developed among one-half of the recipients of purchased blood.[6]

It is generally agreed that the greatest hazard comes from blood sold by the poor of the slums and the derelicts of skid row. These individuals not only have a pressing need for the money they can get by selling their blood, they are also uncommonly susceptible to infectious diseases.[7] The same hazard has been found among prison donors, who give blood in return for a few dollars or a few days off their sentences.

The risk is especially high when blood is drawn from narcotics users who may have infected themselves by the use of unsterilized hypodermic needles. In fact, the rising incidence of hepatitis in the United States is attributed partly to the increased use of drugs and partly to the increase in forms of surgery requiring large amounts of blood. The Department of Health, Education and Welfare has reported the incidence of hepatitis has risen during the past five years. In the first 10 weeks of 1971 it was 32 per cent higher than in the same period a year earlier.

Commercial Blood Banking: Practices and Issues

Donors from the ranks of narcotics users and the urban poor are likely to patronize "bucket shops." These are commercial blood-collecting stations set up in rundown sections of the big cities to attract those in need of easy money. Bucket shops appeared when the growing demand for blood was not matched by voluntary giving. The main hazard lies in the opera-

---

[6] John H. Walsh, Robert H. Purcell, Andrew G. Morrow, Robert M. Chanock, and Paul J. Schmidt, "Posttransfusion Hepatitis After Open-Heart Operations," *Journal of the American Medical Association*, Jan. 12, 1970, p. 261. The patients in this study were all hospitalized at the Clinical Center of the National Institutes of Health, a branch of the U.S. Public Health Service, at Bethesda, Md.

[7] Among conditions that disqualify donors are alcoholic habituation, drug addiction, history of malaria, syphilis, tuberculosis, heart disease, kidney and liver diseases, chronic asthma, or diabetes.

tion of a "drop-in" facility—one in which any passing stranger may drop in to sell his blood. Blood Services, with its chain of 27 nonprofit blood banks in 11 states, claims it avoids the problem by maintaining its own roster of donors. "These are not casual drop-in donors, but...donors who have successfully given blood before." If a donor's first blood offering has been used without mishap, his name will go on a donor file. "Thereafter the [blood] bank calls him in as often as he is needed to make additional donations, and each time offers him a check or other incentive for his trouble."[8]

The issue of bought-versus-volunteer blood came to the fore in a case that concerned the right of hospitals to jointly refuse to buy blood from commercial blood banks. The case started in July 1962 when the Federal Trade Commission issued a complaint against the Community Blood Bank of Kansas City, three local hospitals, the area's hospital association, and several hospital administrators and pathologists. The FTC charged that they had acted in restraint of trade by agreeing to obtain all of their blood from the non-profit community blood bank. The complaint originated with two commercial blood banks that had supplied the hospitals' needs before the community bank was established in 1958. After protracted hearings, the FTC commissioners ruled by a 3-2 vote, in October 1966, that the institutions and individuals cited had in fact engaged in a conspiracy to restrain commerce in blood. More than two years later, in January 1969, the ruling was set aside by the U.S. 8th Circuit Court of Appeals in St. Louis, and no further appeal was made.

During the years the outcome was in doubt, the issues were widely debated. On two occasions, in 1964 and again in 1967, the Antitrust and Monopoly Subcommittee of the Senate Judiciary Committee held hearings on proposed legislation to exempt nonprofit community blood banks from antitrust laws and to declare that blood was not an article of commerce as defined by laws applying to interstate commerce. Much of the material on which Richard Titmuss based his critical study of American blood banking came from these hearings. Leading American blood bankers, however, have challenged his interpretation of the material. They said he distorted the relationship between donor and blood bank in attempting to prove the alleged "commercialization" of blood banking and "exploitation" of blood donors in the United States.

---

[8] Paul M. Roca, general counsel of Blood Services, in a speech at Phoenix, Ariz., Nov. 20, 1970, before the Newcomen Society in North America.

Blood Banking

Three recent developments show promise of reducing the hepatitis risk associated with transfusion: (1) the development of increasingly sensitive methods of identifying hepatitis carriers, (2) indications that "transmembrane" red-cell washing may render frozen blood cells free of contaminating viruses, and (3) new prospects for developing an anti-hepatitis vaccine. A major breakthrough was made in 1963 with the discovery of the Australia antigen. This unusual protein substance was found in the blood of an aborigine by scientists of the Institute for Cancer Research in Philadelphia while engaged in a genetics-research project in Australia. Later studies indicated that the substance was an antigen[9] and that it is found at certain times in the blood of patients with infectious hepatitis.

New Ways to Identify and Screen Virus Carriers

It was also shown that the antigen could be transmitted by transfusion from donor to recipient, who in some cases would come down with the disease. Since then techniques have been developed to check the blood of donors for presence of Australia antigen. Tests indicate that if such techniques could be universally applied, they would reduce the incidence of transfusion-induced hepatitis by 20 to 25 per cent. If some specialists are correct in estimating that 150,000 cases of hepatitis are induced by blood transfusion, this screening would eliminate up to nearly 40,000 cases a year.

Screening of donor blood for Australia antigen began to move ahead after the Division of Biologics Standards of the National Institutes of Health approved the manufacture of a reagent. Screening procedures depend on a reaction between the reagent[10] in the blood given by donors and plasma containing antibodies directed against these antigens. New York became the first state, as of Jan. 1, 1971, to require the screening of all donor blood for Australia antigen. The state supplied laboratories with the reagent until it could become available from commercial producers. Massachusetts soon followed with a mandatory rule. The Red Cross announced on March 2, 1971, that it was instituting this type of screening in all of its 59 regional blood banks across the country—banks

---

[9] A substance in the body that triggers the formation of antibodies against particular foreign agents—a part of the body's natural immunological process. Some scientists believe the Australia antigen is in fact the hepatitis virus.

[10] Known as Hepatitis Associated Antibody, or simply HAA. See *Journal of American Medical Association,* March 8, 1971, p. 1570. A reagent is something added to a complex solution to determine, by the resulting chemical action, the presence or absence of a certain substance.

that collect approximately one-half the blood used for transfusions today. Many non-Red Cross banks took similar action.

The Red Cross is also developing a register of prospective donors whose tests showed them to be suspected carriers of hepatitis. A Red Cross worker told Editorial Research Reports in mid-April that a number of carrier suspects had already turned up and that the donors were being informed that their names were going on the register. Blood taken from hepatitis virus carriers may be used for research purposes.

Hope for perfecting a vaccine has risen. Researchers at New York University Medical Center announced on March 23, 1971, that they had apparently been successful in immunizing a small group of children against serum hepatitis, a type of hepatitis that can be transmitted by an unsterilized hypodermic needle or by contaminated blood.[11] At least five years will be required according to Dr. Saul Krugman, head of the research team, for further development. Dr. Krugman said he announced the tentative results in the hope of discouraging a threatened cutback of federal support for research on this and other diseases.

Evolution of Blood-Banking Systems

DESPITE OCCASIONAL MISHAPS, the transfusion of blood from one human being to another stands as one of the great benefits of modern medicine. Few healing agents can compare with it in the breadth of its application to the sick, the injured and the debilitated. A boundless number of conditions call for its use. The prompt infusion of blood, or one of its components, may determine life or death in cases of injury. Some forms of life-saving surgery would be futile without massive amounts of fresh blood.

Blood has always figured importantly in the medical arts, but it is more than a medicine. It carries an emotional overload that is absent in the use of other medications. This feeling about blood began with primitive man, who learned from experience that it had vital properties. Thus he attached magic powers to it. Blood sacrifices to propitiate the gods, blood pacts to bind members of tribal brotherhoods, and

---

[11] *Stedman's Medical Dictionary* (1962 edition) lists a dozen types of hepatitis.

blood feuds to avenge blood brothers all signify the powerful feelings associated with blood. "The history of every people assigns to blood a unique importance," Titmuss wrote, adding:

> Symbolically and functionally, blood is deeply embedded in religious doctrine; in the psychology of human relationships; and in theories and concepts of race, kinship, ancestor worship and the family. From time immemorial it has symbolized qualities of fortitude, vigor, nobility, purity and fertility. Men have been terrified by the sight of blood; they have killed each other for it; believed it could work miracles; and have preferred death rather than receive it from a member of a different ethnic group.

Some of the emotions have clung to the modern practice of sharing blood. In World War II, the Red Cross provided blood to white and Negro servicemen that had been donated only by persons of their own races. Its Central Committee said in 1947 that "on the basis of recorded scientific and medical opinion, there is no difference in the blood of humans, based upon race or color" but, nevertheless, "chapters will collect and hold blood in such manner as to give the physician and the patient the right of selection at the time of the administration." Three years later, the Committee on National Blood Program of the Red Cross Board of Governors said that "racial designation on donor cards should be withdrawn." The official *Red Cross Manual of Policy and Instruction,* dated May 6, 1963, contains this statement: "The Red Cross Program does not require the segregation of blood. The official medical records will not bear the racial designation of the blood donor."

In other ways, too, some ancient emotions about blood still prevail. "Among the noblest acts of personal generosity is the gift of one's blood for the benefit of another," President Nixon said in a proclamation[12] designating January 1971 as Blood Donor Month. The American Association of Blood Banks quotes from Kahil Gibran in *The Prophet:* "You give but little when you give of your possessions. It is when you give of yourself that you truly give."[13]

Blood Transfusions From 17th Century Onward

The idea of transfusing blood—transferring it from one person to another—took hold after the British physician, William Harvey, published in 1628 proof of his discovery of the circulation of the blood. A number of scientists in the 17th century experimented with animal and human transfusion. The first

---

[12] Issued on Oct. 26, 1970.
[13] Quoted in AABB program of ceremony on first day of issuance by the U.S. Post Office Department of commemorative stamp honoring America's volunteer blood donors, March 11, 1971.

authenticated transfusion was performed by a British physician, Richard Lower, between dogs in 1665. Samuel Pepys commented in his *Diary* that "This technique may be of mighty use to man's health for the amending of bad blood by borrowing from a better body." The next year Jean Baptiste Denis, physician to Louis XIV of France, attempted to transfuse the blood of a lamb into a 15-year-old boy. The boy died and Dr. Denis was charged with murder. Although he was acquitted, France—then England—prohibited further transfusions. Pope Innocent XI forbade them in 1678.

Interest in blood transfusion revived in 1818 when Dr. James Blundell of London invented a direct transfusion device which he used successfully to control hemorrhage of women after childbirth. Some 347 transfusions were recorded by 1875. The great breakthrough came in 1901 with the discovery by Karl Landsteiner, an Austrian physiologist, that there are different types of blood and that if the donor's type and the recipient's are ill-matched the red cells clump and disintegrate. This solved the mystery of why some transfusions had been successful while others resulted in death. Another great advance was made in 1914 with the development of anti-coagulants to prevent clotting and a few years later with the introduction of a syringe-valve device that made it possible to regulate the flow of blood during a transfusion.

As new knowledge in the biosciences developed, the unsuspected complexity of the blood system unfolded.[14] The great diversity of blood types and the variable combinations of blood factors became known. A major milestone was the discovery in 1940 of the Rh factor. More knowledge was gained of the way the body's immunization mechanism functions through the production of antibodies in the blood stream.

All of these advances of knowledge had important applications to blood therapy. They led to refinements of blood-handling techniques, improvements in separating blood components, and the production of derivative products. While the new knowledge disclosed some previously unrealized hazards of blood transfusion, they also led to the development of measures that increased the safety of transfusions. "Now

---

[14] "The mysteries of the blood have been a microcosmal repetition of the greater mysteries of the universe. Each has yielded its secrets through the diligent and inquisitive resourcefulness of the minds of men."—Dr. Dermont W. Melick, professor of surgery, University of Arizona College of Medicine, address to the Newcomen Society of North America, Phoenix, Ariz., Nov. 20, 1970.

one may shudder to think of all the risks imposed on the recipient of a transfusion" several decades ago, Dr. Louis K. Diamond wrote in his history of blood banking. The following is his description of the earlier hazards:

> The bottles, connecting rubber tubes, and steel needles for venipuncture had usually been rewashed and resterilized after each use but were often not completely free of foreign particles. Old blood or heat-precipitated serum proteins occasionally remained in the needle, or tubing, or bottle. Febrile reactions were therefore frequent. The open system of connecting tubing to bottle and the needle...allowed many points of contamination.[15]

As respect for the hazards grew, discrimination in use of transfusions increased. "The hazards of the administration of blood are so great that only when there is urgent need is blood prescribed," the College of American Pathologists said in 1964.[16] A manual published by the American Medical Association that is in current use warns that the physician "must...recognize that the transfusion of blood carries with it a *risk of hepatitis* and other disease entities in addition to the danger of incompatability that may result in disaster for a particular patient. As with any drug with known side effects, the physician must weigh the potential danger against the expected benefit before ordering a blood transfusion. *A blood transfusion should never be ordered or given unless it is worth the risk.*"[17]

Beginning of Nonprofit Blood Banks in America

In the early days of modern blood transfusions, the giving of blood was usually an impromptu affair. When a patient needed blood, friends or relatives were called on to provide it. The transfusion was direct, with the blood flowing from the vein of one person into the vein of the other. The idea of a blood bank did not become feasible until a method of storing blood was developed. This came about with the introduction in 1914 of sodium citrate as an anti-coagulant. It permitted the storage of refrigerated blood for five to seven days. During World War II an improved mixture of acid-citrate and dextrose extended storage time to 21 days.[18]

[15] Louis K. Diamond, "History of Blood Banking in the United States," *Journal of the American Medical Association,* July 5, 1965, p. 43.

[16] Statement submitted to the Antitrust and Monopoly Subcommittee of the Senate Judiciary Committee in hearings on blood banking, Aug. 20, 1964.

[17] American Medical Association, *General Principles of Blood Transfusion,* June 1970, p. 1. Italics in quoted matter are the AMA's.

[18] "Developers of a citrate-phosphate-dextrose formula developed in 1961 claim that it extends the effective shelf-life of the red cells to 28 days, lessens the acid load on the recipient, and improves oxygen release."—National Academy of Sciences-National Research Council, *op. cit.,* October 1970, p. 4.

The first blood depot was set up by the British army at casualty stations during World War I to meet the needs of wounded soldiers. During the 1920s in the United States a number of "walk-in donor" banks were opened but they were regarded as unsafe. The first nonprofit blood bank was established in 1937 at Cook County Hospital in Chicago and the first community-wide blood bank came along four years later —the still-functioning Irwin Memorial Blood Bank of the San Francisco Medical Society.

The American National Red Cross entered blood banking in the year before Pearl Harbor at the request of the U.S. Army and Navy, which wanted blood to be processed into dried plasma for military needs.[19] Over the war period, 1941-1945, the Red Cross procured 13 million units of dried plasma and serum albumin (derived from plasma) for overseas use, and provided more than 300,000 pints of whole blood or plasma to military hospitals in the United States. The Red Cross program for the armed forces was terminated when World War II ended but was reactivated after the outbreak of the Korean War.

Size and Makeup of Nation's Blood-Bank Network

A strong demand had risen in the meantime for a steady blood supply in civilian medicine. A number of hospitals established their own blood banks during World War II and in the early postwar years, or else they cooperated with founders of community or regional blood banks to serve hospitals in their areas. The Red Cross, with the blessing of organized medicine, initiated a program in 1947 to establish regional blood centers throughout the United States.[20] In the same year a group of hospital and community banks, motivated in part by fear of a Red Cross "invasion" of their territories, formed the American Association of Blood Banks[21] to serve as a trade association for its members. Since then both major blood banking systems have expanded and the two organizations have developed extensive programs of donor recruitment, professional education and research support.

---

[19] The Red Cross was called on because it had several thousand local chapters for recruitment of donors and because it had gained some experience in donor recruitment in a special program in Augusta, Ga., in 1937. In the months before Pearl Harbor, the Red Cross conducted "Blood for Britain" drives.

[20] The first Red Cross regional blood bank opened in Rochester, N.Y., in February 1948.

[21] The AABB was founded in Dallas in 1947 and incorporated in 1958 in Illinois as a nonprofit corporation.

BLOOD GROUPINGS AND RH FACTORS

There are four inherited *basic blood groups:* A, B, AB, and O. The letters A and B refer to the kind of antigen found in an individual's red blood cells. An *antigen* is a substance that triggers an immunological reaction, such as the formation of antibodies. The terms anti-A and anti-B were given to antibodies in the blood plasma that would act against the red cells with A or B antigens.

The effect if anti-A antibody met A antigen (or if anti-B antibody met B antigen) would be a dangerous, possibly fatal, *clumping* of red cells. To prevent such clumping, blood for transfusion is matched with that of the recipient for its ABO group. The four basic groups in the ABO system are:

Group A—has A antigen in red cells, anti-B in its plasma
Group B blood—has B antigen in red cells, anti-A in its plasma
Group AB blood—has both A and B antigens in the red cells but neither anti-A nor anti-B in its plasma.
Group O blood—has neither A nor B antigens in red cells, and both anti-A and anti-B in the plasma.

AB blood cannot cause the clumping of red cells of any other group and therefore those with AB blood are called universal recipients. Group O blood cannot be clumped by any human blood and therefore those with Group O blood are called universal donors.

Blood types in the United States, according to the American Association of Blood Banks, occur as follows:

Blood type and Rh	How many have it	Frequency
O Rh positive	1 person in 3	37.4%
O Rh negative	1 person in 15	6.6
A Rh positive	1 person in 3	35.7
A Rh negative	1 person in 16	6.3
B Rh positive	1 person in 12	8.5
B Rh negative	1 person in 67	1.5
AB Rh positive	1 person in 29	3.4
AB Rh negative	1 person in 167	0.6

The *Rh factor* pertains to an inherited condition of the blood cells. This condition is present in most people, hence they are *Rh positive.* Those persons in which it is missing are *Rh negative.* A transfusion of Rh positive blood may cause kidney damage, even death, to an Rh negative person. The Rh factor may be a problem in pregnancy when an Rh-negative mother carries an Rh-positive baby.

One of the complaints often directed at the blood-banking enterprise is that it fails to produce comprehensive national-scale data. The Red Cross now operates 59 regional blood banks in 34 states, the District of Columbia and Puerto Rico.

Some 1,700 of its 3,300 chapters participate directly in its blood-banking program. The AABB has more than 1,300 institutional members (hospital and community blood-gathering, distributing, and utilizing facilities) as well as the administrative, professional and technical personnel of these institutions. Among members of the association are regional systems that comprise a number of blood banks. Institutional members of AABB, like the Red Cross blood banks, must be nonprofit enterprises. This does not bar a member-institution, however, from purchasing blood from a donor when needed. Both the AABB and Red Cross have taken the position that blood banking would be better off if buying and selling could be eliminated. However, not all members of AABB agree that this is a feasible prospect.

In addition to these two large systems, there are an unknown number of other banks, both commercial and nonprofit. The AMA's latest *Directory of Blood Banking and Transfusion Facilities and Services* (1969) lists 5,625 facilities in 50 states and U.S. territories. However, 32 per cent of the 10,443 places questioned by AMA did not respond. Of those that replied, 20 per cent said they did not have a blood banking-transfusion facility. By applying that percentage to all, and on the basis of other studies, it has been estimated that in 1966-68 some 9,000 central, regional and local blood banks existed in the United States.

The growth of commercial activity in blood banking has been encouraged by the development of plasmapheresis, the process by which red cells are separated from the plasma and returned to the donor—all within the space of less than an hour. This method permits a person to give plasma several times a week, whereas he should give whole blood not more than five times a year. Large pharmaceutical firms depend on a plasma supply for the production of blood derivative products, such as gamma globulin for immunization purposes. Virtually all plasma drawn by plasmapheresis comes from "a changing pool of an estimated 100,000 donors, most of whom are paid for their services."[22]

---

[22] National Academy of Sciences, *Safeguards for Plasma Donors in Plasmapheresis Programs,* June 1970, p. 6. These safeguards were developed after it was realized that some plasma donors had suffered serious physical harm in unregulated plasmapheresis centers. "Because the procedure involves the reinfusion of formed elements into the donor, there are risks of incompatible transfusions, air embolism and sepsis." Some donors are immunized or "hyper-immunized" so that their plasma will become suitable for the production of immunoglobulins that protect against whooping cough, tetanus, smallpox, mumps or other conditions.

Tensions between the Red Cross and AABB in the early days of blood banking have eased considerably and the two organizations now cooperate amicably in many ways, though a trace of rivalry persists. A reciprocity agreement, concluded in May 1961, facilitates the exchange of blood and blood-replacement credits between agencies in the two systems. Files for locating rare blood donors without delay have been coordinated with each other and with a similar file maintained by the International Society of Blood Transfusion. This permits rapid worldwide exchange of rare blood as need arises. A woman in Boston, for example, recently received rare blood from Bangkok.

Both organizations have taken steps recently to extend their cooperation. The Red Cross Board of Governors voted on Feb. 15, 1971, to expand the organization's cooperation with other nonprofit blood-collecting agencies. Red Cross chapters were asked to make available to other agencies "specialized assistance in the control and management of blood and blood components inventories, hepatitis surveillance, coordination of a fractionation[23] program for plasma, the use of Reference Laboratories, the Rare Donor Registry service and training programs." In addition, the 1,600 chapters not involved in Red Cross regional blood programs were asked to assist other blood-collecting agencies in their areas by helping them recruit volunteer donors and giving other "appropriate volunteer service support."

This was a historic move, for the original tension between the Red Cross and other nonprofit blood-collecting agencies had centered on competition for the limited supply of donors. Generally speaking, the Red Cross tries to recruit donors from large groups—business concerns, civic clubs, local unions, and the like—while other nonprofit banks tend to rely on friends and relatives of patients in hospitals.

Approaches Toward Solving Blood Needs

THE PROBLEM of maintaining an adequate supply of blood is never-ending. There is no source except in the body of living human beings. Although every person may at some time be in acute need of a blood transfusion or a medication derived from a human blood component, only a tiny fraction of the

---

[23] Dividing up of the blood into its parts. See chart, page 329.

population ever makes a blood donation. It is believed that one-half of the American people are qualified by age and health to donate blood.[24] Yet no more than three million a year actually do so. Three-fourths of them give repeatedly.

Blood-banking officials often wonder how and why they have failed to awaken in more Americans a sense of obligation to give blood at least once a year. They know that one big reason is simple fear—a lot of people are afraid of the needle, or they think erroneously that bloodletting will damage their health. Many do not want to take the time, or else simply never think about it.

Individual 'Credits' to Meet Future Requirements

In these circumstances, it is no wonder that the blood-collecting agencies have felt it necessary to offer various kinds of inducements. In many cases, this means cash payments. But putting a monetary value on blood is believed to confirm a widespread feeling among people that if they or their families ever require blood, all they need is the money to pay for it—or insurance to cover the cost. Some health insurance policies pay for blood. A preferred form of compensation is "replacement credit." Under this plan a patient will not be charged for the blood he receives if he can get someone to replenish the bank. Some hospitals require two or even three units of replacement for each transfusion to compensate for the loss of outdated blood. Some hospitals charge more than the market cost of the blood precisely to encourage blood replacement.

This practice has helped to keep blood banks replenished but it can be hard on patients who have large blood requirements. The Medicare program requires the aged patient either to pay cash or find replacements for his first three blood transfusions. Only one-fourth of the blood provided Medicare patients in the first 18 months of the program was replaced in lieu of payment.[25] Titmuss caustically referred to the replacement-fee system as evidence of an "assumption that waste in the system (over which the sick have no control) should be paid for by the sick."[26]

---

[24] AABB standards limit donors to ages 18-66 (the Red Cross puts the top age at 61). Donors who weigh under 110 pounds give less than the standard amount, one pint. The average man has 12 pints of blood, the average woman 9 pints. Blood accounts for 7 to 8 per cent of the body weight.
[25] National Academy of Sciences-National Research Council, *op. cit.*, p. 35.
[26] Titmuss, *op. cit.*, p. 80.

Some foresighted individuals prepare for their blood needs by contributing blood in advance of scheduled surgery. Some hospitals already require donations prior to certain surgery and, according to Stephen M. Morris, president of the American Hospital Association, other hospitals may be impelled to adopt that practice.[27] Blood-bank leaders have sought to popularize so-called assurance plans whereby donors can build up blood credits for themselves and their families—a form of protection in which both the premium payment and the benefit are in blood, not money. Some blood banks will give credits to an entire organization, thus protecting all of its members, in return for a stated volume of blood contributions from the group.

Blood banks vary in the amounts they pay for blood. Payments may range from $5 to $25 per pint for common types, while very rare blood commands a great deal more. Similarly, hospitals vary in their charges to the patient, which will include not only the cost of the blood itself but the services involved. The National Research Council reported preliminary studies indicate the following are average fees to patients:

Processing fee	$11.75
Typing and cross-matching fee	13.50
Infusion fee	6.75
Donor replacement fee	24.50

The donor-replacement fee represents the cost to the hospital of replenishing the blood used by the patient—unless the patient finds a donor who will replace the blood without charge.

Reducing Waste by Fuller Use of Blood Parts

Another approach to solving the supply problem is to reduce waste. A certain amount of blood is bound to become outdated. The AMA directory of blood banking shows that in the year its survey was conducted, 1968, some 4.35 million pints of blood were used in transfusions but that 6.6 million pints were collected. Compilers of the listing said the difference in numbers did not represent waste because outdated blood is usually returned to the agency that provided it and then is processed into recoverable components. But the National Research Council's panel has noted that "to take full advantage of the potential of a unit of whole blood, the decision to separate the red cells from the plasma must be made shortly after collection."

---

[27] *In Vivo* (quarterly publication of Blood Services), January 1971, p. 3.

Waste has been attributed to the insufficient use of component therapy—the use of blood components rather than whole blood for specific medical purposes. A group of professional and business men concerned with this problem formed the Component Therapy Institute in Washington in 1967 to press for greater application of new medical knowledge and technology on the utilization of blood parts. It was at the institute's behest that the National Research Council appointed a committee to study blood utilization. The committee's report found that "the therapeutic benefits derived from the human blood collected in the United States fall far short of its potential value to the patient population." The reasons given were many, arising largely from the lack of an integrated system uniting the various segments of the blood-service complex. But also at fault was "the failure of a substantial portion of the medical profession to adopt proven concepts of component therapy."

The AMA has sought to overcome a persistent feeling among doctors that whole blood is nearly always better than part-blood. The AMA Committee on Transfusion and Transplantation stated categorically in 1970 that "the routine use of whole blood can no longer be justified." By more extensive use of components, a single donation could serve several patients instead of only one. Medically, a component would be more serviceable than whole blood in most cases, and sometimes much less hazardous.

All the while, a search goes on for a substitute for human blood. In experiments conducted at the University of Cincinnati Medical School, Harvard School of Public Health and elsewhere, animal blood has been replaced with a fluid composed of the synthetic substances fluorocarbons and polyols. Animals survived the exchange; their bodies eventually restored their normal blood supply. *Reuters,* the British news agency, reported in April 1971 that Japanese scientists at Kobe University had produced artificial blood from similar ingredients and had tested it successfully on animals, although there were still problems to be ironed out.

If artificial blood should prove itself in tests on human beings, the problem of recruiting donors would no longer be vital. But until that day arrives, the supply of blood must depend on the small handful who, prodded by unavoidable circumstance, by self-interest, or by simple benevolence, come forward to give.

ARCHEOLOGY BOOM

by

Ralph C. Deans

WIDE RANGE OF RECENT DISCOVERIES
Surging Interest in Archeology on All Continents
New Evidence in Search for Origins of Mankind
Recent Discoveries Concerning Biblical Period
Controversy Over First Arrivals in New World

DEVELOPMENT OF ARCHEOLOGICAL RESEARCH
Dispute Over True Source of Modern Archeology
Success of Dilettantes and Gentlemen Amateurs
Discovery of Egyptian, Mesopotamian Cultures
Mysteries of Mayan, Aztec and Toltec Indians

AIMS AND METHODS OF MODERN ARCHEOLOGY
Reappraisal and Questioning of Old Assumptions
Techniques to Determine the Ages of Remains
Impact of Physical Sciences on Archeologists
Problem of Protecting Invaluable Ancient Sites

1 9 7 1
July 14

ARCHEOLOGY BOOM

A JAWBONE FRAGMENT discovered in Africa indicates that man-like creatures existed a million and a half years earlier than previously supposed. In France, man's earliest known habitation, a temporary hunting camp 300,000 years old, has been unearthed. In Israel, archeologists have uncovered the first physical evidence of Roman crucifixions. In Egypt, an ancient shrine visited by Queen Cleopatra has been discovered. In Greece, the court where Socrates was tried and condemned to death has been found. These are only a few of many dramatic archeological finds in recent years.

New knowledge of the distant past is thus being unearthed on an unprecedented scale. The tempo of activity reflects the surging interest in archeology as a basic tool in man's quest for self-understanding. From artifacts and fossils, he is trying to piece together a portrait of his ancestry from the dim recesses of prehistory. His painstaking efforts also shed further light on events that have taken place since the dawn of recorded history.

Recent discoveries and new thinking about old ones are beginning to challenge many hallowed concepts—including those concerning the origins of man, the birth of civilization and the migration of ancient man from the Old World to the New. The vast and constantly growing body of archeological knowledge has led to specialization of a high order. Archeologists draw heavily on the physical sciences, including biology, physics, computer technology and zoology, in their work. The influx of new methods, while placing great demands on scholarship in the study of ancient artifacts, has helped to whet the interest of a rising generation of scholars.

Archeology courses, frequently over-enrolled in universities, now have been introduced in many high schools. T. Patrick Culbert, secretary of the Society of American Archeologists, told Editorial Research Reports that between 60 and 70 field courses in archeology are being conducted each summer by various American universities and involve more than 2,000 students. In Culbert's opinion, "the romance of the past" and the "lure of the unknown" are still the major factors in the rising interest in archeology. He felt that leisure time had helped bring about the boom.

At least a part of the new interest in archeology can be ascribed to the high value placed on it by new and emergent countries. Anthropologist Clyde Kluckhohn, noting that Mussolini excavated Roman ruins to stimulate Italian pride, has said "new states...developed archeology as a means of nation-building and self-expression."[1] The study of African archeology has quickened in the last decade, another archeologist wrote, "in response to an increasing awareness of the importance of Africa's past, and also as a result of the demand for national histories and intellectual justification for newly independent states."[2] The same may be said of Israel, where enormous archeological tasks are being carried out. George A. Petit, an anthropologist at the University of California at Berkeley, asserts that society "needs a background for its self-pride."[3]

Curiously enough, a great many archeological finds come to light not as the result of a highly scientific and well-organized endeavor but by accident. Excavation for a subway system in Rome has been delayed because work crews frequently stumble across valuable artifacts. Work on the Mexico City subway has similarly turned up important traces of the Aztec culture and the Spanish conquest of the 14th century. In New York State, archeologists found the remains of an early Dutch stockade in the path of an urban renewal project.

There are two matters of general concern to all archeologists today. One is the ever-growing pressure to construct roads and buildings over areas that may abound in priceless artifacts. Many archeologists feel they are in a race with the bulldozer. The other is the plundering of ancient sites by thieves who smuggle the loot out to the world market.

New Evidence in Search for Origins of Mankind

The fragment of a jawbone found in 1967 near Lake Rudolf in Kenya has provided man with one of the most interesting clues about his origins. The fragment was discovered by

---

[1] Clyde Kluckhohn, *Mirror for Man* (1949), p. 47.
[2] Brian M. Fagan (professor of archeology at the University of California, Santa Barbara), "Archeology and History: Can the Gap Be Bridged?" *Africa Report*, March 1970, p. 22.
[3] George A. Petit, *Prisoners of Culture* (1971).

Archeology Boom

Arnold D. Lewis, head of the preparation laboratory of Harvard's Museum of Comparative Zoology, while on an expedition led by Bryan Patterson, professor of vertebrate paleontology at Harvard. After careful study, it was announced in February 1971 that the fragment was 5.5 million years old and came from a creature closely related to *Australopithecus*— an animal known to be about five feet tall and believed to be an ancestor of man. It was offered as evidence that the evolutionary line of man went back a million and a half years beyond previous estimates.

It is no longer questioned that man and ape share a common ancestry, but the point in evolution where the hominid or human line diverged from the pongid or ape line has not been established. The oldest man-ape that authorities admit to the hominid line is *Ramapithecus*, which is believed to have existed 14 million years ago. The skull of an infant primate, found in South Africa in 1924 named *Australopithecus*, was dated at 750,000 years old and considered to be the oldest form of man-like life. Thirty-five years after that discovery, Dr. Louis S. B. Leakey, one of the most controversial archeologists today, found in Tanzania's Olduvai Gorge a similar specimen which he concluded was 1.75 million years old. The timetable was pushed even further back in 1967 when an arm bone fragment from the Lake Rudolf area was dated at about four million years old.

Then came the jawbone discovery. Scientists guessed the jawbone fragment was from a mature female. The single remaining tooth on the fragment indicated that the creature's diet consisted of both meat and vegetation. Patterson, in announcing the discovery, said modern man would recognize the creature "as closely related to ourselves if we were to see it alive today."[4] The find suggests that man was evolving during the Pliocene period, which began about 13 million years ago, not just in the Pleistocene period which began about two million years ago and was marked by the advance and retreat of glaciers. "It is now evident, the paleontologists say, that the human lineage was evolving throughout this stretch of time and that the beginnings of the hominid line occurred still earlier."[5]

Alexander Marshack, a research associate at the Peabody Museum of Archeology and Ethnology at Harvard, came for-

---

[4] Quoted in *The Washington Post*, Feb. 19, 1971.
[5] "More Complete View of Man's Ancestors," *Science News*, Feb. 27, 1971, p. 141.

ward early in 1971 with evidence that ancient man may have had a system of notation and a lunar calendar during the last Ice Age, some 34,000 years ago. Marshack first suggested in 1964 that mysterious markings on fragments of bone found throughout Europe may have been more than just decoration. After seven years of study—using techniques employed in a criminal ballistics laboratory—he showed that at least 24 different points were used to make the tiny indentations. He believes that the notations were a calendar of the phases of the moon and that they could have been used as an aid to hunting, to anticipate phases of the moon or female menses.

Dr. Hallam Movius, curator of Old World archeology at the Peabody Museum, said the decipherment of the system had "thrown revolutionary new light on the intellectual level attained by our Upper Paleolithic forebears." Ralph Solecki of Columbia University, one of the world's leading authorities on Ice Age archeology, said the bone calendars were a "logical springboard" leading to the calendars of ancient Sumeria, Babylon and Egypt. Solecki said the discovery was "a milestone in archeology comparable to the discovery of carbon 14"—a modern technique of dating ancient artifacts.[6]

Between January and July 1966, excavators in France unearthed the earliest known habitation built by man. Writing in the *Scientific American* about the find, French paleontologist Henry de Lumley said it revealed that prehistoric men visited a beach on the Mediterranean for 11 successive years, building temporary shelters and then moving on after hunting and gathering food. A footprint found at the site, near Nice in the south of France, indicated that the individuals may have been about five feet tall. The archeologist's ability to make judgments on the slightest evidence is well known. De Lumley deduced that the small band arrived at the site during the late spring. Analysis of the pollen in fossilized human feces at the site showed that it came from plants that "shed their pollen at the end of spring or the beginning of summer."[7]

Recent Discoveries Concerning Biblical Period

The discovery of the first Dead Sea Scrolls in 1947 was one of the major archeological finds resulting from serendipity rather than scientific method. The original lot of seven scrolls

---

[6] Quoted in *The New York Times*, Jan. 20, 1971.

[7] Henry de Lumley (director of the Laboratory of Human Paleontology and Prehistory in Marseilles), "A Paleolithic Camp at Nice," *Scientific American*, May 1969, p. 43.

was discovered by a bedouin boy who, out of curiosity, entered the caves at Qumran, in what is now Israeli-occupied Jordan. The scrolls included two copies of the Book of Isaiah from pre-Christian centuries, a collection of Thanksgiving Psalms and a Manual of Discipline to regulate the life of a Jewish sect whose members—probably Essenes—believed they were living in the last days of the world.

Further scrolls were found in nearby caves throughout the 1950s. The scrolls, some in fragmentary form, now number about 500 manuscripts. They contain every book of the Old Testament except Esther. They also contain extensive remains of records, receipts, contracts and letters. William F. Albright, a professor of Semitic languages at Johns Hopkins University, has written that it is "hard to exaggerate" what the documents mean for historians of the religious ideas in the obscure age between the fifth century B.C. and the Christian era.[8] Floyd V. Filson, professor of New Testament Literature and History at McCormick Theological Seminary of Chicago, has said that for his area of scholarship, "the discovery of the Dead Sea Scrolls is the most important manuscript find ever made."

Archeologists are working intensively throughout the biblical lands. During the 1970 season, archeologists worked on 25 major sites and the period was described by Peter Grose of *The New York Times* as "the most intense period of archeological exploration ever known in Israel." Work during the 1971 season promises to be even greater. Among many finds to date are these:

—A tomb uncovered by roadbuilders in Upper Galilee is believed to be Canaanite or Amorite, from about 1900 B.C., a relic of the peoples who lived in Galilee before Abraham and the Hebrew patriarchs arrived.

—A wall 21 feet thick found in an inner courtyard of a row of ancient houses on the way to the Wailing Wall in Jerusalem is now believed to be the remains of the New Church of Mary, built by the Byzantine emperor Justinian in 543 A.D.

—In the Old Quarter of Jerusalem, the complete residence of a patrician of the first years of the Christian era was found with the rooms, closets, walls and pottery in place.

—At Tel Dan in northern Israel, only a few hundred yards from the Lebanese border, a team of archeologists found the "High Place" cited often in the Bible as the site of worship of pagan

[8] William F. Albright, "Impact of Archeology on Biblical Research" in *New Directions in Biblical Archeology* (1971), edited by David Noel Freedman and Jonas C. Greenfield, p. 15.

Canaanite gods. This may be the place, archeologists speculate, where a golden calf was erected by the first Israelite king, Jeroboam, in the ninth century B.C. to lure his people away from the Temple in Jerusalem.

—On Jan. 15, 1971, professor Nahum Avigad of Hebrew University announced the uncovering of a building destroyed in 70 A.D. in the Jewish Quarter of old Jerusalem by Roman conquerors. This was the first physical evidence of the razing of Jerusalem 1,900 years ago.

—On March 2, 1971, Greek archeologists unearthed the remains of the basilica that the Roman Emperor Constantine built in Jerusalem in the fourth century B.C.

A major archeological find was recorded when researchers found a foot bone with a nail driven through it on the northern outskirts of Jerusalem. The find, announced Jan. 3, 1971, immediately aroused emotional speculation that the remains could possibly be those of Jesus. Archeologists involved turned this speculation aside. The inscription on the stone coffin containing the remains included the common first name "Jehohanan," and Avraham Biran, director of the Israeli Government Department of Antiquities, cited this as proof that the remains were not those of Jesus of Nazareth.

Controversy Over First Arrivals in New World

New archeological evidence indicates that man is older, developed intellectual skills much earlier, and arrived in the western hemisphere earlier than was previously supposed. A decade ago, archeologists surmised that man came to the New World no more than 10,000 or 12,000 years ago. Current estimates are that Mongolian tribesmen arrived in Alaska— either by boat or over a land bridge crossing the Bering Strait—some 20,000 years ago. Work now in progress in California and Peru indicates that man may have made the journey 40,000 or even 100,000 years ago. But if this is true, most archeologists say, it was not *homo sapiens* but an earlier ancestor of man who made the trek.

A controversial project indicates the earlier crossing. It is an archeological "dig" in the Calico Mountains near San Bernardino, Calif. A team put together by Dr. Louis S. B. Leakey —famous for his work in the Olduvai Gorge in Tanzania—and Ruth Dee Simpson of the San Bernardino County Museum has found what are claimed to be man-made stone tools 50,000 to 100,000 years old. Many archeologists do not accept the theory that the chipped rocks found at the Calico site are man-made, however. "The only one who believes that is S. B. Leakey," one archeologist told Editorial Research

Reports. "But Leakey has been thought wrong before and was proved right." The evidence remains in doubt because there are no deposits at Calico that can be dated by carbon-14 or by potassium-argon tests.

Further confusing the issue, some geologists say the deposit is at least 500,000 years old and perhaps 1.5 million years old. One of the questions raised by the Calico findings is how man crossed the glacier-blocked north, if he arrived in the New World 50,000 to 100,000 years ago.

> The theory, says Miss Simpson, is that there were successive progressions and recessions of the glacial ice in the millennia preceding the last great opening out about 12,000 years ago. As the glaciers formed, they drew water from the Bering Strait, exposing a land bridge between the continents. But they also drew water from along the continental shelf of the West Coast, exposing a strip of land several miles wide, down which animals, and man living off them, gradually extended their range throughout America. There was no need for them to come through the ice-blocked passage down the center of the continent.[9]

A shift in thinking has recently taken place because of solidly dated archeological finds in South America. Richard S. MacNeish, director of the Robert S. Peabody Foundation for Archeology at the Phillips Academy in Andover, Mass., has discovered stone tools in highland Peru which indicate that men lived there 22,000 years ago, almost double the old estimate. MacNeish has written that tools found in the Peruvian caves he studied in 1969 and 1970 were characterized by "large corelike choppers, large side-scrapers and spokeshaves and heavy denticulate implements" which he described as belonging to the "Core Tool Tradition." He said this tradition "may just possibly be represented in North America by the controversial finds at the Calico site...."[10]

Development of Archeological Research

IN THE SIXTH CENTURY B.C., Nabonidus, the last king of Babylon, engaged in antiquarian research and even conducted excavations at Ur, the birthplace of Abraham. Greeks made anthropological studies of various cultures that are still

---

[9] "Early Man in America," *Science News*, Nov. 7, 1970, p. 364.
[10] Richard S. MacNeish, "Early Man in the Andes," *Scientific American*, April 1971, p. 44.

valid today. "The Athenians in the fifth century, carrying out a religious purification of Delos, opened some graves," a scholar recounted. "These they judged from the make of arms buried with the bodies and from the fashion of the burial to be those of Carians, who still used arms and burial of the same kind," E. D. Phillips wrote in 1964 in the British magazine *Antiquity*. The antiquities—if not the cultural development—of older societies thus appear to have been a matter of interest from early times.

The origins of modern archeology are in dispute. Some historians contend that archeology was born during the Renaissance with the rise of intellectual curiosity which spread from Italy about the year 1400. Renaissance scholars directed their attention to Greece, Rome and the Bible lands. Art historian Johann Winckelmann is frequently cited as "the father of archeology" for his 18th century studies of Greek and Roman fine arts. When the word "archaeology"[11] came into use in the 17th century it meant merely the study of the classic art of antiquity.

There are other historians who believe true archeology arose only after the appearance of Charles Darwin's *The Origin of Species* in 1859. Still another view is expressed by Glyn Daniel, who regards Jens Jacob Asmussen Worsaae as the real father of modern archeology. In 1843, fifteen years before Darwin wrote his book, Worsaae wrote the first exposition of the principles of excavation and stressed the need to interest the public in archeological matters. "No words are too lavish in trying to estimate Worsaae's contribution," Daniel said.[12]

Success of Dilettantes and Gentlemen Amateurs

In any case, the beginnings of modern archeology can be glimpsed in the activities of numerous antiquarians who in the 15th century started to collect, catalogue and speculate about the artifacts that came to their attention. They include William Camden, who wrote the first general guide to the antiquities of Britain, *Britannia*, in 1586; John Aubrey, who studied the Stonehenge ruins in the mid-1600s; Edward Lhwyd, who compiled the first catalogue of fossils, *Archeologia Britannica*, in 1707, and William Stukely, who in the mid-18th

---

[11] Spelled with an "a" before the "e"—"archaeology"—a spelling that has fallen out of favor with many archeologists in recent years though it has not been discarded entirely.

[12] Glyn Daniel, *The Origins and Growth of Archeology* (1967), p. 97.

century attributed the building of Stonehenge to the Druids—
a theory that has since been challenged.[13]

Perhaps the most famous of all the dilettantes and traveling
antiquaries was Giovanni Belzoni, an Italian who went to
Egypt in 1815 to sell hydraulic machinery but stayed on to
explore and rob the tombs of the Pharaohs. Among many
other things, Belzoni discovered the tomb of Sethos I, the
predecessor of Ramses. This find paved the way for a series
of other important discoveries in the Valley of the Kings along
the Nile.

Heinrich Schliemann, a wealthy German businessman of
the 19th century, is one of the great names in the development of archeology. Schliemann, an amateur archeologist,
overcame the scoffing of scholars who believed ancient Troy
was a fictional city. Relying on clues found in the writings of
the Greek poet Homer,[14] Schliemann located and in 1871-82
excavated the remains of nine cities near present-day Hissar-
lik, Turkey, each one built on the ruins of the previous city.
His only error in this regard seems to have been in assuming
that the second layer of ruins he uncovered was Troy's. It is
now generally believed that the remains of Troy are those
found in the seventh layer down.

Schliemann's extensive reading of classical Greek writing
also led him to another major discovery—that of graves at
the ancient ruin at Mycenae, a citadel set on the northeastern
part of the Argive Plain in Greece. Still later, in 1885, this
remarkable amateur found the dead city of Tiryns, the legend-
ary birthplace of Hercules. His excavations, and those made
later (1900-25) on Crete by the famous English archeologist
Arthur Evans, established the existence of the Minoan-
Mycenaean culture.

The foundations of Egyptology, a prolific branch of archeology, were laid by Napoleon's expedition to Egypt in 1798.
Though a military failure, it was an artistic success, exposing
European scholarship to the wealth of Egyptian antiquity
which had been preserved in the desert climate. The French
leader took along 175 scholars, the most notable being

---

[13] The official guidebook, issued by the British Ministry of Public Buildings and
Works, makes this opening statement: "The ordinary visitor to Stonehenge will in all
likelihood have been taught from his earliest years that Stonehenge was built by the
Druids. He can clear his mind of this statement, which is quite incorrect...."

[14] According to legend, as related by Homer in his *Iliad*, Troy stood near the
Hellespont and was the capital of King Priam. It was besieged for 10 years by the
Greeks in their attempt to recover Helen, wife of the king of Sparta, who had been
carried off by Paris, son of Priam.

Dominique Vivant Denon. He copied Egyptian hieroglyphics, sketched the pyramids and collected papyrus. It was not a scholar, however, but a soldier who discovered the Rosetta Stone, a basalt slab about three and a half feet long, two feet wide and nearly a foot thick.

Discovery of Egyptian, Mesopotamian Cultures

On one highly polished side were three columns of writing— one in hieroglyphs, the picture writing of Egyptian priests, another in the related demotic script, and the third in Greek. The Greek was promptly translated as a decree of the Egyptian priesthood, issued in 196 B.C., praising Ptolemy Epiphanus. But the hieroglyphs eluded the best efforts of scholars to decipher them until 1822 when a youthful French scholar, J. F. Champollion, finally cracked the hieroglyphic code. This was one of the most important developments in the long study of Egyptian culture. It enabled investigators to read the inscriptions on royal tombs and give precise dates to events in Egyptian history as far back as 3000 B.C.

Many important discoveries were made in Egypt after Napoleon's invasion and Belzoni's excursions. A German, Richard Lepsius, found 30 unknown pyramids between 1843 and 1845; French archeologist Auguste Mariette uncovered 140 sphinxes; Englishman William Matthew Petrie, in years of excavating beginning in 1881, dug the Temple of Ramses out of the hills at Nebesheh and found the plundered tomb of Amenemhet III.

The most dramatic find in Egypt was undoubtedly the discovery of the tomb of King Tutankhamen by Lord Carnarvon and Howard Carter in the Valley of the Kings—a desert region of the upper Nile near the site of ancient Thebes. After six years of excavation, Carter in 1922 found the entrance to the tomb. The burial vault yielded not only Tutankhamen's mummy but fabulous amounts of gold and jewelry as well as a wealth of inscriptions and artifacts.

A sensational sidelight of the Tutankhamen find was the so-called "curse of the Pharaohs." Some 20 persons connected with the unsealing of the tomb died later under mysterious circumstances. Lord Carnarvon, the financial backer of the hunt for the tomb, died of pneumonia six weeks after the tomb was opened. An inscription, "Death shall come on swift wings to him that toucheth the tomb of Pharaoh" was widely reported to have been inscribed somewhere in the tomb. It was not. And many of those connected with the opening lived to ripe ages.[15]

"The discovery of the tomb of Tutankhamen represents the very summit of success in archeological effort," wrote C. W. Ceram in his well-known history of archeology, *Gods, Graves and Scholars*. As a result of the find, "Egyptology ceased once and for all to be a random striking-out into an unknown terrain and became a sort of cultural surveying process, marked by the strictest sort of adherence to method....Carter was able to make the very most of scientific exactitude and discipline [and] it was this combination of sweep and minute thoroughness that made Carter one of the greatest figures in the history of archeology."[16]

Another giant in archeological endeavor was Sir Leonard Woolley, who excavated the royal tombs of Ur in 1928. Woolley's findings filled in many details of the ancient Sumerian civilization, which pre-dated the Babylonians and Assyrians and was believed to be the first civilization of the "fertile crescent" where man's history begins. While archeologists had long suspected the existence of a civilization much older than either the Babylonian or Assyrian, the first evidence of the Sumerians was found late in the 19th century by a French consular agent, Ernest de Sarzec, who excavated mounds between the Tigris and Euphrates rivers. The way had been paved early in the century for study in this region, Mesopotamia, when Georg Grotefend and Sir Henry Creswicke Rawlinson had independently deciphered cuneiform writing.

In excavations at Ur, Woolley concluded that he had found physical evidence of the Great Flood described in the Bible and, before that, in the Gilgamesh Epic, a Babylonian poem dating from the seventh century B.C.[17] Woolley was neither the first nor last to claim such a discovery. History is rich with tales of Noah's Ark being found. Marco Polo mentioned the existence of an ark near the summit of Mount Ararat in Turkey in the year 1300. As recently as 1955, a French engineer, Ferdinand Navarra, announced he found remains of the ark near the summit and pressed his claim in a book

---

[15] Ronald Harrison, professor of anatomy at Liverpool University, recently added a postscript to the King Tut story. He suggested that the youthful king may have died violently. X-rays of the mummy disclosed skull injuries.—Washington *Sunday Star*, Oct. 26, 1967.

[16] C. W. Ceram, *Gods, Graves and Scholars* (1951), p. 174.

[17] The epic written on 30,000 clay tablets and unearthed in the 19th century relates how Ut-napishtim, a prehistoric man, built an ark to escape a flood sent down by primitive gods. Many other ancient cultures also have legends that correspond roughly to biblical accounts of the flood.

entitled *I Found Noah's Ark*. Airplane pilots have persistently reported seeing a ship-like shadow on the mountain's surface.

Mysteries of Mayan, Aztec and Toltec Indians

When Cortez arrived in Mexico in 1519, the Spanish conquistadores he led found the highly developed Aztec culture, which soon disappeared under their rule and was largely forgotten until the 19th century. John Lloyd Stephens discovered the ancient Mayan civilization in 1839 and his fellow American, William H. Prescott, revived interest in the Aztecs about the same time with his book *History of the Conquest of Mexico* (1843). It became clear that the two cultures were connected in some way though the Mayan and Aztec languages were different. Prescott theorized that an even older culture antedated the two. But it was only in the 1940s that Mexican archeologists began to find traces of the preceding Toltec culture, and the even older and more primitive Olmec and Zapotec cultures.

Edward Herbert Thompson, a young American archeologist, discovered the largest and mightiest of the Mayan cities —Chichen-Itza—in 1885. Thompson also discovered and undertook the difficult excavation of the Sacred Well of Chichen-Itza in Yucatan, where Mayan priests threw human sacrifices and precious objects to propitiate the gods. Chichen-Itza is probably the most intensively studied site of the Mayan culture. Giles Greville Healey, who led an expedition in 1947 financed by the United Fruit Company and sponsored by the Carnegie Institute, uncovered 11 temples of the Old Empire of the Mayas, beginning in some unknown time and ending about 610 A.D.

The Mayan, Aztec and Toltec cultures all built pyramids, though not as tombs. Archeologists believe the structures were temple complexes. They had steps leading up to the gods of the sun or the moon and were located according to astronomical lines of sight. All three societies were believed to have been led by priests under whom religious practices became progressively more cruel. But the origins of these advanced cultures, and the interaction between them, are still a mystery.

A frequently expressed belief is that the high civilizations of Middle America may have sprung from one of the lost tribes of Israel. Professor Alexander von Wuthenau of the University of the Americas in Mexico City, in a recent series of lectures at Brandeis University in Waltham, Mass., suggested that Mediterranean people may have discovered America centuries before Columbus. He said a recently excavated Mayan Stela—a tall sculpted stone—shows a squat scowling man wearing an earring with the Star of David in its center. Furthermore the figure wears a torpedo-shaped hat. "In my opinion," von Wuthenau said, "it depicts a ship seen from above, which, surprisingly enough, has great affinity with ancient Egyptian papyrus boats."[18] Brandeis professor Cyrus H. Gordon believes that a stone found in a burial mound in Tennessee shows that Jews were in America in ancient times. The inscription on the stone, Gordon says, is in the writing style of the "promised land" of the Israelites between the Jordan River and the Mediterranean.

Aims and Methods of Modern Archeology

ARCHEOLOGY TODAY is in an age of ferment. Old concepts are falling under the onslaught of new discoveries and new interpretations. The body of knowledge is now vast but it is shot through with mysteries which continue to fascinate and baffle scholars. How did man evolve from the primitive *Ramapithecus* of 14 million years ago? When and how did man come to the New World? C. W. Ceram wrote that "As we get to know more about the history of mankind, the time comes when we begin to feel the faint breath of the eternal wafted to us across the great gap of the years."

We begin to see [he continued] glimmerings of evidence that little human experience during five thousand years of history has actually been lost.... It is frightening to realize in full depth what it means to be a human being: that is, to realize that we are all embedded in the flux of generations, whose legacy of thought and feeling we irrevocably carry along with us. Most of us never become aware that man alone of all mammals lugs forward through time.[19]

Such is the high drama that many archeologists feel their discipline is imbued with. In few disciplines, however, do bedrock assumptions change so thoroughly or so rapidly.

Even the hallowed theory that western civilization sprang

---

[18] Quoted in the Baltimore *Sun*, March 23, 1971.
[19] C. W. Ceram, *op. cit.*, p. 298.

up in the Mediterranean basin is now being challenged. Colin Renfrew, a senior lecturer on prehistory and archeology at the University of Sheffield in Britain, contends that civilization may have begun in northern Europe a thousand years before it appeared in the Mediterranean littoral. "According to Renfrew's calibrations, Stonehenge now becomes older than Mycenae. The megalithic tombs of Brittany and Iberia are a millennium older than the pyramids—and may even be the earliest of man's surviving monuments."[20]

Relics unearthed in Thailand have led another group of archeologists to dispute the theory that agriculture was first practiced in the fertile crescent—the valley between the Tigris and Euphrates rivers. Seeds and tools found in two widely separated areas of Thailand suggest that Thais were farming long before the Babylonians and were using bronze for tools 1,000 years before the Chinese were. William G. Solheim 2nd of the University of Hawaii said domesticated seeds found in Thailand caves were 11,000 years old—"the oldest found by modern man."[21]

Techniques for Determining Ages of Remains

The archeologist is supremely concerned with dating his finds relative to others and to the current concept of time. The most widely used and satisfactory method of chronological dating is the radiocarbon system, developed in the late 1940s by Willard F. Libby at the Institute for Nuclear Studies at the University of Chicago. This method determines the age of anything that absorbed particles of the atmosphere during its lifetime. Carbon-14, a radioactive unstable form of carbon, is among those particles. It penetrates all living organisms and begins to decay at their death; since its rate of decay is constant, scientists can measure the rate and convert their measurements into years.

However, there are limits to radiocarbon dating. For one thing, carbon-14 has a mean lifetime of about 8,000 years and some objects are too old to be dated by it. For another thing, the carbon of the distant past was more radioactive than that of the recent past, Philip Morrison reported in the July 1971 issue of *Scientific American* magazine. "The explanation is not certain, but an admirable fit has been proposed by V. Bucha of Prague that ascribes the entire effect to the known slow decline in the earth's magnetic field...."

---

[20] "Pyramids From France?" *Newsweek*, April 5, 1971, p. 50.
[21] Quoted in *The New York Times*, Jan. 12, 1970.

Archeology Boom

Bucha contends that the reduced size of the magnetic field allowed low-energy cosmic ray protons to enter the earth's atmosphere and that these had the effect of increasing the radioactivity of carbon nearly 50 per cent at the maximum of 10,000 years ago—up to about the present level of the nuclear age.[22]

The thermoluminescence technique used in the dating of pottery relies, like radiocarbon dating, on a knowledge of physics and chemistry. The technique, perfected in 1965 after years of development, makes use of the fact that thorium and uranium in clay bombard other substances and raise electrons to unstable levels. When the clay is fired in the kiln, each electron falls back to its stable position and emits a photon of light. If a fragment of ancient pottery is reheated in the laboratory, it gives off small amounts of light which can be measured in terms of years of age.

One highly specialized method of dating glass artifacts is by counting the variegated scales which build up in iridescent layers on glass that is kept under water or buried in the ground for a long time. "For historic sites on land or underwater sites of shipwrecks, that are only a few hundreds of years old, the method of dating by glass can scarcely be surpassed," George L. Quimby noted in the *Encyclopedia Americana*.[23]

Glacial-varve chronology is another highly accurate means of dating many deposits. Glacial varves are thin layers of clay deposited yearly in basins of water created by the retreating ice of continental glaciers. Cultural remains associated with late glacial and post-glacial features can often be dated by this method. It has been particularly successful in the Baltic region of northern Europe.

"Counting annual layers and reading written documents remain the most reliable means of dating the past," Morrison wrote in *Scientific American*, though he added that they are neither simple nor infallible. "We moderns date the dynasties of Egypt most securely by our astronomy; we know when Sirius rose, and we are lucky enough to have read some Egyptian documents that tell us the day it was by their civil calendar." As for annual layers, "the bristlecone pine forests

---

[22] Bucha's theory is expounded in *Nobel Symposium 12: Radiocarbon Variations and Absolute Chronology* (1971), the record of a symposium held in Uppsala, Sweden, in 1967.
[23] Vol. 2 (1968), p. 196.

of the dry White Mountains of the Sierra Nevada provide a marvelous run of tree rings with little error."

Impact of Physical Sciences on Archeologists

As well as helping to date archeological finds, the physical sciences have had a tremendous impact on the ability of archeologists to find and analyze artifacts. One of the most recent developments was the perfection of a technique which would be applied more naturally in mineralogy than archeology. The method can determine whether ancient bones were those of domestic or wild animals by their crystal structure. Crystals in the bones of domesticated animals tend to be aligned near the joints, while in wild animals they are scattered. The difference is observed by viewing thin slices of bone under a microscope in polarized light filtered through a gypsum plate. Columbia University scientists who developed the technique said it gave archeologists "an important tool in understanding the culture of prehistoric man."[24] An X-ray examination of royal mummies at the Cairo Museum in 1969 indicated the existence of jewels and other artifacts imbedded in or placed on the bodies.

High-sensitivity magnetometers are now being used to locate ancient brick walls, kilns and stone monuments. The recently developed devices, using cesium vapor, can detect anomalies in the static magnetic field of the earth. The neutron activation of pottery makes it possible to determine over 40 different elements in pottery samples and this in turn should make it possible to characterize clays of a particular area. "The analysis of pottery is valuable because clay is one of the most ancient materials and is universally associated with man," notes anthropologist Fred H. Stross of the University of California at Berkeley. "People settling in a new area seem to have brought their pottery with them, but they soon started to use the clay that was available locally, while continuing to make ceramic ware in the style of their country or province." "Therefore, analysis together with comparison of style provides good evidence of migration and transplanting of groups of people."[25]

Scientists from many other disciplines work with archeologists. A group of archeologists and natural scientists from the

---

[24] Quoted in *The New York Times*, Jan. 23, 1971. The discovery was announced in the Jan. 22, 1971, issue of *Science* magazine.
[25] Fred H. Stross, "Application of the Physical Sciences to Archeology," *Science*, Feb. 26, 1971, p. 831.

Universities of Chicago and Istanbul recently made the discovery that men settled in villages in southeastern Turkey 9,000 years ago, even before they learned how to farm. Some 30 geologists, botanists, zoologists and other natural science scholars worked in this group.

All this activity bears on an old question which is being debated anew: Is archeology a science? Archeology has traditionally been considered a branch of the humanities by some of its practitioners, and as such has derived part of its inspiration and methodology from classical scholarship. Other archeologists, especially in the United States, have considered their work to be closely related to anthropology. But a number of young archeologists have been trying to develop what they think of as the "science of archeology."

Having recounted these differences within the ranks of archeologists, Allen L. Hammond suggested in *Science* magazine that the "new archeology" might be closer to the social sciences than the physical sciences. "One of the more outspoken and influential proponents of the new archeology, Lewis Binford of the University of New Mexico, has insisted that archeology can in fact be science, and that it can make a major contribution to the understanding of cultural processes only if it becomes a science."[26] Patty Jo Watson of Washington University in St. Louis, a pro-science archeologist, considers both the traditional and new approach valid. "They're both important and I'd hate to see the field polarized."[27]

Problem of Protecting Invaluable Ancient Sites

Archeologists of all schools can unite on one common cause—that of finding ways to protect ancient sites from being plundered and destroyed. "We are witnessing one of the worst periods of archeological destruction ever seen," said Froelich Rainey, director of the University of Pennsylvania Museum in Philadelphia. "This is the murder of man's history, and it's a tragedy," commented Machteld Mellink, an archeology professor at Bryn Mawr College.[28]

Thievery of art objects is prevalent in India, Thailand, and nations in the Middle East, Africa and Latin America. Many are poor countries unable to afford extensive networks of

[26] Allen L. Hammond, "The New Archeology: Toward a Social Science," *Science* (magazine of the American Association for the Advancement of Science), June 11, 1971, p. 1119.
[27] Quoted in *Behavior Today*, July 5, 1971.
[28] Rainey and Mellink were quoted in *The Wall Street Journal*, June 2, 1970.

guards for archeological sites. Guatemala, for instance, abounds in ancient Mayan ruins in dense jungles hundreds of miles away from modern cities. In these areas, the *huaqueros* or "grave robbers" cut up inscribed stones and cart them away with little fear of detection. In hacking out one relic, they invariably deface and mutilate others. Moreover, once the art object has been removed in this fashion it can rarely be traced to its place of origin even by experts. It may continue to be admired for its beauty but its archeological significance is perhaps lost forever.

Clemency Coggins, an authority on the topic,[29] wrote that in the Americas the theft of artifacts is a big-time business that extends southward as far as Peru, "with the most serious plundering taking place in Mexico and Guatemala." "This plunder has been financed by the international art market, by collectors and by most museums." Turkey, a country particularly hard hit by art looters, has asserted that three American museums—the Boston Museum of Fine Arts, the Metropolitan Museum in New York, and Dumbarton Oaks in Washington—acquired treasures smuggled out of the country.[30] It threatened in 1970 to bar all American expeditions in reprisal for alleged thefts of antiquities.

An international agreement aimed at stopping or slowing the trade in stolen art won approval in the United Nations Educational, Scientific and Cultural Organization (UNESCO) on Nov. 3, 1970, but it represented a compromise between countries from which the treasures are taken and the countries to which they usually go—principally the United States, Britain, France, Switzerland, and West Germany. The agreement put the signatory nations under obligation to help recover stolen art objects, to fight theft from archeological sites, and help to block the purchase of illicit art. Other efforts have been made to curtail the looting of archeological sites but as long as there is a market financed by wealthy private collectors and some museums, robberies are likely to continue.

[29] Coggins, a member of the Panel on the International Movement of National Art Treasures, specializes in pre-Columbian art and archeology. See "The Maya scandal: how thieves strip sites of past cultures," *Smithsonian* (magazine of the Smithsonian Institution), October 1970, pp. 8-16.

[30] As reported from Ankara by Alfred Friendly Jr. in *The New York Times*, Nov. 3, 1970.

Nuclear Power Options

by

Ralph C. Deans

1 9 7 1
Aug. 4

NUCLEAR POWER OPTIONS

NUCLEAR ENERGY currently provides only a fraction of the electrical energy consumed in the United States— about three-tenths of one per cent. But it is expected to produce 20 per cent of the nation's electricity in 1980 and perhaps 50 per cent by the year 2000. America is greeting this age of nuclear energy with a curious mixture of resignation, fear and hope. The atom promises to deliver us from the "energy crisis" that has resulted in summer "brownouts" and cascading power failures. But every nuclear plant has what physicist Ralph E. Lapp calls a "Hiroshima halo" around it. For atomic power is still associated with the awesome explosion that ended World War II and launched the world into the nuclear era.

"Nuclear power came just in time," according to Glenn T. Seaborg, the outgoing chairman of the Atomic Energy Commission. "Civilization as we know it will grind to a halt unless we can develop nuclear power."[1] The thrust of his argument is that the nation's supply of fossil fuels—petroleum and coal, the traditional sources of energy—is finite and nonrenewable. Estimates vary and are highly controversial. Commission officials frequently say that reserves of oil and gas will run out in 30 years and coal in 80 years. At the other extreme, a study commissioned by President Kennedy reported in 1966 that fuel reserves were sufficient to last for some seven thousand years.[2]

The energy crisis involves many factors but it is not yet the result of an over-all insufficiency of fossil fuel. In some cases it is the result of a maldistribution of power. In other cases it stems from an inadequate fuel-supply system. In still others,

---

[1] Quoted in *Penthouse* magazine, June 1971, President Nixon on July 21, 1971, accepted Dr. Seaborg's resignation from the AEC which he had served as chairman for a decade. No date was set for his departure from that office.

[2] *Energy R & D and National Progress: Findings and Conclusions*, September 1966, p. 8. The report's estimate of fuel reserves included those that are prohibitively expensive to recover with present technology—such as low-grade oil shale at depths of 20,000 feet.

it is the result of an insufficient generating capacity.[3] At least in part, shortfalls in the supply of electricity result from the successful efforts of environmentalists to block the construction of pollution-causing generating facilities. Here again, nuclear power is cast in the role of a rescuer because it is widely admitted that the atom is the cleanest source of power available—at least in the sense that a nuclear plant does not belch sulfur dioxides and other pollutants into the air as many fossil-fuel plants do.

A high-power nuclear reactor, however, burns as much nuclear fuel in a year as is consumed in the detonation of more than 1,000 Nagasaki-type A-bombs. AEC officials dislike that analogy because, they contend, atomic explosions are impossible in the 22 nuclear plants now generating electricity in the United States. Nevertheless, men like Lapp worry about the possibility of freak accidents that could release large amounts of lethal radiation. Lapp considers the accumulation of immense amounts of radioactivity in the reactor core "a potential threat without precedent in urban life."[4]

Moreover, nuclear power plants produce more waste heat than fossil-fuel plants and this is a form of pollution. Reactors also constantly "leak" tiny amounts of radiation. While this leakage constitutes only a small fraction of the "background" radiation received from the environment, there is a bitter scientific controversy about how dangerous these small doses are. There is a growing debate about how much protection should be built into nuclear reactors and about who should decide that question. Some critics complain that the AEC's function as the promoter of nuclear energy conflicts with its role as the watchdog over nuclear safety. There are questions about safeguards in the mining, manufacture, transportation, use and disposal of atomic fuels. All of these questions are posed within the context of a larger argument about whether America should—or even can—cut down on its consumption of energy, now doubling about every decade.

New Federal Effort to Develop Breeder Reactors

In a message delivered to Congress on June 4, 1971, President Nixon committed the United States to a further development of nuclear power over the next 10 years. Among his pro-

---

[3] See "Electric Power Problems," *E.R.R.*, 1969 Vol. II, pp. 939-956, and "Fuel Shortages," *E.R.R.*, 1970 Vol. II, pp. 955-972.

[4] Ralph E. Lapp, "How Safe Are Nuclear Power Plants?" *The New Republic*, Jan. 23, 1971, p. 19.

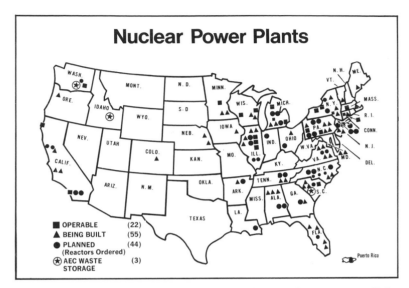

Nuclear Power Plants

■ OPERABLE (22)
▲ BEING BUILT (55)
● PLANNED (44)
(Reactors Ordered)
✪ AEC WASTE (3)
STORAGE

posals "to ensure an adequate supply of clean energy," he requested supplemental appropriations of almost $100 million in fiscal 1972—more than $75 million of it for various aspects of nuclear power. Over the coming decade, he envisioned an investment of $3 billion in federal funds to develop "clean energy" from all sources. Two billion dollars of that would be the federal contribution to the construction of a commercial "fast breeder" nuclear reactor by 1980. Nixon described the advanced reactor as "our best hope for meeting the nation's growing demand for economical clean energy."

Without the breeder reactor, nuclear fuels would not significantly enlarge the nation's fuel supply. AEC officials estimate the rare form of uranium consumed in present-day nuclear power plants will run out in about 40 years. The breeder, however, creates more fuel than it consumes.[5] Estimates vary but breeders are expected to extend the supply of nuclear fuel from decades to centuries.

The Atomic Energy Commission has placed a high priority on the breeder reactor for years. Since 1949, the federal government has spent some $600 million on the development of a liquid-metal fast breeder.[6] The only operating commercial

[5] "Breeding" is accomplished by the careful selection and placement of fissionable and "fertile" elements into the reactor core. Neutrons not necessary to keep the nuclear reaction going are absorbed by fertile material to produce more fissionable material. Thus, when the isotope of uranium (U-238), a fertile material, absorbs neutrons, it is converted into fissionable plutonium.

[6] "Liquid metal" refers to the primary coolant used to transfer heat from the reactor core—in this case, liquid sodium.

173

power plant using a breeder reactor is the Enrico Fermi plant on Lake Erie, about 35 miles southwest of Detroit. It is owned jointly by the Detroit Edison and the Power Reactor Development companies. The Fermi plant is capable of generating nearly 61,000 kilowatts of electricity while at the same time converting "fertile" material into nuclear fuel.

Yet the plant has been a disappointment, Seth Lipsky wrote in *The Wall Street Journal*, July 27, 1971. "The Fermi plant's unhappy history is one of foulups, miscalculations and accidents. It was supposed to cost some $50 million. Close to $130 million actually has been spent. When it was first switched on eight years ago, it was supposed to have grossed, by 1970, $48.6 million from the sale of steam to Detroit Edison and $43.5 million from the sale of plutonium to the government. But during that eight years the plant has operated at full power for the equivalent of only 21 days. It has taken in a paltry $65,000 from the sale of heat, and it has sold not a dime's worth of plutonium."

> Instead of leading the world in breeder reactors [Lipsky continued], the U.S. now is racing to catch up with England, France and Russia where bigger breeders are being built partly because the prototype Fermi plant has failed. Indeed, government-supported efforts are now under way to get a brand-new demonstration breeder that really works....

President Nixon, in his request for supplemental appropriations, specified that $27 million would be spent "so that the necessary engineering groundwork for demonstration plants can soon be laid." AEC officials have talked in terms of a plant capable of producing 300 to 500 megawatts of electrical power—five to eight times more than the Fermi plant.[7] Seaborg told a news conference on the day the President's message went to Congress that a plant of that size would cost $400 million to $500 million—of which about half would be supplied by utility and reactor manufacturers. General Electric, Westinghouse, and North American Rockwell are the three major candidates vying to build the demonstration breeder.

Some scientists consider the breeder to be a temporary facility to provide energy only until they can develop thermonuclear fusion—an experimental but potentially much more powerful source of energy. At present, atoms are split to produce heat in a nuclear plant. In a fusion reactor, atoms would

---

[7] A kilowatt is a unit of power equal to one thousand watts. A megawatt is one million watts.

be welded together to achieve the same result. *(See pp. 602-605.)* Since fusion power would use hydrogen atoms easily extracted from the sea, it promises to supply abundant, cheap electricity for millions of years. A number of fusion-minded scientists have urged an "Apollo-type" commitment to develop fusion power quickly. Its low priority has been called a "disgrace" by some, who predict that the Soviet Union may be the first to perfect the fusion technology. High-energy physicists predict that fusion power will be available in the 1980s, even though it has never been achieved experimentally. Most scientists, however, doubt that fusion will be practical until after the turn of the century and some question whether it will ever produce electricity.

Current State of Nuclear Energy; Cost Drawbacks

Since the first full-scale commercial nuclear power plant went into operation in 1957,[8] 21 others of various types have started functioning. As of mid-1971, 55 additional plants were under construction and 44 more were being designed. The generating capacity of existing, rising and planned facilities totals nearly 100 million kilowatts. Contracts were signed in the first six months of 1971 for 13 new atomic reactors, compared with 14 during the entire previous year.

This growth has taken place despite opposition from some environmentalists and despite rising costs of nuclear-plant construction. Capital costs of a large nuclear plant have risen in the last three years from about $120 per kilowatt of capacity to more than $200—in some cases to more than $300. Construction costs for fossil fuel plants are usually much less. Operating costs are also usually less for fossil fuel plants. The compensating factor is that the cost of nuclear fuel typically is less. The enrichment and fabrication of fuel *(see box, p. 605)* are currently the largest factors in the cost of atomic fuel. Uranium concentrate is worth about $9 a pound. The enrichment process increases the value to about $115 per pound. And the fabrication process increases the cost to $165 a pound.

Costs have been a major drawback to private investment. Jeremy Main described in the November 1969 issue of *Fortune* a "premature rush" that had then been under way on the part of many major utilities to build nuclear plants ahead of their rivals. "The stampede set off after Jersey Central Power & Light Co....accepted in 1963 a $60 million...offer

---

[8] At Shippingport, Pa. The 90,000 kilowatt facility is jointly owned by the Duquesne Light Company and the AEC.

TYPES OF NUCLEAR POWER PLANTS

Boiling Water: Water is heated as it passes up through the reactor. Steam is drawn off the top of the reactor vessel and forced into a turbine. The least expensive to build but the plant must be relatively large.

Pressurized Water: Water passing through the reactor is kept under pressure which allows it to be heated without boiling. This highly pressurized water is forced through a coil or tubes in a steam generator which converts water to steam.

Gas-Cooled: Instead of water, a gas like helium or carbon dioxide is circulated by blowers through the reactor. The gas is superheated and produces steam as hot as 1,000 degrees Fahrenheit in a steam generator.

Heavy Water: Deuterium, an isotope of hydrogen, is commonly called heavy water. It is a very efficient moderator for slowing down neutrons so they can split U-235 atoms. Usually, the fuel elements are placed in tubes running through a tank of heavy water. The coolant—either gas, water or heavy water—passes up the tubes past the fuel and into a steam generator.

Liquid Metal Fast Breeder: The coolant in this case is a metal, usually sodium. It is heated to about 1,000 degrees Fahrenheit and is forced through a "heat exchanger" where it transfers its heat to another "loop" or closed-system of liquid metal. The intermediate loop transfers heat to a steam generator. Liquid metal efficiently removes heat from the reactor but it has a violent reaction on contact with air or water. Thus an intermediate loop is necessary to protect the reactor from any explosion caused by an accidental leakage of sodium into the steam generator.

by General Electric to build a 640-megawatt plant at Oyster Creek [N.J.]," he wrote.

"According to an analysis published soon after the contract was signed, Oyster Creek would produce electricity by mid-1967, at a cost of four mills per kilowatt-hour, which was cheaper than the cost of a comparable coal-burning plant and half the average current cost. The estimate aroused such enthusiasm for nuclear power that almost half the total generating capacity ordered by private utilities in 1967 and 1968 was nuclear."

"But Oyster Creek did not work out," Main continued. It first produced power in the autumn of 1969, more than two years behind schedule. He said the experience at Oyster Creek might be an extreme example of the utilities' problems with nuclear plants. But he added that of the 47 nuclear plants then under construction, most of them expected to start producing electricity from six months to two years late.

Writing in the same issue of the magazine, Harold B. Meyers said: "The long-awaited transition of the U.S. electric-power industry into the nuclear age has been slowed by a number of factors, including technological difficulties and public resistance. But a special and unexpected cause for delay has been one company's crucial failure to deliver a single vital component of nuclear plants." He identified the company as Babcock & Wilson, a maker of steam-generating boilers for conventional power plants. It received contracts to make nuclear pressure vessels: huge steel pots—some are more than 70 feet long and weigh more than 700 tons—to contain atomic reactions.

B & W built a plant at Mount Vernon, Ind., to make them "but nothing seemed to go right," Meyers recounted. "Plagued by labor shortages and malfunctioning machines, the plant produced just three pressure vessels in its first three years of operation.... Last May [1969], B & W was forced to make a humiliating disclosure. Every one of the 28 nuclear pressure vessels then in the Mount Vernon works was behind schedule, as much as 17 months. For the utility industry, the news from B & W meant intolerable delays in bringing 28 badly needed nuclear plants into service."

Development of Nuclear Energy Policy

THE ATOMIC ENERGY ACT of 1946 gave the government monopoly control over nuclear materials, making it impossible for a private company to build and operate a nuclear power plant. This condition presented no problem at first because nuclear production of electricity on a large-scale basis was not then a reality.[9] By 1953, however, the possibility of future production was in sight and arrangements had to be made for future methods of production and sale. The following year Congress changed the law after lengthy hearings. The issue was the future structure and development of the nation's civilian atomic energy program and, in particular, whether

---

[9] The first electric power in the world generated from nuclear power was accomplished on Dec. 20, 1951, with an experimental reactor at the National Reactor Testing Station in Idaho.

private firms should be permitted to market commercial nuclear power when it finally became available.

Advocates of public power, such as the American Public Power Association and the National Rural Electric Cooperative Association, saw the atom as a way to bring really low-cost power to consumers. With this in mind, they favored a government development program to solve technical problems and put nuclear energy on the market quickly. They contended that since the secrets of the atom had been revealed at great expense to the government in World War II,[10] the whole nation should enjoy any financial benefits to arise from atomic power. They argued that if sales were left to private companies, power prices would be higher and private industry would reap the benefits.

Public vs. Private Development of Atomic Power

The private electric industry, through the Edison Electric Institute and the National Association of Electric Companies, pressed for an amendment to the 1946 act authorizing private industry to build and operate commercial nuclear plants. The industry opposed government construction and operation of nuclear electric plants on a commercial basis. President Eisenhower took a position similar to that held by the private power industry.

As finally enacted, the Atomic Energy Act of 1954 let private industry enter the nuclear power business. However, private companies would be subject to AEC regulations, and the ownership of special nuclear materials would be retained by the government. The public power forces also won a number of important victories on patent rights, regulation of nuclear power rates by the Federal Power Commission, and public power preferences in the granting of licenses to non-federal applicants desiring to build nuclear plants.

But on the question of government-owned plants, the public power advocates lost. The final bill contained a provision forbidding the AEC to engage in commercial production; it could only build and operate experimental plants. Starting in 1955, private companies were invited to build, own, and operate atomic reactors. The AEC stood ready to assist by waiving use charges for special nuclear materials, performing

---

[10] For an account of the wartime Manhattan Project which developed the atomic bomb, see "Balance of Terror: 25 Years After Alamogordo," *E.R.R.*, 1970 Vol. II, pp. 293-297.

A.E.C. LICENSING PROCEDURE

1. Utility files a detailed Preliminary Safety Analysis Report which undergoes intensive study by the AEC's regulatory staff. This review takes up to a full year and involves conferences between the AEC and the utility and the construction company.

2. The utility's report, plus the AEC regulatory report on the project, undergoes examination by the Advisory Committee on Reactor Safeguards, a group of 14 nuclear experts. The Advisory Committee has the power to force changes in the reactor design and to deny a construction permit.

3. Public hearings are held prior to the issuance of a construction permit. These are held before a three-man board selected from a 23-man Atomic Safety and Licensing Board panel. Intervenors can argue against the issuance of a permit.

4. After the reactor has been constructed, it is monitored by AEC officials to ensure that it has been built to specification. Another public hearing is held before the awarding of an operations license. Several reactors have been held back from operating because of interventions at this stage.

research and development work at no charge, and contracting for additional research and development.

The Eisenhower administration and particularly AEC Chairman Lewis L. Strauss believed that private industry would do the job. But industry was slow to respond because of high costs, and Democrats in Congress began complaining that the nation's nuclear program was lagging. By late 1957, the AEC indicated that the government itself was prepared to undertake the building of certain reactors if proposals from private industry were not submitted or could not be negotiated.

Under the cooperative program as it developed, the AEC provided virtually every conceivable type of assistance except actual construction of private reactors. A commission suggestion that it provide construction funds for investor-owned power reactors was turned down by Congress in 1959. But in 1962 the commission was authorized to provide funds for engineering design of private demonstration reactors.[11]

---

[11] See "Atomic Power Development," *E.R.R.*, 1964 Vol. I, pp. 433-436.

Questions About Safety and Pollution

WITH THEIR POTENTIAL, however remote, for inflicting death and destruction, nuclear power plants are subject to especially searching criticism. The hazards associated with nuclear power plants usually fall into three categories: 1) low levels of radiation leaked from the plant; 2) the dangers involved in transporting nuclear fuel and radioactive wastes to and from nuclear plants; and 3) the possibility of an accident —a non-nuclear explosion or a melting of the reactor core— which could release large amounts of radiation.

The AEC can point out that in 14 years of operating experience, there is no record of anyone having been injured by radiation from any commercial reactor. And since World War II, there has been no recorded major transportation accident in which people have been injured as a result of the radioactive contents of a shipment. However, critics draw little comfort from these statistics. While commercial reactors have taken no lives, there have been seven deaths connected with the operation of experimental reactors in the United States. And the transportation of all hazardous substances—not just nuclear cargo—is a matter of rising concern. Several train derailments in recent years have involved poisonous chemicals and explosives.[12]

Expressing a prevalent AEC viewpoint, Seaborg has spoken of "excessive fear" by the public. He is convinced that nuclear reactors are safe but acknowledges that an accident could release radioactivity. He explains that this is why nuclear plants are constructed and designed with back-up safety systems to prevent or minimize the possibility of an accident. But other prestigious scientists have voiced concern about the risks involved. Dr. Edward Teller, a former director of the Livermore laboratory, notes that an economical breeder "needs quite a bit more than one ton of plutonium...I do not like the hazard involved."[13] And Walter Jordan, assistant director of

---

[12] A train derailment at Laurel, Miss., in January 1969 ignited tons of liquified petroleum gas in explosions that killed two persons and injured 33. Some 1,400 buildings and homes were destroyed or damaged. See Eric Albone and Julian McCaull, "Freighted with Hazard," *Environment* (magazine of the Scientists' Institute for Public Information, St. Louis), December 1970, pp. 18-27. See also "Chemical-Biological Warfare," *E.R.R.*, 1969 Vol. I, pp. 455-456.

[13] Quoted by Gene Schrader, "Atomic Doubletalk," *The Center Magazine,* January/February 1971, p. 48.

the Oak Ridge National Laboratory, wrote recently that "the $64 million question still remains, and this is whether we have succeeded in reducing the risk to a tolerable level—i.e., something less than one chance in 10,000 that a reactor will have a serious accident in any year."[14]

Critics contend that if risks must be accepted in nuclear plants, they must be balanced against the benefits. Furthermore, the risks must be outlined and the public must be polled in some way as to whether it is willing to take them. The benefit-risk argument leads to a number of additional questions: Who takes the risks and who gets the benefits? Since some risk can be eliminated by additional but expensive safety systems, who will pay for the added measure of safety?

The fears of some critics would be eased if the regulatory role of the AEC were split off from its function of encouraging and promoting nuclear power. For them, this dual role represents conflict of interest which could lead to tragedy. Ralph E. Lapp made several suggestions for a more widely based safety review system for nuclear reactors in a series of articles published early in 1971 in *The New Republic*. He urged that a high-level review of reactor safety be made "by some independent group such as the National Academy of Engineering, funded by the Environmental Protection Agency."

Lapp also suggested that a permanent Nuclear Power Safety Board be established, that reactor size be limited to reduce the conuequences of an accident, that no nuclear plants be placed near metropolitan areas, and that research into reactor safety be intensified. Robert Gillette of *Science* magazine wrote recently that while AEC officials continually assure the public that nuclear reactors are safe, "a number have been discreetly appealing for more money—preferably much more money—to support research on the safety of conventional, water-cooled nuclear reactors."[15] AEC expenditures for safety research reached a peak of $37 million in fiscal 1970 and dropped to $36 million in 1971.

Safeguards to Protect Against Reactor Mishaps

One of the worst accidents that could occur in the light-water plants operating in the United States today is a "blowdown"—

---

[14] Walter Jordan, "Benefits vs. Risks in Nuclear Power," Oak Ridge National Laboratory *Review.* One chance in 10,000 would mean that if there were 100 nuclear reactors operating in the United States, an accident could be expected every 100 years.

[15] Robert Gillette, "Nuclear Reactor Safety: A New Dilemma for the AEC," *Science,* p. 126.

a sudden loss of coolant which would allow the reactor to over-heat. If a reactor core heated to the melting point a number of things could happen. Within a minute, the core would begin to collapse to the bottom of the containment vessel. The molten radioactive material would eventually burn its way through the vessel and then through the slab of concrete under it. The ball of atomic fire would burn itself into the earth in what is known, perhaps tongue-in-cheek, as the "Chinese syndrome"—meaning it may continue to sink into the earth in the direction of Asia. One guess is that the molten material would sink to the water table where it would cause the venting of highly radio-active steam. Another possibility is that the material would congeal into a fiery glob that would burn for a decade or more.

Under normal conditions, reactor-maintenance experts could take action before the core melted. Control rods—as many as 90 of them—can be inserted into the core to soak up neutrons and thus shut down the nuclear reaction. Some experts say there still might be enough residual heat in the core, along with heat released by decaying fission products and by chemical reaction between metal and remaining water, to melt it. Thus one of the key safety features of most reactors is a "backup coolant system" which is designed to flood the reactor vessel if the primary bath of coolant is lost.

This system became a controversial matter in 1971 when the congressional Joint Committee on Atomic Energy released a letter that Seaborg had written to Sen. John O. Pastore (D R.I.), the committee chairman. Seaborg said new techniques for calculating the temperature of cladding in the reactor indicated that "the predicted margins of emergency core cooling systems performance may not be as large as those predicted previously." His letter stemmed from a series of small-scale experiments which the AEC conducted at its National Reactor Testing Station near Arco, Idaho.

Those experiments indicated, but did not prove, that high steam pressures inside the reactor vessel might delay the entry of the backup coolant. On the basis of an "interim policy statement" issued by the AEC June 19, 1971, five reactors were ordered to modernize their backup systems by July 1, 1974, and in the meantime to triple their inspections of pipes, pumps and valves. Four new power plants were ordered to hold their peak operating temperatures below normal.

Concern over nuclear reactor safety arose in the late 1950s. In response to a request by the Joint Committee, the AEC commissioned its Brookhaven National Laboratory on Long Island, N.Y., to study nuclear safety. The laboratory report, *Theoretical Possibilities and Consequences of Major Accidents in Large Nuclear Power Plants,* presented to the committee in March 1957, said an accident might cause no casualties or that conceivably it might leave as many as 3,400 persons dead and 43,000 injured. Property damages might range from $500,000 to $7 billion. A revision of this study was undertaken in 1965 but the results were never made public. In a letter dated June 18, 1965, Seaborg told Rep. Chet Holifield (D Calif.), who was then the committee chairman, that the larger reactors being planned at that time would produce greater damage in the event of an accident but that the likelihood of an accident was "still more remote" than before.

Sheldon Novick, in his book *The Careless Atom,* described the situation in 1957 this way: "The Joint Committee had reached an impasse; in order to have private industry participate in the reactor program, utilities would have to own and operate their own reactors. This they declined to do unless adequate insurance was available. But insurance companies refused to assume the risk."[16] The result was the Price-Anderson Act, which Congress passed that year, stating that the aggregate liability for a single nuclear accident "shall not exceed the sum of $500 million." The act requires the operator of a reactor to obtain as much private insurance as possible. The Atomic Energy Commission then provides the remaining insurance up to a total of $500 million protection. As the matter worked out, private insurance now provides some $82 million in coverage for each nuclear reactor and the government provides $418 million.

Controversy Regarding Radioactive Discharges

Dr. Ernest J. Sternglass, a professor of radiation physics at the University of Pittsburgh, asserted in 1969 that 400,000 infant deaths and more than two million fetal deaths had occurred in the United States as the result of strontium-90 fallout from nuclear weapons tests.[17] Two AEC scientists, Dr. John W. Gofman and Dr. Arthur R. Tamplin, challenged his findings. But, in the process of digging into the statistics, they became convinced that radioactive emissions from nuclear power plants are potentially dangerous. Their findings, in turn, have been challenged. Robert W. Holcomb reported

[16] Sheldon Novick, *The Careless Atom* (1969), p. 70.
[17] See "Infant Health," *E.R.R.,* 1970 Vol. II, pp. 909-910.

RADIATION: ITS CAUSES AND EFFECTS

Radiation is a wave of sub-atomic particles—photons, electrons, protons or neutrons—which penetrates matter, including living tissue, at nearly the speed of light. Radiation causes damage because the particles can either physically tear away molecules within a cell or ionize (split into charged fragments) tissue atoms. In either case, radiation injures, sometimes permanently, the body's atomic chemistry.

The most widely used unit of radiation measurement is the *rad*, an acronym for radiation absorbed dose. It is defined as that quantity of radiation which delivers 100 ergs of energy to one gram of substance. The *rem*, acronym for Roentgen equivalent man, is a measure that includes an estimate of the biological impact of different types of radiation. It is usually roughly equivalent to a rad. A *millirem* is one-thousandth of a rem.

Science has not yet been able to determine a "threshold" below which radiation is always harmless. The lowest absorbed dosage at which medically significant damage to humans has been observed lies somewhere between 50 and 100 rad, according to some experts. A whole-body single dose of 100 rad induces vomiting in about 10 per cent of people so exposed. A whole-body single dose of 450 rad causes death to half those so exposed.

in *Science* magazine: "An analysis of their work reveals that most of the assumptions they use in making predictions can be neither proved nor disproved, but the consensus of their peers is that at least some of the assumptions are wrong."[18] Gofman had asserted that if all Americans were irradiated to the AEC maximum, 32,000 extra cancer deaths would result yearly.

Radiation is a normal hazard of all earth dwellers. The "background" radiation—from cosmic rays, the earth and food—ranges from 90 to 200 millirems per year. Until recently, the Atomic Energy Commission standards for nuclear plants set maximum individual exposure at 500 millirems per year.[19] These were the upper limits when the commission, on Dec. 3, 1970, published design and operating requirements for nuclear power reactors to keep levels of radioactivity emissions "as low as practicable."

---

[18] Robert W. Holcomb, "Radiation Risk: A Scientific Problem?" *Science*, Feb. 6, 1970, p. 855.

[19] These exposure guides are established by the Federal Radiation Council, formed in 1959 to provide "guidance for all federal agencies in the formulation of radiation standards." The guidelines were originally formulated in the 1950s by the International Commission on Radiological Protection and the National Committee on Radiation Protection and Measurements.

New operating criteria for nuclear reactors, issued by the AEC on June 9, 1971, would limit the exposure to people living nearby to no more than 5 per cent of the amount of radiation that they normally receive from natural sources. Harold L. Price, AEC director of regulations, said all but two or three of the 22 power plants now in operation emit less than one per cent of the radioactive wastes that they are allowed to release under the new guidelines.[20] Price added that the companies would be allowed three years to make the necessary improvements to cut down on the emission of radioactivity.

The commission was quick to point out—and its critics to note—that the new regulations applied only to those nuclear plants cooled by ordinary water. Thus breeder reactors, isotope processing plants and waste disposal sites were not affected. The disposal of wastes has created a furor. These wastes are now kept in liquid form in large underground steel tanks near Richland, Wash., Idaho Falls, Idaho, and Aiken, S.C. Exactly how much radioactive waste is under storage is classified information. The AEC told Editorial Research Reports the amount was "in excess of 80 million gallons." Some experts suggest it is in excess of 200 million gallons.

Issue of Waste Disposal, Environmental Criticism

In budget hearings before the Joint Committee in March 1971, the AEC requested funds to buy a salt mine near Lyons, Kan., to dump radioactive wastes on a trial basis. Commission officials explained that studies indicated salt beds were the safest dumping grounds—better than placing waste-filled metal containers in the earth or burying them in the deep sea. They said salt seemed to be the ideal receptacle because it withstands high temperatures, becomes plastic and flows to heal fractures and to fill holes. It has compressive strength and radiation shielding properties comparable to concrete.

Many Kansans, including some state officials and congressmen, remained unconvinced that the wastes could be safely sealed away for all time. They questioned whether seepage might not occur and enter underground streams. Richard S. Lewis, writing in the *Bulletin of the Atomic Scientists,* noted that one question "unasked and unanswered during the budget hearing, but implicit in the debate was: Who would be responsible if 50, 500, 5,000 or 50,000 years from now, deforma-

---

[20] Those not conforming were identified as the Pacific Gas and Electric Company's Humboldt Bay reactor, near Eureka, Calif., and Commonwealth Edison's Dresden-1 reactor and possibly its Dresden-2 facility, both near Chicago.

tion from heat, an earthquake or a new ice age ruptured the rock seal and allowed the radioactivity to percolate through the aquifers into the environment?"[21]

While the AEC's request for these funds was being considered by Congress, the U.S. Court of Appeals in the District of Columbia told the commission that it had failed in its duty to guard against environmental damage at the Calvert Cliffs plant being constructed on the Chesapeake Bay by the Baltimore Gas & Electric Company. The commission's interpretation of the National Environmental Policy Act of 1969 "makes a mockery of the act," the court said in a decision in the Calvert Cliffs case on July 26, 1971. "The very purpose of the...act was to tell federal agencies that environmental protection is as much a part of their responsibility as is protection and promotion of the industries they regulate," the court said. It ordered the AEC to revise its procedures for environmental protection.

Critics of the commission long have accused its members of insensitivity to environmental issues. President Nixon's new appointments to the commission were being viewed in Washington as a reflection of his awareness of this criticism. On announcing Seaborg's resignation, July 21, 1971, Nixon named as his successor—pending Senate confirmation—James R. Schlesinger, an assistant director of the White House Office of Management and Budget. In that role, Schlesinger was credited with persuading the administration to reverse itself and return federal land to the Taos (N.M.) Indians—a cause that conservationists endorsed. Nixon also named William O. Doub, a Baltimore lawyer who has served on the presidential Air Quality Advisory Board, to replace commission member Theos J. Thompson, who was killed in a plane crash.

Promise of Fusion Power in The Future

THE BREEDER reactor promises to provide energy for hundreds of years. But another source of energy—fusion—holds the promise of providing electricity for millions, perhaps billions, of years. According to physicists, fusion power will be inherently safe, cheap and abundant. Nuclear fission releases energy by the splitting apart of U-235 atoms. Fusion releases energy by fusing hydrogen nuclei together. It is the source of the power

---

[21] Richard S. Lewis, "The Radioactive Salt Mine," *Bulletin of the Atomic Scientists,* June 1971, p. 27.

PRINCIPLES OF NUCLEAR REACTION

As physicists now know, not all atoms of an element are similar. The nucleus of one atom may have more or fewer *neutrons*—uncharged particles—packed into it than another does. Such different physical forms of the same chemical are called *isotopes.* Uranium atoms usually have 238 *protons*—positively charged particles—plus neutrons in their nucleus. But a few uranium atoms, about seven-tenths of one per cent of naturally occurring uranium, have three fewer neutrons. This is the isotope U-235 and it is the basic fuel for all atomic power. U-235 is *fissionable,* or capable of being split apart when struck by a slow-moving neutron.

If a neutron is moving too quickly, it is absorbed by U-238 atoms—converting them to *plutonium* atoms. If it is moving slowly enough to split a U-235 nucleus, it produces an enormous multiplication of energy. In the act of splitting apart into flying fission fragments, the destroyed nucleus of a U-235 atom releases two or three new neutrons. The fragments, containing some 200 million electron volts of kinetic energy, produce heat when they collide with surrounding atoms. The new neutrons, meanwhile, if sufficiently slowed down by passage through a "moderator" like graphite, are capable of splitting other U-235 atoms. When this occurs and goes on occurring, the result is a *nuclear reaction.*

A nuclear power reactor concentrates U-235 into a precise geometrical configuration so that: 1) the released neutrons can be slowed down and have a better chance of striking new U-235 atoms; 2) so that the reaction can be controlled by the placement of rods which absorb neutrons and 3) so that a *coolant*—either water, gas, or liquid metal—can be passed by the reaction to transfer heat from it. This heat is eventually converted into steam which is used to drive the blades of a turbine. In light-water plants such as are used in the United States, the water circulated through the reactor also acts as a moderator, slowing down neutrons.

in an H-bomb explosion and it is the energy source of the sun and the stars.

Research on fusion power began secretly about 20 years ago and almost simultaneously in the United States, the United Kingdom and the Soviet Union. Secrecy was ended by international agreement in 1958 and there has been a high degree of international cooperation since then. According to Richard F. Post of the Lawrence Radiation Laboratory, around $210 million is being spent annually on worldwide fusion research—"about 50 per cent in the Soviet Union, 25 per cent in the United States and the rest in the U.K., Western Europe and Japan."[22]

In some respects, starting up a self-sustaining fusion reaction is similar to building a wood fire. The flame from a match must be hot enough and be held long enough, and the fuel must be closely spaced to heat itself and keep the fire going. The fuel in a fusion reactor, however, is deuterium and tritium, two isotopes of hydrogen.[23] "The general idea," wrote Francis F. Chen, "is to heat a plasma, or ionized gas, of deuterium or tritium to more than 100 million degrees centigrade and hold the plasma together by means of a magnetic field long enough for some of the ions to fuse, releasing energy in the form of radiation."[24]

A plasma is sometimes described as the "fourth state of matter." It is a hot gas—so hot that it "ionizes" or breaks up into a mixture of negatively charged electrons and positively charged nuclei. Physicists say that if a deuterium-tritium plasma can be made dense enough, contained long enough and heated high enough, the deuterium and tritium nuclei would be forced to collide with each other, even though they are usually held apart by their positive charges. The fusing would create the nucleus of helium-4 which would instantly release a highly energized neutron. The neutron—in one theoretical design—would produce heat when it penetrated a "lithium blanket" surrounding the reaction. The lithium, in turn, would siphon the heat off to a steam-generating plant.

Actually, only part of the energy would be transferred to the lithium. The rest would be used to keep the fusion process going by heating the incoming gas to fusion temperature. Fusion theorists calculate they need to confine a plasma to a density at least as high as one hundred trillion particles per cubic centimeter. Furthermore, this density must be held for a full second and the plasma must be heated at the same time to about 40 million degrees centigrade. If all three conditions are met, the result would be a self-sustaining fusion reaction that would continue as long as more deuterium was fed into the reaction.

---

[22] Address before the National Academy of Science's Symposium on Energy for the Future, Washington, April 26, 1971. Cooperation in fusion research may soon be matched in the field of U-235 enrichment, a closely held U.S. secret. Victor Cohn of *The Washington Post* reported July 30, 1971, that the AEC was ready to begin talks with 10 friendly nations about sharing the secrets of nuclear enrichment.

[23] Deuterium, or "heavy water," is easily separated from ordinary sea water. Tritium, the radioactive form of hydrogen, is rare.

[24] Francis F. Chen, "The Leakage Problem of Fusion Reactors," *Scientific American,* July 1967, pp. 76-88.

Conversion of Uranium Ore Into Fuel

ENRICHMENT: Since the isotope of fissionable uranium is rare, an enrichment process is necessary before the element can be used in the light-water reactors employed in the United States. Uranium ore is milled to produce a crude concentrate known as *yellow cake* which contains 70 to 90 per cent uranium oxide. Yellow cake is refined into pure uranium trioxide, a fine orange powder known as *orange oxide*. This is subsequently converted to uranium dioxide by hydrogenation and the dioxide is finally converted into uranium tetrafluoride, called *green salt*, by reaction with hydrogen fluoride gas.

Green salt is shipped to one of the three large uranium enrichment plants at Oak Ridge, Tenn., Paducah, Ky., and Portsmouth, Ohio. There it is reacted with fluorine gas to produce *uranium hexafluoride*, a volatile compound which changes into a gas at low temperatures. This gas is fed into a series of chambers, a "cascade," which concentrates the fissionable U-235 isotope. Once enriched, the hexafluoride is converted back, step by step, to uranium dioxide powder.

FABRICATION: This powder is compacted into cylindrical pellets and loaded into long zirconium alloy tubes. The tubes, called *cladding*, protect the fuel material from corrosion and lock in the radioactive fission products. Some 200 tubes are arranged in a *fuel assembly* which may weigh 1,300 pounds. The assemblies, perhaps 200 of them, are loaded into chambers in the reactor equipped with fittings which permit the coolant to enter and leave the assembly. The "geometry" of the core must be precise because a certain spatial relationship is necessary for a controlled chain reaction to take place. Also, hundreds of thousands of gallons of coolant must flow through the reactor each minute.

One or two of the three parameters of fusion—density, time and heat—have been achieved in various experiments. But no experiment has yet achieved all of them at the same time. The closest experiment to achieving these requirements was the Tokomak-3, a Russian fusion experiment in Moscow. The Tokomak features a torodial (doughnut-shaped) containment vessel in which the plasma is confined by means of powerful magnetic fields. In the summer of 1969, Russian scientists announced that the Tokomak had held a plasma of 50,000 billion particles per cubic centimeter at a temperature of 10 million degrees for 0.05 seconds—that is, the temperature and density were near what would be required for sustained fusion and the time was about a tenth of the minimum. The containment time, however, was 10 times as long as was previously attained.

Richard F. Post has suggested that fusion power could make it possible to convert heat directly into electricity without the intermediate step of producing steam. Dr. Bernard J. Eastlund and Dr. William C. Gough of the AEC have suggested a plan whereby the intense heat of a fusion "torch" could be used to reduce garbage and other wastes to chemical elements which could later be recovered and reprocessed.[25] While far from fanciful, these advanced concepts must await proof that fusion power is possible.

David J. Rose, director of long-range planning at Oak Ridge, wrote recently that no fusion experiment had exhibited "scientific feasibility." Yet, he added, "I will bet a modest amount of even money on success of the Tokamak in the next few years...My own guess is that fusion power will be available in appreciable quantity by 2000."[26] Rose noted that more optimistic persons believe fusion power will be available by 1990 while "pessimists propose never."

T. J. Thompson, a member of the AEC, cautioned recently that the fusion reactor will not be hazard-free. While it will not involve a "critical mass" as nuclear power plants do, "it will not do away with the need to be careful of radioactivity and the need to store radioactive waste products."

> I would caution [he continued] that all of us try to be as realistic as possible in our approach to this new field. One thing that the recent concern about the environment has reemphasized is that it is necessary to approach each technology with a realistic assessment of its detrimental effects as well as its beneficial effects.[27]

Thompson suggested that scientists direct all their energies to a demonstration of the feasibility of controlled fusion, "even at the expense of some other desirable but non-vital elements of the program." This indeed appears to be the next step in the evolution of fusion power. With the United States having embarked on a program to vastly increase the number of nuclear power plants in the country, it will take the best efforts of politicians, industrialists and scientists to ensure that nuclear power, whether fission or fusion, is safe, reliable and kind to the environment.

---

[25] See William C. Gough and Bernard J. Eastlund, "The Prospects of Fusion Power," *Scientific American,* February 1971, pp. 50-64.

[26] David J. Rose, "Controlled Nuclear Fusion: Status and Outlook," *Science,* May 21, 1971, pp. 797-808.

[27] Address before the Annual Meeting of the Division of Plasma Physics, American Physical Society, Washington, D.C., Nov. 5, 1970.

LIGHT:
Physical and Biological Action

AMERICAN INSTITUTE OF BIOLOGICAL SCIENCES

and

U. S. ATOMIC ENERGY COMMISSION

MONOGRAPH SERIES ON

RADIATION BIOLOGY

JOHN R. OLIVE, *Series Director*

AMERICAN INSTITUTE OF BIOLOGICAL SCIENCES

ADVISORY COMMITTEE

MONOGRAPH TITLES AND AUTHORS

RADIATION, RADIOACTIVITY, AND INSECTS
  R. D. O'BRIEN, *Cornell University*
  L. S. WOLFE, *Montreal Neurological Institute*
RADIATION, ISOTOPES, AND BONE
  F. C. McLEAN, *University of Chicago*
  A. M. BUDY, *University of Chicago*
RADIATION AND IMMUNE MECHANISMS
  W. H. TALIAFERRO, *Argonne National Laboratory*
  L. G. TALIAFERRO, *Argonne National Laboratory*
  B. N. JAROSLOW, *Argonne National Laboratory*
LIGHT: PHYSICAL AND BIOLOGICAL ACTION
  H. H. SELIGER, *Johns Hopkins University*
  W. D. McELROY, *Johns Hopkins University*

(IN PREPARATION)

MAMMALIAN RADIATION LETHALITY: A DISTURBANCE IN CELLULAR KINETICS
  V. P. BOND, *Brookhaven National Laboratory*
  J. D. ARCHAMBEAU, *Brookhaven National Laboratory*
  T. M. FLIEDNER, *Brookhaven National Laboratory*
IONIZING RADIATION—NEURAL FUNCTION AND BEHAVIOR
  D. J. KIMELDORF, *Radiological Defense Laboratory*
  E. L. HUNT, *Radiological Defense Laboratory*
TRANSPLANT IMMUNITY AND RADIATION
  J. F. LOUTIT, *Radiobiological Research Unit, Harwell*
  H. S. MICKLEM, *Radiobiological Research Unit, Harwell*
TRITIUM IN BIOLOGY
  L. E. FEINENDEGEN, *Services de Biologie, Euratom*
PHYSICAL ASPECTS OF RADIOISOTOPES IN THE HUMAN BODY
  F. W. SPIERS, *University of Leeds*
SOIL-PLANT SYSTEMS
  M. FRIED, *International Atomic Energy Agency, Vienna*
MUTAGENESIS
  I. I. OSTER, *Institute for Cancer Research, Philadelphia*

LIGHT:
Physical and Biological Action

HOWARD H. SELIGER

WILLIAM D. McELROY

*McCollum-Pratt Institute
and
Department of Biology
The Johns Hopkins University
Baltimore, Maryland*

Prepared under the direction of the American Institute of Biological Sciences for the Division of Technical Information, United States Atomic Energy Commission

 1965

ACADEMIC PRESS ● New York and London

ACADEMIC PRESS INC.
111 Fifth Avenue, New York, New York 10003

United Kingdom Edition published by
ACADEMIC PRESS INC. (LONDON) LTD.
Berkeley Square House, London W.1

LIBRARY OF CONGRESS CATALOG CARD NUMBER: 65-21329

PRINTED IN THE UNITED STATES OF AMERICA

FOREWORD

This monograph is one in a series developed through the cooperative efforts of the American Institute of Biological Sciences and the U. S. Atomic Energy Commission's Division of Technical Information. The goal in this undertaking has been to direct attention to biologists' increasing utilization of radiation and radioisotopes. Their importance as tools for studying living systems cannot be overestimated. Indeed, their applications by biologists has an added significance, representing as it does the new, closer association between the physical and biological sciences.

The association places stringent demands on both disciplines: Each must seek to understand the methods, systems, and philosophies of the other science if radiation biology is to fulfill its promise of great contributions to our knowledge of both the normal and the abnormal organism. Hopefully, the information contained in each publication will guide students and scientists to areas where further research is indicated.

The American Institute of Biological Sciences is most pleased to have had a part in developing this Monograph Series.

JOHN R. OLIVE
Executive Director
American Institute of Biological Sciences

PREFACE

Research in the field of photobiology is extremely diverse and requires a concurrent knowledge of the physics of electromagnetic radiation, molecular structures, and the biology of the organisms. Any effort, therefore, to summarize such areas as the measurement of molecular excitation by light, photosensitization, photosynthesis, phototropism, phototaxis, vision, photoperiodism, bioluminescence, diurnal rhythms, and effects of ultraviolet and X-ray on cells, tissues, and organisms in one volume is obviously difficult inasmuch as each subject could fill an entire volume. We have been encouraged to do this, however, even with these limitations in mind, in order to give beginning students a general introduction to the important problems which are usually considered in photobiology. It is evident, because of the large amount of detailed information available, that the student would have considerable difficulty in trying to obtain an over-all view of the broad field of photobiology using current reviews and advanced monographs. This is particularly true in the physical aspects.

Because of the nature of the subject, therefore, it has been necessary to condense, or even omit, many significant investigations, and for this we apologize to our colleagues in the field. In addition, because of the nature of the backgrounds of the authors, it is to be expected that some subjects have been covered more fully than others. In trying to condense and summarize, however, we hope that where other viewpoints could have been presented but were not, we have made this clear in the text and referred the student to other publications.

In addition to the purely biological aspects, there is a great need for a general introduction to physical concepts and their practical application that is not adequately covered in general and intermediate physics and chemistry courses. We have tried to summarize this material in the first two chapters. We hope that we have left the reader with at least one important concept, that all problems of photobiology or all processes which are initiated by light are quantum phenomena and should always be considered as such even though there may be enzymatic amplification after the initial photochemical events.

We have had numerous discussions with many colleagues in the field, some of whom have read certain parts of the manuscript. We would like to thank Dr. M. Kasha and Dr. S. Udenfriend for their comments, Miss Marie Pierrel who drew many of the illustrations in the monograph, and

Mrs. Mary E. Backer who very patiently typed her way through masses of very roughly written drafts.

We would like to take this opportunity to express our deepest appreciation to two very understanding women, to whom this book is sincerely dedicated.

HOWARD H. SELIGER
WILLIAM D. McELROY

March, 1965

*

CONTENTS

CHAPTER 1

Measurement and Characterization of Light

1. The Nature of Light

The whole of nature is a trillion, trillion chemical machines, squirming, twisting, swimming, crawling, floating, flying, and sometimes walking—in the image of Man.

What a spectacle is this vast proliferation of green light-traps, utilizing in a special still-secret process the energy of sunlight to convert carbon dioxide, water, and inorganic salts into more complex molecules at the rate of 375 billion tons of carbohydrates per year. And picture still further the voracious horde of plant-eating animals and animal-eating animals that somewhere in the dim past have given up this basic light-energy conversion process and are utterly dependent upon these green alchemists and, ultimately, upon the light from the sun. For aside from the minor heating of the earth's crust by the naturally radioactive elements, volcanic eruptions, cosmic rays, and fusion reactions, all of the energy for life on earth is contained within a narrow spectral range of electromagnetic radiation received from the sun. The total solar radiation (cal/cm²-min) at normal incidence outside the atmosphere at the mean solar distance is $2.00 \pm 2\%$ (*Smithsonian Physical Tables*, 1959, Table 808). This corresponds to a total incident energy of 1.34×10^{24} cal per year or an irradiance of 0.14 watts/cm². In the equivalent units of mass and energy this is a total of 68,400 tons of sunlight falling on the earth per year. Table 1.1 gives a summary of the distributions of this incident sunlight in the different wavelength regions of the spectrum.

A large variety of physical phenomena are defined either subjectively or physiologically and in most cases with very good reasons. The definition of light is intimately associated with its physiological response. The smallest child knows that "light is to see." This simple definition tells him much more than "light is radiant energy in that portion of the transverse electromagnetic spectrum with frequencies between 3.8×10^{14} and 7.7×10^{14} cycles per second or with vacuum wavelengths between 7800 and 3900 A, respectively, capable of stimulating, in the normal

1

human eye, the sensation of vision." In fact the former definition by virtue of its brevity is less subject to contradiction than the latter. For example, the early workers in radioactivity were able to observe a dull glow upon holding a strong source of radium near their eyes in a dark room; likewise the early X-ray workers, upon exposure to an X-ray beam—a type of scientific "derring-do" that makes one shudder at their intrepidity or naivité. It is also a well-known fact that persons who have suffered from cataracts, who have therefore had their lenses removed, are able to "see" in ultraviolet light below 3800 A; the normal lens apparently absorbs all wavelengths below this value. Honeybees have a component of color vision in the ultraviolet. Many color-blind persons,

TABLE 1.1

SPECTRAL DISTRIBUTION OF SUNLIGHT INCIDENT ON EARTH'S ATMOSPHERE

Wavelength region (A)	Per cent of incident energy	μwatts/cm^2 ($\times$ 10^3)	Einsteins per second over total surface[a] ($\times$ 10^9)
Below 2000	0.1	0.136	0.255
2000–2500	0.8	1.09	2.63
2500–3000	2.2	2.99	8.77
3000–3500	3.5	4.76	16.5
3500–4000	5.4	7.34	29.5
4000–7000	36.0	49.0	288
7000–10,000	24.0	32.6	295
Above 10,000	28.0	38.1	—

[a] 1 Einstein = 6.025 $\times$ 10^{23} light quanta.

although they cannot distinguish red light, have dark-adapted vision which is just as acute as that of persons with normal eyes. A sharp blow on the head, the action of certain drugs, or mental aberrations can produce the sensation of brilliant colors.

With these facts in mind let us now proceed to discuss light from a purely physical and historical viewpoint, and then perhaps it will be easier to relate these physical concepts to biological applications.

Let us begin our story with *Sir Isaac Newton*, born in Woolsthrope, England, in 1642, only 11 years after the "heretic" Galileo had for the second time been called before the Inquisition because of his vocal opposition to the Ptolemaic, earth-centered solar system. In addition to Newton's magnificent contributions of the Differential Calculus to Mathematics and the Universal Law of Gravitation to Mechanics, he discovered in 1666 that white light is made up of the various spectral colors which could be "sorted-out" by means of a prism and which he explained on the basis of light corpuscles ("multitudes of unimaginable small and swift corpuscles of various sizes springing from shining bodies . . . ").

His fame and stature as a monarch of science overshadowed the work of another genius of that era, *Christian Huygens*, who in 1678 had developed a wave theory of light. It was not until 1801 when *Thomas Young*, in a paper before the Royal Society, presented an undulatory theory of light and proposed the phenomenon of interference to explain refraction and the diffraction grating, that the wave theory again dared to lift its head. Even at this late date Young was subjected to a storm of derision and abuse by some of his more dogmatic scientific peers. It was only after the crucial experiments of *Jean Leon Foucault* in 1850, who showed that light travels more slowly in a dense medium such as water than in a rare medium such as air, as predicted by the wave theory, that the wave theory became respectable.

Paralleling these developments in experiments with light were the discoveries in electricity and magnetism. In 1820 *Hans Christian Oersted* showed that there is a magnetic field associated with the flow of electric current. In 1831 the self-taught experimental genius *Michael Faraday* discovered the converse—that there is an electric current associated with a *change* of magnetic field, the principle of the dynamo. Then in 1865, *James Clerk Maxwell*, Professor at King's College, London, published his "Dynamical Theory of the Electromagnetic Field," which clearly ranks with Newton's "Principia" as one of the tremendous landmarks in science. In this treatise Maxwell combined the discoveries of Newton, Huygens, Young, Foucault, Oersted, and Faraday into a unified theory of electromagnetic phenomena which included light. Light consisted of transverse electromagnetic waves (an assumption made by *Auguste Jean Fresnel* in 1818 to explain polarization) whose frequency of vibration ν and wavelength λ were related by $\nu\lambda = v$, where v is the speed of propagation of light in the medium. In vacuum in the Maxwell theory $v = c$, a universal constant, dependent only upon the ratio of the electromagnetic to the electrostatic unit of charge. The value of this ratio even at that time agreed to better than 1% with the experimental value for the speed of light, measured in 1849 by *Hippolyte Louis Fizeau*, and represents one of the great triumphs of theoretical physics.

One of the corollaries of Maxwell's theory was that light was pictured as a series of rapid alterations of electric and magnetic fields, each perpendicular to the other and both transverse to the direction of propagation.

In 1887 *Heinrich Hertz*, in Germany, showed that an electric discharge or an oscillating electric dipole radiated energy in the form of transverse waves which traveled with the speed of light and possessed all of the wave properties predicted by Maxwell's electromagnetic theory. It is interesting to note that during the course of these experiments he accidentally discovered the Photoelectric Effect, one of the fundamental phenomena which require the quantum (corpuscular) theory and which cannot be explained by the very electromagnetic wave theory whose existence he was at the same time demonstrating.

Maxwell's theory gave a mechanism of an oscillating charge giving rise to electromagnetic waves and from *L. Lorenz'* mathematical picture in 1880 of a quasi-statically bound charged particle, the idea was developed that very rapid oscillations of charged particles can give rise to electromagnetic waves of very high frequency, i.e., light.

In 1897, *Joseph John Thompson* showed that cathode rays consisted of particles of negative charge, or electrons. In the same year the Normal Zeeman Effect of a magnetic field on the emission of light was discovered by *Pieter Zeeman* and interpreted by *Hendrik Antoon Lorentz* and Zeeman on the basis of an electron linear oscillator as a model for the radiating atom. The wave theory appeared firmly established. But there were rumblings. Some of the experimental facts would not fall neatly into place. First, there was Hertz' photoelectric effect, the emission of electrons by a material irradiated by light, which had a threshold wavelength above which it did not occur no matter how much energy was absorbed, a result at variance with the wave theory. Then, there were other more complicated theoretical problems. Last, based on thermodynamics and the Law of Equipartition of Energy, the wave theory predicted that the radiant energy emitted from a blackbody should increase rapidly to infinity as the wavelength approaches zero, referred to as the "infinity catastrophe." Since experimentally the spectral energy distribution was found to be a peaked curve with a maximum predicted by the Wien Displacement Law and the total energy emitted was finite and followed the Stefan-Boltzmann Law, it was evident that there was something incorrect in the expanding wave-front picture of the wave theory which could not account for either the absorption of all of the energy in the wave front by a single oscillator or the spectral distribution of energies from a group of oscillators. In 1901 *Max Planck*, in one of the most startling breaks with the whole of scientific tradition, reluctantly put forward the hypothesis that radiation is discontinuous. If, in the expression for the average energy of an oscillator, the assumption was made that an oscillator could not emit all possible energies but could emit one of the *discrete* values, $h\nu_i$, where h was a universal constant and ν_i was the frequency of the light, the predicted energy distribution from a blackbody was in exact agreement with experiment. The "infinity catastrophe" was removed and the Wien Displacement Law and the Stefan-Boltzmann Law, both based on pure thermodynamic reasoning, were unchanged. Although we now accept this hypothesis of discrete energy jumps as one of the fundamental laws of nature, it is important to point out that Planck, whose idea this was, and although it correctly explained the energy emission from a blackbody, could not conceive of how this discreteness could explain the *absorption* of energy and therefore originally modified his theory to permit oscillators to *absorb continuously* but *radiate discontinuously*. It was not until 1905 that *Albert Einstein*, in order to explain Hertz' Photoelectric Effect, made the break complete by assuming that energy was also absorbed in discrete steps and that radiation was composed of discrete packets of energy or quanta, Newton's original "multitude of swift corpuscles." In 1911 *Ernest Rutherford*, in a classic article describing his experiments on the scattering of alpha particles by thin films of matter, developed a picture of the atom as composed of a positively charged central nucleus less than 10^{-12} cm in diameter, surrounded by the negative atomic electrons. It was at this point, in 1913, that *Niels Bohr*, combining the ideas of Planck, Einstein, and Rutherford, postulated the existence of stationary states of allowed energy for the atomic electrons in circular orbits about the nucleus, leading to the prediction of the absorption and emission spectrum of the hydrogen atom. To round out our necessarily brief outline, in 1924 *Louis de Broglie*, with an intuitive feeling that Nature exhibits symmetry, postulated that since radiant energy possessed both wave and corpuscular properties, possibly matter itself possessed both wave and corpuscular properties. Thus the wavelength λ associated with a material particle was given by h, Planck's constant, divided by the momentum (mass times velocity) of the particle. This concept of duality, as well as the Bohr postulates, was incorporated into a self-consistent mathematical theory by *Erwin Schrödinger* in 1926 and forms the basis, with modifications contributed by *Max Born*, *Werner Heisenberg*, *Wolfgang Pauli*, *Paul Adrian Maurice Dirac*, and others, of our present-day Wave Mechanics or Quantum Mechanics.

This then is the nature of our light. It is transverse electromagnetic radiation. It can be treated as a quantum of energy $h\nu$ or as a wave with a wavelength $\lambda = c/\nu$. In its interaction with matter it is always absorbed as whole quanta. In diffraction by a grating, even a single photon will be diffracted according to the laws of interference of continuous waves. It is the ultimate source of free energy for life.

2. Photometry

The absorption of light by a biological system and the emission of light from excited molecules are quantum phenomena, and therefore a proper description of the light either absorbed or emitted during a reaction should contain the number of photons per second per unit wavelength interval. Many workers in photobiology have used photometric units of light *intensity* and *illumination* in dealing with photochemical processes, and unless one is quite careful this can lead to serious errors and ambiguity in the interpretation of experimental results. It is the purpose of this section to describe these photometric units and to correlate them with photon intensity.

2.1.   True, Color, and Brightness Temperature

We have already described the origin of the quantum theory and stated that the Planck formula accurately describes the spectral energy emission of a blackbody at any temperature T_0. A blackbody, or perfect radiator of electromagnetic energy, is seldom realized but is approached in the *Hohlraum* of classical physics. For all other radiators which are not ideal blackbodies there are three characteristic temperatures which describe their radiant emission. These are the *true temperature*, T_t, the *color temperature*, T_c, and the *brightness temperature*, T_b.

Only in the case of a blackbody whose emissivity is unity over the entire spectral range is the energy distribution given exactly by the Planck formula. For non-blackbodies the power radiated is modified by the emissivity, $\epsilon(\lambda, T)$, which for most metals is a slowly varying function. Since the emissivity does not vary greatly *in the visible region*, it is possible to color-match the emission of a non-blackbody with that of an ideal blackbody operating at a blackbody temperature T_0. The *color temperature*, T_c, of a non-blackbody is therefore defined in this way, i.e., the temperature at which it is necessary to operate an ideal blackbody in order that its radiation in the visible region will match in integral color that of the non-blackbody.

In the measurement of the radiant flux from non-blackbodies with an optical pyrometer, the results are equated with the radiant flux that an ideal blackbody would emit. Since non-blackbodies radiate less power in any wavelength interval than blackbodies at the same temperature, the ideal blackbody temperature at which the fluxes are equal will always be lower than the true temperature of the non-blackbody. This ideal blackbody temperature is defined as the *brightness temperature*, T_b. The relation between the brightness temperature and the true temperature of the non-blackbody is given by

$$\frac{1}{T_t} = \frac{1}{T_b} + \frac{\lambda}{1.438} \ln \epsilon(\lambda, T) \tag{1.1}$$

where λ is the wavelength in centimeters,* $\epsilon(\lambda, T)$ is the emissivity of

---

* The constant 1.438 is defined as the "second radiation constant." In the Planck equation the spectral distribution of blackbody radiation is given by $J_\lambda = C_1 \lambda^{-5}/$

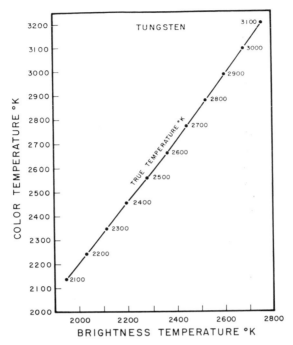

Fig. 1.1. Relations of true, brightness, and color temperature for tungsten. Graph of the color temperature of a tungsten filament versus brightness temperature as measured by an optical pyrometer. The true temperature in °K is given to the right of each point. The relation between the true temperature T_t and the brightness temperature T_b, measured with the optical pyrometer, is given by

$$\frac{1}{T_t} = \left(\frac{1}{T_b}\right) + \left(\frac{\lambda}{1.438}\right) \ln \epsilon(\lambda, T)$$

where λ is the wavelength in centimeters (6.5×10^{-5}) and $\epsilon(\lambda, T)$ is the emissivity of the tungsten source (0.43) at 6500 A and temperature T. The color temperature T_c is then the blackbody temperature which gives the same integral color as $W(\lambda, T_t)\,\epsilon(\lambda, T_t)$ over the visible region where $W(\lambda, T_t)$ is calculated from the Planck radiation formula. In all cases $T_b \leqslant T_t \leqslant T_c$.

the non-blackbody in the particular wavelength band used in the optical pyrometer, and T_b is the observed *brightness temperature*, the pyrometer having previously been calibrated against a standard blackbody. Thus we see that for all cases $T_b < T_t < T_c$. A calibration curve relating T_t, T_b, and T_c for a tungsten filament is given in Fig. 1.1 (*American Institute of Physics Handbook*, 1957).

[exp $(C_2/\lambda T) - 1$]. The numeric for C_2 must have the unit of wavelength times absolute temperature. If the wavelength is expressed in microns, the numeric becomes 14380.

2.2. Standards of Luminous Intensity and Luminous Flux

In the measurement of the emission of radiation from hot bodies, the basic physical technique involves placing a completely absorbing surface in the radiation field and measuring the temperature increase of the detector due to the absorption of this radiant energy. The thermopile which utilizes the Seebeck effect (the development of a potential difference dependent on temperature at the junction of two dissimilar metals) to detect these small temperature differences and later the Golay cell, which depends upon the change in pressure of an enclosed gas, have been used in conjunction with spectrometers to measure the spectral energy distribution of blackbody radiation. The Planck Quantum Hypothesis was developed to explain the divergence of the experimental results from those predicted by the classical wave theory. Both the theory and the primary method of measurement of radiation are based on energy. For example, the Stefan-Boltzmann Law gives the total energy radiated by a blackbody at a given temperature. Per unit of surface area, a blackbody emits into a 2π solid angle $5.673 \times 10^{-5}\ T^4$ ergs/sec.* The Planck equation for the blackbody spectral power distribution into a 2π solid angle, per 0.01 μ wavelength interval, per square centimeter of surface at a wavelength of λ (in microns) is

$$J_\lambda = \frac{3.740 \times 10^9}{\lambda^5\{[\exp\ (14380)/kT]-1\}}\ \text{ergs/sec}\dagger \qquad (1.2)$$

It is therefore natural that the primary standard candle should be specified in terms of the energy emission from a blackbody.

The units in photometry are based on the spectral energy distribution of blackbody radiation, however, *modified by the visual sensitivity of the human eye.* This is an extremely important point. The average relative radiant spectral sensitivity of the normal, light-adapted human eye is defined by the curve of Fig. 1.2 (Preston, 1961). The luminous energy is therefore the integral of the product of the radiant energy per unit wavelength interval, $J_\lambda\ d\lambda$, and the relative photopic luminous efficiency, V_λ,

---

* The constant 5.673×10^{-5} is the Stefan-Boltzmann constant with the units erg cm^{-2} deg^{-4} sec^{-1}, for emission into a 2π solid angle of a blackbody radiator.

† See footnote on page 6. The constant 3.740×10^9 is the "first radiation constant," C_1, in the Planck equation and has the units,

$$C_1 = \text{numeric}\ \frac{(\text{wavelength unit})^5 \times \text{power unit}}{\text{area} \times \text{wavelength interval} \times \text{solid angle}}.$$

In the case of Eq. (1.2) the area is in cm^2, the wavelength is in microns and the wavelength interval is in 0.01 μ. Thus the units of C_1 are erg sec^{-1} cm^2.

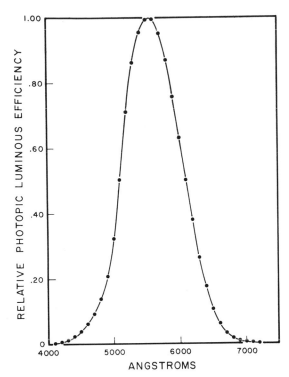

Fig. 1.2. Photopic relative luminous efficiency of "standard eye." The photopic relative luminous efficiency of radiation, normally referred to as the "Photometric Standard Observer," showing a peak efficiency at 5560 A. At this wavelength the mechanical equivalent of light is 680 lumens/watt. While referred to as the wavelength of maximum visibility, it is that only for the light-adapted eye. Actually the wavelength of maximum sensitivity is that corresponding to the scotopic peak for the dark-adapted eye and is a different curve with a peak further toward the blue, around 5070 A. To give some idea of the transition region between photopic (cone) vision and scotopic (rod) vision, with a close-by source having a brightness of 0.25 millilambert or for an illumination of 0.23 foot-candle, the peak visibility lies between 5070 and 5560 A. The full moon has a maximum brightness of 0.8 lambert, corresponding to an illumination of 0.06 foot-candle.

over the region of photopic vision, roughly between 3900 and 7800 A,

$$\text{luminous energy} = \int_{.39\,\mu}^{.78\,\mu} J_\lambda V_\lambda \, d\lambda \qquad (1.3)$$

As a consequence of the emitting properties peculiar to tungsten, *over the visible range*, the relative spectral power distribution of tungsten is very close to that of a blackbody. Therefore we can assign to a tungsten filament a *color temperature*, T_c, as defined previously. This is the temperature at which an ideal blackbody would have to be operated so that the light *color-matches* that of the filament. It is this visual equating of colors that involves the definition of the "Standard Eye," given in Fig. 1.2. The specification of standard light sources in photometry is further based on the visual equating of the *illumination* of a surface by a blackbody at 2042.15°K and by a color-matched tungsten filament. The temperature of solidification of pure platinum is 2042.15°K, which is a convenient and

TABLE 1.2
PHOTOMETRIC UNITS AND DEFINITIONS

Luminous flux	The total *visible energy* emitted by a source per unit time
Lumen	Flux emitted per unit solid angle by a point source of 1 candela. Therefore 1 candela emits 4π *lumens* in all directions
Luminous intensity (of a source)	Property of a source of emitting luminous flux
Candela	One-sixtieth of the intensity of 1 cm² of a blackbody radiator at the temperature of solidification of platinum (2042°K)
Brightness (of an emitting surface)	Flux emitted per unit emissive area as projected on a plane normal to the line of sight
Lambert	1 *lumen*/cm² emitted by a perfectly diffusing surface
Stilb	1 *candela*/cm² = π *lamberts*
Illumination (of a surface)	Luminous flux incident on unit area of a surface
Foot-candle	1 *lumen*/ft²
Lux	1 *lumen*/m²
Phot	1 *lumen*/cm² = 929 *foot-candles*
Luminous efficiency	Ratio of total luminous flux to the total power consumed. *Lumens/watt*

internationally agreed upon standard of blackbody temperature. The brightness of a blackbody radiator at 2042.15°K is defined as 60 candelas/cm². Therefore, if at equal distances from a surface normal to the line of sight the visual *illumination* due to a color-matched tungsten filament is the same as that due to the standard blackbody, the tungsten lamp is *defined* as having a luminous intensity of 60 candelas. Any other tungsten lamp operating at a color temperature different than 2042.15°K is then compared with the primarily standardized tungsten lamp, again by its normal *visual illumination* of a surface, and the *luminous intensity* of the secondary lamp is also specified in *candelas*. Thus the photopic sensitivity of the eye is intimately involved in the specification of the photometric units of *intensity* and *illumination*. The lumen is then defined as the luminous (visible) flux emitted within a unit solid angle of 1 steradian by a point source having an intensity of *1 candela*. Thus a source of *1 candela* emits in all directions a total of 4π *lumens*. The outputs of some types of lamps are specified in lumens because the filament arrays or focusing arrangements do not approach a point source geometry. In these cases a large diffusely reflecting "integrating sphere" is used which integrates over all directions of emission. The illumination of a portion of the sphere surface is then compared with that produced by the standard tungsten lamp and the calibration is then *mean spherical intensity in candelas*. The total luminous flux in lumens is then $4\pi \times$ *mean spherical intensity*. Thus a tungsten lamp with a stated output of 300 lumens has a *mean spherical intensity* of 23.9 candelas previously referred to as 23.9 candlepower. *Brightness* in candelas per square centimeter, *intensity* in candelas, and *illumination* in lumens per square centimeter are specifically related to the illumination of a surface by the visible radiation from a standard blackbody as seen by a standard observer whose relative spectral sensitivity is given by Fig. 1.2. The various photometric units are given in Table 1.2. At the present time photoelectric photometry is used for these standardizations rather than visual photometry in order to eliminate observer differences. However, the principle is exactly the same. The relative spectral response of a filter-phototube combination is adjusted to that of the "Standard Eye" (Crawford, 1962).

2.3.   ENERGY OF VISIBLE LIGHT

The standard blackbody is defined as a perfectly diffusing source (Fig. 1.3). From Lambert's Law the angular distribution of radiant energy emitted from the plane surface is therefore

$$J_\theta = J_0 \cos \theta \tag{1.4}$$

where θ is the angle as measured from the normal to the surface. In this way the brightness of the source, defined as the intensity emitted per unit area of the source in the projection normal to the line of sight, is a constant and is equal numerically to J_0. If we now integrate the radiant power emitted per square centimeter by a standard blackbody over the entire 2π solid angle, correcting the radiant energy distribution for the relative photopic efficiency of the "Standard Eye," we obtain the total luminous flux in a 2π solid angle:

$$\int_0^\pi \int_{.39\,\mu}^{.78\,\mu} J_0(\lambda) \cos \theta \cdot 2\pi \sin \theta \, d\theta \, V(\lambda) \, d\lambda \; lumens \tag{1.5}$$

The order of integration is not important and we now set

$$\text{total luminous flux} = \pi \int_{.39\,\mu}^{.78\,\mu} J_0(\lambda)\,V(\lambda)\,d\lambda = \pi B \text{ lumens} \qquad (1.6)$$

where B is the *brightness* of the source in *candelas per square centimeter*. The total luminous flux emitted by 1 cm² of a standard blackbody into a 2π solid angle is therefore 60π lumens.

Since $J_0(\lambda)$ is the power per square centimeter emitted in the direction normal to the surface and the Planck formula constants used in Eq. (1.2)

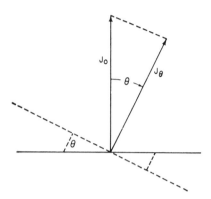

FIG. 1.3.   Lambert Law emission from blackbody. Intensity of light emitted by a perfectly diffusing surface. In this case since the brightness is defined as the intensity emitted per unit area *in the projection normal to the line of sight*, the effective area of source included in this unit area projection is $1/\cos\theta$. Therefore at any angle θ, the brightness is the product of J_θ and the effective source area. This gives

$$J_0 \cos\theta \times 1/\cos\theta = J_0$$

and therefore a perfectly diffusing source has a constant brightness, independent of θ.

are for a 2π solid angle, the factor π is required to convert $J_0(\lambda)$ to $J(\lambda)$. If now the integral of Eq. (1.3)

$$\int_{.39\,\mu}^{.78\,\mu} J(\lambda)\,V(\lambda)\,d\lambda$$

is evaluated numerically using Eq. (1.2) and Fig. 1.2, the result is 2.772×10^6 ergs/sec for the effective luminous power emitted by 1 cm² of a blackbody. Equating this power to 60π lumens, we arrive at what is termed the "least mechanical equivalent of light," 0.00147 watts/lumen. Thus 1 watt of monochromatic power at a wavelength of 5560 A, the maximum of the photopic sensitivity curve of Fig. 1.2, will produce a response equivalent to 680 lumens.

It must be remembered that so far we have described not the total radiant power emitted by blackbodies but the integrated product of this radiant power and the relative photopic efficiency of the "Standard Eye"; the *lumen* refers only to visible flux. For example, 1 cm² of a standard blackbody at 2042.15°K emits into a 2π solid angle a total power of 98.65 watts, from the Stefan-Boltzmann equation. Of this only 0.2772 watts are effective in producing a visible response, resulting in an absolute visual efficiency of 0.28% or a luminous efficiency of 1.91 lumens/watt. As the temperature of a blackbody increases, a larger fraction of the total radiant power is emitted within the visible region. A blackbody at 2700°K has a luminous efficiency of 12.36 lumens/watt. A blackbody at 6000°K, the approximate temperature of the sun's surface, has a luminous efficiency of 93.4 lumens/watt.

2.4. DIFFICULTIES INHERENT IN PHOTOMETRIC UNITS IN BIOLOGY

The preceding sections have dealt with the spectral distributions of the total power radiated by blackbodies and with the integral of these power distributions modified by the relative spectral sensitivity of a "Standard Eye." We therefore have a physiological definition for radiant energy, the *lumen*. It is thus possible for sources of entirely different spectral power distributions to produce equal *illumination* of a surface in *lumens per square centimeter* or in the more generally used units of *foot-candles*. It follows that the characterization of an irradiating beam from a hetero-chromatic source in foot-candles or even in microwatts per square centimeter can be ambiguous, since the information given is incomplete. This can be seen more easily in the following example. In experiments in photosynthesis many experimenters use a large variety of incandescent bulbs or fluorescent lamps or a combination of both. In those cases where illumination is specified, one of the commonly used units of illumination is the foot-candle. Thus the compensation point in photosynthesis is taken as 100 foot-candles, and bright June sunlight at noon on a clear day is 10,000 foot-candles. However, photosynthesis, like all other photochemistry, is based on the absorption of quanta and what we are really interested in is the number of quanta per second per square centimeter incident within the absorption bands of the chlorophyll or the accessory pigments and not the integrated effect of the source radiant emission on the human retina. This is shown in Fig. 1.4 which is a plot from the data of Barbrow (1959), showing the spectral power emission into a unit solid angle of two sources of different color temperatures when the illumination due to each of the sources is the same. Thus at 4500 A, a 2854°K color temperature source contains 78% more quanta/sec than a 2300°K color temperature source, while at 6800 A, the 2854°K source contains 28% fewer quanta per second than the 2300°K source. Even more important for some interpretations, the ratio of quanta emitted at two different wavelengths,

6800 A/4500 A, for 2854°K is 5.6, while the 6800 A/4500 A ratio for 2300°K is 14, a factor of almost 3. When fluorescent lamps are used, the situation becomes even more confused since the spectral distributions of fluorescence have no relation to those of blackbody radiation. Figure 1.5 gives the relative spectral energy emissions between 4000 and 7000 A of a standard candle at 2042°K, a tungsten lamp at a color temperature of 2900°K, and a standard 40-watt daylight fluorescent lamp respectively. The curves are normalized at the wavelengths of maximum intensity.

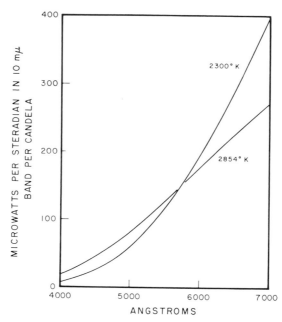

Fig. 1.4.   Blackbody spectral power emission at 2300° and 2854°K. Spectral power emission into unit solid angle of sources of 2300° and 2854°K for equal visual illumination. Data from Barbrow (1959).

From this and the photopic sensitivity curve of Fig. 1.2 it can be seen that the characterization of incident intensity in foot-candles or in photometric units of illumination may bear little relation to the quantitative description of the photochemical absorption phenomenon and can therefore lead to serious errors in the interpretation of experimental results either in the same laboratory, or in different laboratories trying to confirm particular experiments.

The data become even more confusing when phototubes, which are mainly blue-sensitive, are used to measure quantities like the transmission of sunlight through sea water. At the sea surface, bright sunlight of 10,000 foot-candles corresponds to approximately 100,000 μwatts/cm². The spectral power distribution of sunlight integrated over the phototube spectral sensitivity will give a response to which all other measurements are normalized, these measurements being expressed in μwatts/cm². However, as we go deeper and deeper into the sea, the yellows and reds are absorbed to a greater degree and the color of the transmitted sunlight

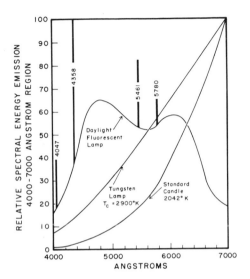

FIG. 1.5. Relative spectral energy emissions of various sources. Comparison of the relative spectral energy emissions of a low-pressure Hg "daylight" fluorescent lamp, a blackbody at 2042°K, and a tungsten lamp with color temperature 2900°K. The 6800 A/4500 A ratios are 0.42, 24, and 5.8, respectively. The continuous distribution of the emission of the phosphor coating of the fluorescent lamp has superimposed on it the strong lines of the Hg arc spectrum. The large relative excess of blue light over that of the black body, to which the tungsten filament lamp is comparable, gives a color temperature of 5600°K for the fluorescent lamp. The phosphor coating of the glass walls is 28% white $ZnBeSiO_4$, 25% pinkish-white $ZnBeSiO_4$, and 47% blue-white $MgWO_4$. The "soft-white" fluorescent lamp coating has only 14% $MgWO_4$, giving a higher red peak, a correspondingly lower blue peak, and a color temperature of about 3500°K.

becomes blue-green. Since most phototubes are not very sensitive in the yellow and red regions, this change in spectral quality would appear as a minor effect and, since even in the region of maximum sensitivity of the phototube the spectral composition is shifting, it would be extremely difficult[1] to assess much quantitative significance to data taken in this way unless the spectral distributions are known.

It would appear obvious at this point that the proper description of a light beam must include the spectral distribution as well as the number of quanta per second per square centimeter. If the spectral data are given in microwatts per centimeter squared, we can relate this to the number of quanta per second by the relation

$$1 \text{ quantum/sec} = \frac{1987}{\lambda \text{ (Angstroms)}} \times 10^{-18} \text{ watts*} \tag{1.7}$$

If we know the spectral composition of a light source, we can find a correlation between photometric units and quantum intensity. For example, at 5560 A we found that 1 lumen = 0.00147 watts. Therefore, the photon equivalent of light, P, is obtained from

$$0.00147 = P \times \frac{1987}{5560} \times 10^{-18}$$

so that

$$P(5560 \text{ A}) = 4.11 \times 10^{15} \text{ quanta/sec-lumen} \tag{1.8}$$

At wavelengths other than 5560 A, more quanta will be required because of the lower "Standard Eye" sensitivity. Since at 6800 A the

TABLE 1.3
Relation between Photometric Units and Wavelengths of Various Chemiluminescent Reactions

Luminescence	Lumen (flux) (μwatts)	Photometric units stilb (brightness) (μwatt/cm²)	Foot-candle (illumination) (photons/sec per cm²)
Luminol chemiluminescence (4300 A)	126,700	398,000	295×10^{12}
Bacterial bioluminescence (4900 A)	7062	22,200	18.75×10^{12}
Firefly bioluminescence (5620 A)	1490	4670	4.54×10^{12}

luminous efficiency is only 0.017, 1 lumen at 6800 A is the equivalent of $(4.11 \times 10^{15})/0.017 = 242 \times 10^{15}$ quanta/sec while at 4500 A one lumen is equivalent to 108×10^{15} quanta/sec. These conversions are particularly relevant in bioluminescence where many workers have expressed the brightness of fireflies or of luminous bacteria or dinoflagellates in photometric units of candelas or lamberts (see Table 1.2 for definitions of photometric units). In Table 1.3 are shown the correlations between the

---

* The number 1987×10^{-18} comes from the values of the constants hc where h is Planck's constant 6.624×10^{-27} erg-sec and c is the speed of light 3.00×10^{10} cm-sec⁻¹. The product has the units erg-cm and to convert to joule-Angstroms we multiply by 10.

photometric units of lumens and foot-candles and the energies and quantum intensities for three of the most common light-emitting reactions, the chemiluminescence of luminol, the bioluminescence of luminous bacteria and dinoflagellates, and the bioluminescence of the firefly. In these cases the peak emission wavelengths were used as approximations in the calculations although, in principle, an integration over the entire emission spectrum is required. As can be seen from the table and reference to Eq. (1.7), 1 lumen of a firefly bioluminescence reaction involves the emission of only one-quarter as many quanta per second as 1 lumen of a bacterial bioluminescence reaction and only one-sixty-fifth as many quanta per second as 1 lumen of a luminol chemiluminescent reaction. Again it is evident that photometric units are not the most suitable for a quantitative description of light emission and that *quanta per second per unit wavelength interval* is the proper characterization of light in biology.

3. Light Sources

3.1. ENERGY DISTRIBUTIONS AND INTENSITIES

The light from the sun has been used quite often in irradiation studies and for many experiments it is still a reasonable choice. The high temperature of the sun produces reasonably high intensities of radiation in the visible and near UV regions of the spectrum. For example, by reference to Table 1.1, in the 3500–4000 A region there are incident from the sun 7340 watts/cm² of earth surface, assuming negligible absorption by the atmosphere. For comparison, at 1 meter from a high pressure 1000-watt Hg arc lamp, within the same spectral region there are only 460 watts/cm² incident. However, for most studies the changes in atmospheric absorption, the low intensities below 3000 A and the unalterable diurnal variations in azimuth and intensity present difficulties conveniently overcome by the use of laboratory sources of high intensity radiation. These sources fall into one of two general classifications, incandescent or electric discharge. In Table 1.4 the various types of light sources that are used are compared in units of surface brightness. In the second column are given the luminous efficiencies in lumens per watt i.e., the ratio of the total visible energy emitted to the total power consumed. These data have been collected from various specifications and while the values of brightness in lamberts are of direct concern to illumination engineers, they are less informative to the researcher interested in the spectral dependence of quantum absorption phenomena. In the case of incandescent sources it is possible from a knowledge of the brightness to calculate an approximate photon spectral distribution, owing to the parallelism with blackbody spectral distributions. However, for electric discharge tubes there is no correlation between brightness, color temperature, and photon spectral distribution. This is shown in Fig. 1.5 by comparing the relative spectral energy emissions of incandescent sources with that of a fluorescent lamp. By virtue of the relatively larger blue component of the

TABLE 1.4
EFFICIENCY AND BRIGHTNESS OF VARIOUS LIGHT SOURCES

Source	Efficiency (lumens/watt)	Nominal brightness (lamberts)
Candle flame	—	3.1
Sun at meridian	90	519,000
Sun near horizon	—	1885
White surface in bright sunlight	—	9.42
Clear sky	—	2.5
Bright moon	—	0.8
Extended source at absolute visual threshold	—	3×10^{-10}
Zirconium concentrated arc	10	31,400
(100-watt, 1.5-mm spot)		(3000°K color temp.)
High intensity carbon arc	18.5	125,000–314,000
Incandescent lamps:		
First commercial carbon filament	1.6	165
Tungsten filament in vacuum (25 watt)	10.6	650
Tungsten filament gas filled (1000 watt)	21.6	3800
Fluorescent lamps:		
40-watt daylight	54	2.1
40-watt standard warm white	64	2.1
High pressure continuous discharge:		
G.E. Type A-6, 1000-watt Hg	65	94,000
Osram HBO 107/1, 100-watt Hg	20	440,000
Osram HBO 500, 500-watt Hg	43.5	63,000
Osram XBO 162, 150-watt Xe	21	23,500
Flash lamps:		
EG and G Type FX-33 Xe (100 joules in 150 μsec)	25	(6000°K color temp.)
AMGLO 1TZ Xe (1000 joules in 2000–4000 μsec)	approx. 25	(6000°K color temp.)
Shapiro and Edwards Model 110 (1–5000	30	(10,000,000 peak
flashes/sec, 5 joules/flash in 1.5 μsec)		candle-power)
Ruby laser, $\lambda = 6943$ A (2000 joules in 1 μsec)	—	—
(7×10^{27} quanta-sec in 1 μsec burst)		

total light, the fluorescent lamp has a *color temperature* of 6500°K, although the surface brightness is only 2.1 lamberts. A blackbody at the same *color temperature* has a brightness of 10^6 lamberts.

In the case of incandescent tungsten filament lamps the following calculations should be useful to those interested in reasonably quantitative estimations of spectral distributions.

Consider a 100-watt tungsten filament lamp operating at a color temperature of 2800°K. At this temperature the luminous efficiency is approximately 15 lumens/watt so that the lamp emits 1500 lumens. Since by definition 1 candela emits 4π lumens, the tungsten lamp is equivalent to a point source of $1500/4\pi = 120$ candelas. At a distance of 1 foot normal to the plane of the filament the illumination will therefore be 120

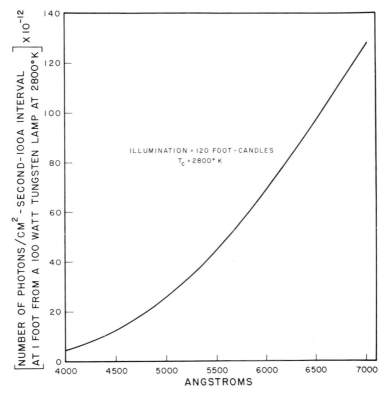

FIG. 1.6.   Photon spectral distribution from tungsten lamp. Number of photons/cm²-sec per 100 A wavelength interval as a function of wavelength corresponding to a visual illumination of 120 foot-candles from a tungsten lamp at a color temperature of 2800°K.

foot-candles. The incident photon spectral intensity distribution, that is, the number of photons per square centimeter per second per 100 A wavelength interval corresponding to an *illumination* of 120 foot-candles, is plotted in Fig. 1.6. At a wavelength of 4500 A we have 12.6×10^{12} photons/sec per cm², while at 6800 A we have 115×10^{12} photons/sec per cm², approximately a factor of 10 more red quanta than blue quanta. If this curve is integrated over the absorption spectrum of the particular pigment involved in the photochemical process, it should be possible to obtain reasonable estimates of the yields of the reactions.

In the case of sources emitting radiation in the UV region, it is, of course, impossible to use photometric units which are defined only in the visible region (4000–7000 A). In these cases lamps are usually specified in terms of a curve or table showing microwatts per square centimeter

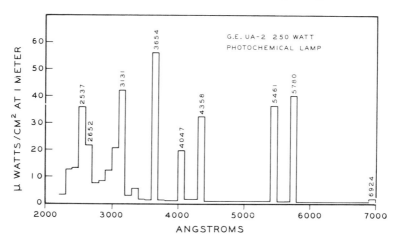

FIG. 1.7.   Line spectra from a photochemical lamp. Histogram of the spectral energy distribution of the radiation from a typical photochemical lamp, calculated from data supplied by the manufacturer. In terms of number of photons per square centimeter per second, the conversion factor from microwatts per square centimeter is (λ in Angstroms/1987) $\times 10^{12}$. Thus 56 μwatts/cm^2 of 3654 A corresponds to 1.03×10^{14} photons/cm^2-sec. For schematic purposes the lines are drawn wider than they actually are.

incident at 1 meter as a function of wavelength. These data are then easily convertible to photons/sec-cm^2 by the relation

$$\text{Energy flux of 1 photon/sec} = \frac{1987}{\lambda[\text{A}]} \times 10^{-12} \ \mu\text{watt} \qquad (1.9)$$

Figure 1.7 shows such a spectral energy distribution from a low pressure 250-watt Hg photochemical lamp, calculated from data supplied by the manufacturer. Low-pressure discharge lamps emit mainly the individual lines characteristic of the vapor while high-pressure discharge lamps emit more or less continuously with the characteristic emission lines broadened and superimposed on the continuum. Figure 1.8 shows absolute spectral energy emissions of several high-pressure discharge lamps. There is a much more intense continuum in the high-pressure lamp. The advantage of the high-pressure lamp is that the arc current

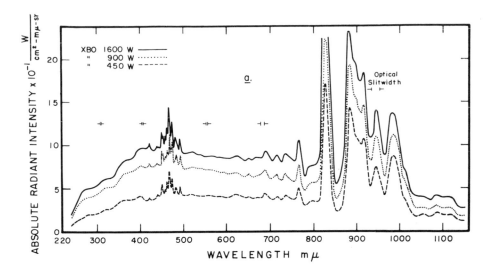

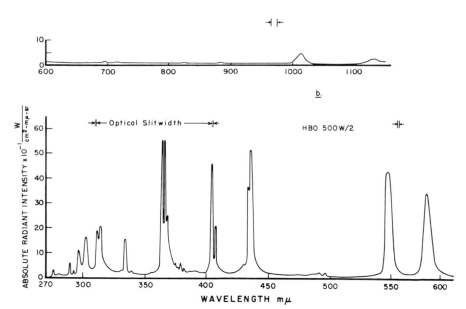

Fig. 1.8. Absolute Spectral Energy Distributions of High Pressure Lamps. (a) 450-watt, 900-watt, and 1600-watt xenon lamps, (b) 500-watt mercury lamp.

These curves were supplied by the Osram Company. They were obtained using a double monochromator with a constant slit width of 0.5 mm in the xenon measurements and 0.25 mm in the mercury measurements. The units are watts per square centimeter per millimicron per steradian. In the cases of the 100- and 200-watt mercury lamps the relative spectral distributions are the same as for the 500-watt lamp and the absolute values are reduced by factors of 0.1 and 0.4, respectively.

density is much higher and consequently the source is much brighter. This is an important point in obtaining sufficiently high intensities when a monochromator is used in irradiation experiments.

3.2. ACHIEVEMENT OF "MONOCHROMATIC" LIGHT

3.2.1. Types of Radiation

(1) *Temperature or Thermal Radiation.* Thermal radiation is emitted by every material system by virtue of its temperature alone. The kinetic energy of translation of the heat motion of the molecules of a substance is converted into electronic, vibrational, and rotational motion of the molecules of the lattice and this energy is dissipated either by further collisional transfer or by emission of a continuum of radiation. Most solids and liquids begin to emit visible radiation ($\lambda < 7800$ A) at temperatures above 500°C.

(2) *Radiation Produced by Electric Discharges.* Here are included arc and spark spectra produced by the passage of electric discharges through gases under reduced pressures and the radiations emitted by gases, liquids, or solids under bombardment by ionizing radiation such as cathode rays, X- and gamma-rays, or corpuscular radiation from radioactive sources or particle accelerators. In all of these cases the system is not at temperature equilibrium. Individual atoms or molecules are raised to extremely high energy levels compared with those accessible by virtue of the Maxwell-Boltzmann distribution of thermal energies and the radiation emitted is characteristic of the electronic, vibrational, or rotational energy levels of the excited atoms or molecules. A portion of the energy delivered to the substance is degraded to thermal levels, and the total radiation has therefore a temperature continuum on which the characteristic radiation is superimposed.

(3) *Chemiluminescence.* This third large group includes radiation emitted as the result of an exergonic chemical reaction. This is the reverse of the photochemical absorption of energy. Again this is a nonequilibrium process and the radiation is characteristic of the energy levels of the excited product atom or molecule, superimposed on the temperature continuum due to the thermal degradation of chemical energy. This type of radiation will be discussed more completely in Chapter 3.

(4) *Photoluminescence.* This type of radiation is again characteristic rather than thermal and differs from (2) and (3) above only in the method of production. In this case radiation with energy below the ionization energy absorbed by matter can be re-emitted as *resonance radiation, fluorescence,* or *phosphorescence.* These will be discussed in more detail in Chapter 2.

(5) *Radiation Produced by High Local Electric Fields.* The basic mechanism of this type of radiation is similar to (2) above. Into this class fall the following phenomena:

Triboluminescence—The faint "sparks" seen when crystalline or microcrystalline structures are broken by mechanical means. Presumably the rupturing of bonds can give rise to high local fields. Cane sugar and uranyl nitrate show this effect quite markedly.

Sonoluminescence—The emission of light as a result of cavitation produced by sound waves, presumably as a result of frictional forces giving rise to electrostatic fields.

Crystalloluminescence—The weak emission or flashing occurring during the crystallization of compounds, i.e., strontium bromate.

Galvanoluminescence—The emission of light at the cathode of an electrolytic cell as reduction takes place, possibly distinguished from chemiluminescence due to radical formation.

Electroluminescence—The emission of light by crystals with specific impurity levels upon application of alternating electric fields.

Recombination radiation—The emission of light by atoms or molecules upon neutralization of their respective ions. In nuclear physics the annihilation radiation produced by the combination of a positron with an electron can give rise either to a single 1.02 Mev quantum or to two 0.51 Mev quanta. In the X-ray region K, L, and M characteristic radiation are emitted, and in the low-energy region of ionization potentials the characteristic fluorescence and phosphorescence spectra are observed.

One of the major steps in the investigation of the action of light on biological systems is the determination of the effective wavelength region, the action spectrum. The proper criteria for, and examples of, action spectra will be presented in Chapter 5. In the present section we shall discuss only the various methods of obtaining "monochromatic" or sharply defined wavelength intervals.

3.2.2. *Line Sources*

A low-pressure electrical discharge tube containing the appropriate vapor is a convenient source of a monochromatic or line spectrum. Figure 1.9 shows the principal arc spectrum lines emitted from Hg, Ne, and Xe arc lamps, together with the principal lines in the Fraunhofer solar absorption spectrum. These lines are ideal for the calibration of monochromators over the entire UV and visible spectrum. In addition, with the use of appropriate band-pass filters, many of these intense emission lines can be isolated and used in irradiation experiments. This is particularly true for the Hg spectrum in the UV (2537, 3130, 3663 A), blue (4358 A), green (5461 A), and yellow (5770, 5791 A) and for Ne in the red and far-red regions. In the far-UV region the Hg 1849 A line is also very intense. However, even some types of quartz begin to absorb strongly in this region and it is best to use fluorite optics.

The data of Fig. 1.7 give some idea of the irradiation intensities available from a low-pressure Hg photochemical lamp.

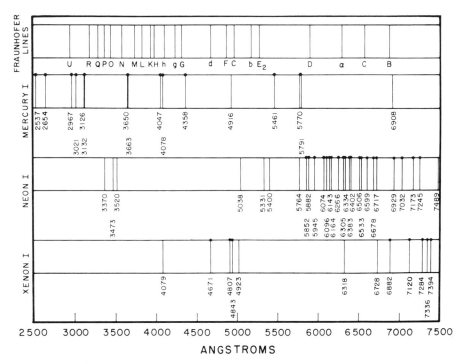

FIG. 1.9. Significant emission lines for calibration. Line spectra which can be used for the calibration of spectrometers or monochromators over the UV and visible range. The Fraunhofer lines are due to absorption of the continuous solar spectrum by cooler gases in the sun's atmosphere. The mercury, neon, and xenon lines shown are spectra such as are obtained in low-pressure discharge tubes. In these spectra the lines identified by the solid dots are the most intense.

The LASER (Light Amplification by Stimulated Emission of Radiation) can be of significant use in biological irradiation experiments because of the tremendous intensities and extreme monochromaticity which characterize its operation. A description of the principles of operation of the ruby, He-Ne gas, and the newly developed semiconductor lasers is given in Appendix III, together with a table describing presently available lasers and their emission characteristics. Aside from its use as an intense microbeam in the "welding" of detached retinas and in carcinoma therapy, the laser properties have not yet been used in any sophisticated experiments in biology. The intense monochromatic lines available in the infrared suggest the capability of exciting specific vibrational modes in molecules and the examination of very specialized reactions. This has promise in the fields of far-red effects and mutation effects, both pre- and postirradiation.

3.2.3. *Filters*

In the visible region both Corning and Chance colored-glass filters are suitable for isolating particular regions or narrow bands centered about particular emission lines of the various vapor discharge tubes. In many cases when the source exhibits a continuum, as from a high-pressure discharge lamp or from an incandescent lamp, narrow band-pass interference filters can be used. Biological pigments are rather large conjugated molecules and therefore exhibit reasonably broad absorption bands in the UV or visible region. Therefore the requirement of monochromaticity required for biological action spectra is many orders of magnitude less stringent than that to which the ordinary physical spectroscopist is accustomed. In most cases a 50–100 A wavelength interval is quite sufficient since the FWHM (full width at half maximum and commonly called the half-width) of the absorption band is about 600 A. Figure 1.10 shows transmission curves of several representative glass combination filters and typical interference filters used for the isolation of specific wavelength bands.

Caution should be exercised in the use of both colored-glass and interference filters for isolating specific wavelength regions. In the case of glass filters the transmission may vary from a few tenths of a per cent to several per cent in regions removed from the main transmission band. It may be in this very region, where the experimenter assumes zero transmission, that other photochemical events can occur and possibly mask the effects of the main transmission band. This is particularly true in biological systems where there are very wide ranges of threshold intensities for various effects. In addition, in many cases glass filters have a second or even several transmission peaks, usually at longer wavelengths. In many filter sets a combination of filters is used to eliminate these secondary transmission bands. The correction, even if 99% effective, may still lead to unwanted light of sufficient intensity to interfere with the main portion of the experiment. As a general rule, the use of a double filter will show up these effects. For example, if a filter with a main peak transmission of 30% is used for irradiation, it may be that a 1% transmission at some other wavelength is interfering with the experiment. Introduction of a second filter, identical to the first, should reduce the main beam intensity to 30% again. The unwanted radiation, if it is really important, will be reduced to 1%. Under these conditions if the decrease in response or rate of product formation follows the main beam attenuation and does drop off by a factor of 30, we can be reasonably sure that our isolation filter technique is satisfactory.

In the case of interference filters one must be extremely careful that the beam in which the filters are used is parallel. The principle of the

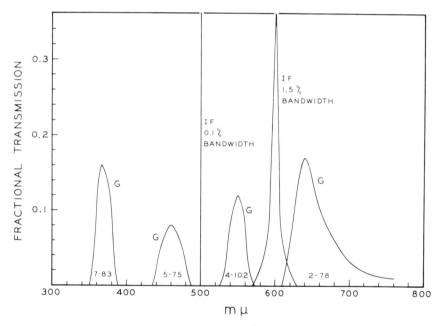

Fɪɢ. 1.10. Types of transmission filters are typical isolation filters available commercially. The curves labeled G are glass combination filters. Those labeled IF are representative of narrow band pass interference filters.

filter is such that reinforcement or transmission occurs at specific wavelengths determined by the length of light path through the reflecting layers of the filter. If light enters the filter at an angle other than normal in addition to the normal light, the light path between the reflecting layers becomes longer. Reinforcement, or transmission, will also occur at shorter wavelengths and the filter is no longer monochromatic. In addition, there is always a small transmission outside of the "stated" transmission band. This can be tested for in the same manner as for the colored-glass filters or in a sensitive spectrophotometer.

Ordinary glass filters cannot be used for the isolation of short wavelength UV lines obtained from a Hg discharge lamp. In this case quartz cells containing various solutions or gases can be employed for the selective absorption of various regions. Kasha (1948) has developed several excellent transmission filters for the ultraviolet which are superior to the glass filters. These consist of combinations of solution filters and glass filters. In addition, several can serve as short wavelength cut-off filters. These, together with other filter data for isolation or absorption, are given in Appendix IV.

3.2.4. *Monochromators*

The monochromator is a more general and versatile instrument for obtaining narrow wavelength intervals for irradiation experiments and has the advantage that it can also be used directly for the determination of emission spectra of fluorescence or chemiluminescence. References to the theory and application of prism and grating monochromators can be found in Sawyer (1944), Clark (1960), and Udenfriend (1962), as well as in the standard classical physics references of Baly (1924, 1927) and Sawyer (1944).

One of the pitfalls in the use of "monochromatic" light in biological experiments is the possible effects of unwanted scattered light of wavelengths different from the narrow bands under consideration. Sawyer (1944) has a section on the types of stray scattering occurring in various types of spectroscopes. However, it must be emphasized that while the physicist is concerned with scattered light because it may mask other very weak emission lines such as may be encountered in forbidden spectra or in Raman spectroscopy, the biologist must be concerned with scattered light to sometimes even a higher degree. Table 1.5 shows that the threshold intensities for various biological phenomena vary over many orders of magnitude. Some of these photoprocesses may be intimately related and it should not be assumed a priori that experimental action spectra have separated these effects. Commercially available grating monochromators specify tolerances of scattered light under the *best conditions* of the order of 0.1% which may not be sufficient for some experiments. A schematic drawing of an f/3 grating monochromator with 50-mm slit height that we have used in our laboratory for several years is shown in Fig. 1.11. The linear dispersion in first order at the exit slit is 20 A/mm. Figure 1.12 shows a photometric record of the relative emission lines of a low-pressure Hg "Penlite" discharge lamp obtained with this monochromator with slit widths set at 0.1 mm in combination with an RCA 7326 phototube. The monochromator grating is blazed at 5000 A. Although the dispersion is good, when using an incandescent source and observing below 4000 A, *the intensity of scattered radiation is equal to or greater than the "true" intensity.* It is therefore desirable to use any monochromator in

TABLE 1.5
Irradiation

Scale (log μwatts/cm^2)	Photoprocesses
5	Bright June sunlight (approx. 10,000 foot-candles) Saturation of photosynthesis (approx. 2000 foot-candles)
4	
3	Compensation point for photosynthesis (approx. 100 foot-candles)
2	Photoperiodic control of flowering (approx. 10 foot-candles)
1	End of twilight
	Lettuce seed germination; 8-min exposure, red light (Withrow, 1959) (Threshold for photoperiodism)
0	Flower bud induction
	Full moon
1	Threshold for positive phototaxis, *Platymonas*, 405 mμ (Halldal, 1961)
−2	
−3	

TABLE 1.5 (*Continued*)
IRRADIATION

Scale (log μwatts/cm^2)	Photoprocesses
-4	*Phycomyces*, sporangiophore curvature; 20-min exposure, blue light (Withrow, 1959) (Threshold for phototropism)
-5	
-6	Threshold of rod vision, human eye; 1 quantum/sec -830 rods (Pirenne, 1958)
-7	(Threshold for photomorphogenesis) Bean hypocotyl; 6-day exposure, red light (Withrow, 1959)
-8	Threshold of electrophysiological response in cat retina; 15-sec exposure (Pirenne, 1958) Threshold of dark-adapted eye—white light (Pirenne, 1962)
-9	Threshold of completely dark-adapted human eye, 5.6×10^{-10} erg/sec, 507 mμ (Marriott *et al.*, 1959)
-10	
-11	

conjunction with narrow band interference filters and infrared absorbing filters at the exit slit to reduce the stray light effect or to use a double monochromator.

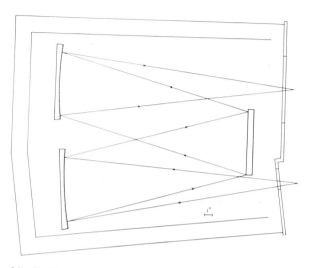

Fig. 1.11. f/3 Grating monochromator. Scale drawing of f/3 grating monochromator used either as a monochromator or spectrometer. The grating is blazed at 5000 A and has 600 lines/mm. The linear dispersion at the exit slit is 20 A/mm.

3.3. Measurement of Light Intensities

In principle, all that is required for the measurement of light intensity is an absorber of known area which absorbs completely all incident radiant energy and which uses this energy to produce a physical or chemical change which can be measured. A knowledge of the detailed energy conversion process will then permit absolute intensity measurements. Otherwise, relative measurements can be made with reference to some absolute standard light source of known spectral composition. In actual practice there are often severe limitations imposed either by the conditions of the experiment or by the types of radiation detectors available. In the following discussion both the physical and chemical methods of radiation measurements will be summarized. For a more complete description of the theoretical and practical aspects the reader will find sufficient references with each topic.

3.3.1. *Physical Methods*

The measurement of radiant intensity implies that some or all of the incident photons must be absorbed. Upon absorption the energy may be

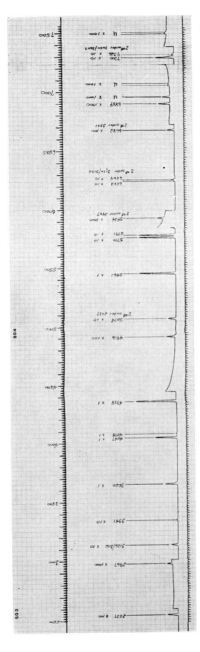

Fig. 1.12. Emission lines of Hg "Penlite" source. Relative photometric record of the emission lines from a Hg "Penlite" source made with an f/3 grating spectrometer. The entrance and exit slits were 0.1 mm or 2 A wide. The photo-tube was an RCA 7326 with a S-20 response. Grating blazed at 5000 A. The phototube has a Pyrex glass face, accounting for the UV absorption. The lines marked u have not been identified. The increase in dispersion for higher order spectra is readily apparent in the separation of the 3126/3132 A pair in second order at 6252 and 6264 A.

converted directly into heat as in a blackbody absorber. Electrons may be raised to excited electronic levels, giving rise to resonance radiation, fluorescence, or phosphorescence. In some cases atoms or molecules may become ionized. In some metals and semiconductors photoelectrons may be ejected giving rise to photoelectric currents. In others, electrons from filled bands can be raised to conduction bands where, under the action of an external electric field, they give rise to photocurrents. In still other solids, particularly the alkali halide crystals containing impurity or trapping centers, the energy may be trapped, giving rise to thermoluminescence. In silver halides we have the photographic effect of reduction of silver ions. In barrier-layer semiconductors there occurs a separation of the electron raised to a conduction band and the "hole" that it leaves behind. This charge separation produces a photovoltaic potential difference across the detector which can be measured. The rectification properties of the barrier-layer semiconductor permit easy flow of charge in one direction and thus give rise to photogalvanic currents through external resistors.

In all cases with the exception of the blackbody radiation detector there is a spectral selectivity for absorption. Therefore equal response at different wavelengths of incident light does not mean equal incident energy flux or equal number of incident photons per second. Representative data on radiation-sensitive detectors are summarized in Fig. 1.13.

(1) *Thermocouple.* Melloni (1833) was the first to use a number of thermocouple junctions or "thermopile" as a radiation detector. When alternate junctions, usually of Bi-Sb, are exposed to radiation, a small temperature difference is produced between the irradiated and protected junctions. This temperature difference produces a thermoelectric potential difference, first discovered by Seebeck, which can be measured with a sensitive galvanometer. When T is small, the Seebeck potential is given by

$$E = P_{ab} \Delta T \qquad (1.10)$$

where P_{ab} is the thermoelectric power of the metal couple. If we measure the open-circuit potential this is the value observed. However, if the thermocouple is connected to a galvanometer there is current flow, i, producing a Peltier *cooling* of the couple such that the amount of heat that flows is

$$\Delta W = -\pi_{ab} i \qquad (1.11)$$

where π_{ab} is the Peltier coefficient and the temperature change is always reduced by the Peltier effect. The responsivity of the thermopile is then very strongly dependent on the method of construction. An excellent discussion of these factors and descriptions of various thermopiles are to be found in Smith *et al.* (1960). Briefly, with a moving coil galvanometer of 100 ohms resistance and a current sensitivity such that 10^{-10} amp is detectable, powers of the order of 10^{-9} watt may be detected. With chopped light sources and tuned a-c amplifiers it is possible to detect,

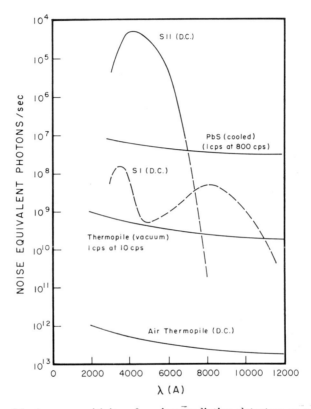

Fig. 1.13. Maximum sensitivity of various radiation detectors as a function of wavelength. Photon flux expressed as log number of quanta per second-square centimeter which will produce in the detector a signal equal to the RMS noise. For chopped light sources this is expressed for a 1 cps band width. In practice it must be remembered that the tuning band width is usually 10%. Thus at 1000 cps a PbS detector may have an equivalent noise input of 10^{-4} watt. Since this figure is normalized to 1 cps, the minimum experimentally detectable noise can be as low as 10^{-12} watt.

under careful experimental conditions and operation in vacuum, of the order of 10^{-10} watt. This differs by a factor of only 2 or 3 from the ultimate sensitivity of the thermocouple detector due to Johnson noise and thermal noise. If we define a quantity W_m, minimum detectable power, so that the RMS signal voltage arising from this absorbed power is just equal to the

RMS fluctuation voltage due to all noise sources, then $W_m \approx 3 \times 10^{-11}$ watt for a 1 cps band width.

In order to obtain a high responsivity for the thermopile the materials composing the couple must have a high value of thermoelectric power, P_{ab}. For pure metals, antimony and bismuth with $P_{ab} = 100 \ \mu V/°C$ are normally used. In order that a given amount of absorbed radiation produce the largest T and therefore the largest E, the transfer of heat through the thermocouple wires and by thermal convection must be a minimum. For this reason very thin wires are used and the thermopile is usually evacuated. Evacuation alone increases the responsivity by between 10 and 100 times, depending on the structure of the couple. It might be assumed that since certain combinations of semiconductors can give higher values of P_{ab}, a higher responsivity can be obtained. However, their resistivities are also high and they are not as stable as the metallic couples. The Bi-Sb couple is about the best combination, although a combination of Bi + 10% Sb and Sb + 25% Cd gives a slightly higher figure of merit (Geiling, 1950).

The total heat capacity of the absorbing couple must be small in order to achieve the highest ΔT. This means that the absorbing layer must be as thin as possible consistent with blackness. The method of deposition of the black surface is therefore critical and very thin layers, giving small heat capacities and consequently faster response times, are subject to deviations from blackness. We have found that some of our gold-blacked fast-response thermopiles are not uniformly "black" from the UV through the infrared region. A proprietary compound, Parson's black, is uniformly black but increases the response time by about a factor of 10 compared with gold-black. The experimenter must make a choice between these two effects. For spectrophotometry with chopped light sources, slight deviations from blackness are cancelled out by the comparative nature of the measurement and fast response time is of primary importance. However, for standardization and for accurate absolute light intensity measurements, uniform blackness is essential. Instructions for the use of the thermopile for the determination of absolute light intensities are given in Appendix V.

(2) *Bolometer.* Very much the same arguments presented for the thermopile apply to the bolometer. The fundamental difference between the bolometer and the thermopile is that the bolometer must be fed from an external power source. This introduces, in addition to Johnson noise and thermal fluctuations, a current noise. For an ideal bolometer of area 1 mm^2 at 300°K, $W_m \approx 5 \times 10^{-11}$ watt for 1 cps band width. In practice, for both the thermocouple and the bolometer, the minimum energy measurement is at least an order of magnitude higher, corresponding, for light

of wavelength 5000 A, to an incident intensity of several $\times$ 10^{10} photons/sec-cm^2.

(3) *Golay Cell.* The Golay cell is a pneumatic detector in which an air- or gas-filled cell is covered by a thin conducting film receiving the radiation. The resulting increase in gas temperature displaces a flexible mirror-membrane at the rear of the cell, light-coupled to a photocell-optical amplifier system. The resultant reading is proportional to ΔT or ΔP, from the gas laws. The novel feature of this detector is that the surface resistance or "resistance per square" of an evaporated aluminum film on which the radiation is to be incident is adjusted during evaporation to match the impedance of free space for electromagnetic radiation, i.e., 377 ohms. In this way the sensitivity is uniformly flat across the entire electromagnetic spectrum from the UV all the way to microwave frequencies. Any changes in the surface due to aging or to oxidation will reduce the impedance match, changing the reflectivity of the surface; however, the flatness of spectral response will be maintained.

(4) *Photoconductive Cells.* Most semiconductors are photoconductive, i.e., their electrical resistance decreases upon exposure to light. For a more complete discussion of photoconductivity the reader is referred to Rose (1956) and Dekker (1958). An absorbed photon excites an electron from the normally filled valence band to a conduction band in which it is free to move through the crystal lattice under the action of an applied field. Conduction can be through electron mobility, "hole" mobility, or a combination of both. If the absorbing surface is an evaporated layer of a semiconducting material, the film absorption spectrum will not necessarily bear any relationship to the bulk material, since the thin film is extremely sensitive to the presence of impurities, particularly oxygen. Under certain conditions the decrease in resistance is not linearly proportional to light intensity and so care must be taken to ensure that in the region of intensities to be measured the slope of R versus I gives a straight line.

If the above simple picture of the photoconductive process were to hold in all cases, the photoconductive cell would be an ideal photon detector for energies within the conduction band limits. The change in resistance would be proportional only to the number of quanta absorbed and not to the energy of the photon, except that the long wavelength limit would be given by the minimum energy separation between the top of the valence band and the bottom of the conduction band. This relation holds for some commercial RCA germanium p-n alloy junction photo conducting cells, over the wavelength region from about 4500 A to 1.4 μ (Radio Corporation of America, 1960), characterized as S-14 response.

TlS, PbS, and PbTe cells are used frequently in the infrared regions. Their sensitivities and long wavelength cut-offs are strongly temperature dependent. Upon cooling and with proper radiation shielding, a value for W_m for a PbS cell of 6.2×10^{-14} watt for a 1 cps band width has been reported by Watts (1949). Simpson *et al.* (1948) reported for a PbTe cell at liquid air a value of W_m of 2×10^{-14} watt for a 1 cps band width.

(5) *Electron Multiplier Phototubes.* The photo-emissive cathode presents a mechanism similar to that of photoconductivity except here the electron is given sufficient energy to escape from the light-absorbing surface. The kinetic energy of the emitted photoelectron is given by the Einstein photoelectric equation.

$$K.E. = h\nu - e\varphi \qquad (1.12)$$

where φ is the surface work function of the material and $h\nu$ is the quantum energy. Here the long wavelength cut-off is determined by the vacuum work function of the film rather than the energy difference between valence and conduction bands. Most photocathode surfaces have a very pronounced photoelectric spectral sensitivity with a peak around 4000 A and a long wavelength cut-off around 6000 A. By special treatment of the photocathode such as the introduction of Ca or K or O atom impurities this spectral sensitivity may be changed quite markedly (Sommer, 1956). This is shown in Fig. 1.14 in which the per cent quantum efficiencies of three representative photocathode surfaces are plotted as functions of wavelength. These data were taken from the typical equal-energy response curves issued by phototube manufacturers and have been converted to quantum efficiencies. As can be seen, the S-20 response of an RCA 7326 at 6500 A is more than 20 times higher than the S-4 response of an RCA 1P21, and only for wavelengths greater than 7800 A would an S-1 photocathode be advisable. However, cooling the S-1 photocathode to liquid nitrogen temperatures will increase the signal-to-noise ratio by several orders of magnitude.

The electron multiplier phototube is the most sensitive detector of visible light. It is more sensitive than the thermopile by a factor of at least 10^5. However, the phototube has its associated problems when it comes to making absolute measurements or even precise relative measurements of light intensities. The nature of the photocathode surface is not well understood, especially in the multi-alkali cathodes. The following can affect the accuracy of measurements:

(a) Variation of photoelectric efficiency with position over the cathode.

(b) Variation of spectral sensitivity with position over the cathode.

(c) Variation of both photoelectric efficiency and spectral sensitivity of the cathode with time, temperature, light intensity, and previous history.

(d) Dynode fatigue whose magnitudes and time dependence vary with dynode materials, current, temperature, construction, and previous history. Figure 1.15 illustrates the results of recent measurements under carefully controlled external conditions of the relative change in over-all phototube gain as a function of time and at different temperatures, taken from the data of Cathey (1958).

The above is not intended to counsel the abandonment of the phototube as a radiation detector. It is rather a caution that constant checking

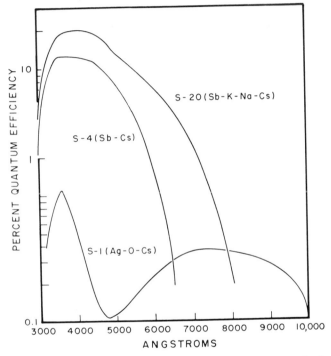

FIG. 1.14.   Per cent quantum efficiencies of various photocathodes.

experiments should be made during use to insure that these effects are minimal. Under ordinary measurement conditions with a liquid-air-cooled photocathode it should be possible to detect 10^{-16} watt or about 250 "blue" photons/sec incident on the photocathode. Under very special conditions Engstrom (1947) has reported being able to detect an average of approximately *1 quantum of light per second incident on the photocathode.* This involved the use of broad-band amplifiers and oscilloscope display to observe single events, liquid-air-cooling, and integrating times of 300 sec.

In the description of phototube characteristics manufacturers often specify the photocathode spectral sensitivity for equal values of radiant flux at all wavelengths rather than for equal numbers of photons. Thus the same radiant sensitivity at 3400 A and at 5800 A means, in effect, that the photocathode is 5800/3400 or 1.7 times more efficient for UV photons than for yellow light, since 34 photons at 3400 A have the same energy as 58 photons at 5800 A. We are mainly concerned, however, with the spectral analysis of light absorbed or emitted by biological organisms or with photochemistry. It is, therefore, the relative *photon spectral sensitivity*

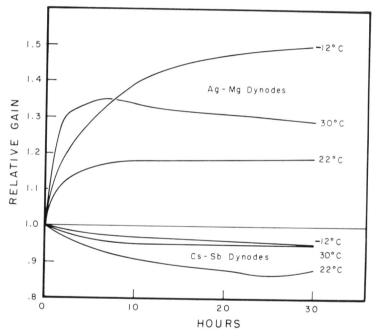

FIG. 1.15.   Fatigue in electron multiplier phototubes (Cathey, 1958).

which is important and it is necessary to correct the manufacturers' curves by weighting each given value by the reciprocal of the wavelength. This is shown in Fig. 1.16 where the standard S-11 radiant sensitivity curve is plotted together with the derived photon spectral sensitivity. The shift is always to the blue in the photon spectral sensitivity curve. The S-11 curve is almost identical with the S-4 response except for the little dip at 4000 A. Therefore this type of correction is valid for most of the blue-sensitive phototubes used at present, i.e., RCA 1P21, 931A, 929, 5819, 6199, 6342, DuMont 6292, and most EMI phototubes.

One very interesting type of semiconductor photocell has an S-14 response. Between 4400 and 12,300 A the relative radiant sensitivity can be expressed as $\epsilon_{radiant}$ (per cent) $= \lambda(A)/123$. Since the radiant sensitivity is related to the photon sensitivity by $\epsilon_{radiant} = \epsilon_{photon} \times \lambda$, we see that between these two limits the photon sensitivity is a constant, independent of wavelength. Therefore the observed photocurrent is directly proportional to the number of photons striking the sensitive area of the photocell and is therefore extremely useful for comparing the intensities of light beams of differing spectral quality. The use of this type of photocell should greatly simplify the extensive calibration now required for the

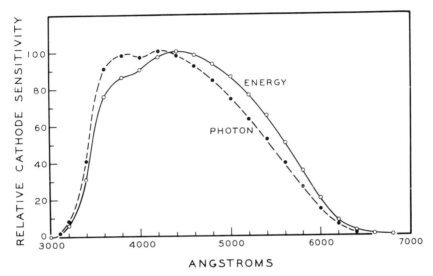

FIG. 1.16.   Spectral energy and calculated photon sensitivity of an S-11 phototube photocathode. The energy sensitivity curve is usually given in commercial specifications.

measurement of biological action spectra and would provide a simple and direct means of doing absolute spectrophotofluorimetry.

(6) *Photographic Film.* Although it has only been about 15 years since the electron multiplier phototube was introduced as an efficient detector of low-level light, particularly in the field of nuclear physics particle detectors, the phototube has virtually displaced the photographic plate as a light-measuring device. Of course the phototube has several distinct and insurmountable advantages over photographic film. It is strictly linear with light intensity over wide ranges, whereas the film has the familiar S-shaped optical density versus log exposure curve. Even more important is the fast response of the former, permitting measurements of intensity as a function of time instead of only the integral of this value. It should be remembered, however, that for certain applications the use

of photographic film can be extremely valuable. For example, in the blue region Kodak Spectroscopic Plate Type 103-0 has a log sensitivity of approximately 2.4 for an optical density of 0.6 above gross fog. At 4500 A this corresponds to a total of 0.9×10^9 photons/cm² incident on the plate. Thus a 3-hour continuous exposure will permit observation of an average incident photon flux of 10^5 photons/sec-cm². Therefore the minimum detectable average flux will be about 10^4 photons/sec-cm².

The dye-sensitized photoreduction of silver ions in silver halide crystals will be deferred until the next chapter where mechanisms of photooxidation and photoreduction will be discussed.

3.3.2. *Chemical Methods*

At the present time there are two reasonably sensitive techniques for the measurement of light intensities. These are the ferrioxalate actinometer developed by Parker (1953) and Hatchard and Parker (1956) (see Appendix VI), and the uranyl oxalate actinometer developed by Forbes and co-workers (Leighton and Forbes, 1930; Brackett and Forbes, 1933); and recently improved by Porter and Volman (1962). The quantum yield of ferrous ion produced by radiation in the ferrioxalate actinometer as a function of wavelength is given in Fig. 1.17. In this case the value of 1.26 obtained by Lee and Seliger (1964) at 365/6 mμ has been used to

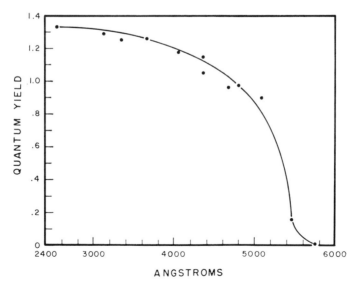

FIG. 1.17.   Quantum yield of the ferrioxalate actinometer. The quantum yield of Fe^{++} in the ferrioxalate actinometer as a function of wavelength. The data are taken from Hatchard and Parker (1956) and Lee and Seliger (1964).

normalize the data of Hatchard and Parker (1956). The use of gas chromatographic methods for determining CO evolution in the uranyl oxalate actinometer (Porter and Volman, 1962) should make this latter technique even more sensitive than the ferrioxalate technique. However, for general use and for convenience and reproducibility the ferrioxalate technique is to be preferred. The sensitivity is essentially 3×10^{14} quanta absorbed in 3 ml of solution. At 4500 A a 0.15 M $K_3Fe(C_2O_4)_3$ solution with a 10-mm path length will absorb about 98% of incident radiation. Assuming a 3-hour exposure, this corresponds to an average flux of 10^{10} photons/sec-cm². This is slightly lower than the minimum detectable flux measurable by the best thermopiles and therefore in the region of spectral sensitivity should be preferable to the thermopile. The important advantages of the chemical actinometer over the thermopile are (a) that it can substitute exactly for any solution being irradiated therefore correcting automatically for geometry, and (b) its ease of preparation and calibration relative to the more complicated thermopile accessory electronic equipment.

4.   The Light from the Sun—Some Biological Implications

Since the sun is the "driving force" for life on earth it would be expected that the spectral distribution of this radiant energy should be directly related to photochemistry.

In the primary photochemistry and in the evolution of more complex photochemistry the most efficient systems were those whose absorption of radiant energy was most efficient. Therefore at the earth's surface the spectral distribution of the incident radiation was presumably the original selection mechanism for molecular species. Those molecular structures for doing photochemical work were favored whose absorption bands were in the region of maximum incident photon intensity. Since photochemistry is a quantum absorption phenomenon, we shall be concerned with the photon spectral distribution of the incident sunlight rather than with the energy spectral distribution.

Figure 1.18 shows the relative *photon* spectral distribution from the sun (black dots). The ordinate is relative number of photons per unit wavelength interval and was calculated from the Planck formula

$$N_\lambda(T, \lambda) \, d\lambda = \frac{2\pi C}{\lambda^4} \frac{d\lambda}{\epsilon(14380/\lambda T) - 1} \tag{1.13}$$

assuming the sun is a blackbody at 5773°K. In this case $\lambda_{max}T = 3670$ and the wavelength at the maximum number of photons is 6360 A. This is to be compared with the wavelength 5020 A which is the wavelength of maximum *energy* emission. From Table 1.1 we see that 3.1% of the solar

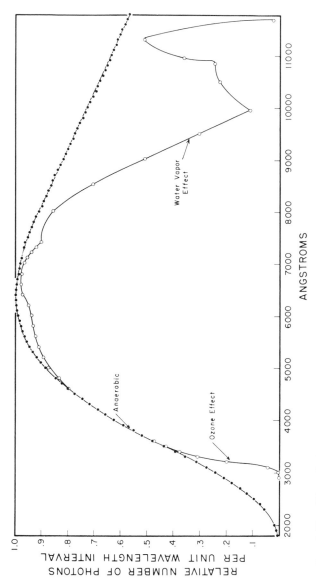

FIG. 1.18. The relative photon spectral distribution from the sun. The black dots indicate the photon distribution incident on the upper atmosphere of the earth. This was calculated from the Planck formula, assuming the sun is a blackbody at 5773°K. The effect of atmospheric ozone absorption in the visible and UV regions and that of atmospheric water vapor absorption in the infrared are shown by the open circles. At the surface of the earth there is a broad peak in photon density at the chlorophyll absorption band and a secondary sharp peak at 1.13 μ.

energy is below 3000 A, equivalent to 11.65×10^9 Einsteins/sec on the earth's surface, providing the atmosphere is anaerobic. With the advent of free oxygen a protective ozone layer was formed. Also shown in Fig. 1.18 is the modification of the incident photon distribution due to ozone

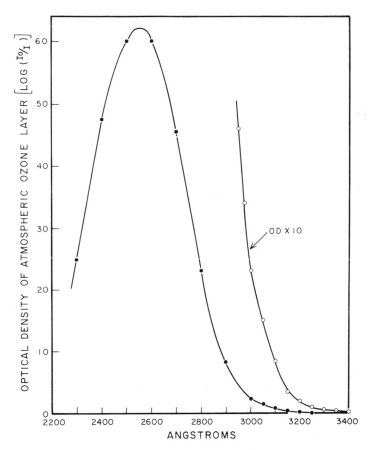

FIG. 1.19. Absorbance of atmospheric ozone layer as a function of wavelength. The open circles are a $10\times$ expansion of the long wavelength portion of the curve showing the very sharp filtering effect below 3100 A.

absorption at the present time. The curve is derived by normalizing the data of Ladenburg and Lehmann (1906), Fabry and Buisson (1913), Dutheil and Dutheil (1926), and Colange (1927) for the extinction coefficients of ozone and assuming Fabry and Buisson's estimate of an ozone content of the atmosphere equivalent to a 5-mm path length at *NTP*. In Fig. 1.19 the optical density of the atmospheric ozone layer is plotted for the short wavelength ultraviolet. Figure 1.20 gives the per cent transmission of the atmospheric ozone layer for the region from 3000 to 8000 A. At 3000 A the atmospheric transmission is 1/200 and at 2900 A it is less than 1/100,000,000. The maximum intensity below 3100 A is approximately 40 μ watts/cm^2 at the present time with seasonal variations of a factor of 10. The short wavelength limit for blackening of a photographic plate varies from 3070 A in January to 2970 A in June.

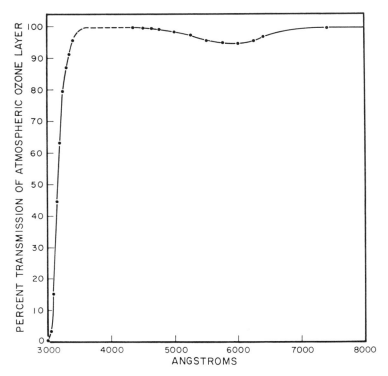

FIG. 1.20.   Per cent transmission of atmospheric ozone layer from 3000 to 8000 A, showing a dip around 6000 A.

The further modification of the solar spectrum incident on the earth's atmosphere (Fig. 1.18) is by water vapor absorption. For the purposes of our calculations we have assumed a path length of 3 cm of precipitable water. The mean concentration of water vapor in the earth's atmosphere is equivalent to 2.6 cm of precipitable water or 0.25% by weight (*International Critical Tables*, 1929). B. Mason (1960) quotes variations of from 0.02 to 4%, depending upon latitude.

The resultant curve, the combination of the ozone absorption in the ultraviolet and visible and the water vapor absorption in the infrared, shows a broad window from 3000 to 10,000 A, peaking at 6800 A and a narrow window between 1.1 and 1.17 μ, peaking at 1.13 μ. This, then, is the range within which we should find all biological photochemistry taking place. A significant argument in a hypothesis of the development and continuance of life on earth is to be found in the results of action spectra for bactericidal action. Coblentz and Fulton (1924) found absorption spectra for *Escherichia coli* in the range of 2960 to 2200 A and for *Staphylococcus aureus* in the range of 2960 to 2380 A which coincided directly with the ranges for lethal effects. The lethal action of rays shorter than 2800 A was orders of magnitude higher than those longer than 3000 A. From their data we have estimated that between 1700 A and 2800 A the energy required to kill a single bacterium was 19×10^{-12} joule.

Bactericidal action occurs completely within the ozone absorbance curve, and a good argument can be made that the protective action of the ozone layer in the upper atmosphere is specifically responsible for the *uninterrupted* replication of proteins and nucleic acids on the earth's surface. Gates (1928) reported that the reciprocal of the bactericidal energy curves matched more closely the absorption curves for cytosine, thymine, and uracil than those of the aromatic amino acids tyrosine, tryptophan, and phenylalanine, presenting the first indication that the killing effect of UV radiation was due to specific DNA (deoxyribonucleic acid) damage rather than to general cytoplasmic protein inactivation. The striking protection of the nucleic acids afforded by the ozone layer is shown in Fig. 1.21 where the relative absorbance of thymus nucleic acid (DNA) is plotted together with the relative absorbance of ozone. With the very close similarity between these two curves it is tempting to speculate a *causal* rather than a *casual* relationship between the absorptive protection by ozone and the nature of the molecular species and chemical bonding involved in the accumulation of primitive nucleic acids. The 2600 A peak absorption of nucleic acids does not normally dissociate the nucleic acid. This is analogous to the activation energy in photopolymerization reactions. We can imagine therefore that under anaerobic conditions in the primitive soup, radiation including short wavelength ultraviolet was the driving force in the polymerization and modification of all combinations of long-chain molecules and that exact template replication of complex structures, owing to the high probability of excitation and subsequent rearrangement, was very improbable. This gave rise to the widest range of variation with little competition owing to the high nutrient concentrations at that time. However, with the advent of the ozone layer, radiation of wavelengths below 3000 A were essentially

eliminated, specifically the nucleic acid absorption range, and exact replication became the probable event, with the "mutation" rate very much
lower. With more exact replication came the accumulation of previous
structure and the subsequent competition and natural selection. In its
general aspects this picture is similar to that presented by Gaffron (1962).
The difference here is that it is proposed that the appearance of the ozone

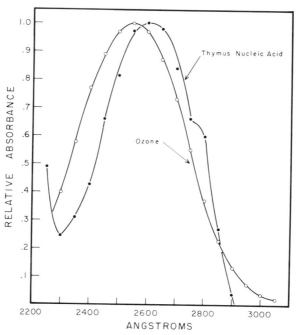

Fig. 1.21.   Correspondence between absorbance spectra of ozone and nucleic acid,
indicative of the radiation protection afforded by the atmospheric ozone. The nucleic
acid data are taken from Gates (1928), and the slight shoulder at 2800 A is probably
due to a slight amount of bound protein.

layer signaled the beginning of exact replication owing to the lowering of
the mutation rate and that at this time the evolutionary competition
began, with success going to those organisms which could develop systems
to utilize visible light energy for reducing power and for driving the
activating enzyme systems. The broad peak of the photon spectral distribution at 6800 A corresponds to the major chlorophyll action spectrum
for green plants.

CHAPTER 2

Excitation of
Molecules by Light

1. Basic Photoprocesses in Atoms and Simple Molecules

A system receiving radiation from an outside source may utilize the energy absorbed in several ways. It may simply convert the absorbed energy into heat. It may reradiate the energy as resonance radiation, fluorescence, or phosphorescence. It may utilize the energy to accelerate certain reactions which proceed in the dark at immeasurably slow speeds or to inhibit other reactions. While "dark" reactions always involve a decrease in the free energy of the reacting system, most of the photochemical reactions with which we shall be concerned result in an increase of free energy.

There are several laws in photochemistry which appear almost obvious but on closer inspection have a deeper significance. The first is the generalization called the Grotthuss-Draper Law which states that *only radiations which are absorbed by a reacting system are effective in producing chemical change.* The second is the Stark-Einstein Law of the Photochemical Equivalent: *each molecule taking part in a chemical reaction induced by exposure to light absorbs one quantum of the radiation, causing the reaction.* The quantum efficiency Q of a primary photochemical reaction is thus defined by

$$Q = \frac{\text{Numbers of molecules undergoing chemical reaction}}{\text{Number of quanta absorbed}} \quad (2.1)$$

In many cases the experimental values measured are

$$Q = \frac{\text{Rate of chemical reaction}}{\text{Number of quanta absorbed per second}} \quad (2.2)$$

In order to develop some of the concepts of the excitation of molecules by light and the subsequent reactions of the excited states of molecules resulting in re-emission, energy transfer, or photochemistry, we shall first consider the relatively well-defined reaction mechanisms of the mercury atom in the gas phase and then use those mechanisms as a basis for discussing the more complex reactions of aromatic molecules in solutions. We shall be concerned with electron configurations of molecules and therefore as a starting point it would be instructive to see how, in the case of the building up of the periodic system, the application of the Pauli Exclusion Principle gives rise to the electron distribution of the ground states of atoms. A similar type of reasoning is employed to picture the formation of molecules from atoms, although some of the direct physical interpretation of the elementary quantum numbers is lost.

1.1   Historical Background of the Nomenclature of Spectral Terms

The apparent regularity of the series of absorption lines in the spectrum of hydrogen was first mathematically formulated by Balmer. He found that the frequencies of spectral lines could be represented to about one part in ten million by a series of terms such as

$$\nu = \frac{R}{2^2} - \frac{R}{m^2} \ (m = 3, 4, 5, \ldots)$$

Later other series were discovered by Lyman, Paschen, Brackett, and Pfund and these series could be represented by similar terms:

$$\nu = \frac{R}{1^2} - \frac{R}{m^2} \qquad (m = 2, 3, 4, \ldots) \qquad \text{Lyman series}$$

$$\nu = \frac{R}{3^2} - \frac{R}{m^2} \qquad (m = 4, 5, 6, \ldots) \qquad \text{Paschen series}$$

$$\nu = \frac{R}{4^2} - \frac{R}{m^2} \qquad (m = 5, 6, 7, \ldots) \qquad \text{Brackett series}$$

$$\nu = \frac{R}{5^2} - \frac{R}{m^2} \qquad (m = 6, 7, 8, \ldots) \qquad \text{Pfund series}$$

The first term of the series representation was called the *series limit*.

In the study of the alkali metal spectra, a series was found corresponding to the Lyman hydrogen series, however, displaced to longer wavelengths. This series therefore was named the Principal series and in order to express the terms of the series in the Balmer representation a constant correction was required for the value of *m*. This correction, called the *Rydberg correction*, was denoted by *p* for the Principal series which could then be described by

$$\nu = \text{Series limit} - \frac{R}{(m + p)^2} \qquad \text{Principal series}$$

Upon more careful observation and with improved experimental techniques two other series of lines could be isolated. In one series the absorption lines on the photographic plate were sharply defined, but in the other series the lines were diffuse. These were therefore named the *Sharp* and *Diffuse* series, respectively, and their representations were

$$\nu = \text{Sharp series limit} - \frac{R}{(m+s)^2} \qquad \text{Sharp series}$$

$$\nu = \text{Diffuse series limit} - \frac{R}{(m+d)^2} \qquad \text{Diffuse series}$$

The *Rydberg corrections* were different for each series and were given the letters s and d, coming from the first letters of the names of the series. In a shorthand notation these series were represented as

$$(\text{Principal series}) = \text{Limit}_{ps} - m\text{P}$$
$$(\text{Sharp series}) = \text{Limit}_{ss} - m\text{S}$$
$$(\text{Diffuse series}) = \text{Limit}_{ds} - m\text{D}$$

Empirically it was found that $\text{Limit}_{ps} = R/(1+s)^2$, $\text{Limit}_{ss} = R/(2+p)^2$, and $\text{Limit}_{ds} = R/(2+p)^2$. Therefore in the shorthand notation the series could be represented as

$$\nu(\text{Principal series}) = 1\text{S} - m\text{P} \qquad (m = 2, 3, 4, \ldots)$$
$$\nu(\text{Sharp series}) = 2\text{P} - m\text{S} \qquad (m = 2, 3, 4, \ldots)$$
$$\nu(\text{Diffuse series}) = 2\text{P} - m\text{D} \qquad (m = 3, 4, 5, \ldots)$$

When the Bohr theory was extended to the alkali metals, it was found that the running letter m corresponded with the principal quantum number n and the terms having the shorthand notations S, P, D corresponded to values of the orbital or azimuthal quantum number $l = 0, 1, 2$, respectively, that is, the orbital angular momentum of the outer electron. Empirically it was found that transitions occurred only for $\Delta l = 0, \pm 1$.

In the case of a number of electrons in the atom an l value was ascribed to each electron. Thus electrons with $l = 0$ were called s electrons; those with $l = 1$ and $l = 2$ were called p and d electrons, respectively. The vector addition of all of the l's of the electrons in a particular shell was then the capital letter L, and an atom with $L = 0$ was described as being in an S state, an atom with $L = 1$ was in a P state, and an atom with $L = 2$ was in a D state. When higher resolution spectrometers became available, it was found that many of the original lines were in reality multiple lines. Still further splitting was observed in electric and magnetic fields under still higher resolution. In addition to the n and l quantum numbers there are required m_l and the m_s, the magnetic and spin quantum numbers, in order to account completely for the observed multiplicity and field splitting.

A more detailed description of the operation of these quantum numbers or of the theoretical justification will not be given. The historical summary of the origin of the lettering system used to describe energy levels and the states of electrons should serve mainly as an introduction to what is now to follow, the building-up of simple atoms containing s, p, and d electrons and their counterparts in simple molecules.

1.2 SPECTRAL TERMS AND ELECTRON ORBITALS

In the theory of atomic structure we completely specify the energy of a single electron system by a set of four quantum numbers, n, l, m_l, and m_s, where n is the principal quantum number, l is the orbital angular momentum quantum number, m_l is the component of l in the direction of an applied magnetic field, and m_s, the spin quantum number, equals $\pm\frac{1}{2}$. The Pauli Exclusion Principle states that in one and the same atom, no two electrons can have the same set of values for the four quantum numbers n, l, m_l, and m_s. In quantum mechanics this is equivalent to the requirement for an antisymmetric total wave function, that is, the product of the spatial coordinate function ψ and the spin function β must be antisymmetric. It is this requirement that gives rise to the prohibition of intercombinations of singlet and triplet states as well as to the building up of the electron shells in atoms. For example consider an atom with two electrons 1 and 2. The total wave function ϕ can be assumed to be the product of two independent functions, one involving the spatial coordinates and the other involving the spin. Thus $\phi = \psi$ (coordinates) $\cdot \beta$ (spin). From the four possibilities of the combinations of the $\pm$ spins of electrons 1 and 2 we can construct four spin functions:

$$\beta_{\mathrm{I}} = \beta_1{\uparrow}\beta_2{\uparrow}$$
$$\beta_{\mathrm{II}} = \beta_1{\uparrow}\beta_2{\downarrow} + \beta_1{\downarrow}\beta_2{\uparrow}$$
$$\beta_{\mathrm{III}} = \beta_1{\uparrow}\beta_2{\downarrow} - \beta_1{\downarrow}\beta_2{\uparrow} \tag{2.3}$$
$$\beta_{\mathrm{IV}} = \beta_1{\downarrow}\beta_2{\downarrow}$$

Only β_{III} is antisymmetric in the electrons, that is, interchange of the electrons gives the negative of the original function. The others are all symmetric. Thus the total wave functions ϕ that can be constructed so that ϕ is antisymmetric are

$$
\begin{aligned}
\phi_{\mathrm{I}} &= \psi(\text{symmetric})\,\beta_{\mathrm{III}}(\text{antisymmetric}) \quad \text{Singlet system} \\
\phi_{\mathrm{II}} &= \psi(\text{antisymmetric})\,\beta_{\mathrm{I}}(\text{symmetric}) \\
\phi_{\mathrm{III}} &= \psi(\text{antisymmetric})\,\beta_{\mathrm{II}}(\text{symmetric}) \\
\phi_{\mathrm{IV}} &= \psi(\text{antisymmetric})\,\beta_{\mathrm{IV}}(\text{symmetric})
\end{aligned}
\left.\vphantom{\begin{aligned}\\\\\\\end{aligned}}\right\} \text{Triplet system} \tag{2.4}
$$

Here then is the origin of the multiplet structure of spectral lines. There are three states represented by the combination of the symmetric spin functions but only one state for the antisymmetric spin function. Therefore a two-electron system can give rise to both singlet lines and triplet lines. An empirical rule first formulated by Hund but which has a solid quantum mechanical basis is briefly that the level with the greatest multiplicity has the lowest energy. This is related to the concept of resonance of forms in quantum mechanics and is the reason why in both atoms and molecules the triplet state is usually lower in energy than the singlet state. We shall come to this point later on in the discussion of singlet and triplet levels in the mercury atom and in molecules. So long as the coupling between the spin states and the orbital angular momentum is weak, the approximation $\phi = \psi \cdot \beta$ is valid. Since the transition probability between a ψ (symmetric) and a ψ (antisymmetric) state is zero, this therefore precludes intersystem crossing. From Eq. (2.4) above, singlet wave functions or states cannot give transitions to triplet states and vice versa. However, for higher Z atoms where the spin-orbit coupling becomes strong, it is no longer meaningful to separate, for example, the triplet wave function ϕ_T into the separate products ψ_T and β_T. The total wave function must include a mixture of singlet and triplet excited states. Thus $\phi_T = \phi_T^0 + \lambda \phi_{S^*}$. In these cases the singlet-triplet transition probabilities are no longer zero and intersystem crossings can occur. This will be seen in the next section for the case of the mercury atom.

The problem of determining the energy levels of the hydrogen atom has been solved completely by quantum mechanics. The mathematical solutions give information which was not available simply from the empirical quantum numbers. This is particularly true in describing the position of the electron outside of the nucleus. The solutions of the wave equations for the hydrogen atom give probability distributions for the electron position. It is usual to think of the electron as a point particle in any interaction but for its position relative to the nucleus it is easier to consider it as a cloud of negative charge density surrounding the nucleus. Then the density of the cloud is proportional to the probability in any series of experiments of finding the electron within a certain finite volume in space. The higher the energy, the further does this cloud extend from the nucleus. The mathematical solutions show also that some states have peculiar charge density distributions for the electrons and by virtue of these distributions the chemical reactivity can be inferred or bond angles of stable molecules can be explained. This is shown in Fig. 2.1 in which the spatial distributions of electron densities about the nucleus are drawn for several of the energy levels of the electron of the hydrogen atom. The distributions are drawn approximately to scale. In order to give a better

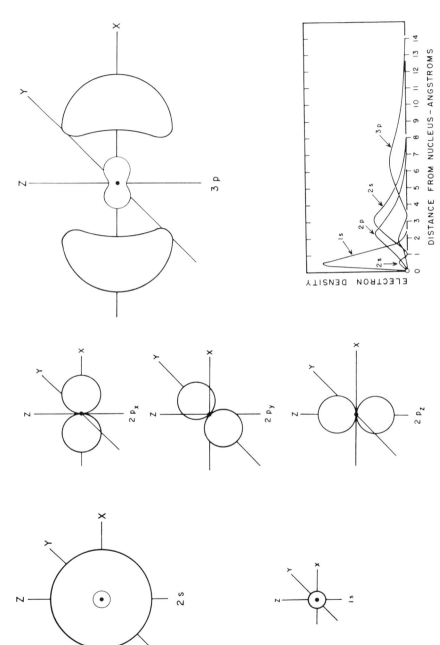

FIG. 2.1.　Electron cloud or probability density distribution of the hydrogen atom or hydrogen-like ions in different states. The electron can be considered to be smeared out over these regions. The relative extent of the regions enclosed are drawn to the same scale. In the lower right-hand portion are drawn the actual electron probability densities as functions of the distance in Angstroms from the nucleus. This is the probability of finding the electron in a spherical shell of thickness dr at a distance r from the nucleus.

idea of the extent of these electron clouds, the curves at the bottom of the figure show the radial probability density distributions of the $1p$, $2s$, $2p$, and $3p$ orbitals. The important point shown here is that only s electrons ($l = 0$) have spherical symmetry. Electrons with higher angular momentum ($l = 1, 2, 3, \ldots$), in addition to being located further from the nucleus, have, at most, axial symmetry. This is shown by the drawings of the $2p_x$, $2p_y$, and $2p_z$ cloud distributions in the figure. The axes in the absence of a magnetic field are entirely arbitrary and an electron can be pictured as being in any one of the p_x, p_y, or p_z orbitals. The electron cloud distribution for a $3p_x$ electron is also axially symmetric about the X-axis as can be seen from the drawing and also from the radial distribution curve for the 3_p orbital. There are intermediate regions in which the probability of finding the electron is essentially zero.

In the building-up of atoms we can imagine a process of adding orbital electrons to surround the nucleus of charge Z. We start at the first major shell, $n = 1$. The second shell is $n = 2$ and so forth. If we treat the angular momentum vectors l as separate from the spin vectors m_s, that is, adding each separately to obtain the total L and S for the atom, the picture will be simpler. We define a closed shell or subshell by $L = S = 0$. The hydrogen atom has only one electron. Therefore the lowest orbital would be $1s$, i.e., $n = 1$, $l = 0$. Since $l = 0$, $m_l = 0$. The electron has a spin of $\frac{1}{2}$ so that in the absence of external fields $L = 0$, $S = \frac{1}{2}$. $L = 0$ corresponds to an S state. A second electron could fit into the first shell only if the spin is $-\frac{1}{2}$, since the other three quantum numbers are the same. Thus for helium with two electrons $L = 0$, $S = 0$ and we have a closed shell. A third electron must then be promoted to $n = 2$. The lowest energy orbital outside of the closed shell would therefore be a $2s$ electron, i.e., $n = 2$, $l = 0$, $m_l = 0$, $m_s = \frac{1}{2}$. Again $L = 0$ so that the ground state of lithium is also an S state. In shorthand notation the electronic configuration of lithium can be written

$$[1s1s]2s; \,^2S \quad \text{or} \quad [1s^2]2s; \,^2S$$

indicating two $1s$ electrons forming a closed shell and a third $2s$ electron, which, since it is a single electron outside a closed shell, will have a *hydrogen-like* orbital. The superscript to the left of the term symbol indicates a doublet for the multiplicity, since the electron spin can be either up or down. It can be calculated from the value of $(2S + 1)$. Now let us consider the carbon atom, since it figures so prominently in all organic molecules. With six electrons we have the following distribution:

$$[1s1s](2s2s)2p2p = [1s^2](2s^2)2p^2$$

The round brackets indicate a closed subshell of the $n = 2$ main shell for which $L = 0$, $S = 0$. Here $n = 2$, $l = 0$, $m_l = 0$, $m_s = \frac{1}{2}$ for the third electron and $n = 2$, $l = 0$, $m_l = 0$, $m_s = -\frac{1}{2}$ for the fourth. The fifth electron must be promoted to the $n = 2$, $l = 1$ next higher orbital. It is therefore a $2p$ electron. Similarly the sixth electron will also be a $2p$ electron. For these two $2p$ electrons L, the *vector sum* of the individual l's, can have the values 0, 1, 2, giving the possibility of S, P, and D terms for the ground state. From an analysis of the various combinations of n, l, m_l, and m_s for these two electrons it can be shown that these terms are ^{1}S, ^{3}P, and ^{1}D. From the Hund Rule the state of highest multiplicity has the lowest energy and thus the ground state of the carbon atom is a triplet level. In line with the drawings of $2p$ orbitals in Fig. 2.1 we can imagine that one of the $2p$ electrons is $2p_x$ and the other is $2p_y$ or $2p_z$. In order for them both to be $2p_x$ the spins would have to be antiparallel in order not to conflict with the Pauli Principle, giving $S = 0$ or a singlet state. From the Hund Rule the lower energy state would be that corresponding to parallel spins ($S = 1$ giving higher multiplicity) and therefore to electrons in different axial distributions. When we come to oxygen we already have placed three electrons in $2p_x$, $2p_y$, $2p_z$ orbitals, so that the fourth $2p$ electron must be added to one of those axial distributions. If the electron is added to a $2p_x$ orbital, the two electrons must have opposite spins, leaving the other two electrons on $2p_y$ and $2p_z$ with unpaired spins. Thus $S = 1$ and for the oxygen atom the lowest state is also a triplet state.

1.2.1. Spectrum of the Mercury Atom

The normal energy level of the Hg atom $[1s^2][2s^22p^6][3s^23p^63d^{10}]$ $[4s^24p^64d^{10}4f^{14}](5s^25p^65d^{10})(6s^2)$ corresponds to the singlet level ^{1}S$_0$. The next higher singlet level is ^{1}P$_1$, giving rise both in absorption and in emission to the intense UV line at 1850 A, ^{1}S$_0$–^{1}P$_1$. Although empirical observation and exact quantum mechanical calculations have shown that for low atomic number elements intersystem crossings or intercombinations of states of different multiplicities are strongly forbidden, this selection rule becomes much less important as the atomic number increases, owing to the stronger coupling of spin angular momentum and orbital angular momentum. Thus in Hg, intersystem crossing because of the strong spin-orbit coupling is no longer improbable and, in fact, gives rise to several of the principal emission lines observed; specifically the intense 2537 A UV line, the 4078 A blue line, and the 5770 A line of the yellow doublet (Herzberg, 1944). These transitions are 6^3P$_1$–6^1S$_0$, 7^1S$_0$–6^3P$_1$, and 6^3D$_2$–6^1P$_1$, respectively. Because of the selection rule on electron angular momentum ($\Delta = \pm 1$) the transitions 6^3P$_0$–6^1S$_0$ and 6^3P$_2$–6^1S$_0$ are still forbidden. The levels 6^3P$_0$ and 6^3P$_2$ are truly *metastable* states. Although they represent high energy levels of the atom, the probability of spontaneous emission to the $^1S_0$ level is so small (the lifetime is so long) that the energy will remain in the atom. For the dissipation of this excess energy, it must depend upon collisions with other atoms or molecules. Transitions giving rise to the lines at 2967, 3650, 3655, 3663, 4047, and 5461 A produce these metastable states in Hg. It is possible then to make a distinction between a *labile* level of a triplet such as 6^3P_1 from which intersystem crossing is possible and a *metastable* level such as 6^3P_0 and 6^3P_2, from which transitions are strongly forbidden. It is possible by collisions to effect the internal transitions 6^3P_1–6^3P_0 and 6^3P_1–6^3P_2. For example, it has been observed that the absorption line 4047 A is absent in Hg vapor with or without the presence of N_2. However, in the presence of low partial pressures of N_2 and *simultaneous irradiation* by 2537 A, the 4047 A absorption line can be seen. From reference to the energy level diagram of Fig. 2.2 the reaction sequence is as follows:

$$Hg(6^1S_0) \xrightarrow{\text{2537 A}} Hg\ (6^3P_1)$$

$$Hg(6^3P_1) + N_2(P_{N_2} = 0.1 \text{ mm}) \rightarrow Hg(6^3P_0) +$$
$$N_2 \text{ (higher kinetic energy by 5 kcal/mole)}$$

$$Hg(6^3P_0) \xrightarrow{\text{4047 A}} Hg(7^3S_1)$$

The nitrogen molecule has been able to absorb in a collisional encounter the small energy difference between 6^3P_1 and 6^3P_0 and therefore populate the 6^3P_0 state which could not ordinarily be reached by photoexcitation alone.

For an isolated atom in a metastable state the probability of transition by spontaneous emission is excluded and the lifetime is essentially infinite. The appearance of weak transitions can usually be ascribed to the perturbation of the energy level by strong electric or magnetic fields. In the absence of strong external fields the lifetime of the metastable state is usually influenced by absorption of radiation as in the example presented above or by collisions with other atoms or molecules or with the walls of the reaction container. As a result of these external influences the atom can either gain energy or lose a part of its internal energy and effect a transition to a labile state. In the case of Hg, the lifetime of the 6^3P_1 level is 10^{-7} sec. By contrast the maximum experimental lifetime of the 6^3P_0 metastable level has been found to be 5×10^{-3} sec. Small traces of impurities, for example, H_2, reduce this observed lifetime considerably. The presence of impurities can also affect the observed lifetime of the 6^3P_1 labile level, as determined by measurements on the quenching of the 2537 A resonance radiation. The measurement of the quenching by O_2 of the 2537 A Hg resonance radiation is able to give, although indirectly, some idea as to the increase in the reactive size of the excited atom. For example, kinetic theory calculations of the collision rate between Hg (6^3P_1) atoms and O_2 molecules, in order to agree with the observed quenching, give a collision radius for the excited Hg (6^3P_1) atom approximately 3.4 times the collision radius of the normal Hg (6^1S_0) atom.

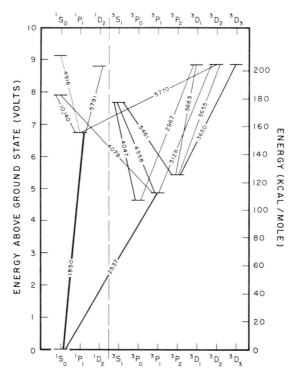

Fig. 2.2. Energy level diagram for the Hg atom. The relative intensities of the transitions are denoted by the thicknesses of the lines joining the energy levels. The energies are given both in electron volts and kcal/mole and for ease of visualization the various states, S, P, D, are separated laterally. Singlet states are to the left of the dashed vertical line so that intersystem crossings are obvious. As described in the text the strong 2537 A emission line is a triplet-singlet transition.

Oxygen is found to have a 100% probability of de-exciting the Hg (6^3P_1) level during a collision. It is quite possible that a chemical transformation is involved in this case.

1.2.2. The Absorption of Light

For simplicity let us consider the absorption of light by the $1s$ electron of the H atom. Let us assume that the incident light is plane polarized with its electric vector in the Z direction and consider only the **E** vector. This is shown schematically in Fig. 2.3a.

In the ground state the electron is $1s$ with spherical symmetry. The absorption of light with the electric field vector in the Z direction will induce an electric dipole in the atom along **E**. Therefore the electron will be disturbed from its spherical symmetry and be forced to oscillate along the **E** direction. In order for the light quantum to be *absorbed*, it must have

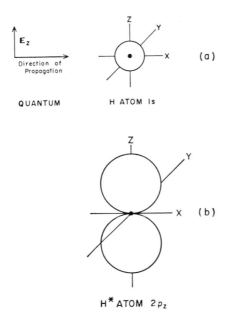

FIG. 2.3. Absorption of plane polarized light by the H atom. (a) The ground state H atom prior to the absorption event. The vector diagram to the left shows schematically the plane of polarization of the electric vector of an incident quantum. (b) The excited H atom in a $2p_z$ state. The induced electric dipole is in the same direction as the incident electric vector.

exactly sufficient energy to raise the electron to the next higher orbital, which in this case is a $2p$ orbital. Therefore, in this case with light polarized as shown, the absorption process corresponds to the transition

$$1s \overset{h\nu}{\to} 2p_z \quad \text{or} \quad 1^2S \overset{h\nu}{\to} 2^2P$$

In the absence of external influences the re-emission of light in the transition $2p_z \to 1s + h\nu$ would be plane polarized with the electric vector still in the Z direction. For ordinary unpolarized light all directions of the **E** vector are possible and therefore all directions of $2p$ excited states are possible. The re-emitted light will therefore be unpolarized. If the energy of the quantum is not exactly equal to the energy difference between the 1s and 2p orbitals, the molecule is distorted by the field during the passage of the light quantum (corresponding to the phenomenon of dispersion), but *no net work is done on the atom*. In a very qualitative sense we can say that the lifetime of this *distorted state* is the time required for the quantum to pass by,

$$\tau \sim \frac{\text{wavelength of light}}{\text{speed of light}} \sim \frac{1000 \text{ A}}{3 \times 10^{10} \text{ cm/sec}} = 0.3 \times 10^{-15} \text{ sec}$$

By contrast the lifetimes of allowed excited states such as the 2P state are of the order of 10^{-8} sec. Thus if the light frequency is a "natural" frequency of the atom the energy can be trapped in a relatively stable excited state, corresponding to absorption.

1.2.3. *Formation of the Hydrogen Molecule*

Let us now consider two hydrogen atoms A and B in the ground state, i.e., with 1s orbitals, and imagine them to be moved slowly toward each other. A stable molecule will result only if there is some distance between the nuclei at which the energy of the system of the two combined atoms is less than that of the two separated atoms. The stability will be determined qualitatively again by the configuration with the greatest number of resonance forms. As we move closer there will be a distance at which the electron cloud of atom B will begin to overlap that of A and vice versa as shown in Fig. 2.4a. Alternatively, this can be considered as equivalent to the structures of Fig. 2.4b.

That is part of the time the two electrons can be considered to be part of the H_A^- negative ion-proton$_B$ complex and part of the time part of the H_B^- negative ion-proton$_A$ complex. In order for the H_A^- or the H_B^- ion to accommodate *both* electrons in 1s orbitals the Pauli Principle requires that the spin of the *added* electron be opposite to the one already there. Otherwise the added electron must be raised to a 2p state. Thus we can have the combinations shown in Fig. 2.4c,d.

In order to obtain the configuration in Fig. 2.4d, it is necessary to add energy to the system. In the case of the H atom the 1s—2p transition is in the short wavelength ultraviolet at about 10 ev, so that form d would not be formed at ordinary temperatures from the ground state H atoms. The configurations of Fig. 2.4c, however, result in a mutual *sharing* of the electrons. Thus each electron can move in a much larger orbit. It is this *resonance* or *delocalization* of electron orbits that can be thought of as providing the stability of these electron pair bonds. In fact, the H_2 molecule of form c is more stable than two separated H atoms by 103 kcal/mole.

Since the H_2 molecule of c is formed from two $1s$ atomic orbitals, the configuration is $1s\sigma^2$. The bond is designated as a σ bond. Greek letters are used to denote for the molecule that which the latin letters denote for the atom. The term designation of the molecule is now the capital greek letter Σ in place of S, since the combination of $1s$ states still has

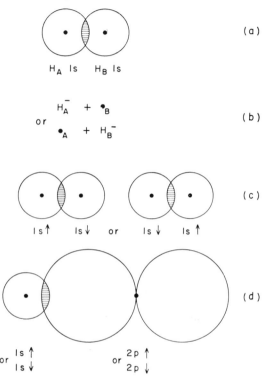

FIG. 2.4. The approach of two H atoms to form the H_2 molecule. (a) The overlapping of the atomic orbitals. The cross-hatched portion is schematic showing the region of overlap. (b) The equivalent structure as a mixture of ionic forms with both electrons on nucleus A or both electrons on nucleus B. (c) and (d) The accommodation of the electron spins based on the Pauli Principle showing the configurations required for paired and unpaired spins, respectively.

$L = 0$. The spins are opposed so that $\Sigma = 0$ corresponding to S = 0 and the multiplicity, $(2\Sigma + 1)$, is singlet. The ground state of the H_2 molecule is $^1\Sigma_g^+$, but these other symbols will not be discussed here.

In a similar way the configuration of the H_2 molecule of Fig. 2.3d is $1s\sigma 2p\sigma$. This type of bond formed is also designated as a σ bond. For simple molecules σ bonds give rise only to Σ terms, so that the state for the two parallel electrons will be $^3\Sigma$. If the spins were antiparallel, we would have $^1\Sigma$. If the second electron were promoted to a $2p$ orbital such that the axis of symmetry of the electron cloud distribution were *perpendicular* to the line joining the nuclei, i.e., the molecular axis, the configuration would be $1s\sigma 2p\pi$. This type of bond is designated as π bond and electrons participating in this bond are π electrons. In the aromatic molecules these π bonds, due to the overlap of p electron orbitals perpendicular to the molecular axis or plane, are the ones that we shall mainly be concerned with, as they give rise to most of the observed chemical and spectroscopic properties.

1.2.4.   *The Ethylene Molecule and the Benzene Molecule*

The carbon atom with six electrons has the ground state configuration $[1s^2](2s^2)2p^2$; ^3P as we discussed before. This corresponds roughly to the divalent carbon atom since only the two $2p$ electrons are available for spin sharing in the formation of chemical bonds. In order for carbon to have quadrivalent character it is necessary that all four electrons be available for bonding. This can be achieved by promoting one of the $2s$ electrons into a $2p$ orbital giving the configuration

$$[1s^2]2s2p_x2p_y2p_z$$

We can therefore qualitatively picture the quadrivalent carbon atom as a combination of a spherically symmetric s orbital and three mutually perpendicular dumbbell-shaped p orbitals. In the ethylene molecule we have the structure shown in Fig. 2.5. In this case in order to avoid the confusion in picturing the overlapping of the s and p orbitals, we have distorted them and effectively separated the two carbon atoms. Figure 2.5a is therefore a schematic top view of the ethylene molecule showing only σ bonds; the formation of a strong σ bond between the $2s$ orbitals of C_1 and C_2 and of four hybrid σ bonds between the $2p_x$ and $2p_y$ orbitals of C_1 and C_2 and four $1s$ H atom orbitals. The Z axis is perpendicular to the plane of the paper. In Fig. 2.5b we have rotated by 90° about the molecule axis and present a side view of the same molecule showing only π bonds. The overlapping of the two $2p_z$ orbitals forms a π bond. For ease in picturing the σ and π bonds only the former have been shown in (a) and then only the latter in (b). One way in which the configuration of ethylene can be written is

$$[1s^2 1s^2]1s2p_x\sigma^2 1s2p_y\sigma^2 2s^2\sigma 2p_z^2\pi$$

indicating that (*a*) two sets of $1s^2$ electrons are in the central cores of the atoms, (*b*) there are two σ bonds of the H atom with $2p_x$ orbitals, (*c*) two σ bonds of H atoms with $2p_y$ orbitals, (*d*) a σ bond composed of two $2s$ orbitals, and (*e*) a π bond composed of two $2p_z$ orbitals. The double bond between C_1 and C_2 is part σ and part π. In the ethylene case we have only two π electrons. For benzene, C_6H_6, as shown in Fig. 2.6 there are six π electrons. In Fig. 2.6a a top view of the planar molecule is drawn schematically showing only the σ bonds forming C—C bonds as well as C—H bonds. Fig. 2.6b is a perspective drawing of the side view of the planar molecule showing only the $2p_z$ π orbitals. These again are distorted

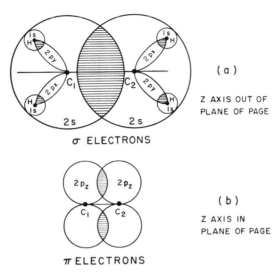

(a)

Z AXIS OUT OF
PLANE OF PAGE

σ ELECTRONS

(b)

Z AXIS IN
PLANE OF PAGE

π ELECTRONS

FIG. 2.5.   Atomic orbitals of the ethylene molecule. (a) Schematic showing σ-bonding electron orbitals. The cross-hatching indicates overlapping of the orbitals but the extent of the overlap is not to scale. (b) π-bonding electron orbitals. The molecule is considered fixed in space and is rotated through 90°. Although both halves of the $2p_z$ orbitals are drawn it must be remembered that there is a difference in phase of the wave. Some authors use + and − signs above and below to indicate the phases of the wave.

in order to show their origin. Actually the overlap among all of the $2p_z$ orbitals would be similar to that shown in Fig. 2.5b. Therefore a $2p_z$ electron from C_2, due to the complete overlap of the $2p_z$ orbitals, can be found with equal probability anywhere within the overlap region, indicated by the "donut-shaped" rings drawn above and below the plane of the molecule. These large overlap regions are what give rise to the delocalization of π electrons. These six π electrons are no longer considered as belonging to six specific carbon atoms but as part of the π orbital system of the molecule as a whole. The "donut-rings" are separated from the σ orbitals in the plane of the molecule and therefore the π electrons are more weakly bound than the σ electrons. It is therefore possible to treat the π electron system of the molecule independently from the σ electrons since most of the lowest energy excited levels of the molecule will be due to π electron transitions.

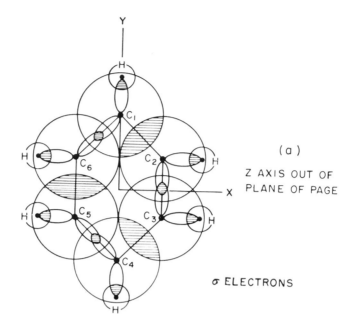

(a)

Z AXIS OUT OF
PLANE OF PAGE

σ ELECTRONS

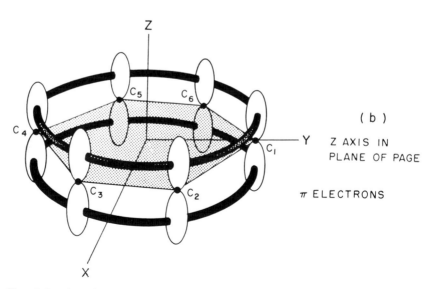

(b)

Z AXIS IN
PLANE OF PAGE

π ELECTRONS

FIG. 2.6.   Atomic orbitals in the benzene molecule. (a) Planar view showing only σ-bonding electron orbitals. The cross-hatching indicates a region of overlap, forming an electron-pair bond. (b) Perspective view of π-bonding electron orbitals. The p_z electron orbitals are much reduced in relative size to show the configuration relative to the molecular plane. Actually the "donut"-shaped rings above and below the plane of the molecule are due to the overlap of the p_z orbitals. It is this overlap that forms the "donut" rings in which the electrons can be considered to be completely delocalized.

1.2.5. n-Orbital Electrons

In the detailed examination of the absorption bands of polyatomic molecules and their displacements due to solvent and substituent effects, there appear to be two separate systems in operation. In one case on the introduction of electron-donating substituents, such as —NH$_2$, —OH . . . , or upon solution in solvents of high dielectric constant, a series of strong absorption bands are shifted toward the red. Other bands showing only weak absorption are shifted toward the blue. Mulliken and McMurry (1940) and McMurry (1941) in a theoretical study of the formaldehyde molecule, proposed the existence of n orbitals as well as π orbitals; that is *nonbonding* or "lone-pair" electrons associated with the oxygen atom in formaldehyde. In 1950 Kasha introduced the concept of $n \to \pi^*$ transitions from these n orbitals as well as the $\pi \to \pi^*$ transitions from the π-orbital electrons. The *nonbonding* electrons are actually physically separate from the delocalized π electrons and therefore retain their atomic character in the heterocyclic molecule. If we consider the planar formaldehyde molecule CH$_2$O in the same type of orbital diagram as Fig. 2.5 for the ethylene molecule we can set down schematically the three types of orbitals shown in Fig. 2.7. The ground state of the oxygen atom is [1s^2](2s^2)2p^4. One of the 2p electrons forms a σ bond with the carbon and another forms a π bond. This leaves an electron pair on the oxygen atom which does not bond to any carbon electrons, or certainly bonds only very slightly. It is this "lone-pair" which is designated as *nonbonding*. The orbitals are p orbitals and have the typical plane of antisymmetry of ordinary p orbitals. The plane of the p orbital contains the C—O axis and is perpendicular to the molecular plane. The orbitals drawn in all of the diagrams are only schematic, with the cross-hatching to indicate the overlap, or most probable region of the electron density. It can be seen that the n electrons, however, having little or no orbital overlap with any other electrons, are in a sense physically separate from the others.

1.2.6. $\pi \to \pi^*$ and $n \to \pi^*$ Transitions

Let us now examine the consequences of the simplified pictures of Figs. 2.5–2.7. It we neglect the n electrons for the moment, we see that the π electrons, being located further away from the nuclei than the σ electrons, are consequently less strongly bound and form the set of lowest-energy electronic excitation levels, the $\pi \to \pi^*$ transitions. The n orbital on the oxygen atom in Fig. 2.7c has an axis $\perp$ to the $2p_x$ and $2p_z$ orbitals forming the σ and π bonds, respectively. Upon formation of the formaldehyde molecule, the σ orbitals can be represented as most strongly bound (and therefore lowest in energy) and the π orbitals next most

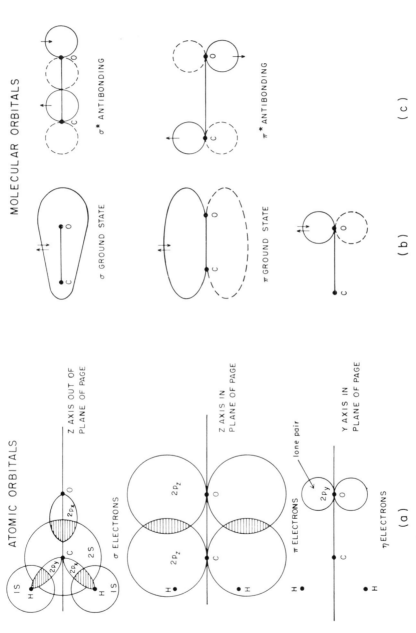

FIG. 2.7. Orbitals in the formaldehyde molecule. (a) σ-bonding and π-bonding electron orbitals similar to the drawings for ethylene and benzene, and the orientation of the nonbonding or lone-pair n electrons of the O atom. (b) Corresponding molecular orbital representation for the CO bonds. (c) Antibonding or excited state molecular orbitals for σ and π electrons.

strongly bound. These are represented in an orbital level diagram Fig. 2.8a. Since the n orbital is nonbonding or very weakly binding, it can be represented in the figure as slightly above the π orbital level. If we consider only the C—O orbitals involving six electrons, we can represent them together with their spin arrows as the black dots in the figure. The lowest-lying excited electronic state will then be π^* and above this will be the σ^* state. Thus an $n \rightarrow \pi^*$ transition or promotion should be the

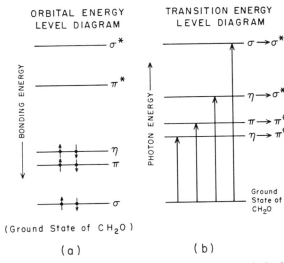

ORBITAL ENERGY
LEVEL DIAGRAM

TRANSITION ENERGY
LEVEL DIAGRAM

(Ground State of CH_2O)

(a)   (b)

FIG. 2.8.   Energy level diagrams for formaldehyde.   In (a) the bonding energy is greatest for the σ orbitals. The n orbital is the highest filled orbital. Thus for the absorption of light, the lowest energy transition would be $n \rightarrow \pi^*$, next $\pi \rightarrow \pi^*$, $n \rightarrow \sigma^*$ and last $\sigma \rightarrow \sigma^*$. These transitions are shown (b) in terms of the photon energy absorbed or emitted during the transition. Thus all transitions start from the ground state of the molecule, containing σ, π, and n electrons. In this case the $n \rightarrow \pi^*$ transition is therefore the lowest in energy and the $\sigma \rightarrow \sigma^*$ transition is highest.

lowest energy transition observed. Figure 2.8b presents the more conventional transition level diagram showing the ground state molecular energy and the relative energies of the various transitions. The spacing of the transitions in Fig. 2.8b are the same as the spacing between levels in Fig. 2.8a. In our discussions of excitation and energy transfer we shall use the convention of Fig. 2.8b, although it is important to keep in mind that aside from the energy difference there is an essential physical difference between an $n \rightarrow \pi^*$ and a $\pi \rightarrow \pi^*$ transition. If we had some way of aligning the benzene molecules of Fig. 2.6 so that they were fixed in space and parallel to one another, all of the $2p_z$ overlapping π orbitals would be parallel to one another. Light with its electric vector *in the plane of the molecule* would therefore have the greatest probability of exciting $\pi \rightarrow \pi^*$

transitions. We say therefore that $\pi \to \pi^*$ transitions are polarized *in the plane* of the molecule.

If now one of the carbon atoms is replaced by nitrogen, forming the pyridine molecule, we will have a "lone-pair" orbital on the nitrogen atom as well as the bonding or overlapping π orbitals. By reference to Fig. 2.7b,c for formaldehyde, the n orbital is perpendicular to the π-bonding orbitals and in the plane of the molecule. On the basis of symmetry considerations for allowed dipole transitions, $n \to \pi^*$ transitions are polarized *out of the plane* of the molecule (Kasha, 1961). This is an extremely important concept relative to a fundamental difference between $\pi \to \pi^*$ and $n \to \pi^*$ transitions and we shall refer to this point again in the discussion of energy transfer in ordered molecules.

As will be discussed in the next section, a nonzero transition probability requires that the integrand $\phi_i \mathbf{M} \phi_j$, where ϕ_i and ϕ_j are the wave functions of the excited and ground states respectively, and $\mathbf{M}$ is the electric dipole moment operator, be symmetric. Without a discussion of group theory we can make the following qualitative arguments. In the pyridine molecule the π orbitals are *antisymmetric* with respect to reflection in the plane of the molecule. Since the π^* orbitals have the same symmetry as the π orbitals, the product of these two wave functions would be symmetric. Therefore only an electric vector which is *symmetric* with respect to reflection in the plane of the molecule will give a *symmetric* integrand and therefore an *allowed* transition. This means that only electric vectors *in the plane* of the molecule can produce $\pi \to \pi^*$ transitions. Conversely the n orbital of the nitrogen atom is in the plane of the molecule and is therefore *symmetric*. The product of the n and π^* wavefunctions is thus *antisymmetric*. Therefore only an electric vector which is *antisymmetric* to reflection in the plane of the molecule will give a *symmetric* integrand. This means that $n \to \pi^*$ transitions are polarized *out of the plane* of the molecule. Because of the lack of overlap between the n orbital and the π^* orbital the transition, though *allowed*, will be weak.

Again by reference to Fig. 2.6 for the benzene molecule, we see that a $\pi \to \pi^*$ transition amounts in effect to a sharing of this excitation energy by the π electron system and leaves the molecule relatively unchanged. However, if again a nitrogen atom is substituted for a carbon atom (i.e., pyridine) the promotion of one of the n electrons to a π^* state leaves a "hole" in the n orbital or an unpaired electron. Since this positive hole is physically separate from the π electrons, an $n \to \pi^*$ transition may give rise to a highly reactive "radical" molecule. This separation of charge from the n orbital will also result in a stronger polarization of the molecule. These may have some importance in the reducing and oxidizing interactions of excited states of dye molecules to be discussed later.

In our discussion of the various bonds involved in molecule formation we have shown only the atomic orbitals associated with the constituents of the molecule. It is also possible to describe the electron orbitals for the molecule as a whole. The overlap among the atomic orbitals leads to the formation of a bond. The resultant molecular orbital is called a bonding orbital. In the case of the hydrogen molecule, formed by the overlap of the two 1s atomic orbitals of the hydrogen atoms, we obtain a bonding molecular orbital which can be represented as

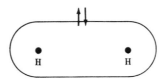

In order for both electrons to be present in the same molecular orbital their spins must be opposite or paired. As can be seen, the spherical symmetry of the 1s atomic orbitals is now a cylindrical symmetry with respect to the axis of the molecule. In the case of the atomic p orbitals, it should be emphasized that the dumbbell or "figure 8"-shaped atomic orbitals represent the total average electron density around the atom. At any one instant because of the node at the nucleus the electron will be on either one side or the other of the nodal plane.

In Fig. 2.7b are shown the molecular orbitals corresponding to the ground state atomic orbitals for formaldehyde shown in Fig. 2.7a. Only the CO molecular orbitals are drawn. The paired arrows represent the two electrons with paired spins in the molecular orbital. There are also repulsive states for the formaldehyde molecule, i.e., energy must be added to the system compared with the free atoms. These states are represented by the antibonding molecular orbitals in Fig. 2.7c. The solid and dashed contours represent the positive and negative regions of the orbitals or the phases of the electrons. The $n \rightarrow \pi^*$ transition will be to a π^* antibonding orbital and can be represented in Cartesian coordinates as

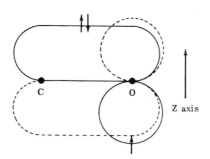

The second n electron with opposite spin remains in its $2p_y$ atomic orbital as drawn in Fig. 2.7a or b. The total spin is still conserved as we are dealing with singlet-singlet transitions. However, since there is no longer a shared n orbital, the spins have become uncoupled and triplet states are possible. In a similar way a $\pi \rightarrow \pi^*$ transition will also be to a π^* antibonding orbital represented by

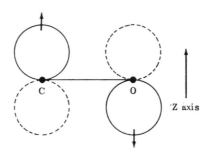

Two alternatives now become possible. First, owing to the bonding of the ground state π electrons, the $\pi \rightarrow \pi^*$ transition is normally to a higher energy state than the $n \rightarrow \pi^*$ transition. Therefore it is energetically possible for one of the nonbonding n electrons to pair with one of the regular π orbitals, resulting effectively as though an $n \rightarrow \pi^*$ transition had occurred. Second, since the electrons are now uncoupled from their shared orbital, triplet states again become possible.

1.2.7. Transition Probabilities and Accessible Energy Levels

Thus far we have described two fundamental differences between $\pi \rightarrow \pi^*$ and $n \rightarrow \pi^*$ transitions, the difference in electric polarization directions relative to the plane of the molecule and the formation of highly polar "radical" states in one type of transition and not in the other. In order to go further it is necessary to introduce the concept of transition probability. In quantum mechanics the transition probability for the absorption or emission of radiation of frequency ν_{ij} has the form for allowed dipole transitions

$$P \sim |\phi_i^* \mathbf{M} \phi_j|^2 \tag{2.5}$$

where ϕ_i^* is the wave function describing the excited state, ϕ_j is the wave function for the ground state, and $\mathbf{M}$ is the electric dipole operator. We can think of the integrand as an element R_{ij}, where the X_{ij} component of R_{ij} is given by

$$X_{ij} = \int_{-\infty}^{\infty} \phi_i^* \sum_n ex_n \phi_j \, dx \tag{2.6}$$

and is the x-component of the average value of the dipole moment over the n electrons making up the system. In order that $X_{ij} \neq 0$, the total expression must be symmetric with respect to any symmetry operation involving the coordinates of the molecule. Therefore since Σex is an odd function, i.e., $f(x) = -f(-x)$, it is necessary that ϕ_i^* and ϕ_j be states of opposite symmetry.

In the case of atoms we stated that it was possible to describe the electron by four quantum numbers n, l, m_l, and m_s. Using the model of the linear harmonic oscillator, it can be shown (Pauling and Wilson, 1935, pp. 82, 306) that owing to the nature of the solutions

$$\int_{-\infty}^{\infty} \psi_n \psi_m x \, dx = 0 \qquad \text{except for} \qquad m = n \pm 1 \qquad (2.7)$$

Therefore the selection rules for principal quantum number n are $\Delta n = \pm 1$. Further from an analysis of the surface harmonic wave functions (Pauling and Wilson, 1935, p. 306), it is found that $\Delta l = \pm 1$ and $\Delta m_l = 0$ or ± 1. We can also describe the state function of an atom as even or odd with reference to changing the signs of all the positional coordinates of the electron. Thus if

$$(x_1, x_2 \ldots x_n; y_1, y_2 \ldots y_n; z_1, z_2 \ldots z_n \ldots) =$$
$$(-x_1, -x_2 \ldots -x_n; -y_1, -y_2 \ldots -y_n; -z_1, -z_2 \ldots -z_n)$$

the function is an even function. If $(x_i, y_i, z_i) = -(-x_i, -y_i, -z_i)$, the function is an odd function. From the expression for the transition probability and since x by itself is an odd function, we see that

$$\int_{-\infty}^{\infty} \psi_n^* \sum ex_i \, \psi_m \, dx = 0 \qquad (2.8)$$

unless ψ_n^* is even and ψ_m is odd or vice versa. Thus transitions are allowed only between states of opposite symmetry. In the atom we speak of this as symmetry with respect to inversion of the electronic coordinates. In the case of a one-electron wave function, when $l = 0, 2, 4, \ldots$ the function is even and when $l = 1, 3, 5, \ldots$ the function is odd. Even functions are given by $s, d, g, \ldots$ orbitals and odd functions are given by $p, f, h, \ldots$ orbitals. Here we see, at least according to our mathematical rules, why the transitions $1s \to 2s$ (even $\to$ even) or $1s \to 2d$ are forbidden.

As we progress from the description of the atom to the diatomic molecule, we introduce a new parameter into the description of the system. There is now a molecular axis and instead of considering the l quantum numbers of the individual electrons it is more meaningful to consider their projections along the axis of the molecule, and for this we use the greek letter λ. The m_l tends to lose its physical significance. In polyatomic molecules the situation becomes further complicated since there are molecular planes as well as axes. Now the symmetry requirements are not with respect to a single nucleus at the center of the coordinate system but with respect to the more complicated symmetry of the molecule. We speak of symmetry with respect to reflection in a plane or with respect to rotation about an axis. For a discussion of this particular aspect of symmetry in more detail, the reader is referred to Reid (1957, pp. 48, 188), Platt (1949), and Eyring et al. (1944).

If the two electronic state functions ψ_i^*, ψ_j have a large spatial overlap, the R_{ij} of Eq. (2.6) will therefore be large and we should expect to observe an intense absorption. This is usually the case for $\pi \rightarrow \pi^*$ transitions since in the π^* orbital the electron coordinates are not changed appreciably. However, the spatial overlap between n and π^* orbitals can be very small, so that if these do occur they will have weak absorption bands. Experimentally, these are at least ten times weaker than $\pi \rightarrow \pi^*$ absorption bands.

With what we have said thus far about the antisymmetry of the state functions with regard to symmetry operations we can see that if the spin functions are included in the total wave functions of the transition moment integral, only $\Delta S = 0$ or like-spin transitions have nonzero probability. However, if there is interaction between spin states and orbital angular momentum, it is no longer possible to consider spin and therefore singlet and triplet states separately. Mathematically, this can be represented as a perturbation of the pure triplet state by an admixture of singlet state, or singlet-triplet mixing. The amount of mixing is strongly dependent on atomic number constituents of the molecule. Thus a singlet $\rightarrow$ triplet* transition is in reality ground singlet $\rightarrow$ small component of excited state singlet mixed with excited triplet. In the expression for the theoretical transition probability for singlet-triplet transitions there is a factor

$$\frac{1}{|E_T - E_S|^2}$$

This inverse square dependence on the energy difference between the triplet state and the mixing singlet state is of extreme importance, for although $|E_{T^*} - E_{S(\text{ground})}|$ is usually large, $|E_{T^*} - E_{S^*}|$ is much smaller. Therefore radiationless transitions from excited singlet levels to triplet levels can occur with high probability. Kasha (1960, p. 254) refers to an intersystem crossing ratio between excited singlet and triplet states as an experimentally accessible method of determining the population of a triplet level. This is just the ratio of the phosphorescence ($T^* \rightarrow S^0$)

quantum yield to the fluorescence $(S^* \rightarrow S^0)$ quantum yield. In this picture the assumption is made that $E_{S^*} - E_{T^*} \gg kT$. The $S^* \rightarrow T^*$ transitions are irreversible and since $T^* \rightarrow S^0$ transitions have low probability (long mean life), there is a much greater probability for T^* states to lose energy by solvent interaction giving rise to a radiationless transition to the ground state than by radiative emission. If now the solution is cooled to prevent solvent interaction, $T^* \rightarrow S^0$ transition can occur and phosphorescence is observed. The cooling does not appreciably affect $S^* \rightarrow S^0$ or $S^* \rightarrow T^*$ transitions.

In polyatomic molecules which have no lone-pair electrons the lowest energy electronic levels will occur in the UV and visible regions and will involve excitations of π electrons only. Kasha (1960) divides these into two classes: dyes with intense visible absorption bands and conjugated hydrocarbons whose absorption bands occur in the ultraviolet and usually with lower intensity (lower f values). In accordance with Hund's rule, if both singlet and triplet levels are possible, the triplet level (having a statistical distribution relative to singlet of $3:1$) will be lower in energy. Thus in naphthalene the fluorescence is in the ultraviolet, while the phosphorescence is in the green, region of the spectrum. As the degree of conjugation increases, i.e., as the number of electrons involved in the delocalized π orbitals and therefore the size of the π orbitals increases, the absorption energy decreases (wavelength increases) from the ultraviolet to the visible and the absorption becomes more intense. Thus benzene absorbs at 2620 Å, naphthalene at 2750 Å, anthracene at 3750 Å, and finally pentacene absorbs so strongly and broadly at 5800 Å in the visible that it has the color of potassium permanganate. This is shown in Table 2.1 where the absorption data for these molecules are given. The characterization of the π orbitals by the nonlocalized electrons has in the limit the graphite structure where the conjugation is so large and the electrons so delocalized that the millions of electronic levels blend in with the vibrational and rotational levels, resulting in a continuous absorption from the ultraviolet through the infrared and therefore the characteristic black color. Radicals are usually strongly colored owing to the "odd" electron which is not as strongly bound as the others. The excitation energy levels of molecules whose electrons are not spin-paired are lower than those with normal spin-paired electrons. The yellow triphenyl radical absorbs at much longer wavelengths than triphenylmethane. In general, a dye has two or more polar groups that can be described by alternative structures. It is this "resonance" between different equivalent structures that effectively lowers the energy levels to the visible region and permits strong absorption. Another way of saying the same thing is that the dye molecules have a large polarizability. It is this polarizability that

TABLE 2.1
ABSORPTION LEVELS OF AROMATIC HYDROCARBONS

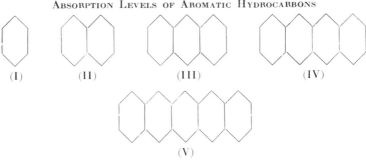

	$E_{1\mu}$		E_{2g}	
Molecule	λ max (A)	log ε	λ max (A)	log ε
(I) Benzene	1840	4.67	2620	3.84
(II) Naphthalene	2200	5.05	2750	3.75
(III) Anthracene	2520	5.30	3750	3.90
(IV) Naphthacene	2780	5.10	4730	4.05
(V) Pentacene	3100	5.45	5800	4.20

favors the aggregation or ordering of dye molecules in concentrated solutions or when absorbed on a surface. Pseudoisocyanine in low concentra-

tions in alcoholic and aqueous solution exhibits broad molecular absorption "M" bands at 4819 and 5297 A. In aqueous solution, if the dye concentration exceeds 10^{-3} M, a new *narrow* absorption "P" band appears at 5727 A (Pringsheim, 1949, p. 359). At 10^{-2} M the absorption and fluorescence due to this 5727 A band are greatly enhanced at the expense of the broader bands and the liquid solution is converted into a gel. The phenomenon is caused by the formation of highly polymerized aggregates consisting of large numbers of parallel planar dye molecules piled upon each other face-to-face so that long chains are built up. The further observation that minute amounts of impurities can quench the fluorescence of this "P" band is extremely important in terms of understanding biological analogs such as the chlorophyll photosynthetic unit. Addition of 10^{-5} M pyrocatechol to a solution of 0.2 M pseudoisocyanine chloride

quenches the fluorescence to 80% of its maximum value. Thus we must assume that the fluorescence of an aggregate can be quenched if any of the dye molecules forming the chain is in contact with a pyrocatechol molecule. In pure pseudoisocyanine solutions the fluorescence yield of the "P" band rises rapidly as gelation occurs and remains constant. With complete dehydration both the "P" band and the fluorescence disappear. As further evidence of the effect of "bonding" water, the position of the "P" band depends on the amount of water present in the gel and neither the "P" band nor fluorescence is observed in alcoholic solution. In a streaming aqueous solution these filaments formed by the face-to-face molecules exhibit a strong dichroism. The "M" bands absorb most strongly when the electric vector is in the plane of the individual molecules (as would be expected for monomeric dye molecules) and the "P" band absorption corresponds to the electric vector parallel to the molecular axis (parallel to the streaming direction). The fluorescence is also polarized parallel to the molecular axis (Scheibe, 1938; Matoon, 1944).

The colors of ions of the transitional elements are due to the fact that there are incompletely filled sets of $3d$ orbitals. Since the electrons can be alloted in several ways among the five possible d orbitals, there are a number of atomic levels with small energy differences, giving rise to visible absorption. For example:

$$Fe^{3+} + e^- \rightleftharpoons Fe^{++}, \quad Mn^{3+} + e^- \rightleftharpoons Mn^{++}$$

yellow   green   violet   pink

$$Co^{3+} + e^- \rightleftharpoons Co^{++}, \quad Cu^{++} + e^- \rightleftharpoons Cu^+$$

red   blue   blue   colorless

In Table 2.2 are shown structures of typical dyes based on the anthracene structure as well as some of the conjugated di- and triphenyls. When hetero atoms are substituted into polyatomic molecules, there is the possibility for "lone-pair" electrons. However, the mere presence of hetero atoms does not always imply the existence of n orbitals. For

TABLE 2.2
STRUCTURES OF ANTHRACENE-LIKE DYES AND PHENYL DYES

Anthracene

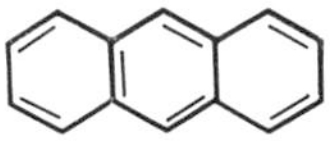

(Anthraquinoid Type)

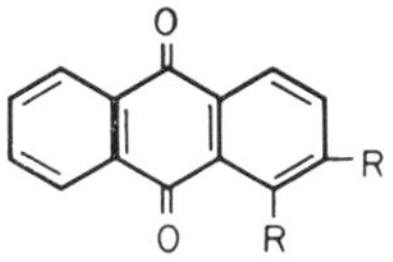

Name	R
Anthraquinone	H
Alizarin	OH

(Xanthene Type)

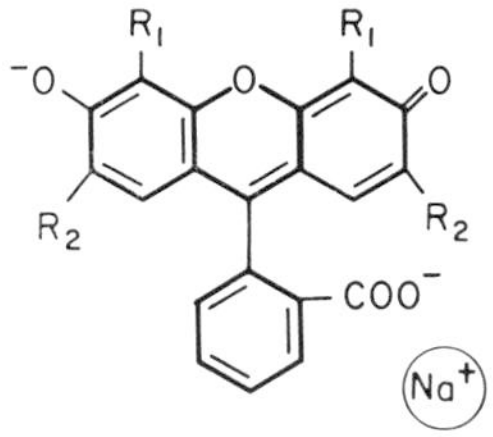

Name	R_1	R_2
Fluorescein - Na (Uranine)	H	H
Eosine I Blue	Br	NO_2
Eosine Yellow	Br	Br

(Xanthene Type)

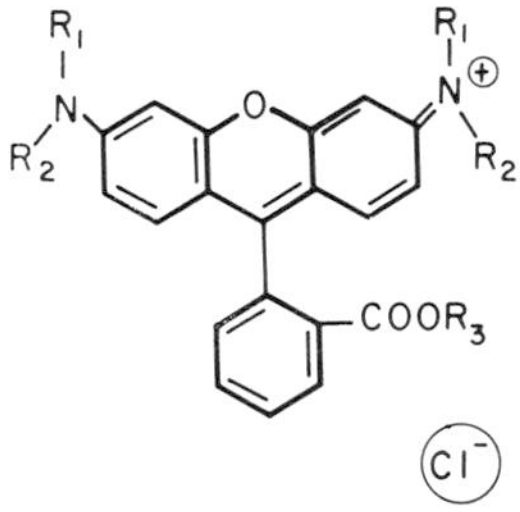

Name	R_1	R_1	R_3
Rhodamine	CH_3	CH_3	H
Rhodamine B	C_2H_5	C_2H_5	H
Rhodamine 3 B	C_2H_5	C_2H_5	C_2H_5
Rhodamine 6 G	H	C_2H_5	C_2H_5

(Acridine Type)
(Cyanine Dyes)

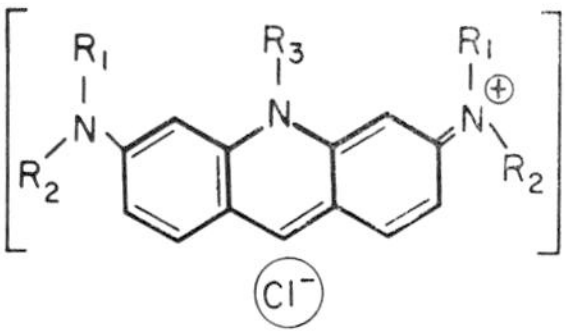

Name	R_1	R_2	R_3
Trypaflavine (Acriflavine)	H	H	CH_3
Acridine Orange	CH_3	CH_3	H

(Oxazine Type)

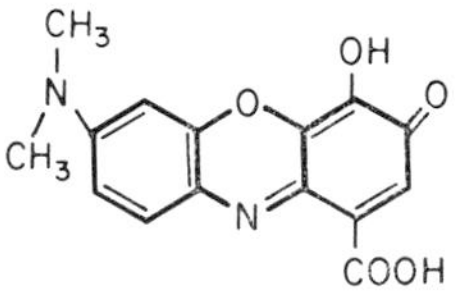

Gallocyanine

TABLE 2.2 *(Continued)*

(Thiazine Type)

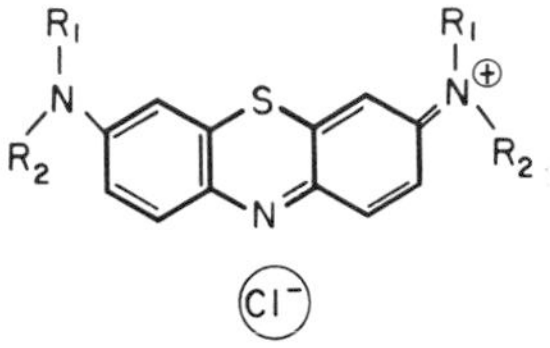

Name	R_1	R_2
Thionine	H	H
Methylene Blue	CH_3	CH_3

(Azine Type)

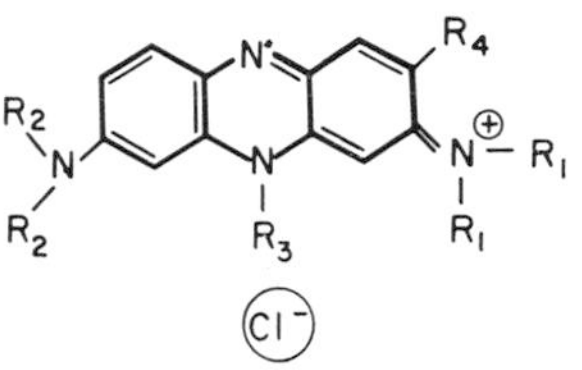

Name	R_1	R_2	R_3	R_4
Safranine	H	H	ϕ	H
Neutral Red	H	CH_3	H	CH_3

(Isoalloxazine Type)

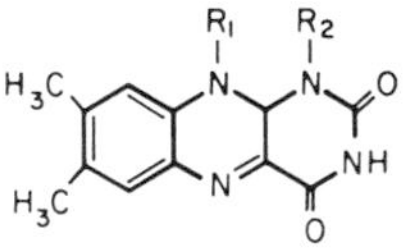

Name	R_1	R_2
Lumichrome	None	H
Lumiflavin	CH_3	None
Riboflavin	D-ribitol	None

(Diphenyl Polyenes)

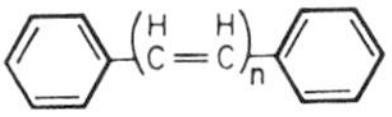

n	Fluorescence Color
1	Violet
2	Blue Violet
3	Blue
4	Yellow Green
5	Yellow

(Triphenyl Methane Type)

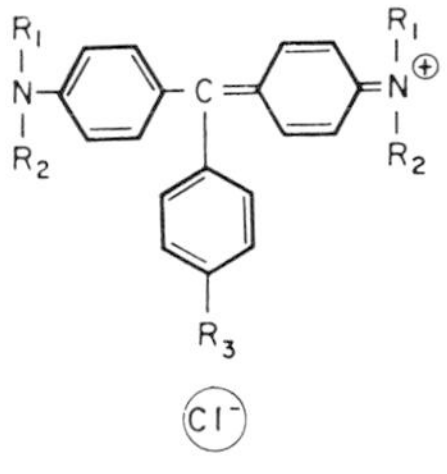

Name	R_1	R_2	R_3
Malachite Green	CH_3	CH_3	H
Crystal Violet	CH_3	CH_3	$N{<}^{CH_3}_{CH_3}$
Ethyl Violet	C_2H_5	C_2H_5	$N{<}^{C_2H_5}_{C_2H_5}$
p-Rosaniline	H	H	NH_2
Brilliant Green	C_2H_5	C_2H_5	H

(Carbocyanines)

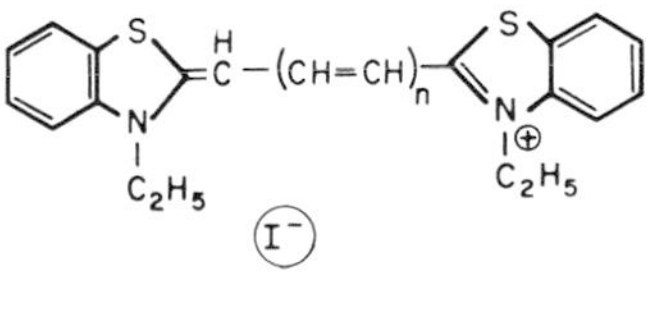

n	λ_{max} (Absorption)
0	4230
1	5570
2	6500
3	7580

example, in the aniline dyes, the electron pair on nitrogen conjugates

with the ring electrons in the planar configuration and therefore contributes to the π electron system, in fact lowering the energy levels of the molecule. In pyridine, however, the lone-pair is in-plane and interacts

very weakly with the π electrons. As we described in the previous section, in pyridine both $\pi \to \pi^*$ and $n \to \pi^*$ transitions are possible.

If we now consider both singlet and triplet states associated with orbital promotions, we can classify several types of molecules. Let us examine physically, however, the meanings of n, π^* singlet and triplet levels. By reference to Fig. 2.8 the physical interpretation of an $n \to \pi^*$ and a $\pi \to \pi^*$ singlet absorption is quite direct. Either one of the lone-pair electrons is excited to a π^* orbital leaving a hole in the n orbital or one of the π electrons is excited to a π^* orbital. However, one must not confuse the lines in Fig. 2.8 with a unique absorption level. These represent only the band peaks and in reality are *broad overlapping regions*. Thus the fact that $n \to \pi^*$ transitions are usually observed experimentally only as long wavelength shoulders on the main $\pi \to \pi^*$ transition indicates a very large overlap in energy. Thus it should be quite probable owing to this energy overlap that, subsequent to a $\pi \to \pi^*$ promotion, one of the lone-pair electrons jumps into the ground-state π system, the atom as a whole losing the energy difference between $\pi \to \pi^*$ and $n \to \pi^*$ transitions. In effect, this would correspond to the transition $^1\Gamma_{\pi,\pi^*} \to {^1\Gamma_{n,\pi^*}}$. There will be a competition between the above transition and the $^1\Gamma_{\pi,\pi^*} \to {^3\Gamma_{\pi,\pi^*}}$ transition. However, in view of the much higher probability of singlet-singlet transitions and the usually smaller energy difference between $^1\Gamma_{\pi,\pi^*}$ and $^1\Gamma_{n,\pi^*}$ states, the most likely transition is to $^1\Gamma_{n,\pi^*}$. The triplet n, π^* level is again lower in energy than the singlet, n, π^* level. It is important to note the difference physically between an n, π^* triplet and a π, π^* triplet. The former corresponds to a "hole" in an n orbital while the latter is a less polarized molecule. The transition from a higher π, π^* triplet state to a lower n, π^* triplet state or vice versa, depending on the molecule, is allowed. By virtue of the small energy difference between the levels it will also be a very intense transition so that effectively all triplet excitation will end up in the lowest triplet state. Figure 2.9 shows typical state levels for the various types of transitions discussed.

In Fig. 2.9a naphthalene is used as an example of an aromatic hydrocarbon with only $\pi \rightarrow \pi^*$ promotions. The straight vertical arrow pointing up indicates absorption with a band centered at 2750 A leading to a singlet π^* state, designated as $^1\Gamma_{\pi,\pi^*}$. Because of the polyatomic nature of the molecule and by virtue of the Franck-Condon principle which we shall

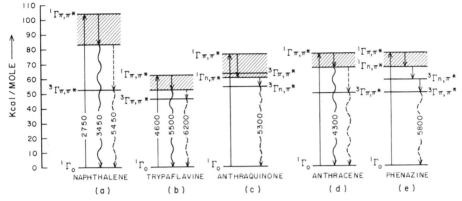

FIG. 2.9.   Transition energy level diagrams for various types of dye molecules. The vertical straight-up arrows indicate the center of the absorption band. The vertical straight-down arrows indicate internal radiationless transitions to lowest excited states. The solid wavy lines indicate observed fluorescence and the dashed wavy lines indicate low-temperature observed phosphorescence. The numbers represent the absorption or emission peaks in Angstroms and the state symbols give the multiplicities and the types of transitions. Energies are given in kcal/mole above the ground state.

discuss in the next section, the fluorescence bands of polyatomic molecules are usually at a lower wavelength than the absorption bands. To illustrate this in the energy-level diagram, the area between these bands has been shaded and the downward-pointing straight arrow indicates a radiationless transition to this lowest singlet excited state. The UV fluorescence emission of naphthalene at 3450 A is drawn as a wavy arrow, indicating light emission. At room temperature in solution the phosphorescence of naphthalene is not observed. According to our picture of the transition probability integral, few of the $^1\Gamma_{\pi,\pi^*}$ states should lead to $^3\Gamma_{\pi,\pi^*}$. First because of the forbiddenness of singlet-triplet transitions and second because of the large energy difference between singlet and triplet excited states. As a general rule, singlet-triplet transitions are about one-one-hundredth as probable as singlet-singlet transitions. The small fraction of $^1\Gamma_{\pi,\pi*} \rightarrow {}^3\Gamma_{\pi,\pi*}$ transitions which do occur are subsequently quenched by solvent interaction because of the long mean lifetime of the $^3\Gamma_{\pi,\pi*}$ state. When the temperature of the solution is lowered sufficiently so that solvent interaction no longer quenches the $^3\Gamma_{\pi,\pi*}$ level or in the solid state, a weak phosphorescence is observed in the green at 5450 A. The $^1\Gamma_{\pi,\pi*} \rightarrow {}^3\Gamma_{\pi,\pi*}$ transition is therefore indicated by a dashed downward-pointing straight arrow. Note that all internal radiationless transitions are drawn as straight arrows. Since solution phosphorescence is observed only at low temperature, the 5450 A $^3\Gamma_{\pi,\pi*} \rightarrow {}^1\Gamma_0$ phosphorescence of the triplet state is drawn as a dashed wavy arrow. In the case of the acridine-type dye trypaflavine or acriflavine (see Table 2.2), the NH_2 groups contribute electrons which conjugate with the π electron system of the acridine molecule, effectively lowering all of the excitation levels in the molecule. This is shown in Fig. 2.9b where the absorption is now in the visible, giving a yellow-colored dye in solution. In addition the singlet-triplet energy separation is much lower (5.75 kcal/mole). Again the internal transition region of the $^1\Gamma_{\pi,\pi*}$ state is shaded, the fluorescence at 5500 A is drawn as a solid wavy line, and the low temperature phosphorescence at 6200 A is a dashed wavy line. Since the singlet-triplet energy separation is small, it might be expected that the overlap between the $^1\Gamma_{\pi,\pi*}$ and the $^3\Gamma_{\pi,\pi*}$ levels would be temperature dependent. Therefore the fluorescence yield would be expected to be temperature dependent, i.e., an increase of $^1\Gamma_{\pi,\pi*} \rightarrow {}^3\Gamma_{\pi,\pi*}$ transitions to triplet levels which are quenched would lower the fluorescence yield. For example, the fluorescence yield of the xanthene-type dye rhodamine B, dissolved in glycerol, decreases from 75% at 16°C to 21% at 90°C (Pringsheim, 1949, p. 319).

In those cases where the introduction of the hetero atom produces lone-pair electrons which do not conjugate with the π orbitals, the levels are not lowered. In Fig. 2.9c–e we consider three cases, anthraquinone

with its carbonyl oxygens, anthracene, and phenazine.

The energy level scheme for anthracene is essentially the same as for naphthalene. Neither anthraquinone nor phenazine is fluorescent at room temperature. In anthraquinone (Fig. 2.9c) the phosphorescence observed in a rigid glass solution at 77°K is in a band centered around 5300 A. This is a $^3\Gamma_{n,\pi*} \rightarrow {}^1\Gamma_0$ transition (Kasha, 1960, p. 266) showing the C-O vibrational structure, characteristic of $n \rightarrow \pi^*$ promotions in carbonyl molecules. The phosphorescence quantum yields of these types of carbonyl or nitro additions to aromatic rings such as benzophenone, acetophenone, anthraquinone, 4-nitrobiphenyl, 2-nitrofluorene, etc., are between 80–100%.

Figure 2.9e is illustrative of a different type of nitrogen heterocyclic where the $^3\Gamma_{\pi,\pi*}$ level is the lowest energy state. In these cases the $^1\Gamma_{n,\pi*}$ state overlaps very strongly with the $^1\Gamma_{\pi,\pi*}$ state. Since the $n \rightarrow \pi^*$ singlet-triplet energy differences are usually smaller than $\pi \rightarrow \pi^*$ singlet-triplet energy differences, it is possible for the $^3\Gamma_{\pi,\pi*}$ level to be lower than the $^3\Gamma_{n,\pi*}$ level. In these cases the radiationless transitions

$$^1\Gamma_{\pi,\pi*} \rightarrow {}^1\Gamma_{n,\pi*} \rightarrow {}^3\Gamma_{n,\pi*} \rightarrow {}^3\Gamma_{\pi,\pi*}$$

provide a very efficient pathway for triplet excitation. Phenazine at room temperature is a nonfluorescent molecule and in a rigid glass solvent at 77°K the phosphorescence quantum yield is approximately 100%.

2. Fates of Excited Molecules

One must be careful in defining just what is meant by the average lifetime of an atom or molecule in any given state of excitation. If the molecule is completely isolated from other matter *and also from radiation* and if it is in its lowest excited state, then the probability of spontaneous emission is formally given by the Einstein coefficient A_{nm}. The mean life τ_0 is therefore $1/A_{nm}$. If the molecule is in a higher excitation level, there will be several possibilities for spontaneous emission and

$$A = A_1 + A_2 + \cdots .$$

In this case the mean life $\tau_0 = 1/\Sigma A$. The complete isolation of the excited state is hardly a practical proposition. In general, there are many mechanisms whereby thermal collisions with other molecules or with the walls of the container or *interaction with radiation* can lead to deactivation or induced emission during the natural lifetime τ_0. In the Einstein derivation of the Planck relation, based on thermodynamics, the assumption is made that there can be an *induced emission* from other excited states, stimulated by the exciting radiation or by extension, by the radiation emitted by an excited atom in reverting to its ground state. This induced emission has been realized in specialized cases in certain crystals, semiconducting surfaces, and gases and is the basis of laser action (see Appendix III).

The intensity of a transition is frequently expressed in terms of oscillator strength or *f-number* of a transition. If the *molar extinction* coefficient ϵ is plotted against the wave number $\bar{\nu}$ in cm^{-1},

$$f = 4.32 \times 10^{-9} \int_{\substack{\text{absorption} \\ \text{band}}} \epsilon \, d\bar{\nu} \tag{2.9}$$

Thus the integrated absorption is a measure of the oscillator strength or the transition probability. The molar extinction coefficient ϵ is defined by

$$\epsilon c l = \log_{10} \frac{I_0}{I} \tag{2.10}$$

where c is the concentration in moles per liter and l is the path length in centimeters. The natural mean lifetime of the excited state (Lewis and Kasha, 1945) can be simplified to

$$\tau_0 = 1.5 \frac{g_u}{g_l} \cdot \frac{1}{f\bar{\nu}^2} \tag{2.11}$$

where g_u/g_l is the ratio of the multiplicity of the upper state to the lower state. In most cases that we shall be concerned with this ratio is 1. If the measurements are made in solution or in a medium of refractive index, Eq. (2.11) becomes

$$\tau_0 = \frac{1.5}{\eta^2} \frac{g_u}{g_l} \cdot \frac{1}{f\bar{\nu}^2} \tag{2.12}$$

To give some feeling for the orders of magnitude of these quantities, for allowed transitions $f \approx 1$. If there are several π electrons involved in the orbital delocalization, f can be greater than 1 but usually is about one-fifth the number of π electrons. The full width at half maximum of an absorption band for a conjugated system is about 3000–4000 cm^{-1}. If $f \approx 1$,

$$\int \epsilon \, d\bar{\nu} = \frac{10^9}{4.32}$$

from Eq. (2.9). Assuming for simplicity a triangular shape, this corresponds to a value of ϵ_{max} of 60,000 to 80,000. From Eq. (2.11) the lifetime for a peak absorption at 4000 Å would therefore be

$$\frac{1.5}{(25000)^2} = 0.24 \times 10^{-8} \text{ sec}$$

Thus molar extinction coefficients of the order of 60,000 and mean lifetimes of 10^{-8}–10^{-9} sec can be used to characterize allowed transitions.

The above discussion holds strictly for the atomic case where absorption and emission take place at the same wavelength with a very narrow line width. However, even for complex molecules the lifetimes calculated by means of Eq. (2.11) have been found to agree with observations within a factor of 2 (Lewis and Kasha, 1945). More recently Strickler and Berg (1962) have given a treatment valid for broad molecular bands and strongly allowed transitions. Their derivation is based on the sharp line transition case. However, since transitions occur over a finite range of vibronic frequencies, it is necessary to integrate over this range. When they do this, the resulting equation for the natural mean lifetime is given by

$$\tau_0 = \frac{1.5}{\eta^2} \frac{g_u}{g_l} \cdot \frac{1}{f'} \langle \bar{\nu}_{fl}^{-3} \rangle_{Av} \tag{2.13}$$

where

$$f' = 4.32 \times 10^{-9} \int \frac{\epsilon(\bar{\nu})}{\bar{\nu}} \, d\bar{\nu} \tag{2.14}$$

$\bar{\nu}_{fl}$ is the wave number associated with the *fluorescence emission spectrum* of the molecule and

$$\langle \bar{\nu}^{-3} \rangle_{Av} = \frac{\int \bar{\nu}^{-3} I(\bar{\nu}) \, d\bar{\nu}}{\int I(\bar{\nu}) \, d\bar{\nu}} \tag{2.15}$$

taken over the *fluorescence emission spectrum*. If absorption and fluorescence occur at the same wavelength as in the case of atomic transitions, the factor $1/\bar{\nu}$ can be removed from under the integral sign in Eq. (2.14), and Eq. (2.13) reduces to Eq. (2.12).

In principle the probability of spontaneous emission of radiation from an isolated atom in a metastable state is zero and the lifetime is theoretically infinite. The fact that these transitions are observed is due to the perturbations of the energy levels by the electric or magnetic fields of the atom, collisions with other atoms or electrons, or the absorption of radiation. As a result of collisions the atom may either lose or gain sufficient energy to bring it to a labile state or to deactivate it completely in a radiationless transition. By absorption of radiation the metastable atom can be transferred to a higher state from which there are other possibilities of emission. We have already described in Section 1.2.1 the appearance of the 4047 A absorption line in Hg.

The lifetimes of metastable states are very strongly influenced by small trace impurities. In the gas phase, hydrogen and oxygen have been shown to exert a remarkable accelerating effect upon the decay of metastable states. Perhaps a similar mechanism acts for excited molecules in solution or on a surface.

So far as we have been able to obtain experimental evidence, there are only two biologically useful reactions that are produced by light. The first, as typified by photosynthesis, is an actual conversion of the light energy into chemical reducing and oxidizing power. The total gain in chemical free energy of chemical compounds formed is supplied by the absorbed quantum. In the second type of reaction, the light acts as a trigger for releasing a large amount of stored chemical energy. This latter process occurs in vision and presumably in all other biological photoreception. At the present time the photodynamic action of light, UV-induced polymerization, lysogenesis, killing, and photorecovery are apparently incidental secondary effects due to the presence of light.

2.1. CHEMICAL REACTIVITY OF THE EXCITED STATE

Ordinarily one thinks of the photochemical action of light as a means of supplying free energy of activation to a reaction, as in a *cis-trans* isomerization or as a means of storing free energy in slow back-reacting processes such as photosynthesis. Both of these concepts are valid. However, light can assume an even more specialized role. Aside from the fact that it is consumed by virtue of being absorbed, it is a "catalyst." Although we are not dealing with systems in thermodynamic equilibrium, it may be instructive, since there are localized concentrations of energy, to assume intra molecular "temperatures" of the orders of thousands of degrees Kelvin. In this case "temperature" is defined as the equivalent blackbody cavity temperature at which there would be an energy density at the particular wavelength corresponding to that being absorbed by the molecule (Einstein, 1912a, b). Therefore reactions should be possible photochemically that are completely inaccessible thermally. For example, Duysens (1959) has calculated that the absorption by chlorophyll of 1000 ergs/cm^2-sec between 6600 and 6800 A corresponds to an equivalent blackbody cavity at 1100°K.

The chemical behavior of molecules depends mostly on their valence or weakly bound electrons. Since these valence electrons form the lowest energy levels in the molecule, these are specifically the electrons excited by light. Raising an electron to an excited state by definition places it in a more weakly bound orbit extending over a larger volume than the original orbit. As an example consider Fig. 2.3 for the $1s \rightarrow 2p$ transition of the hydrogen atom. This new orbital can react more easily with an electrophilic molecule, resulting in a charge-transfer complex or even the actual abstraction of the electron. In effect, light absorption has transformed the molecule into a reductant, and what is more important, into a specific reductant, since specific electron orbitals and therefore specific reaction paths are involved. In addition, the electron in the excited (*)

orbital is no longer paired with the remaining electron in the ground-state orbital. The (*) orbital can then overlap with any other unpaired orbital of a different molecule, forming an electron-pair bond leading to a stable product. In general, one would expect that the pK of the excited state would be quite different from the ground state. An excited acid might much more easily give up a proton owing to the decreased binding of an electron. Excited bases for the same reason would be less able to pick up protons.

The thermodynamic arguments used by Einstein in his proof of the Law of Photochemical Equivalence postulate equivalent blackbody enclosures at such elevated temperatures that for all practical purposes one can assume that the primary photochemical event should be independent of the temperature of the gross solution. Thus photochemistry occurring at room temperature should also occur at liquid N_2 temperature. This, in fact, is the experimental method used to separate the *primary* photochemical event or "light" reaction from the *secondary* chemical or enzymatic processes or "dark" reactions which are strongly temperature dependent. The fact that there should be such two-phase reactions involved in photochemistry was first postulated by Stark (1908).

A further interesting insight into the greater chemical reactivity of light-excited molecules comes from a consideration of the Franck-Condon Principle (Franck, 1926; Condon, 1926, 1928). As an example, consider Fig. 2.10 where are drawn the equilibrium potential energy curves of two states of a molecule, the ground state and the first excited state. The Franck-Condon Principle states that there is little interaction between the loosely bound electrons in a molecule and the nuclei. The latter because of their relatively large mass will not change their internuclear distance or velocity during the short time (10^{-15} sec) during which the electronic transition occurs. Therefore electronic transitions can be represented as vertical lines on the diagram. The upper curve represents the energy of the first excited state as a function of nuclear separation. Since the total energy is greater in the excited state, the equilibrium position (minimum of energy) is displaced to the right. In the ground state in the lowest vibrational level the electron density probability distribution is greatest at a separation r_g. The value of the integral

$$\int_{-\infty}^{\infty} \psi^* e x \psi \, dx \qquad (2.16)$$

(transition probability) will depend on the spatial distribution of the electrons. Therefore absorption transitions from ground vibration level 0 to excited level 2* will be strong and the absorption band peak should be centered about $0 \rightarrow 2*$. During the lifetime of the excited state (10^{-8} sec) the much shorter nuclear vibration periods (10^{-12} sec) can equilibrate this excess vibrational energy with the environment and the nuclei will rapidly assume their equilibrium separation r_e. The greatest electron density in the excited state will now be at r_e and the most intense emission transition will therefore be between the 0* and the 4 vibrational level

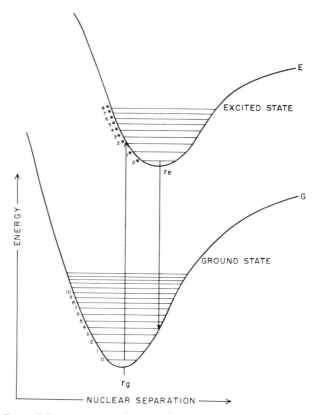

Fig. 2.10. Potential energy curves for two electronic states of a molecule, illustrating the Franck-Condon principle. The absorption from r_g and the emission from r_e represent the centers of the absorption and emission bands, respectively. There is usually an overlap between the long wavelength region of the absorption band and the short wavelength region of the emission band.

of the excited and ground state, respectively. The emission band peak will be centered about 0* → 4.

The net result of this absorption and re-emission is that a large fraction of the absorbing molecules will end up in a highly excited vibrational level of the ground state with therefore a greater chemical reactivity. Since the thermal distribution of energies follows a Maxwell-Boltzmann law, a very high temperature of the ground state would have been required in order to populate this same level directly from the ground state, in turn populating many other levels leading to many other possible reactions. Thus light absorption and emission can, at ordinary temperatures, result in the formation of specific high vibrational levels of ground states leading to reaction rates for specific chemical reactions that could ordinarily occur only at very high temperatures.

It may be, therefore, that many chemical reactions which are accelerated by light are ground-state reactions where the free energy available is just the difference in energy between a particular vibrational level of the ground state that is occupied transitorily subsequent to the emission of fluorescence radiation or a radiationless quenching and the lowest ground vibrational level of the molecule. Chemical reactions accelerated by temperature increases would be most likely to fall into this category. Presumably, therefore, a ground-state chemical reaction which is accelerated by the absorption of light to its first excited state could also proceed by the absorption of infrared quanta of wave number corresponding to the particular bond involved. Hall and Pimentel (1963) have reported what appears to be the first example of this infrared photochemistry in the isomerization of nitrous acid in a nitrogen matrix at 20°K. A narrow range of wave numbers 3650–3200 cm^{-1} isolated by filters from a Nernst glower at 1950°K was found effective for the *cis* to *trans* isomerization of HONO. The height of the potential barrier to isomerization in the nitrogen matrix was estimated to be 9.7 kcal/mole. Both *cis-trans* and *trans-cis* isomerization take place. UV radiation, absorbed by HONO, will also produce the isomerization. The authors also estimate that for infrared energy in the 3650–3200 cm^{-1} region the quantum yield is unity and therefore intramolecular energy transfer is an extremely efficient process.

The reaction of excited states of dye molecules with molecular oxygen, and, in particular, the photodynamic action of light in the presence of oxygen is usually a characteristic of highly fluorescent compounds. It is normally assumed that the reaction proceeds via excited triplet states which can interact with the triplet ground state of the oxygen molecule. The oxygen may itself perturb the excited singlet levels to increase singlet-triplet mixing. Nonfluorescent molecules whose n, π^*, or π, π^* triplet levels are deactivated by crossover to the ground-state singlet may exist for sufficient time to react with other substrates.

2.2. FLUORESCENCE, PHOSPHORESCENCE, AND QUENCHING

For the biological utilization of light energy it would appear that the luminescence of excited molecules is a wasteful process. If most of the absorbed quanta are re-emitted, there is only the Franck-Condon energy difference between absorption and emission which can be available for useful work. It is therefore interesting that many of the biological pigment molecules important in utilizing the energy of light can be classified as fluorescent molecules. This does not mean that light-utilizing reactions are inefficient processes, for in the particular biological system little or no fluorescence may be observed depending on the matrix or perturbing environment. The nature of the reaction, however, whether it be energy transfer or photoreduction or oxidation, apparently requires an electronic configuration conducive to fluorescence in the absence of quenching. It is

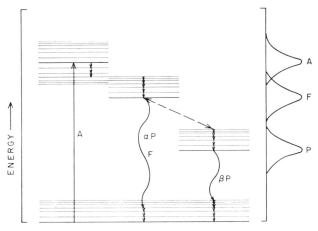

FIG. 2.11   Jablonski-type transition energy diagram showing the relative energy levels of the singlet absorption band, the first excited singlet level from which fluorescence occurs and the triplet level from which phosphorescence occurs. Following convention, the "delayed fluorescence" which has an emission spectrum corresponding to the "immediate fluorescence" and is temperature dependent is called α phosphorescence. The "long-wavelength emission" or triplet-singlet transition is called β phosphorescence.

specifically this latter phenomenon, the quenching of excited states of potentially fluorescent molecules, that permits the effective utilization of absorbed light energy. We shall now proceed to define the characteristic types of luminescence associated with solid and liquid solutions of organic molecules, preparatory to a discussion of the effects of solvents, concentration, and the degree and type of aggregation. From the Franck-Condon diagram in Fig. 2.10 and the previous discussion of triplet states associated with singlet excited states we can draw a modified Jablonski-type diagram (Fig. 2.11). In Fig. 2.11 the straight-up vertical arrow A indicates the center of the absorption band. The small vertical arrows indicate interval vibrational transitions with the fluorescence emission (wavy line

F) originating from the lowest excited singlet state. As can be seen from the diagram there is an overlap between the absorption and the fluorescence spectrum, giving rise to the so-called mirror image symmetry, characteristic of fluorescent molecules. The transition $^1\Gamma_{\pi,\pi*} \leftrightarrow {}^3\Gamma_{\pi,\pi*}$ is shown as the dashed arrow. In the long-lived (small transition probability for $^3\Gamma_{\pi,\pi*} \to {}^1\Gamma_0$) metastable triplet state, one of several events can occur: There may be sufficient thermal energy available so that a $^3\Gamma_{\pi,\pi*}$ state can revert to a $^1\Gamma_{\pi,\pi*}$ state, in which case the emission (αP) observed will be indistinguishable in wavelength from fluorescence. Owing to the long lifetime and the unpaired spins of the electrons, the triplet state will have a greater probability of solvent or quencher interaction and thus the excitation energy will be "lost." Lastly, the triplet state can revert to the $^1\Gamma_0$ state by the emission (βP) of what we normally call phosphorescence. It is important to differentiate between the two types of phosphorescence originating from the metastable triplet state. The one which Jablonski called α phosphorescence is identical in spectral characteristics with ordinary fluorescence and is the origin of the "thermoluminescence" measured by workers on irradiation and subsequent heating of dried proteins, amino acids, DNA, chloroplasts, organic crystals, and metal halide crystals containing a variety of impurities. The second or β phosphorescence is relatively temperature independent the same as fluorescence, except that a lowering of the temperature or adsorption onto surfaces in many cases serves to reduce quenching interactions and therefore will enhance the observed phosphorescence. It must be remembered, however, that the enhancement is due to a removal of competing processes rather than a change in the transition probability either from $^1\Gamma_{\pi,\pi*} \to {}^3\Gamma_{\pi,\pi*}$ or from $^3\Gamma_{\pi,\pi*} \to {}^1\Gamma_0$. The transition to ground state from both the singlet and triplet excited levels will result in bands with vibrational structure as described in the Franck-Condon diagram. These are shown schematically on the right side of Fig. 2.11. If there is very little overlap in energy between the $^1\Gamma_{\pi,\pi*}$ and the $^3\Gamma_{\pi,\pi*}$ states, such as in anthracene or naphthalene, we should expect very little phosphorescence even at low temperature. However, as in the case of phenazine where the overlap between singlet π, π^* and triplet π, π^* and n, π^* states is marked, all of the excitation is channeled into triplet states. At low temperature in the absence of quenching interactions only β phosphorescence is observed, with a phosphorescence quantum yield approaching unity.

According to our picture of the origin of fluorescence, the fluorescence quantum yield, the number of quanta emitted per quantum absorbed. should be constant for any particular molecule. This has been demonstrated experimentally for many substances. However, in some cases on the long wavelength side of the absorption band the fluorescence yield has been observed to drop markedly (Valentinei and Roessiger, 1924, 1925, 1926). In fact, for fluorescein in aqueous solution the quantum yield for excitation by 5461 A is ten times smaller than for excitation by 4358 A. This is in marked contrast to the constancy observed on the short wavelength side of the absorption band down to 2500 A (Vavilov, 1922a,b, 1925, 1927). It is not possible in terms of our simple picture to provide a mechanism for this effect. It may be that for the complex polyatomic pigment molecules the high vibrational levels of the ground state are close in

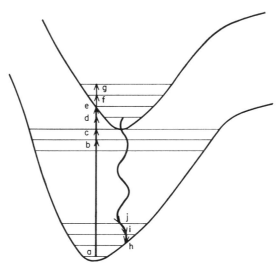

Fig. 2.12.   Possible mechanism for ground-state energy levels with long wavelength excited-state absorption levels to explain the decrease in observed fluorescence quantum yield for long wavelength absorption.

energy to the lowest excited levels of electronic states. For example, consider Fig. 2.12 which is a Franck-Condon diagram showing the difference between the absorption and fluorescence emission, say for the fluorescein molecule. The difference in this case is that there may be accessible vibrational levels in the ground state near low-lying levels in the "first excited state." These levels, while they may contribute to the observed long wavelength absorption, do not necessarily contribute to the fluorescence since the energy is insufficient to reach the lowest excited singlet state.

In Fig. 2.12 the transition ae would be most probable and the transitions ab, ac, ad, af, ag, . . . would, in principle, give rise to other vibrational structure. However, the long wavelength transitions ab and ac are not sufficiently energetic to produce the lowest excited state. The lifetimes of these high vibrational states are very much less than 10^{-9} sec. Fluorescence results from the series of transitions dh, di, dj, ... The net result is that fluorescence can result in the production of very highly excited ground-state molecules as described in the previous section on chemical reactivity and the long wavelength portion of the absorption curve with small extinction may produce extremely reactive ground-state molecules with essentially negligible fluorescence yield.

It would be interesting to examine the temperature dependence of the fluorescence yield as a function of wavelength and also, in the case of certain molecules, to look at the photooxidation yield as a function of wavelength. Either of these might corroborate the existence of these postulated high vibrational levels.

It should also be pointed out that the fine structure observed in absorption is the expression of the *excited-state* vibrational energy distribution, while fine structure in fluorescence should agree with infrared data on the *ground-state* vibrational levels.

In another case it has been shown by Khvostikov (1934, 1936) that in aqueous solutions of magnesium platinocynanide, while the absorption consists of two bands, a strong one at 2500 A and a weak band at 2780 A, only the long wavelength band gives rise to fluorescence. Excitation within the 2500 A band is completely quenched.

A distinction should be made between the quenching of excited states by molecules permanently in contact with the fluorescent molecule such as the solvent and quenching by a third molecule present in solution in low concentration. The quenching efficiency per collision of the solvent need not be high, since during the lifetime of the excited state ($>10^{-9}$ sec) there will ordinarily be the order of 10^3 collisions. However, in the case of the third molecular species, the quenching must be extremely efficient owing to the rarity of collisional encounters. The longer the natural lifetime of the excited state the greater is the probability for quenching. Usually the quenching has a positive temperature dependence, indicating a collisional or diffusion process. It is conceivable, however, that in some types of quenching processes an intermediate complex is involved, in which case it should be possible to observe a negative temperature dependence. More usually this effect is observed as a compensation of the increasing probability of internal conversion in the nonassociated molecules. For example, the fluorescence yield decreases less rapidly with rising temperature in aromatic solvents than in aliphatic solvents. This is ascribed to the tendency of aromatic solvents to form solvent-solute complexes which act to quench the fluorescence. The increase in temperature inhibits the formation of these complexes and thus lowers the contribution of solvent-solute complex quenching. The quenching owing to internal conversion always has a positive temperature coefficient. The potential energy configuration for the approach of the two molecules is not entirely repulsive, but there can exist a shallow trough with a depth of possibly 1–3 kcal/mole owing to dispersion forces or dipole interactions, depending on whether the solvent is nonpolar or polar.

2.2.1 Solvent Quenching

The quinoline molecule in hydrocarbon solvents is nonfluorescent and

at low temperatures shows the $^3\Gamma_{\pi,\pi*} \to {}^1\Gamma_0$ phosphorescence similar to that of naphthalene. In hydroxylic solvents the naphthalene-like $^1\Gamma_{\pi,\pi*} \to {}^1\Gamma_0$ fluorescence appears and the phosphorescence yield of $^3\Gamma_{\pi,\pi*} \to {}^1\Gamma_0$

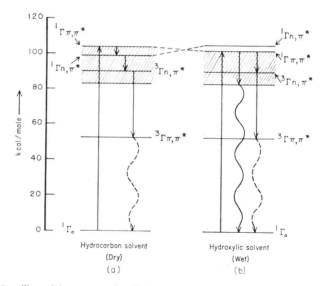

FIG. 2.13. Transition energy level diagrams illustrating the inversion of $\pi, \pi*$ and $n, \pi*$ levels in dry and wet solvents, giving rise to the observed fluorescence of "wet" chlorophyll.

is reduced. This type of shift has been interpreted by Becker and Kasha (1955) and by Platt (1956) as resulting from an interchange of $n, \pi*$ and $\pi, \pi*$ promotion levels, as the result of the blue-shift and red-shift, respectively, of these types upon solvation in polar solvents. Thus in Fig. 2.13 the inversion of the $n, \pi*$ and $\pi, \pi*$ levels is shown schematically.

In Fig. 2.13a the $^1\Gamma_{n,\pi*}$ level lies so close to the $^1\Gamma_{\pi,\pi*}$ level as to be indistinguishable even as a shoulder and yet the pathway

$$^1\Gamma_{\pi,\pi*} \rightarrow {}^1\Gamma_{n,\pi*} \rightarrow {}^3\Gamma_{n,\pi*} \rightarrow {}^3\Gamma_{\pi,\pi*}$$

is so much more probable that the molecule does not emit fluorescence from the lowest $^1\Gamma_{\pi,\pi*}$ level at the bottom of the shaded region. The $^1\Gamma_{n,\pi*} \rightarrow {}^3\Gamma_{n,\pi*}$ splitting is so small that this transition becomes very probable. In a hydroxylic solvent the large blue shift of $n,\pi*$ and the smaller red shift of $\pi,\pi*$ are sufficient to invert the levels. Then the pathway

$$^1\Gamma_{\pi,\pi*} \rightarrow {}^1\Gamma_0$$

will compete with the

$$^1\Gamma_{\pi,\pi*} \rightarrow {}^3\Gamma_{n,\pi*} \rightarrow {}^3\Gamma_{\pi,\pi*}$$

phosphorescence pathway, giving rise to fluorescence and a reduced phosphorescence yield.

Livingston and co-workers (1949) showed, in their absorption curves for nonfluorescent chlorophyll-b in anhydrous benzene, a long wavelength shoulder at 6700 A. In the presence of water or alcohol this shoulder disappears and the solution becomes fluorescent. These observations can be explained on the basis of the $n,\pi*$–$\pi,\pi*$ inversion of Fig. 2.13 and are consistent with the proposal of Franck et al. (1962) that there are two types of chlorophyll involved in photosynthesis; a lipid phase "dry" chlorophyll which is nonfluorescent and an aqueous phase "wet" chlorophyll, responsible for the observed fluorescence.

The sodium salt of phenolphthalein which is nonfluorescent in solution

Na$^+$

is very similar to the sodium salt of fluorescein (see Table 2.2) except for the oxygen bridge of the latter. It is the closing of this middle ring by oxygen that is responsible for the fluorescence. Therefore it should be possible under certain conditions to stabilize the phenolphthalein structure in the fluorescein-type planar configuration, in which case fluorescence is possible. This is actually observed for phenolphthalein in solid solutions of gelatin or sucrose. This same phenomena is observed for solid or gel solutions of malachite green and methyl (crystal) violet (Table 2.2) which in ordinary liquid solution are very weakly fluorescent or nonfluorescent.

It has also been known for some time that even in the case of highly fluorescent substances such as fluorescein, at high concentrations the fluorescence becomes negligible. This is ascribed to the occurrence of deactivating collisions between excited molecules and ground state molecules but depends also upon the solvent. For example, at low concentrations of fluorescein in water and in glycerol the fluorescence yield is the same. However, at a concentration of 9 g/liter (0.027 M) the aqueous fluorescence is only one-tenth the glycerol fluorescence. This is presumably a viscosity effect governing the diffusion of solute molecules.

In those cases where the spontaneous transition probability A determines the mean lifetime of the excited state

$$\tau_0 = \frac{1}{A} \tag{2.17}$$

If we now irradiate a solution so that N_0 quanta per second are absorbed, the emission, N, will, in the absence of perturbing reactions, be equal to N_0 and the quantum yield, Q, defined by

$$Q = N/N_0 \tag{2.18}$$

will be equal to 1. Competing nonradiating processes for the excitation energy will therefore change the rate at which de-excitation occurs and decrease the observed fluorescence. The rate of de-excitation will be

$$N_0(A + \alpha_1 + \alpha_2 + \cdot \cdot \cdot) \tag{2.19}$$

where α_1, α_2, . . . are the respective probabilities for the competing events. Of this only the fraction $A/A + \alpha_1 + \alpha_2 + \cdot \cdot \cdot$ will give rise to fluorescence, so that the *observed* fluorescence will now be

$$N = \frac{N_0 A}{A + \alpha_1 + \alpha_2 + \cdot \cdot \cdot} = \frac{N_0}{\tau_0}\left(\frac{1}{A + \alpha_1 + \alpha_2 + \cdot \cdot \cdot}\right) \tag{2.20}$$

If we define the observed lifetime by

$$\tau = \frac{1}{A + \alpha_1 + \alpha_2 + \cdot \cdot \cdot} \tag{2.21}$$

we see from Eq. (2.18) above that the quantum yield is therefore

$$Q = \tau/\tau_0 \tag{2.22}$$

If we lump all nonfluorescent (quenching) pathways, $\alpha_1 + \alpha_2 + \cdots = q$

$$N = \frac{N_0}{1 + \tau_0 q} \tag{2.23}$$

The "quenching constant," q, is given by

$$q = \frac{1 - Q}{\tau_0 Q} = \frac{N_0 - N}{\tau_0 N} \tag{2.24}$$

The "quenching constant," q, is physically the number of effective quenching collisions per second. Since this number should be proportional to the concentration we rewrite Eq. (2.23) in the form

$$\frac{1}{N} = \frac{1 + c/c^*}{N_0} \tag{2.25}$$

where c is the concentration of quencher in moles per liter, and c^* is the concentration at which $Q = \frac{1}{2}Q_0$.

A plot of $1/N$ versus c should yield a straight line with a slope $1/c^*N_0$ and a zero intercept $1/N_0$. The product of the zero intercept with the reciprocal slope gives c^*. On the average the collision rate of a molecule in a liquid is 10^{12}/sec. If we assume that every collision of an excited molecule with a quenching molecule leads to a radiationless transition to the ground state and that therefore the probability of a quenching molecule being involved in a collision is just the ratio of the molar concentration of quencher to that of the solvent, we find that in aqueous solution (55 M) the natural lifetime τ_0 should be given by

$$\tau_0 = \frac{55}{c^*} \times 10^{-12} \text{ sec} \tag{2.26}$$

Thus for uranine (sodium fluorescein) for which τ_0 has been measured to be 5×10^{-9} sec (Gaviola, 1927), the minimum value of c^* (the most efficient quencher) should be 0.01 M. From data extrapolated from Pringsheim (1949, p. 351) the concentration of fluorescein at which the observed fluorescence is reduced to one-half is 0.0065 M. It is to be noted that for all dyes in liquid solution for which $Q \approx 1$, τ_0 has a value about 5×10^{-9} sec. This is the same as the undisturbed lifetime of the excited molecule in the gaseous state. For the quenching of quinine sulfate in water by phenol, hydroquinone, and resorcinol. The values of c^* are reported as

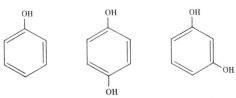

0.013, 0.010, and 0.011, respectively, in excellent agreement with these very qualitative calculations. Conversely it should be possible from the determination of c^* for these compounds to estimate the value of τ_0.

The fluorescence quantum yields obtained for different compounds in the same solvent and for the same compound in different solvents vary over a wide range. Values available in the literature are given in Table 2.3.

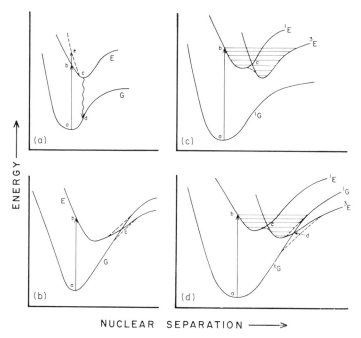

Fɪɢ. 2.14. Franck-Condon-type energy surfaces illustrating several of the mechanisms of overlap and crossing-over of excited and ground-state levels giving rise to delayed fluorescence and internal quenching.

In general, from Eq. (2.22) the observed lifetime τ is related to the natural lifetime τ_0 by

$$Q = \tau/\tau_0 \tag{2.27}$$

Even in the complete absence of "third molecule" quenching there is the possibility for internal conversion processes in the excited molecule to dissipate the energy of the excited state.

In Fig. 2.14 are shown several of the mechanisms that have been observed in terms of potential energy surfaces. Figure 2.14a is a possible explanation of an observed solvent shift of absorption maximum from

TABLE 2.3
FLUORESCENCE QUANTUM YIELDS OF VARIOUS DYES IN SOLUTION

Substance	Solvent	Fluorescence quantum yield	Reference
Acridone	EtOH	0.83	Melhuish (1961)
Acriflavine (trypaflavine, euflavine)	H_2O	0.54	Weber and Teale (1957)
9-Aminoacridine	H_2O	0.98	Weber and Teale (1957)
	EtOH	0.99	Weber and Teale (1957)
Anthracene	Benzene	0.25	Pringsheim (1949)
	Benzene	0.26	Melhuish (1955)
	Benzene	0.29	Weber and Teale (1957)
	EtOH	0.30	Weber and Teale (1957)
	Hexane	0.44	Melhuish (1955)
	Hexane	0.31	Weber and Teale (1957)
	Paraffin	0.36	Melhuish (1955)
	Paraffin	0.23	Pringsheim (1949)
	Acetone	0.21	Pringsheim (1949)
	$CHCl_3$	0.17	Pringsheim (1949)
	Crystalline	1.00	Pringsheim (1949)
Anthranilic acid	Benzene	0.59	Melhuish (1955)
Chlorophyll a	Benzene	0.325[a]	Weber and Teale (1957)
	Benzene	0.29[b]	Weber and Teale (1957)
	Benzene	0.32[c]	Weber and Teale (1957)
	Ether	0.32[c]	Weber and Teale (1957)
	Acetone	0.30[c]	Weber and Teale (1957)
	EtOH	0.225[c]	Weber and Teale (1957)
	MeOH	0.23[c]	Weber and Teale (1957)
	Cyclohexanol	0.30[c]	Weber and Teale (1957)
	Dioxane	0.32[c]	Weber and Teale (1957)
Chlorophyll b	Benzene	0.11[a]	Weber and Teale (1957)
	Benzene	0.122[c]	Weber and Teale (1957)
	Ether	0.117[c]	Weber and Teale (1957)
	Acetone	0.09[c]	Weber and Teale (1957)
	EtOH	0.095[c]	Weber and Teale (1957)
	MeOH	0.10	Weber and Teale (1957)
9, 10-Dimethyl-anthracene	Benzene	0.81	Melhuish (1955)
	EtOH	0.84	Melhuish (1955)
9, 10-Di-CH_2-CH_2-COOH-anthracene	0.1 M NaOH	1.00	Melhuish (1955)
Erythrosin	H_2O	0.02	Pringsheim (1949)
	H_2O-Acetone	0.18	Pringsheim (1949)
	Acetone	0.50	Pringsheim (1949)

TABLE 2.3 (*Continued*)

Substance	Solvent	Fluorescence quantum yield	Reference
Eosin[d]	H_2O	0.15	Pringsheim (1949)
	EtOH	0.40	Pringsheim (1949)
	Glycerol	0.60	Pringsheim (1949)
	0.1 M NaOH	0.19	Weber and Teale (1957)
Fluorene	EtOH	0.357	Bowen and Sawtell (1937)
Fluorescein sodium (uranine)	H_2O	0.71	Pringsheim (1949)
	H_2O, pH 7	0.65	Weber and Teale (1957)
	EtOH	0.71	Pringsheim (1949)
	Glycerol	0.71	Pringsheim (1949)
	0.1 M NaOH	0.92	Weber and Teale (1957)
N-Methylacridinium chloride	H_2O	1.00	Weber and Teale (1957)
Naphthacene	Xylene	0.06	Pringsheim (1949)
	In crystalline anthracene	1.00	Pringsheim (1949)
Naphthalene	EtOH	0.12	Weber and Teale (1957)
	Hexane	0.10	Weber and Teale (1957)
Perylene	Benzene	0.98	Melhuish (1955)
Potassium cyano-platinite	H_2O	0.045	Pringsheim (1949)
	Crystalline	1.00	Pringsheim (1949)
Rhodamine B	H_2O	0.25	Pringsheim (1949)
	EtOH	0.42	Pringsheim (1949)
	EtOH	0.97	Weber and Teale (1957)
	Ethylene glycol	0.89	Weber and Teale (1957)
	Glycerol	0.70[e]	Pringsheim (1949)
Rose bengale	H_2O	0.01	Pringsheim (1949)
	Acetone	0.40	Pringsheim (1949)
Rubrene	Benzene	0.70	Pringsheim (1949)
	Benzene	1.00	Bowen and Williams (1939)
	Acetone	1.00	Pringsheim (1949)
	Crystalline	0.10	Pringsheim (1949)
Sodium-1-naphthyl-amine-4-sulfonate	H_2O	0.80[f]	Bowen and Seaman (1962)
Sodium salicylate	H_2O	0.28	Weber and Teale (1957)
Triphenylmethane	EtOH	0.24	Bowen and Sawtell (1937)
Uranyl sulfate	H_2O	0.01	Pringsheim (1949)
	H_2SO_4	0.26	Pringsheim (1949)
	Crystalline	1.00	Pringsheim (1949)

[a] Exciting wavelength 3663 A.
[b] Exciting wavelength 4358 A.
[c] Exciting wavelength 6438 A.
[d] Most likely eosin (YS).
[e] Strongly temperature dependent; $Q = 0.70$ at 20°C.
[f] Temperature dependent; $Q = 0.80$ at 18°–20°C.

ab to ac, while the fluorescence cd, remains unchanged. Figure 2.14c is an example of the overlap of singlet-excited and triplet-energy surfaces, giving the resultant dashed curve. In this case both delayed fluorescence (α phosphorescence) and phosphorescence (β phosphorescence) would be observed and it should be possible to measure the triplet-excited-singlet activation energy by measuring the delayed fluorescence as a function of temperature. Figure 2.14b,d illustrates the possible internal quenching interactions that can occur owing to a crossing-over of excited and ground state energy surfaces. The nature of the solvent and the temperature often strongly influence the position of any internal conversion crossover point and thus it is ambiguous to separate internal conversion quenching from solvent quenching. However, it is also possible that there exist internal conversion crossover points independent of the solvent interaction in which case the energy dissipation can be by stepwise emission of infrared quanta corresponding to the vibrational structure of the ground state. If a dye cannot be excited to emit fluorescence even though it absorbs light strongly, we say that internal conversion is high. Many dyes which are nonfluorescent in liquid solution become fluorescent and even phosphorescent when frozen, adsorbed on gels and solid surfaces, or when dissolved in a liquid of high viscosity. Let us make a distinction between this effect and the more general observation of the increase in fluorescence yield as the temperature is lowered. In the latter case we are reducing the collisional frequency of the solvent-solute and thus increasing the emission probability. In the former cases we are effectively eliminating rotational degrees of freedom by making the structure rigid as in the case of phenolphthalein described above.

Bowen and Seaman (1962) have shown for several dye molecules a very strong solvent dependence on fluorescence quantum yield and at the same time essentially no temperature effect. This is shown in Fig. 2.15 where the fluorescence yield Q of 1-naphthylamine-5-sulfonate is plotted as a function of temperature for various solvents. They interpret this in terms of two different energy degradation processes; one dependent on temperature and related to the diffusional or collisional quenching of the singlet state and the other independent of temperature, possibly due to a charge-transfer reaction facilitating singlet-triplet conversion and subsequent solvent quenching. It is also possible that the "quenching" can be due to a reversible photochemical reaction such as the transfer of a hydrogen atom between the excited solute molecule and a ground-state solvent molecule. This type of exchange reaction would also be extremely sensitive to the nature of the solvent.

The fluorescence yield of rhodamine B is very sensitive to temperature. We can interpret this in the following way. If in Fig. 2.14d the level of the crossover point c is close to the lowest vibrational level of the excited state of the molecule, a slight increase in temperature will populate levels at the crossover point and thus lead to increased quenching. Therefore from the reasoning leading up to Eq. (2.24), the "quenching constant" will depend upon the vibrational energy distribution due to the temperature. Assuming a Maxwell-Boltzmann energy distribution, the number of

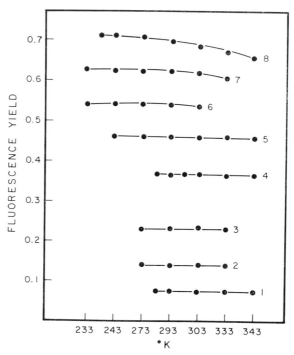

FIG. 2.15.　Fluorescence yield of 1-naphthylamine-5-sulfonate as a function of temperature in various solvents (Bowen and Seaman, 1962).

Key: (1) water; (2) 30% methanol in water; (3) 50% methanol in water; (4) formamide; (5) ethylene glycol; (6) methanol; (7) isopropanol; (8) dimethylformamide.

effective quenching collisions will depend on $e^{-E_q/kT}$, where the height of point c is given by E_q, the "activation" energy for fluorescence quenching. If we plot the logarithm of the quenching constant, q, defined by Eq. (2.24) as a function of $1/T$ we should obtain a straight line whose slope is $-E_q/k$, where k is Boltzmann's constant. An example of this temperature dependence is shown in Fig. 2.16. The data are taken from Pringsheim (1949, p. 319) for rhodamine B dissolved in glycerol. If the value of $\Delta(1/T)$ is measured for a decrease of a factor of 2 in the ordinate, the activation energy for quenching, E_q in calories per mole is given by

$$E_q = \frac{1.39}{(1/T)_{\frac{1}{2}}} \text{ cal/mole} \tag{2.28}$$

In this case all of the data fit a straight line very well and $E_q = 6.76$ kcal/mole. This activation energy for quenching is specifically a result of solvent-solute interaction and can be very different in different solvents.

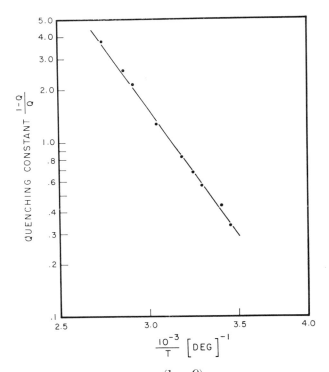

FIG. 2.16. Quenching constant $q = \frac{(1 - Q)}{Q}$ plotted as a function of $1/T$ for rhodamine B on semilog paper, showing good agreement with the behavior predicted by Eq. (2.24). The data are taken from Pringsheim (1949, p. 319).

In fact rubrene dissolved in aliphatic solvents has a value of E_q of approximately 7 kcal/mole, while in hexane the fluorescence quantum yield is constant from $-60°$ to $+60°$C (Pringsheim, 1949, p. 320).

2.2.2. Solute Quenching

In the case of solvent quenching the excited solute molecules are in permanent contact with solvent molecules. However in the case of solute quenching, that is, either self-quenching (quenching of solute excited molecules by normal ground state molecules of the same species) or quenching by "third molecules" or impurities (quenching of solute excited molecules by ground state molecules of a different species), the concentration of quenching molecules is so low that in most cases a specific interaction must be assumed to take place with very high efficiency during each collision. It is perhaps misleading to discuss solvent and solute quenching as though there were two independent and additive mechanisms involved. In fact, the efficiencies of many "third molecule" quenchers show a strong solvent dependence.

It is observed that the fluorescence yields of aqueous solutions of fluorescein, most other dyes, and uranyl salts decrease above some particular optimum concentration, giving rise to the term "self-quenching." Pringsheim (1949, p. 349) reports that the optimum concentration for the phosphorescence of eosin in glycerol is 5×10^{-5} g/ml, whereas for fluorescence it is 100×10^{-5} g/ml. For rhodamine 6G in water, the concentration at which self-quenching begins is 5×10^{-5} M, whereas for the same dye in ethanol or glycerol, self-quenching begins only at 10^{-3} M. The phenomenon of fluorescence depolarization is an even more sensitive function of dye concentration and leads to the concept, which will be discussed in the section on energy transfer, that during the lifetime of the excited state excitation can be transferred to an acceptor molecule removed in space from the original absorbing species. One assumption made is that owing to the identity of the molecules involved there is a large resonance cross section for transfer. The specific mechanism of how the quenching then occurs is not obvious. Rabinowitch and Epstein (1941) have postulated that with increasing concentration ground state dimers or polymers are formed which are nonfluorescent. In the case of thionine in aqueous solution they found that as the concentration was increased from around 10^{-7} M there appeared new and overlapping "D" bands which they ascribed to dimers of the dye molecules. Thus only a fraction of the light absorbed in the region of the two overlapping bands will excite the fluorescent monomers and it will appear as though the fluorescence yield has decreased. The "D" bands decrease with increasing temperature and therefore despite the increased internal and solute quenching at the higher temperatures it has been possible to demonstrate the increased fluorescence yield of a 0.026 M fluorescein solution between 0° and 60°C. While this is a reasonable explanation of some forms of concentration quenching, the mechanism predicts that the lifetime of the excited state should be concentration independent. However, in many cases of concentration quenching the observed fluorescence lifetimes decrease with the decrease in quantum yield such that Q/τ is approximately constant. Two mechanisms that have been proposed which can account for the lifetime decreases are "excimer" quenching or dimer formation in the excited state, leading to radiationless decay to the ground state and "charge transfer" quenching. Both of these require the formation of an encounter complex, although the complex need not be stable. These processes can be represented as

$$A^* + A \rightarrow [A \cdot A]^* \rightarrow A + A + \text{vibrational energy} \qquad (2.29)$$

and

$$A^* + A \rightarrow [A^- \cdot A^+] \rightarrow A + A + \text{vibrational energy} \qquad (2.30)$$

The charge transfer complex is probably very short lived. In any case it has not yet been observed. On the other hand the photodimerization of anthracene in solution is a well-established observation, and Weigert (1912) has shown that the photodimerization rate in phenetole or in anisole solutions can account for the decrease in observed fluorescence yield. This mechanism for explaining self-quenching came about secondarily since originally Byk (1908a,b) and Weigert (1908) were hard put to explain why increasing the anthracene concentration from 0.02 M, at which concentration the light absorption of their solutions was already complete, to 0.16 M had the effect of doubling the photostationary concentration of dianthracene.

The quenching of solution polymers of pseudoisocyanine by minute concentrations of pyrocatechol has already been discussed and is an excellent example of "third molecule" quenching. Kasha (1952) has suggested that high atomic number quenching ions such as I^- increase the spin-orbit perturbation and accelerate singlet-triplet radiationless intercombinations. This mechanism can be represented as

$$^1A^* + Q_z \rightarrow (^3A^* \cdot Q_z) \rightarrow A + Q_z \qquad (2.31)$$

In this case the efficiency will depend on the proximity of the π^* antibonding orbital with the electric field of the heavy atom nucleus and therefore the nature of the solvent may play an important role in determining, by the degree of hydration, the closest distance of approach. From Eq. (2.23) fluorescence intensity should follow the general Stern-Volmer relation

$$N = \frac{N_0}{1 + \tau_0 q}$$

where τ_0 is the lifetime without quencher and q is the number of effective quenching collisions per second. The effective rate of quenching collisions should therefore be proportional to quencher concentration. In solvents of different viscosities the value of τ_0 may change due to the solvent effect on intramolecular quenching.

The cations K^+, Na^+, Ca^{++}, Ba^{++}, and Mn^{++} have very little or no quenching effect on most fluorescent organic solutions. Cu^{++}, Ni^{++}, Fe^{++}, Co^{++}, and Cr^{3+} cations are strong quenchers as are the organic molecules phenol, hydroquinone, resorcinol, and pyrogallol.

A different mechanism of quenching must be put forward to explain the anomalous quenching of dyes such as rhodamine, fluorescein, and eosin by long-chain molecules such as gardinol $[CH_3—(CH_2)_{15}—SO_3]^-Na^+$ and sodium stearate $[CH_3—(CH_2)_{15}—COO]^-Na^+$. As the concentration of quencher is increased, the fluorescence yield decreases. However, at increasing quencher concentrations, starting at 10 volume per cent and ending at 50 volume per cent at which the solution becomes a gel, the fluorescence yield now increases to about 60–70% of the initial value. In addition, the anomalous reactions described occur between positive dye ions such as rhodamine and negative long-chain ions such as stearate, while negative ions such as fluorescein are quenched by positive long-chain ions such as saponin. This is indicative of a specific alignment or binding of the planar dye molecule to the chains or gel-like structures formed by the hydrocarbons.

2.2.3. *Paramagnetic Quenching*

In the absence of perturbations all processes, including energy transfer, which involve a change in multiplicity are forbidden. However, it has been shown that conversion between electronic states of different multiplicities does occur, in many cases with high probability. In view of these apparently contradictory statements it is of interest to ascertain the conditions under which multiplicity changes can occur. In photobiology, owing to the energies of the light quanta involved, there are only two excited electronic states which need to be considered. These are the lowest excited singlet state and the lowest excited triplet state. Since the ground state of most molecules is a singlet state, the $^1\Gamma_0 \rightarrow {}^3\Gamma^*$ absorption probability will be extremely low, compared with the $^1\Gamma_0 \rightarrow {}^1\Gamma^*$ owing to the forbiddenness of the multiplicity change. It should be emphasized that the phosphorescence experimentally observed is related to the *spontaneous emission probability*, $A_{m,n}$, while the absorption is related to the induced emission or Einstein absorption coefficient, $B_{m,n}$. From thermodynamic reasoning these two coefficients are related by the equation

$$A_{m,n} = \frac{8\pi h}{\lambda^3_{m,n}} B_{m,n} \tag{2.32}$$

where $\lambda_{m,n}$ is the transition wavelength. So far as we know, with the possible exception of β-carotene (McGlynn, 1962, p. 282), excited triplet levels are lower in energy than excited singlet levels. In general, then, the triplet level of a molecule (a) is the lowest excited electronic state of the molecule, (b) has a lifetime of the order of 10^4–10^8 greater than that of the singlet state, and (c) exhibits the chemical behavior of a biradical. Because of this relatively long lifetime, the triplet state has often been suggested as an intermediate in energy transfer processes and in biochemical reactions, particularly in photosynthesis and in photodynamic action. Boudin (1930) showed that KI quenches both the fluorescence and phosphorescence of eosin solutions, and Oster and Adelman (1956) have been able to demonstrate the photoreduction of eosin by allyl thiourea with no appreciable change in the fluorescence yield of eosin. Both of these experiments indicate that a long-lived metastable state is involved. Calvin and Dorough (1948) in the photooxidation of zinc chlorin by β-naphthoquinone found that the reaction was of zero order with respect to quinone. In view of the fact that the quantum yield was independent of quinone concentration their interpretation was that a long-lived non-sidereacting metastable state was involved, presumably the triplet state. This is rather surprising in view of the strong quenching of triplet states in solution at room temperature. It would, therefore, be interesting to examine for this reaction the phosphorescence quantum yield of zinc chlorin, if any, at β-naphthoquinone concentration from zero up through the point where the reaction becomes zero order, similar to Boudin's approach. Porter and Windsor (1953) observed triplet state absorption spectra in solution by flash photolysis and were therefore the first to demonstrate that triplet states were readily populated in fluid solution. The difficulty heretofore was that the quenching of the triplet state by oxygen is so efficient that *only in very carefully outgassed* solutions can triplet-triplet absorption be observed. This as well as the relatively short triplet lifetimes in solution (10^{-4} sec) make the detection problem a difficult one. Using the technique of flash photolysis and spectrophotometry for observing the triplet state populations directly, Porter and Wright (1959) have proposed a reasonably consistent mechanism for intermolecular triplet state quenching by paramagnetic molecules such as O_2, NO, . . . which they call "catalyzed spin conservation." They found that while all paramagnetic ions such as Cu^{++}, Fe^{++}, Fe^{3+}, Mn^{++}, . . . would quench the phosphorescence of triplet naphthalene and diamagnetic ions such as K^+, Zn^{++}, Ga^{3+}, and $Cu(CN)_2^-$ had no effect on the phosphorescence, there was no correlation between the degree of quenching by paramagnetic ions and their paramagnetic susceptibility. On the basis of a statistical analysis of spin states available to a postulated collisional complex formed of an excited triplet, ^3A, and the paramagnetic quencher,

^3Q, they state that the reaction

$$^3A^* + {}^3Q \to [^3A^* \cdot {}^3Q] \to {}^1A + {}^3Q \qquad (2.33)$$

has a reasonable probability. For example, consider the allowed spin states for the complex formed by $[^3A^* \cdot {}^3Q]$. Each of the molecules in the complex has $S = 1$. Thus the combination of $\Sigma S = 2$ can have possible spin quantum numbers 2, 1, 0, with statistical weights, $(2S + 1)$, of 5, 3, 1. The products, with spins of 0 and 1, respectively, can have only the spin 1 with a statistical weight, $(2S + 1)$, of 3. Therefore the statistical probability of $^1A_0 + {}^3Q$ emerging from a $[^3A^* \cdot {}^3Q]$ complex is 3/9. For reasonable stable complexes such as O_2 might form, the statistical factor for this spin transition is as high as one-third. The surprising result of these calculations is that all paramagnetic molecules have, on statistical grounds, an *equal* probability of inducing a spin change in $^3A^*$. Differences do arise based on differences in spin-spin coupling between $^3A^*$ and ^3Q in the complex and in the type of complex formed by the paramagnetic molecule. The ^1A, upon dissociation from the complex, is presumably in a highly excited vibrational level of the ground state. On a configurational diagram this would be equivalent to the crossing-over of the collisional complex with the singlet ground state potential well. It is also possible for the ^3Q molecule to emerge in an excited vibrational or electronic level.

The above argument holds only for triplet quenching by a doublet or higher multiplicity ground state quencher. The reaction

$$^3A^* + {}^1Q \to {}^1A + {}^1Q$$

is still forbidden by spin selection rules and on the basis of the statistical argument above. The reaction

$$^3A^* + {}^1Q \to {}^1A + {}^3Q$$

might be possible if the triplet level of Q could be excited by $^3A^*$.

2.3. ENERGY TRANSFER

In this section we shall be concerned with the description of various mechanisms for the transfer of electronic excitation energy between atomic and molecular constituents, both in homogeneous solution and in structured systems. For the most part we will not be concerned with the simple case of the emission of light by one system and its subsequent reabsorption by a second one. Only the nonradiative transfer of excitation energy during the lifetimes of excited states will be discussed.

There is an extensive literature describing experimental observations of energy transfer and the detailed description of each is outside the scope of this section. Therefore only representative samples of the observed effects shall be presented in order to demonstrate the applicability of or the need for any particular theoretical mechanism to account for it.

2.3.1. Sensitized Fluorescence

Cario and Franck (1922) in experimenting with the fluorescence of mercury vapor found that when a mixture of mercury and thallium vapors derived from their metals at 100° and 800°C, respectively, was illuminated by the Hg 2537 A resonance line (which the mercury but not the thallium atoms can absorb) the observed fluorescence contained, in addition to reradiated HG 2537 A, a number of lines characteristic only of thallium, including the visible Tl 5350 A green line. Cario and Franck (1922, 1923) analyzed the *sensitized fluorescence* spectrum of thallium vapor as well as of silver, zinc, and cadmium vapors in combination with that of mercury and in each case were able to demonstrate the transfer of electronic excitation energy from mercury atoms to acceptor atoms, giving rise to a series of characteristic emission lines of the acceptors. They even found evidence that very weak thallium lines could be excited where the excitation potential of the particular thallium level was as much as 14 kcal/mole higher than the 112.7 kcal/mole of the excited Hg atom, requiring the concept that in a collision translational kinetic energy as well as internal electronic excitation energy can be converted to internal energy. Donat (1924) found that the presence of argon or nitrogen greatly enhanced the sensitized fluorescence of the Hg-Tl system and attributed this to partial deactivation of the Hg $^3P_1$ excited level to the slightly lower but metastable Hg (108.5 kcal/mole), thus preventing further collision deactivation by either argon or nitrogen or ground state Hg atoms but not by Tl atoms. Although in this case it was not possible to distinguish between transfer between distant atoms without direct collision and transfer as the result of direct collision, in the mercury sensitization of sodium vapor atoms the energy transfer was observed to occur over distances much larger than those involved in ordinary collisional distances.

Sensitized fluorescence in solution was first observed by Perrin and Choucroun (1929). In order to distinguish between nonradiative energy transfer and the case of emission and reabsorption there are several types of experiments that can be performed. If we irradiate a small volume of solution containing sensitizer (D) and acceptor (A) molecules at a fixed concentration there will be a certain total intensity (I) of sensitizer and acceptor fluorescence that can be measured. We next irradiate this same small volume of solution, but now this volume forms part of a larger volume at the same concentration as before. There should be no change in I_D and I_A if nonradiative transfer is effective but large changes in I_D and I_A if the process is emission and reabsorption, owing to the geometry.

Likewise, if the process is a nonradiative transfer, the observed emission lifetime of the sensitizer should decrease with increasing concentration of acceptor while remaining the same for the case of emission and reabsorption. This acceptor concentration-dependent quenching of sensitizer accompanied by a decrease in sensitizer lifetime is the most general method of distinguishing between the two processes.

Light absorbed in the absorption bands of crystalline anthracene (3500 and 3900 A) is able to excite the fluorescence of very low concentrations of naphthacene dissolved in the anthracene crystals. Consider the case of a naphthacene-containing anthracene crystal in which the normal violet anthracene fluorescence is completely quenched by naphthacene, the latter emitting a green fluorescence with essentially 100% efficiency of energy transfer. If this crystal is heated to the melting point, the green fluorescence of naphthacene disappears completely. If the same crystal is dissolved in benzene, the fluorescence spectrum reverts to the anthracene fluorescence. Much higher concentrations of naphthacene in anthracene-naphthacene liquid solutions are required in order to observe the same degree of sensitized fluorescence, although in both liquid and solid solutions we can observe sensitized fluorescence. These results indicate that the *structure* or degree of coupling strongly affect the degree of, and presumably the mechanism of, energy transfer and that a simple treatment of electrodynamic interaction between isolated molecules in solution may not be sufficient to explain energy transfer in organized structures. In this connection we have already described the quenching of the fluorescence of pseudoisocyanine in Section 1.2.7 of this chapter.

As an example of the quantitative observations to be explained by any mechanism, Bowen and Brocklehurst (1955) have demonstrated significant energy transfer in sensitized fluorescence at concentrations such that the mean distance between donor or sensitizer and acceptor were about 40 A.

2.3.2.   *Fluorescence Depolarization*

If an isotropic molecule is excited with plane polarized light, the radiation emitted must also be plane polarized. If, however, there are interactions with external fields or with other atoms or molecules during the lifetime of the excited state this "memory" can be partially or completely destroyed. If, in a condensed system, where vibronic periods are of the order of 10^{-12} sec as compared with fluorescence lifetimes of the order of 10^{-8} sec, there *persists* a partial polarization of the fluorescent radiation, this can only be due to the fact that there are fixed axes *within the molecules* along which electronic excitation can occur preferentially and that the molecules as a whole are optically anisotropic. The further observation that in some solid solutions the *phosphorescence* also shows a residual polarization (Carrelli and Pringsheim, 1923) proves that there can be a *structural orientation* of molecules. In the experimental observation of the effects of structural orientation we must distinguish two cases. In one the molecules may be adsorbed on a continuous surface so that the alignment over the area of observation is constant. In the other the molecules may be adsorbed, say on silica gel particles. The alignment may be uniform on each face of a particle, but the individual particles may have random orientations with respect to one another. In the former case irradiation with nonpolarized (normal) light will result in an observed fluorescence polarization anisotropy. In the latter case although the "microscopic" anisotropy is present, the macroscopic fluorescence polarization will appear isotropic. In both cases irradiation with plane polarized light will result in observed fluorescence polarization anisotropy. In all cases the simple picture that can be drawn is one of the electric vector of the light wave exciting the stabilized molecule maximally along its axis of maximum polarizability. The fluorescence radiation in the transition to the ground state will then also be emitted with its electric vector along the same axis. In the case of the silica gel particles containing the adsorbed dye, the unpolarized incident radiation has electric vectors in all planes perpendicular to the path and therefore the individual polarizations will contribute in all directions, giving isotropic behavior.

Consider a molecule whose polarizability ellipsoid can be represented as having the major semiaxis b_1, with minor semiaxes $b_2 = b_3$. For an isotropic molecule the ellipsoid becomes a sphere, with $b_1 = b_2 = b_3$. The polarizability, α, is given by $\alpha = \frac{1}{3}(b_1 + b_2 + b_3)$. Linearly polarized light, incident in the x-direction, with the electric vector in the z-direction can be represented as $E_z \rightarrow x$. The direction of observation is the y-direction. Stuart (1934, p. 171) has shown that under these conditions the observed intensity of light polarized in the z-direction is

$$I'_z \rightarrow y = \text{constant } [\alpha^2 + \tfrac{4}{45}(b_1 - b_2)^2] \qquad (2.34)$$

The intensity of light polarized in the x-direction and observed in the y-direction is given by

$$I'_x \rightarrow y = \text{constant } [\tfrac{1}{15}(b_1 - b_2)^2] \qquad (2.35)$$

Therefore for linearly polarized incident light the depolarization ratio, Δ', is given by

$$\Delta' = \frac{I'_x}{I'_z} = \frac{(b_1 - b_2)^2}{3b_1^2 + 4b_1b_2 + 8b_2^2} \qquad (2.36)$$

As can be seen, for an isotropic molecule $I'_x = 0$ and $\Delta' = 0$.

For natural incident light, having components $\mathbf{E}_z \to x$ and $\mathbf{E}_y \to x$, the depolarization ratio, Δ, is given by

$$\Delta = \frac{2I'_x}{I'_x + I'_z} = \frac{2(b_1 - b_2)^2}{4b_1^2 + 2b_1 b_2 + 9b_2^2} \qquad (2.37)$$

For completely anisotropic molecules, $b_2 = b_3 = 0$ so that $\Delta' = \frac{1}{3}$ and $\Delta = \frac{1}{2}$.

Feofilov (1961, p. 38) defines the degree of polarization, P, as

$$P = \frac{I_{\parallel} - I_{\perp}}{I_{\parallel} + I_{\perp}} \qquad (2.38)$$

where the $\parallel$ and $\perp$ refer to the direction of polarization of the observed light intensity, referred to the z-direction, with light incident in the x-direction and observation in the y-direction. In these cases for perfectly anisotropic molecules

$$P' = \frac{I'_z - I'_x}{I'_z + I'_x} = \frac{1}{2} \qquad (2.39)$$

for linearly polarized incident light. For natural light with $\mathbf{E}_z \to x$ and $\mathbf{E}_y \to x$ vectors there will be components of $I_{\parallel}$ and $I_{\perp}$ from both vectors. Thus $I_{\parallel}$ consists of the $\parallel$ component of the $\mathbf{E}_z \to x$ vector and the $\perp$ component of the $\mathbf{E}_y \to x$ vector and the $I_{\perp}$ consists of the $\perp$ component of $\mathbf{E}_z \to x$ and the $\parallel$ component of $\mathbf{E}_y \to x$. From symmetry the $\perp$ component of $\mathbf{E}_z \to x$ is equal to the $\perp$ component of $\mathbf{E}_y \to x$, both contributing to $I_{\perp}$. In addition both the $\perp$ components of $\mathbf{E}_y \to x$ are equal in magnitude, one contributing to $I_{\perp}$ and the other to $I_{\parallel}$. Thus for unpolarized incident light,

$$P = \frac{[\parallel(\mathbf{E}_z \to x) + \perp(\mathbf{E}_y \to x)] - [\perp(\mathbf{E}_z \to x) + \perp(\mathbf{E}_y \to x)]}{[\parallel(\mathbf{E}_z \to x) + \perp(\mathbf{E}_y \to x)] + [\perp(\mathbf{E}_z \to x) + \perp(\mathbf{E}_y \to x)]} \qquad (2.40)$$

Since $\perp(\mathbf{E}_y \to x) = \perp(\mathbf{E}_z \to x)$

$$P = \frac{\parallel(\mathbf{E}_z \to x) - \perp(\mathbf{E}_z \to x)}{\parallel(\mathbf{E}_z \to x) + 3\perp(\mathbf{E}_z \to x)} \qquad (2.41)$$

From the expression for the polarization for linearly polarized light

$$P' = \frac{\parallel(\mathbf{E}_z \to x) - \perp(\mathbf{E}_z \to x)}{\parallel(\mathbf{E}_z \to x) + \perp(\mathbf{E}_z \to x)} \qquad (2.42)$$

so that

$$[\perp(\mathbf{E}_z \to x)][1 + P'] = [\parallel(\mathbf{E}_z \to x)][1 - P'] \qquad (2.43)$$

Substituting Eq. (2.43) into Eq. (2.41) to express P in terms of P' we obtain

$$P = \frac{P'}{2 - P'} \qquad (2.44)$$

All of the above relations were derived on the basis of measurements of light scattering. Exactly the same arguments are made in the case of fluorescence polarization. These represent the limiting cases of light incident on a random aggregate of stationary molecules when the direction of the electric vector of the fluorescence radiation is the same as that of the exciting radiation. If there is an angle θ between the fluorescence dipole oscillator and the absorption dipole oscillator (Levshin, 1925)

$$P' = \frac{3 \cos^2 \theta - 1}{\cos^2 \theta + 3} \qquad (2.45)$$

in the case of perfectly anisotropic molecules.

Thus in viscous solvents, if the relaxation time for molecular reorientation is of the same order as the lifetime of the excited state, an anisotropy of observed polarization of fluorescence will be observed, with the polarization given by the above equations. However, Gaviola and Pringsheim (1924) and Weigert and Käppler (1924) observed very sharp decreases in the polarization of dye solutions in viscous solvents as the dye concentration increased, even at dye concentrations where the fluorescence yield remained unchanged. The concentration of rhodamine G at which the polarization is reduced to one-half of that in a dilute solution is only 1.5×10^{-3} M. The trivial case of emission and reabsorption is not nearly sufficient to explain the strong depolarization. Therefore it appears necessary that molecules other than the one excited can accept the excitation energy within the lifetime of the excited state and, by virtue of the random orientation of molecules, have a fluorescence dipole oscillator in a direction different from the original molecule. F. Perrin (1932) gave a quantum mechanical treatment of such mechanisms of energy transfer and arrived at a critical intermolecular distance

$$R_0 = \frac{\lambda}{\pi} \left[\frac{\bar{l}}{\tau} \right]^{\frac{1}{6}} \qquad (2.46)$$

where $\bar{l}$ is the vibrational collision time, τ is the mean lifetime of the excited state and λ is the fluorescence wavelength. For fluorescein, $\lambda \simeq 5000$ A, $\bar{l} \simeq 10^{-12}$ sec, $\tau = 5 \times 10^{-9}$ sec, R_0 is almost 400 A, giving critical concentrations for depolarization much lower than actually observed.

2.3.3. Förster Mechanism of Energy Transfer

Förster (1946, 1948, 1960) has extended the Perrin theory in terms of the experimentally accessible properties such as absorption and fluorescence spectra and the excited lifetime of the molecule concerned. The mechanism is a dipole-induced dipole resonance transfer from a donor or sensitizer, D, to an acceptor, A. The mechanism is shown schematically in Fig. 2.17. The ground and lowest excited electronic states of the sensitizer and acceptor molecules are shown as solid horizontal lines with the vibrational levels above them as dashed horizontal lines. Light is incident

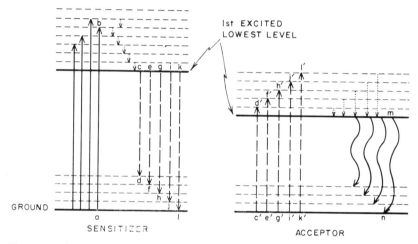

FIG. 2.17. Schematic energy level diagram illustrating the dipole-induced dipole mechanism for Förster-type energy transfer.

within the sensitizer absorption band. Owing to the applicability of the Franck-Condon Principle there will be a series of transitions between the ground state and various vibrational levels of the first excited state which, with the further rotational fine structure, will give rise to a band centering about ab, the peak absorption. The excess vibrational energy will be dissipated extremely rapidly (within the period of the vibrational frequencies of the molecule) and the sensitizer molecule will end up with only the lowest level of the first excited state populated. From this level one would ordinarily observe a fluorescence band corresponding to the transitions cd, ef, gh, ij, and kl. Again by virtue of the Franck-Condon Principle these are intramolecular vibrational transitions and are governed by selection rules. However, if there is some "coupling" between the sensitizer molecule and the acceptor molecule by virtue of their proximity or possibly their environment, there can be included in the calculations a quantum mechanical resonance such that the $c'd'$ dipole absorption transition in the acceptor molecule, whose energy above the ground state is exactly the same as the sensitizer dipole emission transition cd, has a definite probability. Similarly the sensitizer fluorescence transitions ef, gh, ij, kl, etc., have their comparable resonance forms in acceptor absorption transitions $e'f'$, $g'h'$, $i'j'$, $k'l'$, etc., resulting in the transfer of excitation energy from sensitizer to acceptor. In an analogous manner to the sensitizer molecule all of the acceptor vibrational levels except the lowest are depopulated by intramolecular energy dissipation. However, since the fluorescence transitions of the acceptor molecule are much lower in energy than the absorption levels of the sensitizer, the sensitizer can no longer be excited and the acceptor molecule has no recourse but to revert to its ground state with the emission of its characteristic fluorescence, one of whose transitions is indicated by mn. The case illustrated in Fig. 2.17, with the solid absorption lines in the sensitizer, the dashed "emission" transitions, and "absorption" transitions in the sensitizer and acceptor respectively, and the solid fluorescence transitions in the acceptor, corresponds to 100% energy transfer. Förster defines a critical distance, R_0, such that the probability of sensitizer-acceptor energy transfer is just equal to the probability of sensitizer fluorescence emission. In practice, R_0 corresponds with the concentration, C_0, at which the intensity of sensitizer fluorescence is reduced to one-half.

From the type of "virtual transitions" described above, for example, $cd \rightarrow c'd'$, it follows that for continuous bands, rather than isolated vibrational transitions, the summation would become an integral over the bands involved. Since these are dipole transitions, we might expect an energy-transfer probability of the form

$$|\phi_S^* \phi_S \mathbf{M} \phi_A^* \phi_A|^2$$

where the integration is performed over all of the energy levels of both sensitizer fluorescence and acceptor absorbance and $\mathbf{M}$ is the dipole moment operator. In the standard treatment the energy states and wave functions are obtained by adding to the Hamiltonian operator for the unperturbed molecules an intermolecular dipole-dipole perturbation operator of the form

$$\frac{-e}{r_{kl}^3} \sum_{i,j} (2z_k^i z_l^j - x_k^i x_l^j - y_k^i y_l^j)$$

where r_{kl} is the distance between the dipole in the sensitizer and the acceptor and the summation extends over all of the electrons on both molecules. In the Förster derivation, in which the transition probabilities are based on second order perturbation theory, the transition probability enters as the *square* of the dipole-dipole interaction energy. Thus the expression derived by Förster (1959) for the rate constant of the transfer process is

$$n(S^* \to A^*) = \frac{9000 \ln 10 \; K^2}{128\pi^6\eta^4 N\tau_s^0 r_{kl}^6} \int_0^\infty f_s(\bar{\nu})\epsilon_A(\bar{\nu}) \frac{d\bar{\nu}}{\bar{\nu}^2} \qquad (2.47)$$

where $\bar{\nu}$ is the wave number, $f_s(\bar{\nu})$ is the spectral distribution of fluorescence of the sensitizer, $\epsilon_A(\bar{\nu})$ is the molar extinction coefficient of the acceptor as a function of wave number, η is the index of refraction of the solvent, τ_s^0 is the intrinsic lifetime of the sensitizer, r_{kl} is the intermolecular separation, and K is a factor relating the orientation of the transition moment vectors of sensitizer and acceptor. There are apparently some typographical errors in Förster (1959). The $1/\bar{\nu}^2$ factor given in Förster (1948) is reported in the former paper as $1/\bar{\nu}^4$. When $n(S^* \to A^*)$ has equal probability with deactivation of the sensitizer

$$R_0^6 \simeq \frac{9000 \ln 10 K^2}{128\pi^6\eta^4 N\langle\bar{\nu}\rangle^2} \frac{\tau_S}{\tau_S^0} \int_0^\infty f_S(\bar{\nu})\epsilon_A(\bar{\nu}) \; d\bar{\nu} \qquad (2.48)$$

where here $\langle\bar{\nu}\rangle$ is the average wave number between sensitizer fluorescence and acceptor absorption peaks and τ_S/τ_S^0, the ratio of the observed lifetime to the intrinsic lifetime of the sensitizer molecule, is the quantum yield of fluorescence. From Eq. (2.47) and Eq. (2.48) we see that

$$n(S^* \to A^*) = \frac{1}{\tau_S}\left(\frac{R_0}{r_{kl}}\right)^6 \qquad (2.49)$$

Equation (2.48) gives the values for R_0 shown in Table 2.4. The actual measurements of Weber and Teale (1959) shown in the table are in reasonably good agreement with Förster R_0 values.

There are some questions that can be raised regarding the applicability of the Förster mechanism in coupled systems, i.e., for sensitized phosphorescence in rigid media and for triplet-triplet transfer both in rigid media and in solution. Since the Förster mechanism applies only to transfer over distances greater than normal, collision diameters and the transition probability falls off as $(1/r_{kl})^6$, no long range energy transfer due to dipole-induced dipole interactions is predicted.

Recently Bersohn and Isenberg (1963) have observed a phosphorescence quenching of DNA purines at $-196°C$ by much less than stoichiometric amounts of Mn^{++} ions *although the fluorescence is not affected*. In their initial report they postulate a delocalized triplet state in DNA. In our interpretation of these data a delocalized triplet is another way of saying that efficient triplet-triplet energy transfer has been observed to

TABLE 2.4

FÖRSTER CRITICAL DISTANCES FOR VARIOUS SENSITIZER-ACCEPTOR SYSTEMS

Sensitizer	Acceptor	$\tau_S \times 10^8$	$\int_0^\infty f_S(\bar{\nu})\epsilon_A(\bar{\nu})\,d\bar{\nu}$	R_0 (Å)
Fluorescein	Fluorescein	0.5	7×10^{12}	50[a]
Chlorophyll a	Chlorophyll a	3.0	1.1×10^{13}	80[a]
Phenylalanine	Phenylalanine	1.1	4.04×10^6	5.6[b]
Phenylalanine	Tyrosine	1.1	4.1×10^8	12.0[b]
Phenylalanine	Tryptophan	1.1	21.1×10^8	16.0[b]
Tyrosine	Tyrosine	0.91	4.58×10^7	8.3[b]
Tyrosine	Tryptophan	0.91	1.3×10^9	15.0[b]
Tyrosine	red. DPN	0.91	3.4×10^{10}	25.0[b]
Tyrosine	oxid. FMN	0.91	1.2×10^{10}	21.0[b]
Tryptophan	Tryptophan	0.2	3.3×10^7	6.3[b]
Tryptophan	red. DPN	0.2	1.2×10^{11}	25.0[b]
Tryptophan	oxid. FMN	0.2	1.7×10^{11}	26.0[b]
Tryptophan	Dimethylaminonaphthalene-sulfonyl chloride (DNS)	0.2	2×10^{10}	37–53[c]

[a] From Förster (1948).
[b] From Karreman *et al.* (1958).
[c] From Weber and Teale (1959) using fluorescence measurements on labeled heme proteins.

occur over many molecular distances. If this is true, it can not be accounted for on the basis of the Förster mechanism of energy transfer.

2.3.4. *Exciton Mechanism of Energy Transfer*

In the case of dilute solutions of essentially independent molecules the Förster mechanism of inductive resonance provides a consistent physical description of energy transfer, particularly between unlike molecules. However, in molecular aggregates and in molecular crystals where there are intermolecular interactions it is important to consider the resonance interaction between excited states of the coupled molecules. It is more likely that energy transfer in proteins or DNA or between dye molecules in the lamellar structures of biological systems would be described by the exciton mechanism rather than by the Förster mechanism. The exciton, which is used in the sense of a traveling wave packet of excitation in the crystal, was first developed by J. I. Frenkel (1931) and extended by Davydov (1962). In all cases comparisons are made between the spectral characteristics and transition intensities of the individual molecules and the system composed of the same molecules in the aggregate. The theory can explain phenomena such as blue and red shifts, spectral splittings, hypochromism, and polarization of luminescence; and the reader is referred to the extensive papers by McClure (1959), Kasha (1959, 1963), McRae and Kasha (1958, 1964) and Hochstrasser and Kasha (1964). In particular, the papers by McRae and Kasha (1964) and Hochstrasser and Kasha (1964) are excellent treatments, outlining the qualitative aspects of the molecular exciton theory. For strong electric dipole transitions the exciton treatment, similar to the Förster treatment, uses the dipole-dipole potential function

$$V = \frac{-e^2}{r_{kl}^3} \sum_{ij} (2z_k^i z_l^j - x_k^i x_l^j - y_k^i y_l^j)$$

where r_{kl} is the distance between point dipoles in molecules k and l and x_k^i is the x coordinate of the ith electron on molecule k, etc. The summation is over all electrons in each molecule. However, the exciton calculations are based on a first-order perturbation treatment, so that there is only an inverse cube dependence on the distance of separation of the dipoles. For example, in a dimer consisting of two identical molecules A and B, the energy of interaction is given by

$$E = \iint \psi_A^* \psi_B V_{AB} \psi_A \psi_B^* \, d\tau_A \, d\tau_B \tag{2.50}$$

In the case of linearly polarized light the equation can be factored and

$$E = \frac{1}{r_{AB}^3} (\mu_A)(\mu_B) \tag{2.51}$$

where μ_A and μ_B are the transition moment integrals for the transition $A \rightarrow A^*$ and $B \rightarrow B^*$, respectively. The choice of an arbitrary phase factor is made such that $\mu_A = -\mu_B$ will make one linear combination of these degenerate states lower in energy. Thus the energy lowering in the case of the simple dimer corresponds to

$$E = \frac{-\mu_A^2}{r_{AB}^3}$$

and the exciton splitting or band width is $U = 2E$. If we define the resonance lifetime of a locally excited state as

$$\tau = \frac{h}{2U} \tag{2.52}$$

the number of energy transfers per second is $n = 1/\tau$. Therefore the exciton theory predicts

$$n \sim \frac{\mu_A^2}{r_{AB}^3} \tag{2.53}$$

Kasha (1963) has compared the exciton theory with the Förster theory and has pointed out some differences between the predictions of both that may be amenable to experiment. A distinction is made in the exciton theory as to the degree of intermolecular coupling in the aggregate. $\Delta\epsilon$ is the Franck-Condon band width of the molecular electronic transition in the individual molecule. Strong coupling will give rise to large values of E and therefore to large band splitting, i.e., the absorption spectrum of the aggregate is strongly shifted from that of the monomer or individual molecule. In this case $2U \gg \Delta\epsilon$. If the coupling is weak, there will be

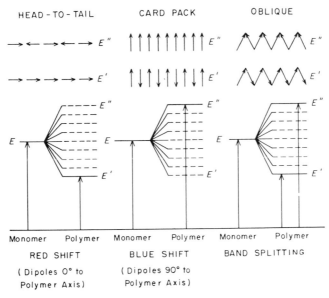

Fig. 2.18.   Energy level scheme for exciton band splitting for different geometrical arrangements of molecules in a linear polymer. After McRae and Kasha (1958).

essentially no band splitting but only a hypochromism or hyperchromism and therefore $2U \ll \Delta\epsilon$. The case for intermediate coupling is not as clear since this gives rise to complex absorption curves. The important point here is that the type of coupling can be determined spectroscopically. In addition, the geometrical arrangement of monomers in a linear molecular polymer in the strong coupling case can be inferred from the observed spectra. Figure 2.18 adapted from McRae and Kasha (1958) shows the types of splitting that are observed for various geometrical arrangements of the transition dipoles associated with the monomer units in a linear polymer. The arrangement of the individual dipoles determines selection rules for exciton transitions so that only certain levels within the band splitting are allowed. For example, in the head-to-tail arrangement where the dipoles are parallel to the linear polymer axis, the in-phase arrangement to give allowed exciton states is also the lowest in energy. This arrangement is shown as E'. The band splitting is $E'' - E'$ and the shift from the monomer band E is shown by the solid vertical arrows. Other arrangements of the dipoles give rise to the forbidden dashed levels between E' and E''. In the figure only nine arbitrary levels are shown. In the parallel card pack arrangement where the dipoles are perpendicular to the linear molecular axis, the in-phase arrangement of dipoles corresponds to the highest energy and we therefore observe a blue shift from the monomer band at E to E''. In the oblique arrangement the E'' configuration gives rise to an in-phase net moment perpendicular to the polymer axis and the E' configuration gives an in-phase net moment parallel to the polymer axis so that both levels are allowed. McRae and Kasha (1964) have also shown that at an angle of 54.7°, intermediate between the 0° head-to-tail and the 90° card pack there should be no splitting even for strong coupling. This type of argument is also of importance in the interpretation of the spectra of helical polypeptides, since it gives information about the arrangement of the molecular chromophores with respect to the helix axis.

The rates of energy transfer are different in the exciton theory and in the Förster theory. In the strong coupling exciton case $n \sim f \cdot (\theta)/r^3$ where f is the oscillator strength for the electronic transition in the monomer, (θ) is the geometry function of the dipole arrangement and r is the distance between dipoles. These transfer rates are 10^{15} sec^{-1} and depend on $1/r^3$. In the weak coupling exciton case $n \sim \dfrac{f \cdot (\theta)}{r^3} \Sigma_{v,v'} g_v^* g_{v'} S_{v,v'}^2$. The additional factor sums the populations and the vibronic overlap integrals over all vibrational levels of the molecules. Even in the case of weak coupling $n \sim 10^{12}$–10^{13} sec^{-1} and the $1/r^3$ dependence remains. In the Förster model

$$ n \sim \left[\frac{f}{r^3} (\theta) \right]^2 \frac{\Sigma_{v,v'} g_v^* g_{v'} S_{v,v'}^4}{\Delta \epsilon'} $$

where now the dependence is on $1/r^6$ and on the fourth power of the vibronic overlap integral. $\Delta \epsilon'$ is the vibronic band width. In this case Kasha (1963) predicts much slower energy-transfer rates of the order of 10^6–10^{11} sec^{-1}. It might therefore be feasible, in the case of certain models, to differentiate between the two mechanisms by virtue of the slower transfer rates predicted by the Förster theory.

2.3.5. "*Addition*" *of Light Quanta*

Recently Peticolas *et al.* (1963) and Kepler *et al.* (1963) reported the double photon absorption and the direct generation of triplet excitons, respectively, in single crystals of anthracene, using ruby laser light, 6943 A. The presence of these triplet excitons was inferred by the latter

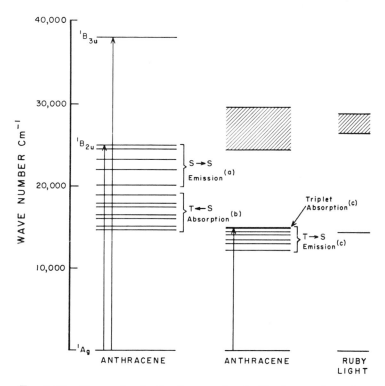

Fig. 2.19. Energy levels of anthracene for double photon absorption.

authors from the observation of the characteristic blue fluorescence presumably produced by the annihilation of triplet pairs. In a further experiment, using incoherent red light (Avakian *et al.*, 1963), these same authors were able to obtain indirectly the triplet excitation spectrum of single-crystal anthracene, shown in Fig. 2.19. This triplet absorption spectrum is obtained from an action spectrum for the blue singlet exciton fluorescence. There is apparently some disagreement in this interpretation in that while for coherent laser light Kepler *et al.* (1963) found a linear dependence of blue fluorescence intensity with incident "red" light, Adelman and Verber (1963) and Singh and Stoicheff (1963) found, for single crystals of the naphthalene-tetrachlorophthalic anhydride complex at 77°K and for anthracene single crystals, respectively, a fluorescence intensity dependent on the square of the incident intensity. In the experiments with incoherent "red" light, Avakian et al. (1963) found a dependence on the square of the intensity. Adelman and Verber interpret their dependence on intensity squared as a two-photon absorption through virtual levels. In view of the agreement between the lowest triplet excitation level found by Avakian and co-workers (14,720 cm^{-1} or 6793 Å) and the solution phosphorescence data of Padhye et al. (1956), it would appear that the triplet level can indeed be involved and possibly a subsequent absorption event or a triplet-triplet annihilation could add sufficient energy to effect the observed blue singlet fluorescence. In the case of incoherent light at the relatively low intensities used (7.3 × 10^{15} photons/ cm^2-sec), a dependence on incident intensity squared would be reasonable. At higher light intensities or *if all the excitation energy were funneled to one long-lived trapping site* the blue fluorescence intensity could be pseudo first order with respect to incident intensity. It is this type of photon addition reaction that has been postulated to account for the "photoionization" of water in photosynthesis. In Fig. 2.19 the shaded area in the second column represents the range of energies available from triplet-triplet annihilation and that in the third column represents the range available owing to a double photon absorption. Ruby laser light (6943 Å) is equivalent to 14,400 cm^{-1} and this level is also shown in the third column.

Although we do not believe that the triplet annihilation mechanism is operative in green-plant photosynthesis (Robinson, 1963), it is instructive to know that such a mechanism is physically demonstrable. Neither do we believe that color vision is the combined effect of different impurity trapping centers in a semiconductor matrix, although a color-sensitive photoconduction in inorganic semiconductors is demonstrable.

3.   Photochemical Reactions

Photosynthesis in plants is a remarkable example of the utilization and chemical storage of the free energy of absorbed sunlight. Theoretically, there is enough energy in sunlight to provide power needs in sunny, rural areas. On a bright, sunlit day the total radiation falling on each square centimeter of earth surface amounts to about 500 cal or almost enough to vaporize 1 gram of water. For comparison, 1 gram of wood on combustion liberates about 4000 cal. If a photochemical reaction could store energy or do work with 10% efficiency, 1 acre of a photochemical reaction can provide in 1 day the equivalent of 2000 kw hours of electricity. A large number of photochemical and photoelectric reactions have been described in the literature. The following summary from Daniels

(1960) shows the types of reactions that are being considered for practical use.

When exposed to sunlight, hydrogen and chlorine react photochemically with a quantum yield approaching one million molecules per photon absorbed. The reaction is exothermic and irreversible.

The photodissociation of iodine to give iodine atoms is endothermic and reversible in the dark. It is produced by sunlight, but the recombination of iodine atoms is so fast that it is useless for storing energy. Even during illumination, the iodine atoms are reforming iodine molecules and the sunlight is instantly transformed into heat.

Brown nitrogen dioxide is dissociated by violet light into colorless nitric oxide and oxygen, which recombine by a third-order reaction to give nitrogen dioxide. The recombination is considerably slower than that of iodine atoms, but it is still much too fast for the storage of energy.

The reaction $AgBr \rightarrow Ag + Br$ takes place in sunlight and it is reversible in the dark; theoretically the bromine could be stored and the silver bromide can be sensitized with dyes to the whole visible spectrum.

Photopolymerization and photoisomerization are endothermic and some of them are reversible in the dark with suitable reaction rates. Their heats of reaction are low, however, and most of them take place only in UV light. The reaction

$$\text{anthracene} \underset{\text{dark}}{\overset{\text{light}}{\rightleftharpoons}} \text{dianthracene}$$

absorbs about 50 cal/gram when illuminated with UV light at 3600 A. The anthracene may be dissolved in benzene or other solvent and the depolymerization, which is sufficiently slow for heat storage at room temperature, may be accelerated by raising the temperature. The photopolymerization of sulfur from $S\lambda$ to $S\mu$ is reversible, but the energy changes are very small.

The reaction

$$\text{maleic acid} \underset{\text{dark}}{\overset{\text{light}}{\rightleftharpoons}} \text{fumaric acid}$$

stores 70 cal/gram when illuminated with UV light and it reverses slowly, even at elevated temperatures. The isomerization requires UV light. The photoisomerization of stilbene to isostilbene absorbs light and stores 57 cal/gram and the photoisomerization of oleic acid to elaidic acid stores 63 cal/gram, but these reactions respond only to UV light.

The reaction

$$CO_2 + H_2O \xrightarrow{\text{light}} 1/n(H_2CO)_n + O_2$$

has an endothermic heat of reaction of 112 kcal/mole or 1800 cal/gram of reactants. The product carbohydrate can be stored indefinitely and reacted to give back this heat at will by igniting the material with the oxygen to start the exothermic reaction. In this case, of course, it is not necessary to provide for the storage of oxygen—only the carbohydrate needs to be stored because the oxygen of the air can be used for the reoxidation process. This reaction requires very short ultraviolet of 2500 A to give 112 kcal/mole. Moreover, the water and carbon dioxide are transparent throughout the visible and absorb only in the very short wavelength ultraviolet. However, in the photosynthesis of growing plants, this reaction is photosensitized by chlorophyll so as to make available most of the sunlight in the visible region. In this reaction photons of low energy are able in coupled reactions to bring about a photochemical change which requires much greater energy.

Nitrosyl chloride (NOCl) is dissociated by sunlight of wavelengths as long as 6500 A and it absorbs about as much of the visible sunlight as does chlorophyll in photosynthesis. The quantum yield in the gas phase is 2 molecules/photon. When it is dissolved in carbon tetrachloride, one of the products, chlorine, remains in solution, but the other product, nitric oxide, being insoluble, escapes from solution and can be stored. The nitric oxide reacts with chlorine in carbon tetrachloride at slightly elevated temperatures to produce nitrosyl chloride, giving back heat equivalent to 137 cal/gram of nitrosyl chloride (NOCl). This reaction seems to meet some of the requirements for storing energy, but the reverse reaction takes place during illumination and chlorine absorbs some of the light, so that a flowing process is necessary. The exothermic heat evolved by the reverse reaction is considerably less than the photochemical energy needed for the dissociation.

The photodissociation of nickel carbonyl dissolved in carbon tetrachloride, hexane, or cyclohexane takes place with UV light with a quantum yield of 2.2 molecules/photon at 3660 A and 2.8 at 3130 A. The carbon monoxide is insoluble in the solvents and can be collected and stored. It can recombine under suitable conditions at 50°C with the nickel suspended in the solvent, thus reforming the nickel carbonyl and evolving 290 cal of heat/gram. It has been reported that the photodissociation of nickel carbonyl can be brought about by visible light with the help of suitable photosensitizers. The heat of formation of 50 kcal/mole corresponds to an upper limit of 5600 A and it is expected that light of longer wavelength would not be effective even if a photosensitizer could be found. Because of its poisonous properties, it must be handled carefully in closed systems.

The photochemical production of hydrogen peroxide from water and oxygen can be brought about with the photocatalyst zinc oxide. It seems possible that the quantum yield might be increased by using zinc oxide which has been purified to a very high degree. The catalytic activity of a zinc oxide crystal depends on its crystal structure, which can be varied by conditions of preparation, particularly purity and temperature.

Most chemical reactions require activation energies of 25–60 kcal/mole and more, which are available in visible light. The energy of infrared radiation, which is half of the sun's total radiation, is less than this, so it does not ordinarily bring about these reactions.

In the Becquerel effect, discovered in 1841, an electrical potential is produced when two electrodes are immersed in a solution of a suitable electrolyte and one of the electrodes is illuminated while the other is kept in the dark. Electrons are liberated at the illuminated electrode and as they go along the circuit through a wire to the nonilluminated electrode they can do useful electrical work. Well-known examples include silver–silver chloride electrodes in an aqueous solution of chloride ions, or copper electrodes with an oxide coating immersed in a solution of sodium hydroxide.

It is not necessary to have the electrodes immersed in solutions; a crystal of semiconductor can be used with electrodes attached, as in the selenium cell, to give a photovoltaic effect. The conversion of light energy into electrical energy in a selenium cell is low.

By far the most efficient photovoltaic cell is the so-called solar battery. It is made of highly purified silicon, grown as a single crystal, which allows electrons released by light to travel all the way to the electrodes. The single crystal of tetravalent silicon is treated with pentavalent arsenic and trivalent boron, in thin layers, and then sawed off in the form of a sandwich. The place of contact between the boron-rich layer and the arsenic-rich layer is only slightly below the surface. A potential of about a half volt is developed at the contact within this sandwich when it is exposed to sunlight; and 14% of the total solar radiation has been converted directly into electricity.

In photogalvanic cells as distinguished from photovoltaic cells, the photochemical effect takes place in the solution surrounding the electrodes rather than on the electrodes themselves. One electrode compartment is illuminated and the other is kept in the dark. If, for example, the solution in the illuminated compartment is converted photochemically into a solution with a higher concentration of oxidized ions, a potential is set up between the electrodes in the two compartments (Ghosh, 1929). If the reactions at the two electrodes are fast enough, some of the electrons which are present in the unoxidized ions or molecules of the dark compartment are transferred through the electrode and wire to the electrode which is surrounded by the oxidized ion in the illuminated compartment—and the flow of electrons can do useful work.

A good example of the photogalvanic cell involves a purple organic dye, thionine, and ferrous ions, which give ferric ions and an altered, colorless form of the dye, called leucothionine, on exposure to light (Rabinowitch, 1940). The reaction reverses in the dark and restores the purple color of thionine. Thus,

$$\text{thionine} + Fe^{2+} \underset{\text{dark}}{\overset{\text{light}}{\rightleftharpoons}} \text{leucothionine} + Fe^{3+} \tag{2.54}$$

The light activates the purple thionine and causes it to accept an electron from the ferrous ion, but there is also a tendency for the electron to go back to the ferric ion thus produced. The reaction is fairly rapid but slow enough to be observed. If two inert electrodes are placed in the ferrous ion-thionine solution and one is illuminated while the other is kept in the dark, the illuminated electrode becomes surrounded by the ferric ions and acquires a potential of about a half volt negative with respect to the dark electrode. Useful electrical work can be obtained when the electrons taken up by the thionine go back through the electrodes and wire to the ferric ions.

Another similar photogalvanic cell involves the following chemical reaction:

$$Fe^{2+} + HgCl_2 \underset{\text{dark}}{\overset{\text{light}}{\rightleftharpoons}} Fe^{3+} + HgCl + Cl^- \tag{2.55}$$

3.1. Photoreactions in the Vapor Phase

Two of the earliest photochemical reactions investigated were the photochemical decomposition of HBr and HI (Warburg, 1916, 1918). Two molecules of HBr or HI are decomposed per quantum absorbed. The reactions in the case of HI are

$$
\begin{aligned}
HI + h\nu &\rightarrow H + I \\
H + HI &\rightarrow H_2 + I \\
I + I &\rightarrow I_2
\end{aligned}
\tag{2.56}
$$

A second and biologically important reaction is the photochemical production of oxygen and ozone in the atmosphere. At the present time the atmosphere contains 78% nitrogen, 20.95% oxygen, 0.9% argon, 0.03% carbon dioxide, and $4 \times 10^{-5}\%$ ozone in addition to other trace amounts of methane, nitrous oxide, hydrogen, and the noble gases. The total mass of oxygen is 11.84×10^{20} g or 0.74×10^{20} moles. There is evidence as stated in B. Mason (1960) that the earth's surface was anaerobic throughout most of Precambrian time, that is about 250 million years ago. There are three theories that have been suggested to account for the evolution of free oxygen:

(a) Photochemical dissociation of water vapor.

(b) Thermal dissociation of water vapor during the cooling period of the earth's surface.

(c) Photosynthesis by green plants.

At the present time the oxygen content of the earth's atmosphere is maintained by photosynthesis.

Igneous activity has probably contributed most of the water vapor and carbon dioxide to the atmosphere. According to H. Brown (1947) ". . . the earth's atmosphere is almost entirely of secondary origin, and it was formed as the result of chemical processes that took place subsequent to the formation of the planet."

Using the figure of 1.3×10^{24} cal/year of solar radiation incident on the earth and a wavelength region between 1400 and 1800 A for the water vapor and oxygen absorption continua, approximately 4.8×10^{-4} of the incident energy is available for the formation of oxygen from the photodissociation of water vapor and for the formation of ozone from oxygen.

Thus 6.2×10^{20} cal/year or 26×10^{20} joules/year have been available. Assuming a mean wavelength of 1600 A this corresponds to

$$\frac{1600}{1987} \times 10^{18} \times 26 \times 10^{20} \times \frac{1}{6.02 \times 10^{23}} = 3.5 \times 10^{15} \text{ Einsteins/year} \quad (2.57)$$

Over the period of the age of the earth (4.5×10^9 years), therefore, aside from the protective absorption by the oxygen thus formed, there would have been available a maximum of 16×10^{24} Einsteins of UV radiation in the 1400 to 1800 A region, certainly sufficient to account for all of the atmospheric oxygen present today. The thermal dissociation of water vapor during the earth's cooling may have contributed some or all of the oxygen and B. Mason (1960) presents a value of 7×10^{20} moles of oxygen produced cumulatively by photosynthesis. It is therefore not possible to choose any one of the three mechanisms as specifically responsible for the present atmospheric oxygen. The photodecomposition of water vapor and the subsequent gas reactions to form oxygen and ozone are shown in Fig. 2.20. Ozone is also decomposed by light absorbed in the red and yellow regions of the visible spectrum as shown by Fig. 1.21.

The H_2–Cl_2 reaction has already been mentioned. Coehn and Jung (1923, 1924) have found that in extremely carefully desiccated mixtures of $H_2 + Cl_2$ where $P_{H_2O} < 10^{-7}$ mm Hg, there was no detectable combination of H_2 and Cl_2 vapor exposure to sunlight for 20 days, although there is a slow combination in UV light. Above $P_{H_2O} \sim 10^{-5}$ mm Hg the light sensitivity of the system becomes independent of water vapor. The reaction is initiated by light absorbed by the chlorine. The nature of the action of the H_2O molecules in catalyzing the combination reaction is not well understood. It has been postulated that the large dipole moment of the water molecule can exert an effect on the electronic configuration of the reactants at distances of many molecular diameters; this is in order to account for the combination rates observed between 10^{-7} and 10^{-5} mm Hg

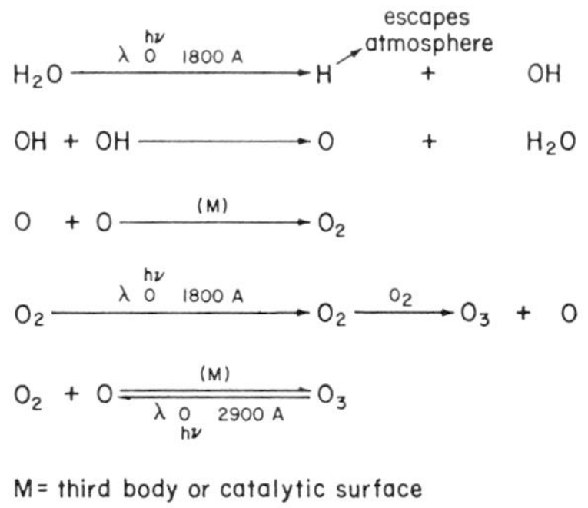

M = third body or catalytic surface

FIG. 2.20. Reaction scheme for the photochemical production of atmospheric oxygen and ozone from water vapor.

(Kornfeld, 1927). Another possibility may be the formation of a hydrate with a chlorine molecule which is then very reactive. However, the exact mechanism is still not known.

3.2. SOLUTION PHOTOCHEMISTRY

Most investigations of quantum yields of photochemical reactions in liquid solution have been confined to either aqueous or CCl_4 solutions. Solution photochemistry in general presents more complications than do gaseous reactions and processes of deactivation are more prevalent. The photochemistry of malic and fumaric acids in aqueous solution will be discussed in Section 3.5. One of the earliest reactions studied was the photodecomposition of chlorine water and hypochlorous acid, HOCl. The reactions are

$$Cl_2 + H_2O \xrightarrow{h\nu} 2HCl + \tfrac{1}{2}O_2 \tag{2.58}$$

$$3Cl_2 + 3H_2O \xrightarrow{h\nu} HClO_3 + 5HCl \tag{2.59}$$

for chlorine water and

$$HOCl \xrightarrow{h\nu} HCl + \tfrac{1}{2}O_2 \tag{2.60}$$

$$3HOCl \xrightarrow{h\nu} HClO_3 + 2HCl \tag{2.61}$$

for hypochlorous acid. In a solution of Cl_2 in water there is an equilibrium

$$Cl_2 + H_2O \rightleftharpoons HOCl + Cl \tag{2.62}$$

The quantum yield for photolysis of pure chlorine water is 2 and the maximum quantum yield for HOCl in the presence of 0.125 M Na_2HPO_4 is 1.72 (Griffith and McKeown, 1929, p. 486). On this basis the reaction mechanisms have been postulated to be

$$Cl_2 \cdot H_2O \xrightarrow{h\nu} 2HCl + O \tag{2.63}$$

$$O + O \rightarrow O_2^* \tag{2.64}$$

$$Cl_2 \cdot H_2O + O \rightarrow 2HCl + O_2^* \tag{2.65}$$

$$Cl_2 \cdot H_2O + O_2^* \rightarrow HClO_3(6\%) + HCl \tag{2.66}$$

$$O_2^* \rightarrow O_2 \tag{2.67}$$

and

$$HOCl \xrightarrow{h\nu} HCl + O \tag{2.68}$$

$$O + O \rightarrow O_2^* \tag{2.69}$$

$$HOCl + O \rightarrow HCl + O_2^* \tag{2.70}$$

$$HOCl + O_2^* \rightarrow HClO_3(86\%) \tag{2.71}$$

$$O_2^* \rightarrow O_2 \tag{2.72}$$

Both reactions are postulated to proceed through O_2^*, to which we refer in Chapter 3. A similar mechanism is postulated for the photolysis of NaOCl in aqueous solution and it is interesting that the catalytic splitting of peroxide by substances such as heme takes the place of the light quantum for the decomposition of NaOCl to produce O_2^*. It was found very early that in the photochemical combination of $H_2 + Cl_2$ the mere presence of minute amounts of water vapor exerted a tremendous catalytic effect on the reaction rate. This question of the influence of water on chemical reactivity in general has great theoretical importance.

Some of the most common types of sensitized photochemical reactions in solution are the sensitization by ferric and uranyl salts. Examples are the decomposition of oxalic acid in the presence of a uranyl salt:

$$H_2C_2O_4 + UO_2SO_4 \xrightarrow{h\nu} CO_2 + CO + H_2O + UO_2SO_4 \tag{2.73}$$

the decomposition of uranyl oxalate:

$$UO_2C_2O_4 + H_2SO_4 \xrightarrow{h\nu} UO_2SO_4 + CO + CO_2 + H_2O \tag{2.74}$$

the decomposition of potassium ferrioxalate:

$$2[Fe(C_2O_4)_3]^{3-} \xrightarrow{h\nu} 2Fe(C_2O_4) + 3[C_2O_4]^- + 2CO_2 \tag{2.75}$$

and the Fe^{3+} catalyzed Eder reaction

$$2HgCl_2(NH_4)_2C_2O_4 + 2FeCl_3 \xrightarrow{h\nu} Hg_2Cl_2 + 2NH_4Cl + 2CO_2 + 2FeCl_3 \quad (2.76)$$

In general the primary reaction is a reduction of ferric to ferrous by light. As described in Appendix VI (Parker, 1953, 1954; Hatchard and Parker, 1956; Lee and Seliger, 1964), chemical actinometry by measuring the amount of Fe^{++} produced can be made extremely precise when, as Parker has found, the Fe^{++} formed in the reaction is bound to 1 10 phenanthroline and is thus removed from further participation.

Photosensitization by fluorescent dyes can be considered to be an analogous process. In effect, the sensitizer must be capable of absorbing light in a manner which does not lead to degradation of the absorbed energy into heat so that the loosely bound electron can participate in an electron transfer. For example, the xanthine series of dyes, fluorescein, eosin, rose bengal, strongly catalyze the Eder reaction and also exhibit photodynamic action in the presence of oxygen. It is not the purpose of this section to treat the various photochemical reactions; these are covered in detail in the chemical literature. In general, we have pointed out that as the result of the absorption of light of sufficient energy, electron-transfer reactions can occur. Upon addition of a suitable sensitizer a complex can be formed, absorbing in the sensitizer absorption band and thus making available in the complex a loosely bound electron, which is another way of describing a highly reducing environment. This aspect will be covered in more detail in Section 3.4.

3.3. CHARGE TRANSFER COMPLEXES

In the case of two molecules A and B in their ground states there may be sufficient interaction such that an equilibrium

$$A + B \rightleftharpoons A^+B^- \tag{2.77}$$

can be maintained (Mulliken, 1952). In this case the formation of the ground-state complex can be observed by the appearance of a new absorption band, usually displaced to longer wavelengths. This is usually observed in concentrated solutions of molecules which are classified as strong electron donors or electron acceptors. If, in solution or in an array in a biological matrix such as the granum of a chloroplast, the ground-state interaction is very weak, the equilibrium of the process

$$[A + B] \rightleftharpoons [A^+ + B^-] \tag{2.78}$$

where the brackets indicate close proximity of A and B either by collision or by physical incorporation in a matrix, may be far toward [A + B]. However, upon the absorption of light by A the equilibrium may now be displaced to the $[A^+ + B^-]^*$ complex and it may therefore be possible, as we have postulated in Fig. 5.15, that the loosely bound electron can participate in a photoreduction reaction. This mechanism, while it can result in a "long-range" electron transfer, may not require the specific crystalline lattice of the analogous mechanism in photographic action described in the next section. The very old suggestion (Baur, 1925) that the equilibrium of

$$[Fe(CN)_6]^{3-} \rightleftharpoons Fe^{3+} + 6(CN)^- \tag{2.79}$$

is displaced to the right upon illumination, in order to account for the photosensitivity of potassium ferrocyanide solution in the presence of *ferricyanide*, is an example of a similar reaction in solution.

The presence of these displaced electrons in biological systems, produced either by enzyme reactions or by irradiation can be observed by electron-spin resonance techniques (Ehrenberg, 1961). This is an extremely powerful and sensitive method of assay for primary photochemical reactions of electron transfer, since one would expect that at liquid nitrogen temperatures dark chemistry would be inhibited. These transient measurements have also been investigated by Grossweiner and Zwicker (1959, 1961; Zwicker and Grossweiner, 1963) by flash photolysis and are discussed in Chapter 5, Section 7 in relation to the mechanism of sensitized photooxidation.

Recently Franck and Rosenberg (1964), along these lines, have proposed a model for the photochemical steps in photosynthesis where the absorbed light energy is delivered to a unique type of chlorophyll-enzyme complex. They postulate only one type of center for photochemical action, a chlorophyll *a* molecule exposed to water and its solutes and complexed with cytochrome *f* and an unknown electron-transport enzyme. The net effect is an electron transfer to form a highly reducing compound through which TPN (triphosphopyridine nucleotide) is reduced. This type of charge transfer complex is discussed more completely in Chapter 5, Section 2.

3.4. MECHANISM OF PHOTOGRAPHIC ACTION

3.4.1. *Silver Halides*

A photographic film composed of silver halide crystals in a matrix of gelatin just after exposure to light shows no change whatsoever to the eye or even to the electron microscope. However, very important changes have occurred, for when the film is placed in a "developing solution," the areas that have been exposed to light are converted chemically to a deposit of finely divided metallic silver while the unexposed portions remain unchanged. For this reason the terminology "latent image" has been applied to the exposed portions. The property of the "latent image" is that it can catalyze the production of atomic silver by the developing solution. In a sensitive emulsion the crystals of silver halide form triangular or hexagonal grains about 1 μ in diameter. After exposure to light, the emulsion is placed in a reducing or developing solution and the positively charged silver ions in the ionic lattice are gradually reduced to elemental silver. *This reduction process occurs only in grains which contain a "latent image," i.e., have been exposed to light.* If samples of the emulsion are removed at successive times during development and the unreduced portions of the silver halide grains are dissolved away by "hypo," the progressive growth of the image, composed of clumps of elemental silver can be observed in the electron microscope. At the beginning of development isolated dark clumps of filamentary silver can be seen, appearing at the edges of what were the original silver halide grains. As development proceeds the clumps extend more and more, until finally the whole grain has been converted to metallic silver. It is assumed that the original clumps are the sites of the "latent images" and that the reduction of the silver ions by the reducing solution is catalyzed by the presence, at the original site, of minute particles consisting of a minimum of 3 or 4 atoms of metallic silver, *produced by the action of the original light.* An average grain of a sensitive emulsion contains of the order of 10^9 silver ions. If these are reduced to metallic silver by the catalytic action of the latent image with less than 10 original silver atoms, we see that the amplification of the system is of the order of 10^8, comparing favorably in this respect with that of the high-gain electron multiplier phototube and close to that of the electron multiplication in the Geiger counter. The effect of light on the silver halide crystal is to produce one or more "reduced centers" or "latent images."

The photoconductivity of alkali halide crystals has been known for some time and has been the subject of extensive research. From the nature of the photoconductivity currents the charge carriers are electrons with high mobility. In addition to this electron conduction produced by light, there is an electrolytic conduction, independent of light but dependent upon temperature, of a small fraction of the positive silver ions.

In the crystal the positive and negative ions at their particular lattice sites effectively neutralize each other's charge. It is, in fact, this neutralization that determines the equilibrium spacing of the lattice. When an electron of a negative bromine ion acquires sufficient energy to be raised to a "conduction band," this is in a sense equivalent to physically removing the electron to a greater average distance from the bromine parent, analogous to the larger orbital of an electron in the excited state of an atom or molecule. Once removed, the electron, no longer being strongly bound to the bromine, can wander away from it, i.e., it is a free electron. Under the action of an external field the electron motion becomes a current. This is the photocurrent that we observe. In the quantum mechanical theory of solid state ionic crystals the separations between bound and free electrons are rather sharp and energetically the allowable levels are separated into filled bands for bound electrons and conduction bands for free electrons. Thus it is customary to talk of an electron from a filled band being raised to a conduction band as the mechanism of photoconductivity. The lack of an electron at some particular bromine site effectively makes that originally neutralized region net positive. There is now a "hole" in the filled band where an electron used to be, since the electron is no longer bound at that site. This "hole" with a net positive charge is therefore a positive hole. Thus in our terminology, light absorbed in an ionic crystal produces a free or conduction electron and a positive hole. At this point one of several things can happen.

(a) The free electron has not moved too far away. It can recombine with the positive hole to reconstitute the original bromine ion. This is analogous to the transition from the first excited state to the ground state in ordinary atoms or molecules. In both cases radiation may be emitted.

(b) The free electron may migrate away from the positive hole and eventually combine with a positive silver ion to form a silver atom, or it may combine with some other positive hole.

(c) An electron from a neighboring negative bromine ion may "jump" into the "positive hole" leaving a positive hole at its own bromine site. In this way positive holes may migrate in the crystal.

Obviously for efficient photographic action we should like to prevent recombination and insure reduction of silver ions. If there were some means of trapping or modifying the positive hole once it was formed, or of trapping the electron, the probability of recombination of the free electron and the hole would be reduced and the only reaction available would be the reduction of a silver ion. Actually, it is found that in very pure silver bromide the recombination of electrons and holes is extremely rapid and that this very pure material is extremely insensitive photographically. In the same way in which transistor action in semiconductors depends upon the nature and concentration of minute "impurities," photographic action depends on the presence of "impurities" in the AgBr lattice. The impurities cause local distortions in the crystal field and can therefore "trap" or bind electrons or positive holes so that they cannot recombine with one another.

It had been realized for a long time that aside from its advantage as a suspension medium, gelatin was an efficient sensitizer of AgBr grains.

It was found by Sheppard of the Eastman Kodak Laboratories that slight traces of sulfur in the gelatin react with AgBr to form traces of Ag_2S at the surfaces of the grains. These "impurities" can then act as "trapping" centers for either the electrons or the positive holes, preventing recombination. At the present time there are two schools of thought as to whether the silver sulfide traps electrons or positive holes. The steps of trapping in the two schemes are shown in Table 2.5.

TABLE 2.5

ELECTRON TRAPPING AND HOLE TRAPPING MECHANISMS IN PHOTOGRAPHIC ACTION

Step	Electron trapping mechanism	Hole trapping mechanism
0	AgBr crystal with a trace of Ag_2S on the surface	AgBr crystal with a trace of Ag_2S on the surface
1	Absorbed photon produces free electron and positive hole	Absorbed photon produces free electron and positive hole
2	High mobility electron becomes trapped at Ag_2S site on surface	Electron is captured directly by a Ag^+ ion in an active site at the surface to form Ag atom
3	$(Ag_2S)^-$ now attracts an interstitial Ag^+ ion to form Ag atom	
4	Second absorbed photon repeats steps 2 and 3. Two Ag atoms now form Ag_2, a more stable diatomic silver molecule	Second absorbed photon repeats step 2. Two Ag atoms now form Ag_2, a more stable diatomic molecule
2a	Positive hole migrates to surface by electron jumping to become an adsorbed Br atom	Positive hole migrates to surface by electron jumping and is trapped at Ag_2S site.
4a	Second positive hole migrates to surface, combines with first Br atom, can leave surface as a Br_2 molecule	Second positive hole is trapped at Ag_2S site

The silver atom itself can react with a positive hole, donating an electron and becoming an ion. However, the Ag_2 molecule is much more stable. Several of these molecules apparently are required to form a stable latent image.

3.4.2. Optical Sensitizers

Vogel (1873) found that a yellow dye that he had added to a silver bromide-collodion emulsion to prevent halation conferred an unusual sensitivity to the emulsion in the green region of the solar spectrum and thereby discovered the process of "optical sensitization." Thus it was possible, by adsorbing certain dyes on AgBr crystals, to obtain latent images with wavelengths of light within the absorption spectrum of the dye, as well as within the absorption spectrum of AgBr.

TABLE 2.6

OPTICAL SENSITIZERS IN PHOTOGRAPHIC EMULSIONS

HETEROCYCLIC DYES:

ACRIDINE ORANGE λ_{max} 520 mμ

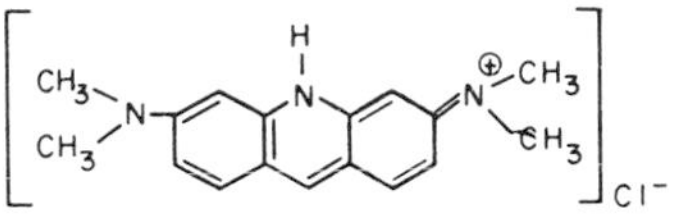

FLUORESCEIN URANINE

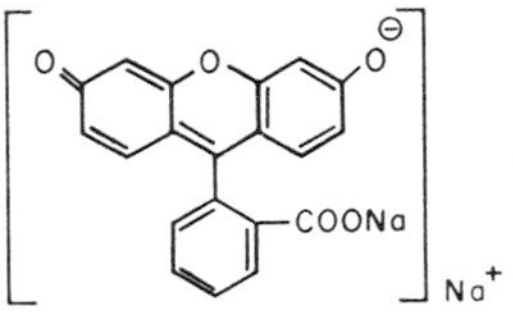

RHODAMINE

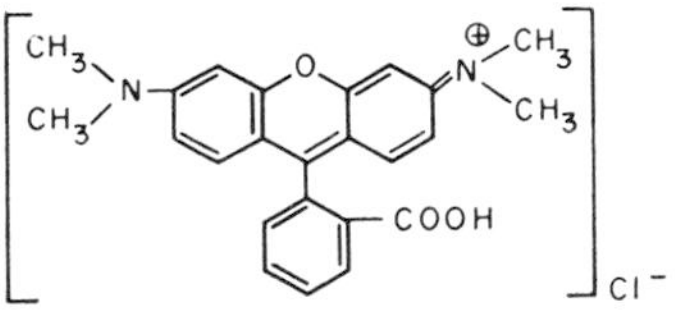

ERYTHROSINE λ_{max} 560 mμ

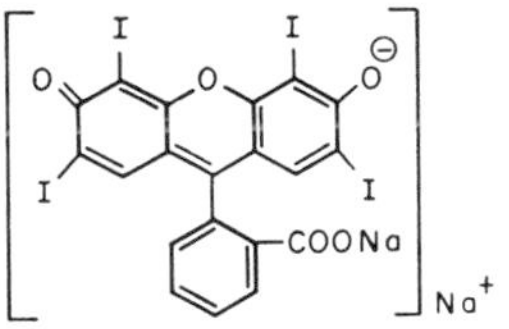

MALACHITE GREEN

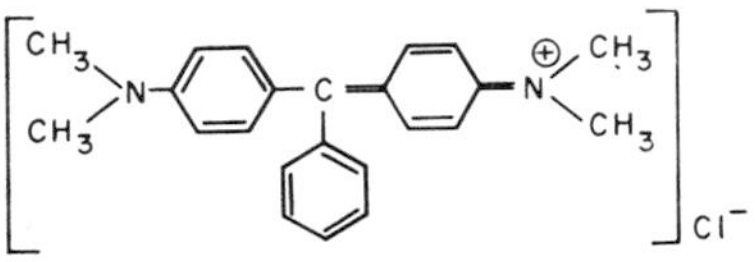

In the very next year E. Becquerel (1874), father of the discoverer of radioactivity first used chlorophyll as a red-sensitizing dye for photographic plates. This makes a question in the back of our minds have an even stronger significance. Does the optical sensitization of AgBr crystals by chlorophyll have anything to do with photosynthesis? Is it photosynthesis? Let us examine what is known of these mechanisms in photography. In Table 2.6 are given many of the known efficient optical sensitizers for latent image formation in silver halide emulsions. The characteristics that all of these dyes possess are (a) planar hetereocyclic rings coupled by (b) conjugated chains of the general form

$$X = C - (C = C)_n - Y \tag{2.80}$$

where X or Y can be oxygen or nitrogen. A more complete discussion is given in Mees (1946). The wide variety and large range of sensitizing wavelength regions has permitted the use of photographic film to beyond 1.2 μ so that effectively the strong water absorption at 1.4 μ presents the limit of photographic sensitivity for long wavelengths.

The wavelengths of maximum absorption are usually obtained from water or ethanolic solutions of the dyes. There is some shift, usually to longer wavelengths, when the dyes are adsorbed on the silver halide crystals, as evidenced from the spectral sensitivity curves. In some cases the aqueous absorption spectrum is strongly modified by adsorption,

TABLE 2.6 (*Continued*)

HETEROCYCLIC NUCLEI
IN ORDER OF LONGER λ_{max}

THIAZOLINE

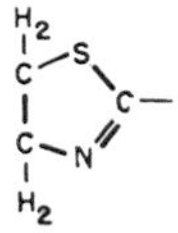

BENZOXAZOLE

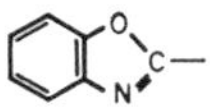

THIAZOLE

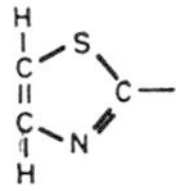

BENZOTHIAZOLE

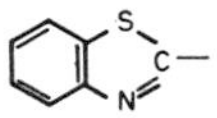

BENZOSELENAZOLE

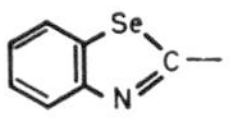

α - NAPHTHOTHIAZOLE

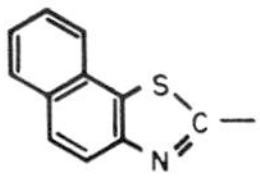

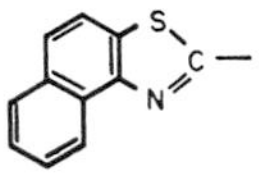

β -NAPHTHOTHIAZOLE

2 - QUINOLINE

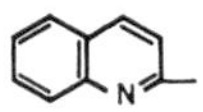

4 - QUINOLINE

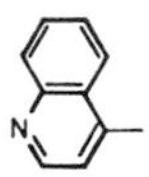

TABLE 2.6 *(Continued)*

SYMMETRIC CYANINES

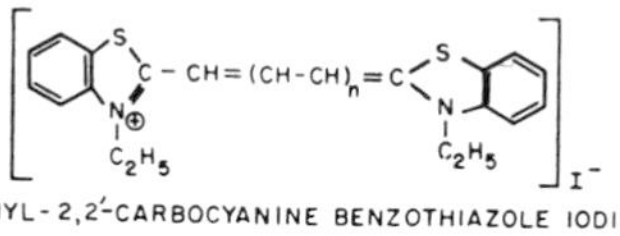

n	$\lambda_{max}\ m\mu$	COUPLING NAME
0	423	MONOMETHINE
1	557	CARBOCYANINE
2	650	DICARBOCYANINE
3	758	TRICARBOCYANINE

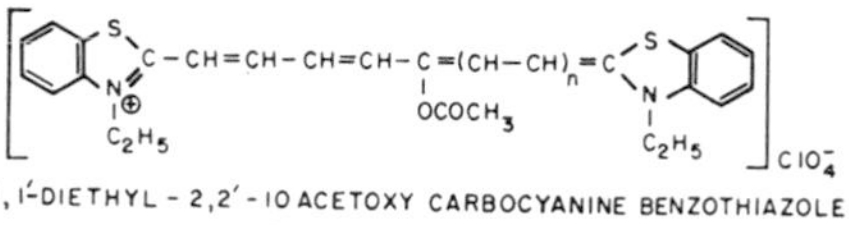

1,1′-DIETHYL-2,2′-CARBOCYANINE BENZOTHIAZOLE IODIDE SERIES

n	$\lambda_{max}\ m\mu$	COUPLING NAME
1	760	10 ACETOXY TRI-CARBOCYANINE
2	870	10 ACETOXY TETRA-CARBOCYANINE
3	995	10 ACETOXY PENTA-CARBOCYANINE

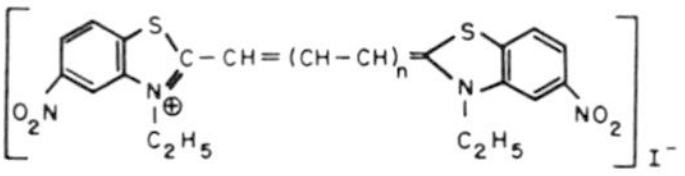

1,1′-DIETHYL-2,2′-10 ACETOXY CARBOCYANINE BENZOTHIAZOLE PERCHLORATE SERIES

n	$\lambda_{max}\ m\mu$
0	430
1	585
2	681

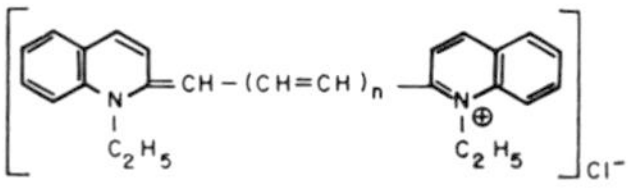

1,1′-DIETHYL-6,-6′ NITRO-2,2′-CARBOCYANINE BENZOTHIAZOLE IODIDE SERIES

n	$\lambda_{max}\ m\mu$	COUPLING NAME
0	522	DIETHYL PSEUDO-CYANINE
1	605	PINACYANOL
2		

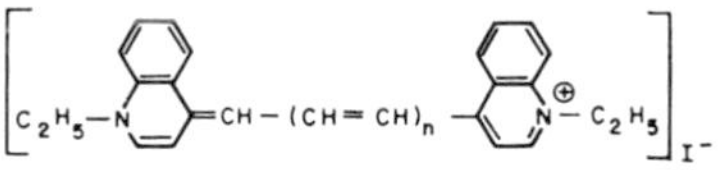

1,1′-DIETHYL-2,2′-CARBOCYANINE QUINOLINE CHLORIDE SERIES

n	$\lambda_{max}\ m\mu$	
0	590	
1	710	(KRYPTOCYANINE)
2	810	

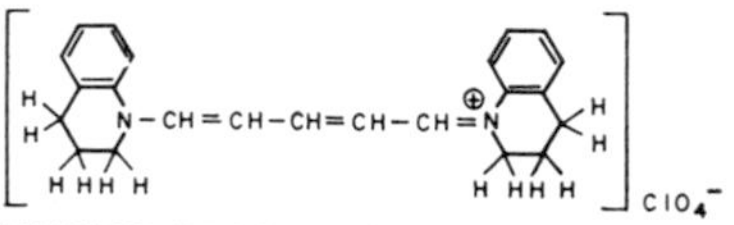

1,1′-DIETHYL-4,4′-CARBOCYANINE QUINOLINE IODIDE SERIES

GLUTACONALDEHYDE DITETRAHYDROQUINOLIDE PERCHLORATE
2,3,4,2′,3′,4′-DIHYDROQUINOLINE-1,1′-DICARBOCYANINE PERCHLORATE

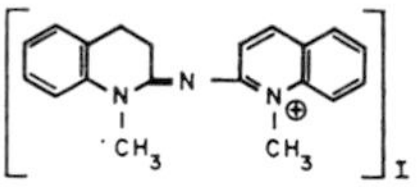

1,1′-DIMETHYL-2,2′-AZACYANINE QUINOLINE IODIDE

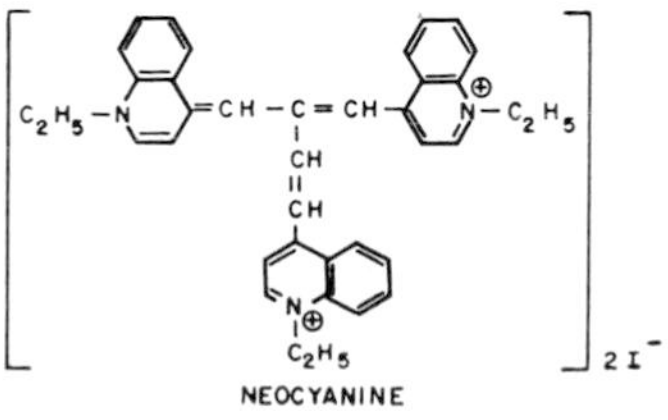

λ_{max}
820 $m\mu$

NEOCYANINE

possibly owing to "polymerization" or simple orderly stacking of the planar dye molecules on the surface of the crystals. Present evidence indicates that the stacking is the important consideration. We have discussed this previously in the experiments of Scheibe (1939) with diethyl pseudoisocyanine chloride. The formation of a very sharp absorption or J band in concentrated solutions in aqueous solution is shown in Fig. 2.21. The corresponding absorption spectrum of diethyl pseudoisocyanine adsorbed as a less than monomolecular layer on a crystal of silver bromide

TABLE 2.6 (*Continued*)

UNSYMMETRIC CYANINES

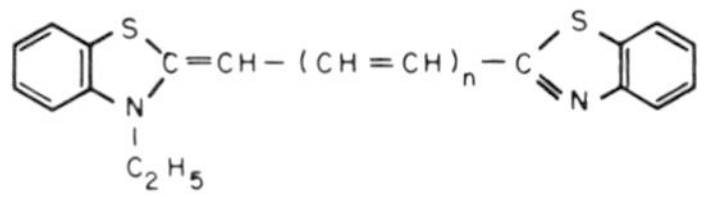

n	$\lambda_{max}\ m\mu$
0	395
1	460
2	490

1-ETHYL - 2,2′-CARBOCYANINE BENZOTHIAZOLE SERIES

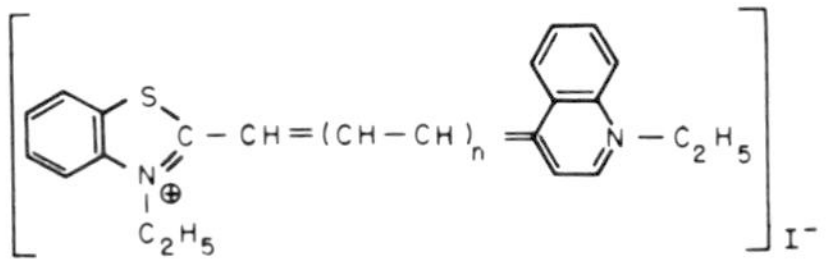

n	$\lambda_{max}\ m\mu$
0	502
1	630
2	728

1-ETHYL BENZOTHIAZOLE - 2,4′CARBOCYANINE-1′-ETHYL-
QUINOLINE IODIDE

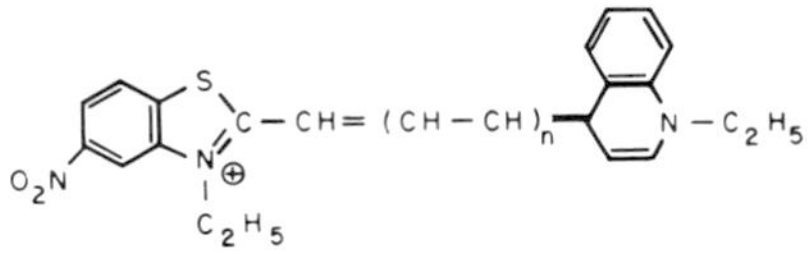

n	$\lambda_{max}\ m\mu$
0	505
1	619
2	680

1-ETHYL-6-NITROBENZOTHIAZOLE-2,4′-CARBOCYANINE
1′-ETHYLQUINOLINE IODIDE SERIES

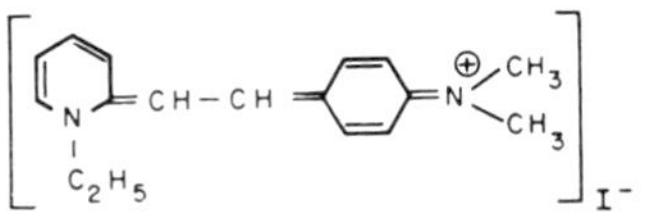

PINAFLAVOL
1-ETHYLPYRIDINE-(1-VINYLENE) PHENYL-P-A-DIMETHYL
NITROIODIDE

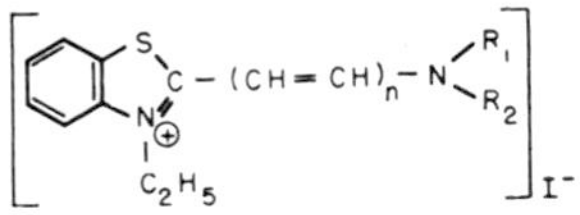

HEMICYANINES

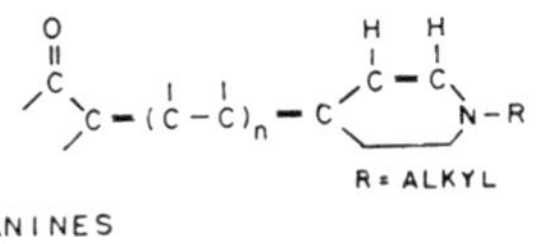

R = ALKYL

MEROCYANINES

as a function of the degree of coverage of the crystal is shown in Fig. 2.22. West and Carroll (1951) have reported that *dimethyl pseudoisocyanine*, while absorbing in alcoholic solution essentially the same as the diethyl derivative, shows only a trace of the J band on adsorption on chlorobromide emulsions. *However, with low concentrations of supersensitizers, the quantum yields for latent image formation at the J band (575 mµ) are*

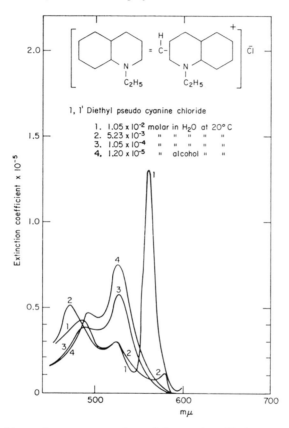

1, 1' Diethyl pseudo cyanine chloride

1. 1.05×10^{-2} molar in H_2O at $20°C$
2. 5.23×10^{-3} " " " " "
3. 1.05×10^{-4} " " " " "
4. 1.20×10^{-5} " alcohol " "

FIG. 2.21. Absorption spectrum of pseudoisocyanine chloride. From Mees (1946).

the same for both the diethyl and the dimethyl sensitizers in spite of the observed absorption spectrum difference at 575 mµ. The action of supersensitizers will be discussed next.

In basic dyes adsorption consists effectively of the electrostatic attraction of the dye cation for an external bromide anion of the AgBr lattice. In acid dyes the anion is held by the silver cation. Thus, in adsorption to polar surfaces the polar portion of the dye is fixed at the crystal lattice with the nonpolar portion extending into the medium. Optimum sensitization is obtained when the surface is covered with the molecules piled leaf-like in as close a packing as is consistent with their own structure. A monomolecular layer of adsorbed dye gives maximum sensitization. Beyond this the sensitization falls off rapidly, in some cases the sensitivity is less than that observed in the complete absence of dye.

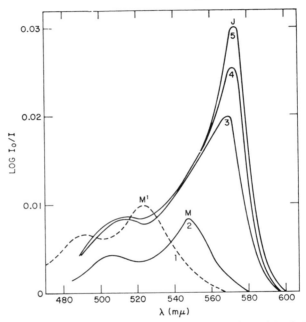

FIG. 2.22. Absorption spectra of the bromide of 1,1′-diethyl-2,2′-cyanine, adsorbed as a monomolecular film on a crystal of silver bromide from a 1 mg/liter aqueous solution of the dye. In this experiment the dye in solution was of an amount sufficient to cover only 75 % of the surface with a monomolecular layer of the aggregate J. After West (1958).

Curve 1: spectrum of a solution diluted with methanol.

Curves 2–5: spectra of adsorbed layers. (2) After 3 min in the dye solution; (3) after 10 min in the dye solution; (4) after 20 min in the dye solution; (5) after 50 min and up to 17 hr in the dye solution; (M′) absorption maximum for molecules in solution; (M) absorption maximum for molecules adsorbed in an isolated condition; (J) maximum of the aggregate layer J. After West (1958).

Mechanisms of optical sensitization are shown in Fig. 2.23 (West, 1962b). Here we have the general picture of the valence (filled) band and the conduction (free electron) band in the AgBr crystal. There is no definite limit to the long wavelength absorption edge. As the temperature increases, levels in the gap between the valence band and the conduction band become occupied and therefore lower energy transitions can be observed.

Optical sensitization occurs because of impurities and crystal defects in the AgBr crystal. These form energy levels which extend between the valence and conduction bands. In the energy-transfer mechanism of Mott (1948) (Fig. 2.23b) the excitation energy of the dye molecule is transferred by resonance induction to a bromide anion at an occupied surface state M. The energy is sufficient to raise the bromide electron to the conduction band, leaving a positive hole. In the electron transfer mechanism of Gurney and Mott (1938) (Fig. 2.23a) the dye molecule itself by virtue of

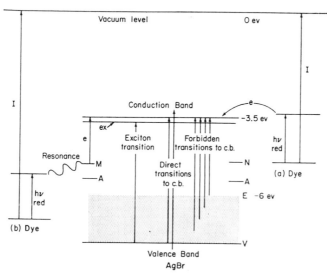

FIG. 2.23. Energy levels for AgBr and for a sensitizing dye, acting (a) by electron transfer and (b) by energy transfer (West, 1962b).

being adsorbed is at an energy level below the conduction band. Excitation by light raises a *dye electron* to the conduction band, leaving a dye positive hole or radical. The level N, representing an energy rich bromide ion site, now acts as an electron donor to the dye molecule. The net result in both cases is a free electron and a positive hole.

3.4.3. *Supersensitization and Antisensitization*

The phenomenon of "supersensitization" occurs when the sensitization produced by a mixture of sensitizing dyes is greater than the sum of the individual effects. West and Carroll (1951) have reported a striking example of this effect on the sensitizing dyes diethyl pseudoisocyanine and dimethyl pseudoisocyanine by the merocyanine dye, 4-[(3-ethyl-2(3H)-benzothiazolylidene)isopropylidene]-3-methyl-1-(p-sulfophenyl)-5-pyrazoline. The absorption band of this dye alone is in the blue-green region. It itself does not confer optical sensitization at 575 mμ, the J band of the pseudoisocyanines. Yet, without increasing the absorption of light by the combination of diethyl pseudoisocyanine-merocyanine the addition of the supersensitizing merocyanine dye produced a 17-fold increase of sensitization at 575 mμ over that of the sensitizer alone. The quantum yield of sensitization at 575 mμ was increased from 0.05 with the dye alone to practically 1.0 in the presence of the supersensitizer.

The inverse phenomenon has also been observed, i.e., a relatively small concentration of certain dyes will markedly decrease the optical sensitization of the sensitizing dye, much more so than they decrease the silver halide sensitivity when they are adsorbed alone. These are termed "antisensitizers."

In general dyes whose absorption spectra when adsorbed on silver bromide resemble their solution spectra are least affected by supersensitizers. Those which exhibit J bands on adsorption are most susceptible. In addition, the sensitizer-supersensitizer combination is quite specific; a supersensitizing dye for one sensitizer need not work on another.

At normal sensitizing concentrations of the order of 10^{-10} mole of dye/cm^2 of surface (Mees, 1946, p. 1069), it is assumed that approximately one-half of the surface of the crystal is covered. Dyes of the cyanine class are in closely packed islands of stacked molecules with relatively open spaces between them. The sharp J band is a sign of exciton migration of energy (Franck and Teller, 1938). In addition, the greatest absorption of polarized light occurs when the light electric vector is parallel to the "fiber" or filament of molecules built up and perpendicular to the N$^+$-N or planar axes of the individual molecules. This exciton migration of energy will therefore extend over the greatest number of individual molecules, so that even a minute singularity anywhere in the regular potential field of the filament could trap the exciton, facilitating transfer to the silver halide lattice. The criterion here is that the supersensitizer must be capable of entering into the adsorbed islands rather than aggregating by itself outside these islands. This mechanism has support in that in adsorbed sensitizing dyes which show, as the concentration on the silver halide is increased, the transition from a molecular absorption spectrum to the J-band absorption spectrum, the supersensitizing effect occurs only when J-band absorption appears. In this same sense there is a marked correlation between the ability of a dye to quench the J fluorescence of an aggregating dye and its ability to produce supersensitization.

Molecular planarity is essential for sensitizing action by a dye. Proceeding along this hypothesis and the above mechanism of supersensitization, it has been shown by West and Carroll (1951) that the nonplanar dye cation 1,1',3,3'-tetramethyl-2,2'-cyanine quinoline strongly depresses

the sensitization by 1,1'-diethyl-2,2'-cyanine quinoline (diethyl pseudo-isocyanine). As a result of the methyl groups on the 3,3' positions the former dye cannot be planar and must be twisted; the nuclei joined by the methine bridge are forced from coplanarity. Again just those dyes which exhibit strong J bands are most affected by the antisensitizing action of nonplanar dyes. In the same type of experiments as above, with sensitizing dyes that show a transition with concentration to a J-band absorption peak, antisensitization occurs with the appearance of the J band and in degree is related directly to the intensity of the J band.

The sensitizer excitation energy must be sufficient to bring about electronic excitation of the antisensitizer. However, there is no loss of efficiency of antisensitization at sensitizer absorption bands considerably shorter in wavelength than those of the antisensitizer. Presumably there is a large range of vibronic energy levels available to the antisensitizer included in the sensitizer island so that excess energy can be dissipated easily.

The mechanisms of latent image formation, sensitization, supersensitization, and antisensitization described above all appear to be consistent with experimental observations.

West (1962a,b) has found that sensitized emulsions, exposed at $-196°C$ and developed at room temperatures and compared with emulsions exposed at room temperature have from 10 to 100 times less sensitivity in the *sensitized region* than they have in the 400 mμ region, where silver halide itself absorbs. He further observed that at $-196°C$ only light absorbed by AgBr itself (blue, 400 mμ) excited a green luminescence. Even with sensitizer light 6000 $\times$ the original blue intensity, no luminescence was observed. The mechanism for the emission of luminescence requires the recombination of free electrons and positive holes. Since both the AgBr absorption and the sensitizer absorption lead to latent image formation requiring the liberation of free electrons, it may be that dye sensitization differs from the primary absorption by AgBr in that in the former, *free* positive holes are not produced, preventing migration as in the case of AgBr absorption.

Both the temperature dependence of the dye sensitization compared with AgBr absorption and the luminescence owing to AgBr absorption point up essential differences between the mechanisms of primary and sensitized formation of latent images. Further work will be required to complete the picture.

No doubt as we describe the various photoreactions in biological systems—photoreduction, energy transfer in structural systems, trapping centers, photoisomerism, the effects of impurities, etc.—the mechanisms of photographic action described here will present intriguing analogies. The mechanism of energy transfer in aggregates is analogous. The adsorption and stacking of pigment molecules is analogous. The trapping of exciton energy by a "supersensitizer" or an "antisensitizer" is analogous. Perhaps these are all that are needed. Perhaps the concept of a protein or a membrane as a semiconductor is not even necessary.

3.4.4. *Photoconductivity in Biological Materials*

In 1941 Dr. A. Szent-Gyorgyi made the rather startling proposal that proteins may have electronic structures analogous to those of semiconductors. This has produced a great many experimental results on semiconductor properties of all kinds of dried proteins (Eley *et al.*, 1953) as well as deoxyribonucleic acid (Liang and Scalco, 1964). Many other workers have found evidence for trapping phenomena such as thermoluminescence and photoconductivity. In particular, Strehler and Arnold (1951) found that *Chlorella, Scenedesmus*, and *Stichococus*, and the higher green plants *Phytolacca americana* and *Trifolium repens* emitted luminescence for some short time after an exciting light was removed. Strehler (1951) repeated the measurements on isolated chloroplasts. The action spectrum for this delayed luminescence in *Chlorella* is shown in Fig. 2.24 and as will be described later in Chapter 5, Section 2, this is essentially the action spectrum for photosynthesis in green plants. Commoner *et al.* (1954) showed that electron-spin resonance signals gave evidence of the existence of free radicals in metabolizing tissues. They also showed an increase in ESR signal amplitudes when green leaves were irradiated by light, demonstrating the existence of stable free radicals caused by light. These observations were later extended to isolated chloroplast suspensions (Commoner *et al.*, 1956). Arnold and Sherwood (1957) then published a paper with the intriguing title "Are Chloroplasts Semiconductors?" in which they described glow curves or thermoluminescence obtained with dried chloroplast preparations exactly analogous with those obtained with the alkali halide crystals and fitting the kinetics of glow curves given by Randall and Wilkins (1945). From their data they intend the question to be rhetorical. Further work on the photosynthetic system (Arnold and Sherwood, 1959; Sogo *et al.* 1957; Tollin and Calvin, 1957;

Arthur and Strehler, 1957; Strehler and Lynch, 1957) as well as a large number of observations on the electrical properties of irradiated dry biological material have extended these ideas.

Terenin and co-workers (1959) have studied the photoinduced transfer of charges through microcrystalline dyes by standard techniques based on dynamic capacitance measurements. They find that in all of the phthalocyanine dyes, metal-free or with Ag^+, Cu^{++}, Mg^{++}, Zn^+, Co^{++}, Fe^{++},

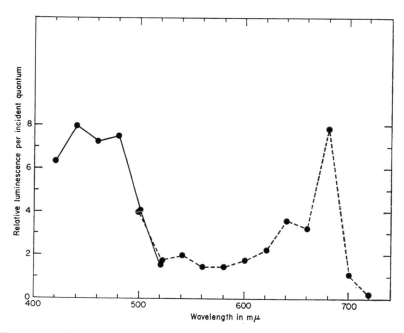

FIG. 2.24. *Chlorella* luminescence action spectrum. Abscissa, wavelength of exciting light; ordinate, relative luminescence per incident quantum. From Strehler and Arnold (1951).

and Fe^{3+}, the charge carriers are positive, i.e., there is a trapping of electrons and the observed photocurrent is due mainly to an electron exchange (hole conductivity) between the molecules at the ground level. In addition, they observed that while the relaxation time for photoconductivity in metal-free phthalocyanines was 5×10^{-5} sec, metal-bound phthalocyanines exhibited an enhanced component with a much longer relaxation time, between $5-7 \times 10^{-3}$ sec, indicating a much more effective trapping of electrons. This latter observation goes to the heart of the present problem, i.e., the sensitized photoeffect or charge separation whose interpretation, particularly in relation to photosynthesis, is still vague. The detachment of an electron from an isolated Mg-phthalocyanine dye molecule or from a layer of the solid dye requires 6.3 ev of energy or a wavelength in the UV region of 1967 Å. However, when this same dye is monomolecularly adsorbed on the surface of ZnO or the alkali halide semiconductors, a photocurrent is observed which has a spectral dependence identical with the optical absorption spectrum of the adsorbed dye. The net effect is a reduction of the photoelectric activation or electron-transfer energy from 6.3 ev to less than 2 ev, in the region of 6800 Å. This is, in effect, a repetition of the Vogel effect (Vogel, 1873), i.e., the optical sensitizing of a silver bromide photographic plate.

The charge transfer complexes exhibit, as do most other organic systems, an exponential temperature dependence of conductivity characteristic of semiconductors. The experimentally observed thermal activation energy E_a is obtained from a plot of σ versus $1/T$ from the expression

$$\sigma = \sigma_0 e^{-E_a/kt} \tag{2.81}$$

Many authors have used a term, E_g, representing the energy gap between the valence band and the conduction band analogous to an intrinsic inorganic semiconductor, where $E_g = 2E_a$ in order to obtain agreement between lowest triplet state energies obtained from phosphorescence measurements and E_g. However, there are strong arguments against the factor of 2. Fox (1961) and Kleinerman et al. (1962) state that triplet states play absolutely no part in dark conductivity.

The use of the lamellar chloroplast structure to provide a separation of the oxidation and reduction intermediates in photosynthesis has been suggested by Bradley and Calvin (1955), using the concept of the barrier layer semiconductor. In a novel series of solid-state photoconductivity experiments Calvin and co-workers (Tollin et al., 1960; Tollin, 1960; Calvin, 1961; Calvin and Androes, 1962) have used phthalocyanine as a model for chlorophyll energy transfer. The phthalocyanine molecule is similar to the tetrapyrrole of chlorophyll except that it is a tetrazaporphyrin with benzene rings fused to the pyrrole rings. The "photosynthetic" cell is a layer of phthalocyanine (donor layer) 1–10 μ thick sublimed onto interlacing aquadag fingers. The alternate aquadag fingers serve as positive and negative electrodes with approximately 100 μ spacing. Next a thin layer of acceptor molecules (o-chloranil) is sublimed on top of the phthalocyanine layer. Both the dark conductivity and the photoconductivity increase by several orders of magnitude as the o-chloranil is added to the surface phthalocyanine layer, in collecting fields of the order of 10^4 volts/cm. On the basis of photoconductivity, polarization, and ESR measurements, Calvin (1961) proposes the following series of reactions to account for the separation and migration of charge in this system.

$$
\begin{align}
\text{(a)} \quad & D + A \xrightarrow{\text{dark}} D^+ + A^- \\[4pt]
\text{(b)} \quad & D + A^- \underset{\text{dark}}{\overset{\text{"red" light}}{\rightleftharpoons}} D^+ + A^{--} \\[4pt]
\text{(c)} \quad & D + A^- \underset{\text{dark}}{\overset{\text{"blue" light}}{\rightleftharpoons}} D^- + A \\[4pt]
\text{(d)} \quad & D^+ + D^- \xrightarrow{\text{dark}} 2D
\end{align}
\tag{2.82}
$$

In the dark the strong electron acceptor, A, and the phthalocyanine electron donor, D, react to form a pair of radical ions. In "red" light a double negative negative ion of o-chloranil (A^{--}) is formed which is reversible in the dark. In "blue" light the anion absorbs, transferring an electron back into the phthalocyanine layer, leading to recombination and a *decrease* in conductivity. This is a photoinduced separation of oxidizing power (positive holes) and reducing power (o-chloranil double negative ions). The presumption is made that a similar mechanism is active in chloroplasts, since photoconductivity and thermoluminescence have been demonstrated in dried chloroplast preparations (Arnold and Thompson, 1956; Arnold and Clayton, 1960).

One assumption made is that a charge-transfer complex is formed between the adsorbed electron acceptor impurity o-chloranil and the host donor phthalocyanine and that under the action of the electric field the phthalocyanine shows p-type conduction. The formation of this charge-transfer complex in the dark is presumed to give rise to the observed ESR signal. All of the effects, such as photoconductivity, light-induced polarization have action spectra corresponding to phthalocyanine absorption. This proposed mechanism has recently been criticized (Franck, *et al.*, 1962).

One of the problems in the interpretation of photoconductivity measurements is the separation of the bulk photoconductivity effects from the surface photoconductivity effects. For example, Schneider and Waddington (1956) and Compton and Waddington (1956) have shown that in the presence of O_2, Cl_2, or NO_2, the dark semiconductivity and photoconductivity of anthracene crystals is increased by many orders of magnitude. Vartanyan and Karpovich (1956) observed this same effect by O_2 for metal-phthalocyanine complexes. Presumably, the general effect of these oxidizing agents or electron-accepting impurity molecules is to increase the mobility of the charge carriers, although Kearns, Tollin, and Calvin postulate the production of charge carriers.

3.5 Photochromism and Photoisomerism

Stereoisomeric changes in organic compounds are very often brought about by light. One of the earliest and most investigated cases is the reversible transformation between malic and fumaric acids. A description of this reaction will produce several analogies in mechanism to light reactions in biological systems. In particular, the reader is referred to Chapter 5, Section 3 for a discussion of the photochemistry of the phytochrome pigment involved in the red far-red light effect in photomorphogenesis and Section 6 for the photoisomerism of 11-*cis* retinene.

In the Krebs (citric acid or tricarboxylic acid) cycle in the metabolism of carbohydrate, the complete oxidation of acetyl-coenzyme A begins with the condensation between oxalacetic acid and acetyl CoA to form citric acid and regenerate CoA. Through a series of further steps involving isomerization, oxidations, and decarboxylations, all catalyzed by enzymes, we arrive at fumaric acid which by the action of the enzyme fumarase is reversibly hydrated to malic acid.

$$
\begin{array}{ccc}
 & & H \\
 & & | \\
H{-}C{-}COOH & \xrightarrow[-H_2O]{+H_2O \atop fumarase} & HO{-}C{-}COOH \\
\| & & \text{\textbar}\,\text{\textbar} \\
HOOC{-}C{-}H & & H{-}C{-}COOH \\
 & & | \\
 & & H \\
\text{fumaric acid} & & \text{malic acid}
\end{array}
$$

Malic acid forms the last step in the cycle, for by the action of malic dehydrogenase it is reversibly dehydrogenated to oxalacetic acid.

This same reversible hydration reaction in aqueous solution can be brought about by short wavelength UV light. Both fumaric and malic acid are sensitive to ultraviolet. When aqueous solutions of either are subjected to UV irradiation, a photostationary state is obtained in which the ratio of malic acid concentration to fumaric acid concentration is about 7 to 3. The equilibrium concentrations are independent of light intensity and wavelength as would be expected. Warburg (1919) measured the quantum yields for these reactions at several wavelengths and these are shown in Table 2.7. Warburg also noted that the quantum yield for F → M increased slightly with increasing concentration of fumaric acid but that the quantum yield of M → F decreased slightly with increasing concentration of malic acid. However, these appear to be small effects. Here we have two different compounds that absorb in the same wavelength region, and the photostationary equilibrium concentration is determined by their respective quantum yields for hydration or dehydration.

Phototropy or photochromism was first observed for tetrachloro-α-

TABLE 2.7
QUANTUM YIELDS OF UV PHOTOCONVERSION IN 0.01 N
FUMARIC AND MALIC ACIDS

	Quantum yield	
λ mμ	F $\rightarrow$ M	M $\rightarrow$ F
207	0.11	0.03
254	0.10	0.04
282	0.13	0.03

ketonaphthalene, a normally white substance which reddens upon illumination and returns to white in the dark or upon heating. More recently Hirshberg and Fischer (1954a,b), and Heiligman-Rim *et al.* (1961) have worked extensively on photochromism in various spiropyrans, and Fischer and Kaganowich (1961) have reported on photoisomerism in the quinoid-1,2-naphthoquinone-2-diphenylhydrazone,

form I

When solutions of this compound in polar (ethanol) or nonpolar (methylcyclohexane-isohexane) solvents are irradiated to photoequilibrium with light at various wavelengths, the absorption spectra change markedly as shown in Fig. 2.25. The well-defined isobestic points throughout the visible and UV regions, indicating the existence of two isomeric states I and II, with long wavelength absorptions in ethanol peaking at 480 and 550 mμ, respectively. Interestingly enough at thermal equilibrium in ethanol there is 98% of I (480 mμ absorbing). By irradiation within this absorption band we can shift the equilibrium. In the dark there is a slow spontaneous return to the thermal equilibrium concentrations. At 19°C in ethanol the half-time for thermal equilibration, I $\rightleftharpoons$ II, starting with an excess of II was 47 min. Unfortunately, Fischer and Kaganowich did not measure quantum yields for these reactions. The reader is again referred to the discussion of the phytochrome pigments in Chapter 5, Section 3. In this present case we have

$$\text{I}_{480} \rightleftharpoons \text{II}_{550} \text{ (predominantly I in dark)}$$

whereas in the phytochrome case we have

$$P_{660} \rightleftharpoons P_{730} \text{ (predominantly } P_{660} \text{ in dark)}$$

A more extreme case of separation of absorption bands in photochromism is given by the reversible reactions

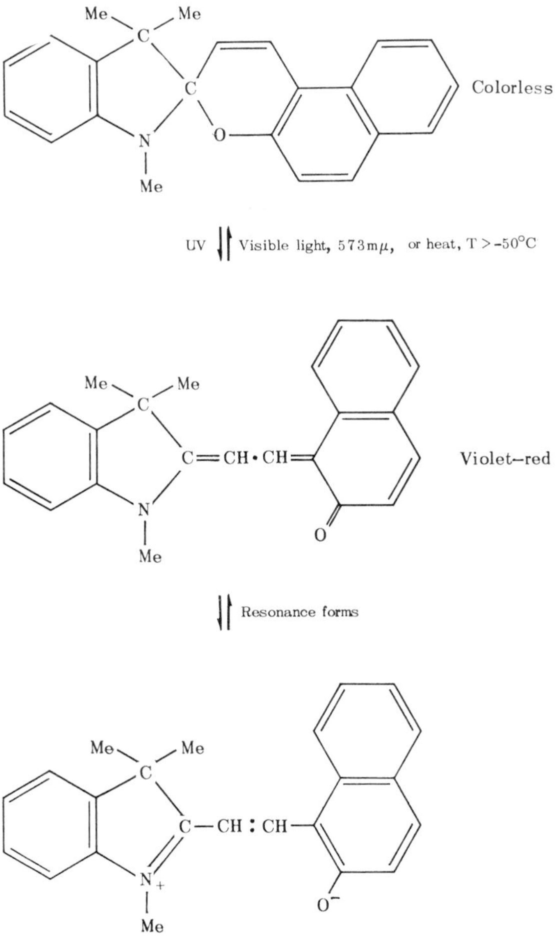

Colorless

UV ⇅ Visible light, 573 mμ, or heat, T > -50°C

Violet–red

⇅ Resonance forms

The absorption spectra upon progressive irradiation with UV light are shown in Fig. 2.26. We go from a colorless form, the normal room temperature isomer, to a highly colored (yellow-absorbing) form, stable only at low temperatures. The effect is modified by the nature of the solvent and the viscosity of the medium. We shall discuss the photoisomerism of rhodopsin in Chapter 5, Section 6.

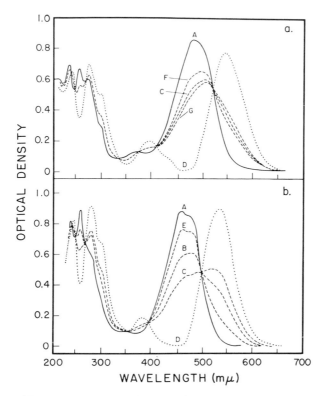

Fig. 2.25. Absorption spectra of 1,2-naphthoquinone-2-diphenylhydrazone at −100°C in (a) ethanol, 3.06 × 10⁻⁴ M; (b) in methylcyclohexane-isohexane, 1.53 × 10⁻⁵ M. Curves "A" and "D" are computed for pure I and II, respectively. In ethanol, A is practically identical with that for thermal equilibrium (98% form I) and therefore further irradiation to photoequilibrium with 578 mμ light where form II only absorbs will not increase I. However, in the nonpolar solvent at thermal equilibrium form I is present to only 84% corresponding to curve E of (b). If this thermally equilibrated mixture is irradiated with 578 mμ light at −100°C the remaining 16% of form II is converted to form I. This is the curve A of (b). For both sets of curves, B, photoequilibrated at 436 mμ; F, photoequilibrated at 510 mμ; G, photoequilibrated at 480 mμ (Fischer and Kaganowich, 1961).

It is possible to conceive, therefore, of some similar type of photochromic dye, adsorbed on some cellular particle or membrane where the restrictions may simulate the effect observed in homogeneous solution only at low temperatures or high viscosities. One must be careful, therefore, when extrapolating from photochemical reactions in homogeneous solution to reactions occurring in biological systems. Conversely the lack of an analogous effect in soluble extracts may not preclude the reaction on a lamellar structure.

A very interesting example of UV- and infrared-induced photoisomerization is given in Section 2.1. Hall and Pimentel (1963) have been able
to isomerize nitrous acid, HONO, with infrared light with essentially
100% efficiency. The frequencies corresponded to the OH stretch frequencies of the molecule. UV irradiation will also produce isomerization.

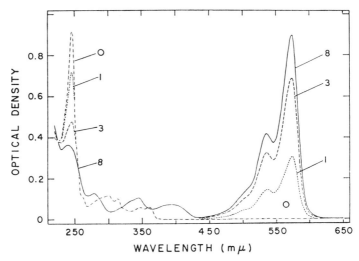

FIG. 2.26.   Spectra of spiran solution 9×10^{-6} M in ethanol-methanol irradiated with 365 mμ UV light at $-100°C$. The numbers on the curves denote the time
of irradiation in minutes. There can be seen several well-defined isobestic points
(Heiligman-Rim et al., 1961).

Presumably the molecule, in returning to the ground state from its first
excited state, passes transitorily through this particular vibration level
of the ground state and owing to efficient intramolecular energy transfer,
the reaction can proceed. Thus the isomerization is a ground-state
reaction.

CHAPTER 3

Chemiluminescence— Reactions and Mechanisms

In Chapter 2 we have described the production of excited states of molecules by light. In the present chapter we shall be concerned with the production of excited states where the energy is the result of a specific chemical reaction. Not all chemically reacting systems which emit light are necessarily emitting *chemiluminescence*. It may be that the energy liberated by a highly exergonic reaction can raise the temperature of the macroscopic system to that of incandescence. Pre-illumination, mechanical or electrical influences must also be discounted.

The process of chemiluminescence can be differentiated from pure temperature radiation in that *over some spectral range* the emission from the system enclosed in a blackbody cavity will be of greater intensity than would be emitted by a blackbody at the same temperature at which the chemical reaction occurred. Ordinarily, even in the absence of a blackbody enclosure for the chemical reaction, the observation of an emission which in some spectral region is more intense than the corresponding blackbody emission is definite evidence of chemiluminescence. For example, blackbody radiation at temperatures less than 500°C is not visible to the naked eye. Therefore any *visible* radiation emitted by a chemical reaction below 500°C is definitely a *chemiluminescence*. We must be careful, however, since the absence of visible radiation does not necessarily eliminate the possibility of chemiluminescence. Only the blackbody cavity experiment described above can demonstrate the absence of chemiluminescence. As in the case of fluorescence, the emitting molecular species is not in a state of thermodynamic equilibrium with its neighbors. In principle, the only requirements for chemiluminescence are a sufficiently exergonic reaction and a suitable emitter which is formed in an excited electronic or vibrational state or which can be raised by the absorption of energy to an excited state.

Many chemical reactions have been observed to emit light. These are:

1. Oxidation of higher alcohols, aldehydes, fatty acids, amides, polyphenols, sugars, urea, albumin, and a large variety of other biological

149

and organic compounds by oxygen, ozone, or hydrogen peroxide in alkaline or alcoholic potassium hydroxide solutions.

2. Oxidation in solution by hypohalogen (iodite, bromite, or chlorite), perhalogens, or acid permanganate as in the case of pyrogallol.

3. Oxidation of organic compounds by heating in air or by heating with anhydrous alkali in air.

4. Action of oxygen or the halogens on vapors of the alkali metals, on ammonia, on molten stearic and palmitic acids and their salts.

5. Hydration of calcium and barium oxides and of calcium chloride.

6. Neutralization of alkaline oxides with concentrated acids.

7. Cold flames; ether vapor, preheated to 260°C and let into air gives a bluish low-temperature flame. Stearic acid at 280°C, oleic acid, olive oil, and paraffin wax at 310°C, and sulfur at 180°C also give low-temperature *visible* flames.

8. Grignard reactions; the reaction between phenylmagnesium bromide or iodide and chloropicrin.

9. Action of halogen vapors on mercury vapor.

10. Reactions between sodium vapor and the vapors of mercuric chloride, iodide, and cyanide, cadmium iodide, hydrogen chloride, and phosphorus trichloride.

11. Oxidation in air of phosphorus, phosphorus trioxide, and freshly cut surfaces of sodium and potassium at 70°C.

12. Oxidation of unsaturated silicon compounds.

13. Recombination of hydrogen atoms and oxygen molecules.

14. Bioluminescence; enzyme-catalyzed oxidation by oxygen of specific substrate molecules in biological systems.

15. Anodic oxidations in electrolytic cells.

16. Thermal decomposition reactions.

Most of the chemiluminescent reactions described above are extremely inefficient in terms of light emission and in some cases the chemiluminescence quantum yield, Q, the number of photons emitted per molecule reacted, can be as low as 10^{-14}–10^{-15} (Audubert, 1939). They can, however, be as high as 1 in the case of enzyme-catalyzed chemiluminescence or bioluminescence (Seliger and McElroy, 1959, 1960b).

The recombination of hydrogen atoms in the presence of alkali metal vapors or anthracene vapor gives rise to emissions corresponding to the line spectra of the metal vapors or the fluorescence of anthracene, respectively (Bonhoeffer, 1925). Similarly, the recombination of hydrogen atoms and oxygen molecules will give rise to an infrared chemiluminescence corresponding to stretching of OH bonds (Charters and Polanyi, 1960). The recombination of sodium and chlorine atoms to form NaCl in the presence of mercury vapor will induce the 2537 A mercury emission line (Kallmann and Fränz, 1925). The above mentioned chemiluminescent reactions involve extremely exergonic chemical recombination reactions and give visible, infrared and ultraviolet light emissions, respectively. Oxygen is not necessarily involved. By far the largest number of chemiluminescences observed, however, including bioluminescence, take place in solution and involve conjugated fluorescent organic compounds. In these cases the energy requirements are such (40–80 kcal/mole) that reactions where direct oxidation by oxygen can occur will be sufficiently exergonic to produce visible light emission. While most chemiluminescent reactions investigated have given a visible light emission and in some cases an emission in the ultraviolet, there is nothing to preclude the possibility that there exists a class of chemical reactions which emit infrared radiation.

The requirements for chemiluminescence are twofold:

1. Sufficient energy in a chemical reaction must become available in a single step to leave the product or intermediate molecule in a state corresponding to its first electronically excited state or a vibrationally excited state.

2. The product or intermediate molecule must be a reasonably fluorescent molecule so that de-excitation by fluorescence is a probable event, in which case we obtain *direct chemiluminescence*, or else efficient energy transfer can occur between the excited product molecule and a second fluorescent species giving rise to a *sensitized chemiluminescence*.

The first clear recognition of the distinction between these two types of chemiluminescences is due to Strutt (1913), who found that when "active" nitrogen, produced by an electric discharge in nitrogen at low pressure, reacts with metal vapors to form nitrides, the chemical reaction is accompanied by a line emission of the resonance radiation of the *free metal* atoms. Hence it is necessary that energy transfer must have preceded the act of emission.

The emission of blue light by an aqueous luminol reaction solution is an example of direct chemiluminescence, the emitting molecule being the excited ion of aminophthalic acid, which has shared the free energy of formation of the nitrogen molecule. If we now add to this reaction solution fluorescein or rhodamine B, the emission will change in color from blue to the green fluorescence emission of fluorescein or to the red fluorescence emission of rhodamine B. The excitation energy in each case has been transferred to the dye, whose absorption spectrum overlaps the emission of the aminophthalic acid ion in the same manner as wavelength shifters are used in liquid scintillation counting. Here is a specific example of a chemiluminescent reaction where we can have either direct chemiluminescence or sensitized chemiluminescence, the difference being that in sensitized chemiluminescence the emitting molecule is not destroyed in the oxidation reaction. These mechanisms can be outlined as follows:

$$A \rightarrow B^* \rightarrow B + \text{light (direct chemiluminescence)} \tag{3.1}$$

$$\begin{aligned} &A \rightarrow B^* \\ &B^* + C(A) \rightarrow B + C^* \text{ or } A^* \\ &C^* \text{ or } A^* \rightarrow C \text{ or } A + \text{light (sensitized chemiluminescence)} \end{aligned} \tag{3.2}$$

$$\begin{aligned} A \rightarrow B^* \rightarrow & B + \text{light} \\ & \Big\downarrow \begin{smallmatrix} \text{dark} \\ \text{reaction} \end{smallmatrix} \\ & C \end{aligned} \tag{3.3}$$

The third mechanism presented above (McElroy and Seliger, 1963) may hold for a number of light-emitting reactions in which the emitter is created during the chemical reaction and is rapidly destroyed subsequent to light emission.

We believe that the chemiluminescence of zinc tetraphenylporphine and of riboflavin reported by Linschitz (1961) and by Strehler and Shoup (1953), respectively, fall into the class of sensitized chemiluminescence, while in the case of firefly bioluminescence the quantum yield and the kinetic data point to the light emission being a direct chemiluminescence.

1. Gas Phase Reactions

When a stream of nitrogen is saturated with sodium vapor at a temperature of around 350°C and is led into an atmosphere of chlorine, the region of the reaction between the sodium and the chlorine is visible as a bright yellow cone 5–10 cm long, starting at the point of introduction of the sodium vapor (Haber and Zisch, 1922). The emission consists of the sodium D doublet superimposed on a thermal continuum. The net result is the formation of NaCl molecules in a highly energetic state and the emission by neutral sodium atoms must be due to energy transfer from reaction products to sodium atoms by collision. Quantum yields of the order of 10^{-4} were observed. Beutler and Polanyi (1925) eliminated nitrogen as a carrier (and also as a diluent) by introducing the reactants at low partial pressures (10^{-2}–10^{-3} mm) and found that by introducing the halogen into excess sodium vapor they could obtain quantum yields as high as 35% in the Na-Cl$_2$ reaction. There takes place the following series of reactions:

$$\text{Na} + \text{Cl}_2 \rightarrow \text{NaCl}^* + \text{Cl}^* \text{ (gas phase)} \tag{3.4}$$

$$\text{Na} + \text{Cl}^* \rightarrow \text{NaCl}^* \text{ (wall reaction)} \tag{3.5}$$

$$\text{Na}_2 + \text{Cl}^* \rightarrow \text{NaCl}^* + \text{Na}^* \text{ (gas phase)} \tag{3.6}$$

$$\text{NaCl}^* + \text{Na} \rightarrow \text{NaCl} + \text{Na}^* \text{ (gas phase)} \tag{3.7}$$

Reactions (3.4), (3.6), and (3.7) take place in every collision between reactants and are responsible for all of the observed chemiluminescence of sodium. Reaction (3.5) leads to quenching and, depending on the partial pressures and the geometry, can seriously reduce the quantum yield.

The total energy contained in the reacting system immediately following the reaction is equal to the sum of the kinetic energies of the reactants and the chemical energy released. This total energy may be shared between the reaction products. The internal energy can have discrete values determined by the allowed energy levels in the products and the kinetic energy will take up the remainder with the proviso that the laws of conservation of linear momentum and energy are followed. Thus for an excess of kinetic energy the products NaCl* and Cl* will share the available translation energy E_{kin} in the proportions

$$E_{NaCl^*} = \frac{58.45}{58.45 + 23} E_{kin} \tag{3.8}$$

$$E_{Na^*} = \frac{23}{58.45 + 23} E_{kin} \tag{3.9}$$

where the molecular weight of the NaCl molecule is 58.45 and that of the sodium atom is 23. We see, therefore, that in order for a sodium atom to utilize the *kinetic energy* of a NaCl* molecule for its own *internal energy* of excitation, a third body must be present in order to conserve momentum.

The internal energy required for the excitation of the sodium yellow doublet is 48.3 kcal/mole. Since the heat of reaction of (3.4) is 34.9 kcal/mole, Beutler and Polanyi (1928) reasoned that only reactions (3.5) and (3.6) with heats of reaction of 93.4 and 75.4 kcal/mole, respectively, could account for the sodium emission. Since reaction (3.5) is primarily a ternary (wall) reaction, only reaction (3.6) can account energetically for the observed emission. It will be informative in establishing how a reaction mechanism is established to follow their reasoning. It may be that reaction (3.6) leads to the direct excitation of the participating Na atom or the NaCl* product can transfer its energy to a Na atom which had played no part in the reaction. In an ingeniously conceived experiment, Polanyi and Schay (1928) introduced N_2 at low partial pressures ($P_{N_2} < 0.1$ mm) into the Na-Cl_2 reaction and found a very marked decrease in quantum yield. However, the same partial pressures of N_2 in pure Na vapor excited by sodium D light had very little effect on the intensity of resonance radiation. Therefore they concluded that reaction (3.7) was responsible for the observed chemiluminescence; the N_2 molecules act to quench the excited NaCl* products before the latter can interact with a Na atom. The reaction is therefore a *sensitized chemilumi-*

nescence rather than a direct chemiluminescence as might have been inferred from reaction (3.6).

In addition to the strong sodium D doublet at 5890–5896 A, other much less intense lines were observed at 3302 and 5688 A corresponding to the 1S-3P and 2P-4D transitions of the Na atom, requiring 86 and 98.2 kcal/mole, respectively, for excitation from the normal 1S state of sodium to the upper quantum level. Since the maximum energy available in reaction (3.6) is 75.4 kcal/mole it follows that ternary collisions must also be possible. For example, the collision

$$\text{NaCl*} + \text{NaCl*} + \text{Na} \rightarrow \text{Na**} + 2\text{NaCl} \tag{3.10}$$

may take place, leaving the Na atom in a very highly excited state. This type of reasoning is used in the explanation of the chemiluminescence of the Na-$HgCl_2$ reaction. In this case the reactions are

$$\text{Na} + \text{HgCl}_2 \rightarrow \text{NaCl} + \text{HgCl} + 24 \text{ kcal/mole} \tag{3.11}$$

and

$$\text{Na} + \text{HgCl} \rightarrow \text{NaCl} + \text{Hg} + 57 \text{ kcal/mole} \tag{3.12}$$

The Hg 2537 A line and the shorter wavelength Na lines which appear in this chemiluminescence require much more than 57 kcal/mole for excitation. The fact that 2×57 kcal/mole is very close to the excitation energy of the Hg 2537 A transition (112 kcal/mole) should favor a high energy transfer efficiency from 2NaCl* molecules to a Hg atom. Again the reactions

$$\text{NaCl*} + \text{Na} \rightarrow \text{Na*} + \text{NaCl} \tag{3.7}$$

$$\text{NaCl*} + \text{NaCl*} + \text{Na} \rightarrow \text{Na**} + 2\text{NaCl} \tag{3.10}$$

and

$$\text{NaCl*} + \text{NaCl*} + \text{Hg} \rightarrow \text{Hg*} + 2\text{NaCl} \tag{3.13}$$

are *sensitized chemiluminescences.*

Other mechanisms have been postulated, such as

$$\text{NaCl*} + \text{NaCl*} + \text{Hg} \rightarrow \text{Hg**} + (\text{NaCl})_2 \tag{3.14}$$

and

$$\text{NaCl*} + \text{N}_2^* + \text{Hg} \rightarrow \text{Hg**} + \text{NaCl} + \text{N}_2 \tag{3.15}$$

However, the basic nature of the reactions remains the same. Kondratjew (1928) has even proposed for the K-$HgCl_2$ chemiluminescence that reactions involving sublimation of product molecules on the wall of the reaction vessel can contribute the heat of sublimation to the available energy of the reaction. This is interesting for it is presumably the first instance where the subject of surface chemiluminescence of vapor molecules is discussed.

2. Condensed Phase Reactions

The most efficient chemiluminescent organic substrates are shown in Fig. 3.1. For historical reasons we shall begin this section with the inorganic oxidation of phosphorus.

2.1. PHOSPHORUS

The slow oxidation of phosphorus is one of the earliest examples of chemiluminescence. Oxygen is required and the reaction is strongly exothermic, the reactions being

$$2P + \tfrac{5}{2}O_2 \rightarrow P_2O_5 + 370 \text{ kcal/mole} \tag{3.16}$$

$$2P + \tfrac{3}{2}O_2 \rightarrow P_2O_3 + 74.8 \text{ kcal/mole} \tag{3.17}$$

and

$$P_2O_3 + O_2 \rightarrow P_2O_5 + 295 \text{ kcal/mole} \tag{3.18}$$

In air or in oxygen under reduced pressure there is a greenish emission in the visible as well as five groups of bands in the ultraviolet at 3270, 2600, 2530, 2460, and 2390 A. There is some question as to whether the emission extends into the 1800 A region where ozone can be formed. Petrikaln (1924, 1928) has examined the UV emission of this chemiluminescent reaction and compared it with the excitation emission of P, P_2O_5, and P_2O_3. He concludes that P_2O_3 is the emitter in the chemiluminescence based on the absence of 3270 A band emission in excited P_2O_5, although in most other respects the emission spectra of P_2O_5 and P_2O_3 are the same. In any case the light emission is most likely a direct chemiluminescence.

There are two significant experimental observations that have not been studied extensively and are quite fundamental to our understanding of the mechanism of these oxidations. The first is the fact that chemiluminescence occurs only between partial pressures of O_2 of 10^{-2} mm and that corresponding to roughly the concentration of oxygen in air. The second and more striking is the specific requirement for the presence of water vapor in order for the oxidation to proceed. This "catalytic" effect of water vapor has been discussed in Chapter 2, Section 3.2 under "Solution Photochemistry." It may be that the chemiluminescence is incidental to a chain reaction photochemiluminescence where the very short wavelength emission around 1800 A is reabsorbed by the system, enhancing the reactivity. The higher partial pressures of oxygen may quench the chain due to the photochemical production of ozone.

2.2. LUMINOL

White and co-workers have recently reviewed the general mechanism of chemiluminescent reactions with particular emphasis on luminol (5-

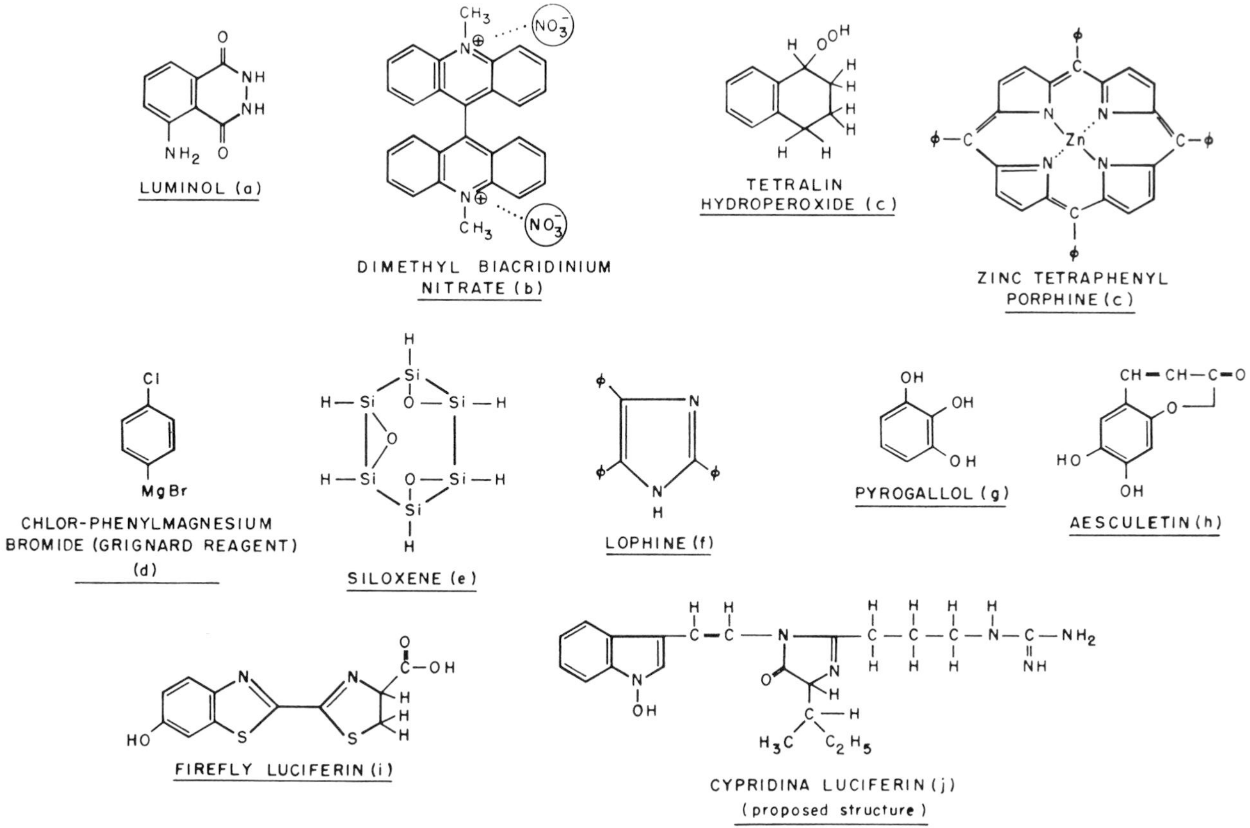

Fig. 3.1. Structures of various chemiluminescent compounds. (a) From Albrecht (1928); (b) from Gleu and Petsch (1935). (c) Helberger (1938) found chemiluminescence of magnesium phthalocyanine and tetralin hydroperoxide; Rothemund (1938) found chemiluminescence of chlorophyll; see also Linschitz and Abramson (1953). (d) From Lifschitz and Kalberer (1922) and Dufford et al. (1923). (e) From Kautsky and Zochar (1922); (f) from Radziszewski (1877a,b); (g) from Eder (1887) and Trautz and Schorigen (1905); (h) from Kearney (1924). (i) White et al. (1961) determined structure; see also Harvey (1952a); Seliger and McElroy (1962) produced chemiluminescence of adenylic acid ester and other esters. (j) Hirata et al. (1959) proposed structure; see also Harvey (1952a). (j) From McElroy and Seliger, 1963b.)

amino-2,3-dihydro-1,4-phthalazinedione). This appears to be the most efficient of the nonenzymic, light-emitting reactions and the one that has been described most completely.

Various mechanisms of oxidation and subsequent light emission have been proposed (Bremer, 1953; Bremer and Friedman, 1954; Dufford *et al.*, 1925; Seliger, 1961; White, 1961; White *et al.*, 1961, 1964; White and Bursey, 1964). The facts that are known definitely concerning the luminol oxidation are: (1) oxygen is required stoichiometrically, (2) the anion of luminol is the reactant, (3) free radicals are involved, (4) nitrogen is produced in the reaction in stoichiometric amounts, and (5) amino-phthalic acid is a product.

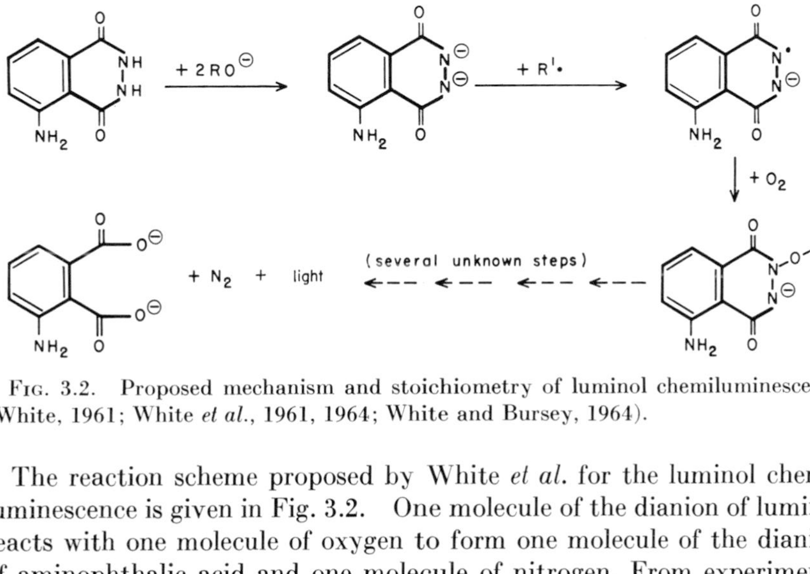

FIG. 3.2.   Proposed mechanism and stoichiometry of luminol chemiluminescence (White, 1961; White *et al.*, 1961, 1964; White and Bursey, 1964).

The reaction scheme proposed by White *et al.* for the luminol chemi-luminescence is given in Fig. 3.2.   One molecule of the dianion of luminol reacts with one molecule of oxygen to form one molecule of the dianion of aminophthalic acid and one molecule of nitrogen. From experiments on the stoichiometry of this reaction in the organic solvent dimethyl-sulfoxide, White has been able to demonstrate a recovery as amino-phthalic acid of approximately 90% of the initial luminol oxidized.

In this connection it would be instructive to describe how, in the case of the luminol chemiluminescence, it has been possible to formulate a mechanism for the reaction by analyzing the fluorescence properties of a proposed product molecule and correlating these directly with the observed chemiluminescence. It is also possible that the emitting species, formed as a result of the oxidation, is a transient, stable compared with the lifetime of the excited state, but subsequently reacting with the solvent or with other components of the reaction solution to produce stable products whose fluorescence properties are different from the chemiluminescence. Fortunately, in the luminol case the emitter is a stable product.

Many workers have assumed that the chemiluminescence of luminol is a sensitized chemiluminescence, since the fluorescence spectrum of the luminol molecule almost coincides with the observed chemiluminescence spectrum. However, the fluorescence quantum yield of luminol is essentially zero at those pH values where a maximum chemiluminescence intensity is observed. Further, the pH dependence of the fluorescence quantum yield of aminophthalic acid closely matches the pH dependence

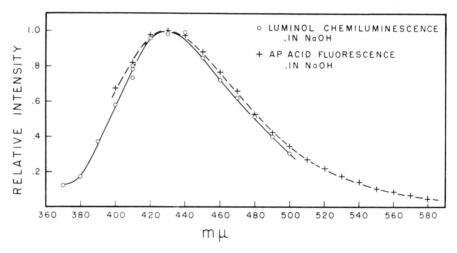

FIG. 3.3.   Chemiluminescence emission spectrum of luminol compared with the fluorescence emission spectrum of aminophthalic (AP) acid. Both the chemiluminescence and the fluorescence were measured in 0.1 N sodium hydroxide (Seliger, 1961).

of the chemiluminescence quantum yield of luminol and, as shown in Fig. 3.3, the fluorescence emission spectrum of aminophthalic acid at the pH values of maximum chemiluminescence efficiency matches even more closely the chemiluminescence emission spectrum than the original luminol molecule (Seliger, 1961). Under suitable conditions in dry dimethylsulfoxide, the product remaining after the total chemiluminescent oxidation of luminol exhibits a fluorescence identical with the chemiluminescence emission. This product has been isolated and identified as aminophthalic acid. Luminol chemiluminescence is therefore a direct chemiluminescence in that the product molecule, formed in an excited state, is also the emitting molecule.

The evidence is not as unambiguous for bacterial bioluminescence, and from the tentative mechanisms summarized in the next chapter it is possible that the emission is a sensitized chemiluminescence. If this were true, then the most likely candidate for the emitter would be a reduced or partially reduced flavin mononucleotide (FMN) molecule bound to the enzyme surface. It is also conceivable that one of the components of a bioluminescent reaction can be modified transiently in an enzyme complex during the reaction so that its light emission is completely different from its fluorescence in the unbound or free state. An example would be the blue shift in the fluorescence of DPNH bound on alcohol dehydrogenase. The reasons for these various suggestions is that flavin mononucleotide (FMN), which in the bacterial bioluminescent system is the only reactant with any reasonable fluorescence properties, has a fluorescence emission peak at 5300 A, while the bioluminescence emission peak occurs between 4750 and 5000 A, depending on the species of bacteria. Unfortunately, a molecular species with these emission properties has not been observed even transiently in the bacterial enzyme system except for a phosphorescence emission at liquid N_2 temperatures.

The bioluminescence and chemiluminescence emission spectra thus far examined have rather broad bands with widths of the order of 700 A. Assuming some overlap between the excitation and emission bands of the emitting molecules, it would be reasonable to expect that the minimum energy requirement for a light-emitting reaction would correspond to a wavelength of approximately 600–700 A less than the observed peak wavelength. Thus in the aqueous luminol reaction with a peak at 4300 A, approximately 77 kcal/mole would be required. Based on bond energies and the reactions as shown in Fig. 3.2, there are approximately 90 kcal/mole available from the formation of the nitrogen molecule. In the firefly and bacterial bioluminescences with peak emission wavelengths at 5620 A and ranging from 4750 to 5000 A, respectively, the energy requirements would be 57 kcal/mole and 65–69 kcal/mole, respectively.

These specific energy requirements can therefore rule out completely those mechanisms in which insufficient energy is released. For example, in the oxidation of firefly luciferin (LH_2), the reaction

$$LH_2 + O_2 \rightarrow L + H_2O_2 \tag{3.19}$$

can be ruled out, since the reaction decomposing H_2O_2 can only make available approximately 25 kcal/mole. It is necessary, therefore, to invoke a reaction of the type

$$LH_2 + O_2 \rightarrow L{=}O + H_2O \tag{3.20}$$

In this case the free energy liberated by the formation of the water molecule should be sufficient to excite the enzyme complexed with the L=O molecule to fluorescence. By a similar argument the bacterial bioluminescence cannot be simply

$$FMNH_2 + O_2 \rightarrow FMN + H_2O_2 \tag{3.21}$$

but must involve a chemical oxidation in which at least 69 kcal/mole can be liberated in a single step.

It is somewhat easier to investigate the mechanism of nonenzymic chemiluminescence than of bioluminescence because of the severe physical restrictions imposed by the enzyme system in the latter. In the luminol chemiluminescent oxidation in water it is quite clear that free radicals are required for the oxygen activation step, based on the inhibition by radical chain stoppers and on the fact that benzoyl peroxide as well as hydrogen peroxide will initiate the light-emitting reaction. The formation of an organic peroxide as an intermediate is thus reasonably certain. This presumably occurs in a radical form which initiates the splitting of the nitrogen-containing ring of the luminol molecule. This same free-radical initiation process is required for the observed chemiluminescence of riboflavin and lucigenin (10, 10′-dimethyl-9,9′-biacridylium nitrate), or DBA.

The same mechanism has been used to predict the conditions for the nonenzymic chemiluminescence of firefly luciferin both in organic solvents and in aqueous solution (Seliger and McElroy, 1962). This appears to be a general property of reactions in which molecular oxygen can be made to combine with organic molecules to form radical peroxides and could be the reaction in bacterial bioluminescence, which is described in more detail in the next chapter. In this latter case the evidence is quite good that the oxygen activation step, that is, the formation of a radical peroxide, occurs prior to the addition of the long-chain aldehyde, which is one of the requirements in the light-emitting reaction. Thus if the enzyme system is reduced in the presence of oxygen and then made anaerobic, subsequent anaerobic addition of aldehyde will result in light emission (Hastings et al., 1963). No light emission occurs, however, if all the reactants are mixed anaerobically. It has been demonstrated by many authors that in oxidation-reduction reactions, particularly with the flavins, free-radical intermediates are involved and that in the nonenzymic cases these intermediates are readily attacked by molecular oxygen. Presumably, in bioluminescent reactions a similar process is directed by the enzyme system in concert with the appropriate substrate molecule.

Recently, Totter et al. (1960a,b) have reported that both luminol and DBA in solution exhibit chemiluminescence during the enzymic oxidation of hypoxanthine by xanthine oxidase. Subsequently, Greenlee et al. (1962), De Angelis and Totter (1964), and Totter et al. (1964) have studied this reaction in detail and have presented evidence that the chemiluminescence is due to the interaction of oxygen and reduced DBA. They observe no apparent loss of DBA during the utilization of xanthine. However, it is doubtful that such a loss could be detected chemically if only one molecule of DBA is destroyed per quantum emitted. From energetic and other considerations we feel that the following over-all reaction is the one most likely to describe the organic mechanism:

$$\text{DBAH} + \text{O—O—H} \rightarrow \text{H}_2\text{O} + (\text{DBA}{=}\text{O})^* \rightarrow \text{products} + \text{light} \qquad (3.22)$$

This is an excellent example of the formation of radical intermediates during an enzyme reaction. Recently, Handler and Beinert (1963) have shown on the basis of ESR spectra that xanthine oxidase is capable of partially reducing both DBA and oxygen. The subsequent interaction of these radicals, presumably on the enzyme surface, leads to light emission.

In connection with the luminol chemiluminescence, it is not H_2O_2 per se that stimulates the observed chemiluminescence but the free radicals generated in the presence of molecular oxygen by the dissociation of H_2O_2. Any catalyst which splits H_2O_2 will stimulate a chemiluminescence intensity several orders of magnitude above that observed with H_2O_2 alone. As mentioned previously, the addition of benzoyl peroxide in the absence of H_2O_2 will also produce a bright chemiluminescence.

2.3. LOPHINE, PYROGALLOL, AND LUCIGENIN

Radziszewski (1877a,b) was the first to study the chemiluminescence of organic solutions. Lophine, shaken in air in alkaline alcohol solution, emits a greenish chemiluminescence, first characterized as peaking at the Fraunhofer E line (5269 A) (Radziszewski, 1880). Chemiluminescence of lophine also results upon addition of H_2O_2 and hemoglobin. The detailed mechanism of this reaction is not known. According to White, some of the oxidation products are benzoic acid and ammonia, indicating a rather violent decomposition.

Eder (1887) found the chemiluminescence of alkaline pyrogallol while using the latter as a developer for photographic plates. The reaction is accelerated in the presence of H_2O_2 and Johnson *et al.* (1954) presume the final product to be

A faint flash of luminescence results on addition of NaOCl or of H_2O_2 plus $K_3Fe(CN)_6$, but a bright luminescence results on addition of H_2O_2 + $K_4Fe(CN)_6$.

In the case of lucigenin the oxidation reaction is assumed to result in the formation of methylacridone

The chemiluminescence is yellowish and takes place only in the presence of H_2O_2 and molecular oxygen. The reaction is accelerated in alkaline solution by the addition of OsO_4 or of reducing agents such as sulfites or stannites. A bright emission is also observed in concentrated ammonia solutions with trace amounts of copper ion. Totter (1964) has reported the quantum yield for dimethylbiacridylium ion chemiluminescence to be between 1.1 and 2.0%, based on the amount of methylacridone formed after reaction with H_2O_2 or in the xanthine oxidase-hypoxanthine enzymatic reaction in 0.01 M Na_2CO_3 at pH 10.4. He states that the methylacridone is the primary emitter and has observed a maximum yield of 0.8 mole methyl acridone per mole DBA^{++}. This reaction, like the luminol reaction, appears to be a direct chemiluminescence, with essentially the same quantum yield. The chemiluminescence emission spectrum of DBA^{++} in alcohol-pyridine-0.01 M KOH (1:1:2) with a small amount of H_2O_2 is significantly different from that in 0.01 M Na_2CO_3 and appears to depend upon the amount of water present.

2.4. OXYSILICONE

Some very intriguing analogies can be drawn for both chemiluminescence and the inverse phenomenon of photosensitization from the experiments reported years ago by Kautsky and Zochar (1922) and Kautsky and Thiele (1925) on the chemiluminescence of the unsaturated silicon compounds. When calcium silicide is treated with HCl, hydrogen is liberated and silicon compounds of various degrees of oxidation are obtained. These include siloxene $Si_6O_3H_6$, which resembles a benzene ring with C replaced by Si and with three oxygen bridges replacing the double bond structure (see Fig. 3.1); silical hydroxide or oxysilicone, $Si_6O_3H_5OH$; other leuco compounds of undetermined composition; and finally silicic acid. Only silical hydroxide is colored, usually being a dark red; its acid salts are bright yellow. The unique feature of these oxygen derivatives of silicon is that they are built up of very thin lamellar flakes in a parallel orientation. The flakes are extremely porous so that the adsorption surface is very large and therefore oxidation can take place readily in solutions throughout the entire crystalline structure. Siloxene can be oxidized in the dark by potassium permanganate in acid solution or by hydrogen peroxide. There is a chemiluminescence accompanying these oxidations, which at first is a low intensity green. As more and more silical hydroxide is formed the intensity increases and finally reaches a maximum when the crystalline structure has a yellow color. During this time the color of the emission has shifted from green to yellow. With a further increase in silical hydroxide content, the intensity decreases and the emission color shifts gradually through orange to red. Both effects are due to increased self absorption. This chemiluminescence emission is identical both in color change and in changes in relative intensity with the fluorescence of silical hydroxide during the oxidation process of siloxene and is therefore considered by Kautsky to be a sensitized chemiluminescence. Exactly the reverse chemiluminescence color changes are observed, starting with pure silical hydroxide and further oxidizing it to its nonchemiluminescent leuco compounds. If silico-oxalic acid, which on oxidation gives no visible chemiluminescence, has absorbed on it basic fluorescent dyes such as rhodamine B or isoquinoline red, a visible chemiluminescence emission is observed which corresponds to the fluorescence emission of these dyes. The chemiluminescence was shown by Kautsky to be due not to dye oxidation but to the oxidation of the surface layer of silico-oxalic acid contiguous to the adsorbed dye. In the case of dye sensitization in the aqueous luminol reaction described previously, in order to sensitize the fluorescence of fluorescein or rhodamine B a reasonably high concentration of the dye was required. However, this may demonstrate one of the functions of a primitive enzyme—the preferential adsorption of a dye molecule at the site of a chemical reaction, producing on a microscopic level the high concentration conditions required for efficient energy transfer in homogeneous solution.

Returning to bacterial bioluminescence, it is therefore possible that the light emission observed is a sensitized chemiluminescence of a bound flavin moiety, the energy coming from the peroxidation of the long-chain aldehyde molecule which we know is destroyed during biolumi-nescence. In this case the energy transfer efficiency could be approximately 100%, the bioluminescent quantum yield (photons emitted per aldehyde molecule destroyed) being equal to the fluorescence yield of the emitting species.

2.5. PHOTOSENSITIZED CHEMILUMINESCENCE

Chemiluminescence accompanying the oxidation of siloxene is accelerated by external light, which is absorbed by the silical hydroxide molecules themselves. Thus, in these unusual compounds we can have simultaneously, sensitized chemiluminescence, sensitized photochemical oxidation, and photosensitized chemiluminescence (see also Serono and Ciuto, 1928). This latter effect has recently been reported by Rosenberg and Shombert (1960a,b) for acriflavine adsorbed on silica gel. In Kautsky's experiments, illumination with diffuse daylight of a nearly colorless mixture of siloxene, ethyl iodide, and water increased the rate of formation of the colored oxysiloxene, the latter acting as a sensitizer for the colorless siloxene. Simultaneous with the photooxidation the system luminesces with light of color varying from green through yellow to red as the amount of oxysiloxene increases. The luminescence is much brighter than the chemiluminescence that would be emitted in the absence of irradiation. In an analogous manner, the photooxidation of moist oxysiloxene to the leuco compounds is accompanied by an emission which changes in color from red through yellow to green with diminishing oxysiloxene content. It has been shown that oxysiloxene is the fluorescent molecule. Upon oxidation of siloxene the product oxysiloxene can be the direct emitter. However, upon further oxidation of oxysiloxene there must be an energy transfer to a nonreacting neighboring molecule. In Rosenberg and Shombert's work the presence of an active oxygen, O_2^*, is postulated, similar to the proposal of Kautsky (1939).

2.6. RIBOFLAVIN

Since riboflavin 5'-phosphate or FMN is the only naturally fluorescent compound involved in the bioluminescence reaction of luminous bacteria, the chemiluminescence of this material is of considerable interest. Strehler and Shoup (1953) observed the chemiluminescence of riboflavin upon the addition of small amounts of 30% H_2O_2. The chemiluminescence emission spectrum that they report is similar to the fluorescence emission of pure riboflavin (5300 A), and not to the bacterial bioluminescence (4900 A). They found a stimulation of luminescence upon addition of cupric and ferrous ions and a very rapid rate of reaction upon addition of OsO_4. The pH dependence of the chemiluminescence intensity shows an optimum between pH 7 and 8. The shape of their curve is narrower than the pH dependence of the fluorescence yield of FMN measured in this laboratory and is probably due to a pH dependence of the reaction rate superimposed on the pH dependence of the fluorescence yield. The products are unknown. Although it was not experimentally verified by kinetic analysis, the emission is most likely a sensitized chemiluminescence.

2.7. LUCIFERIN

The biochemical steps leading to the enzyme-catalyzed emission of light by firefly luciferin are described in the next chapter. Adenosine triphosphate (ATP) is required to form the active luciferin-enzyme complex (E·LH$_2$-AMP), which can then react with molecular oxygen, resulting in the emission of a yellow-green band with a peak at 5620 A. Rhodes and McElroy (1958a,b) have also shown that synthetically produced LH$_2$-AMP will react with the enzyme luciferase to produce light in the absence of ATP.

Recently we have been able to demonstrate the nonenzymic chemiluminescence of LH$_2$-AMP as well as that of the phosphate and methyl

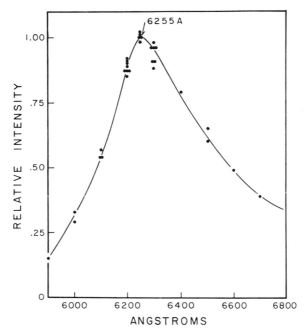

FIG. 3.4.   Chemiluminescence emission spectrum of LH$_2$-AMP in dimethylsulfoxide (Seliger and McElroy, 1962).

esters of luciferin (Seliger and McElroy, 1962). By analogy with the luminescent oxidation of luminol, the abstraction of one of the hydrogen atoms in the 1- or 2-position in basic solution can then permit oxygen attack, resulting in an excited state of the product molecule. Since LH$_2$-AMP is extremely labile in aqueous alkaline solution, we have worked primarily in the strongly hydrogen-bonding organic solvent, dimethylsulfoxide (DMS).

Firefly luciferin was condensed with adenylic acid or metaphosphoric acid in dry pyridine with dicyclohexylcarbodiimide (DCC) using the method described by Khorana (1954), and with diazomethane. Chemiluminescence was obtained in all cases upon making the solution basic. In the case of LH_2-AMP, slightly basic solutions gave a yellow-green emission. Strongly basic solutions gave a red emission, showing a strong pH dependence for the color of the chemiluminescence. This is similar to the pH dependence of color of light emitted in the enzyme-catalyzed reaction discussed later.

The red emission spectrum of LH_2-AMP in strongly basic DMS is shown in Fig. 3.4. The peak emission at 6255 A differs from the 6140 A peak in the enzymic reaction (Seliger and McElroy 1960a,b). However, this is not unexpected, since in the case of luminol there is also a red shift in chemiluminescence emission from 4300 A in aqueous solution to 4840 A in DMS. The unique character of the LH_2-AMP compound is shown by the fact that the methyl ester of luciferin, LH_2-CH_3, gives only a yellow-green chemiluminescence and the phosphate ester, LH_2-PO_4, gives only a red chemiluminescence under the same experimental conditions where LH_2-AMP emission shifted from yellow-green to red. Thus, the oxidized product of LH_2-AMP can exist in either of two fluorescent excited species, dependent on pH. The results indicate that an essential role of the enzyme in the oxidation of firefly luciferin is to permit, by virtue of the binding of the LH_2-AMP, the removal of a proton and the subsequent attack by oxygen. In the absence of enzyme, the aqueous environment would have to be so basic that LH_2-AMP would hydrolyze before ionizing.

2.8. HYDROGEN PEROXIDE

If a 4% NaOCl solution in a hypodermic syringe is squirted into a 30% H_2O_2 solution, a brief red glow is observed. This chemiluminescent reaction was originally observed by Mallet (1927) and independently rediscovered by Gattow and Schneider (1954) and Seliger (1960). Although the red glow is the only visible light, other weak emission bands have been observed in this reaction. There are bands at 5780, 6350, 7030 (Seliger, 1964), 7600, and 8600 A (S. J. Arnold et al., 1964). Khan and Kasha (1963) also found bands at around 6350 and 7030 A. Although this is a very weak chemiluminescence, there is reason to believe that the emitter is active oxygen, O_2^*, originally proposed by Kautsky (1939) as a reactant in the sensitized photooxidations. If this is true, it may be that O_2^* plays an important part in radiobiological reactions and in reactions involving peroxide formation and subsequent oxidations.

The reaction mechanisms in direct and sensitized chemiluminescence have been studied by Rauhut and Hirt et al. (1963a,b, 1964a,b), who have concentrated on the oxalyl chloride-hydrogen peroxide reaction of sensitized chemiluminescence (Chandross, 1963), the phthalhydrizide direct and sensitized chemiluminescences, and the chemiluminescences obtained as a result of oxidations or reductions of aromatic hydrocarbon derivatives. These latter ion radical chemiluminescences furnish examples in the visible region of the release of sufficient energy during a chemical reaction or electron transfer, not including oxygen, to leave the reduced or oxidized molecule in its first excited electronic state. The chemiluminescence emission spectrum corresponds to the fluorescence of the neutral molecule. For example, Rauhut, Hirt *et al.* (1964b) report that the oxidation of sodium-9,10-diphenylanthracenide by chlorine or bromine in tetrahydrofuran to give NaCl plus 9,10-diphenylanthracene plus light proceeds in the absence of oxygen and the emission is typical of 9,10-diphenylanthracene fluorescence. Chemiluminescence was also observed when 9,10-dichloro-9,10-diphenylanthracene was reduced with sodium metal or with sodium naphthalenide. Again the blue chemiluminescence corresponding to 9,10-diphenylanthracene was observed.

Oxalyl chloride reacts vigorously with H_2O_2 to give CO, O, HCl, and a weak light emission. However, in the presence of anthracene or N-methylacridine there is a bright chemiluminescence corresponding to the fluorescence of these hydrocarbons. The reaction at room temperature gives a bright flash. In ether solution at $-78°C$ the intensity is strong and the chemiluminescence lasts much longer. It has been demonstrated (Chandross, 1963; Rauhut and Hirt *et al.*, 1963a) that the *vapors* from the reaction give rise to sensitized chemiluminescence (fluorescence) when they come in contact with anthracene or other fluorescent materials. From this it might be inferred that the excitation energy is carried by the gaseous products of the reaction, analogous to the $NaOCl$-H_2O_2 chemiluminescence previously described.

Solid perhydrates consisting of a mixture of oxalic acid, dicyclohexyl carbidimide, sodium pyrophosphate peroxide, p-toluenesulfonic acid, and 9,10-diphenylanthracene have been found to emit moderately strong chemiluminescence when scratched on a clay plate or when added to an organic solvent such as benzene or ether. The highest sensitized chemiluminescence quantum yield thus far reported is 0.06 (Rauhut and Hirt *et al.*, 1963b). These authors placed 1.7 ml of 0.01 M 9,10-diphenylanthracene in ether and 1.0 ml of 1 M anhydrous ethereal hydrogen peroxide in a 3-ml cell and then rapidly injected 0.02 ml of 0.75 M ethereal oxalyl chloride. The light lasted for about 90 sec. Chemiluminescence as the result of decomposition of photoperoxides of anthracene has been known for some time. Among these are the photoperoxides of 9,10-diphenylanthracene and 1,4-methoxy-9,10-diphenylanthracene.

CHAPTER 4

Bioluminescence—Enzyme-Catalyzed Chemiluminescence

The emission of light by organisms has interest for all with scientific curiosity in the fields of biology, chemistry, and physics; indeed, important contributions to our understanding of bioluminescence have come from investigators representing these various areas.

The emission of light through enzymatic processes occurs among many types of living organisms—from protozoa to fish in animals and in plants from bacteria to the higher fungi. The luminous forms present in the various phyla seem to occur at random. However, the great majority are indigenous to the sea, some inhabiting great depths, while others occur in large numbers at or near the surface where they may give rise to brilliant displays of "phosphorescence" when agitated. Luminous bacteria appear to be present in low concentrations in all marine waters.

Many luminescent organisms have been described and studied from various points of view. Because of the sensitivity of electron-multiplier phototubes it is possible to study many of the enzymatic properties of bioluminescent reactions using concentrations of substrate or enzyme which may be as low as 10^{-15} M. Since luminescence is an enzymatic process, it has been used as a sensitive system for studying the effects of drugs, temperature, and pressure and other factors which affect catalytic processes. In addition, light emission has been used as a sensitive assay system for such biologically important compounds as ATP, DPN, FMN, and other cofactors. Recent reviews have summarized most of the data on these various aspects of light emission (Harvey, 1952a; Chase, 1960; McElroy and Seliger, 1963b). Therefore we will be concerned primarily with what is known about the mechanism and significance of light emission.

Dubois (1885, 1887a,b) demonstrated that light emission by organisms required at least three different substances. One is an *enzyme* which he called luciferase and which catalyzes the oxidation of a low molecular weight *substrate*, luciferin. Molecular *oxygen* is the third requirement for this oxidation. During the enzymatic oxidation of luciferin large amounts of energy (40–80 kcal/mole) become available in a single step so that an intermediate or product of the reaction is left in a highly excited electronic state. The intermediate or product must be a fluorescent molecule and consequently emits light when the excited state returns to the ground state. Biochemical studies have been concerned with the identification of the chemical steps involved in the light-emitting reaction.

It is important to point out that the luciferin and luciferase from one organism are usually quite different chemically from the "luciferins" and "luciferases" isolated from different luminous forms. Thus the bacteria and the firefly have entirely different chemical substances and reactions for promoting light emission. The basic reaction creating the excited state, however, may be quite similar and certainly involves molecular oxygen.

1. Firefly Reactions and Mechanism

Among the insects the largest group of luminous organisms are those found in several families of beetles. The families of Lampyridae, Elateridae, Phengodidae, Drilidae, and Rhagopthalmidae contain numerous species which have developed the capacity to emit light. All the forms that are commonly called fireflies, glowworms, lightning bugs, railroad worms, automobile bugs, blinkies, peeny-wallies, and many other names occur in these groups.

Of all the luminous groups, the true fireflies (Lampyridae) have received the greatest attention. The scientific literature is far too large to review adequately and consequently only certain specific forms will be discussed in order to establish the relationship of the chemistry to the flashing mechanism. Not all genera of Lampyridae contain luminous species. Some may be luminescent for only a brief time during their life history and in a few cases only the larval forms are luminous (Barber and McDermott, 1951).

In some cases the female of the species may be luminous, whereas the male is not. However, according to Harvey, the reverse is never true. In most species the form of the lanterns varies in the two sexes. Buck (1938, 1948) has studied this problem in great detail, particularly as it relates to recognition and mating.

The old discovery that the light of the firefly is a mating device to attract the sexes is now rather universally accepted. In those species of fireflies where both male and female are luminous, a fairly complicated signal system has been developed (Seliger *et al.*, 1964a). Each species appears to have a characteristic flash pattern which the female of the species can recognize (Fig. 4.1).

According to Buck, recognition depends on the time interval between the male flash and that of the female. This interval for *Photinus pyralis* is approximately 2 sec at 25°C. It varies with the temperature. Although

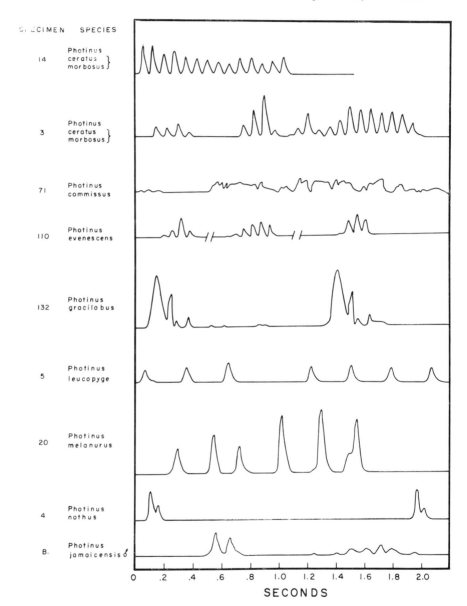

FIG. 4.1.   Representative records of flashing patterns in flight of several species of fireflies, taken in the field, in their natural habitat (Seliger *et al.*, 1964a).

the male may flash at various times, the female invariably responds 2 sec after a male flash. Other species of fireflies have other systems and types of flashes. Synchronous flashing of a number of males to a female has been observed, but it is rare in the United States. However, among tropical fireflies it is not a rare event. In Burma and Siam and other eastern countries, all the fireflies on one tree have been reported to flash simultaneously, while on another tree some distance away this same synchronous flashing would be apparent but out of step with those of the first tree. Unfortunately, very little experimental work has been done on this problem. It is difficult to understand how one female could be located in a tree so that it could trigger the flash of a large number of males uniformly distributed throughout the tree.

The eggs of American fireflies are laid on or near the ground and hatch in about 3 weeks. The larvae differ considerably in habit. They live mostly in damp places among dead fallen leaves, becoming active at night and feeding on slugs, snails, and small insect larvae. The larvae winter, in the case of most species, under stones or a short distance underground, often in specially constructed chambers. Pupation is near the surface and in pupal cells.

The first indication of the formation of the light organ takes place about 15 days after egg development begins. After about 22 days of development, the light organ has become functional and appears as two bright spots of light. The larvae emerge on about the 26th day of incubation and become the well-known glowworms with two small lights at one end. In 1 to 2 years they reach maturity as pupae. During pupation additional light organs develop which will become the adult organs of the firefly. The light organs of both the larva and the adult are formed from fat bodies which become differentiated into the photogenic and reflector layers of the mature light organs. Extensive references to the various studies concerning the development and habits of fireflies may be found in Harvey's book.

The substrates and enzyme required for light production in extracts from the American firefly, *Photinus pyralis*, have now been prepared in a highly purified, crystalline state (McElroy, 1960; McElroy and Seliger, 1961, 1963b). All evidence indicates that these same factors, i.e., luciferase and luciferin, are responsible for light emission in other genera of the Lampyridae and probably other groups in the Coleoptera. The crystallization of firefly luciferase and luciferin has allowed for the first time an extensive quantitative study of the interaction of these substances during the process of light emission. Luciferase, purified by repeated crystallization, is homogeneous as judged by electrophoresis and ultracentrifugation. This method of purification, which eliminates the contaminating enzyme pyrophosphatase, is an important factor in the interpretation of the kinetics of the reactions.

The preparation and some of the properties of firefly luciferin have been described by Bitler and McElroy (1957). Recent studies by White *et al.* have demonstrated the structure of luciferin to be that shown in Fig. 4.2. This structure was proved by total synthesis. In the last step of the chemical synthesis of luciferin, 2-cyano-6-hydroxybenzothiazole is reacted with cysteine. When D(−)-cysteine is used, a luciferin, D(−)-LH₂ is obtained, which has all the properties of natural luciferin. When L(+)-cysteine is used in the synthesis, the resulting luciferin L(+)-LH₂

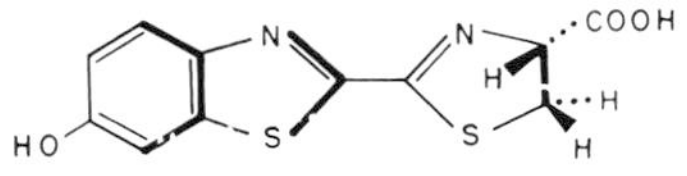

D (−) LUCIFERIN

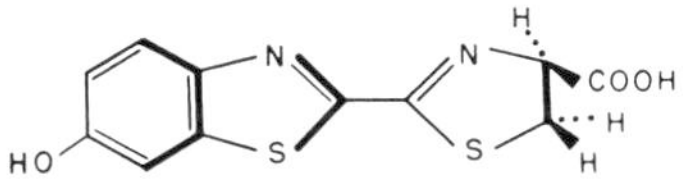

L (+) LUCIFERIN

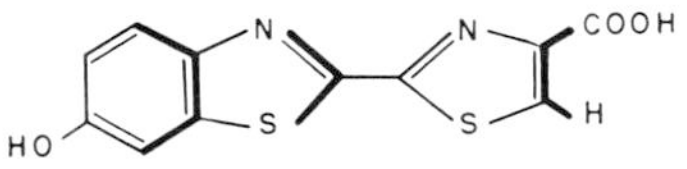

DEHYDROLUCIFERIN

FIG. 4.2.   Structure of firefly luciferin and dehydroluciferin (White, 1961).

is inactive for light production although it is otherwise chemically identical with D(−)-LH₂. Both L(+)-LH₂ and D(−)-LH₂ will react with ATP in the presence of luciferase to liberate pyrophosphate. However, only D(−)-LH₂ will be oxidized and will produce light.

When one starts with free luciferin and luciferase, ATP and magnesium ions are required for light emission. The initial reaction is, in reality, an adenyl transfer to the carboxyl group of luciferin to form luciferyl adenylate with the elimination of inorganic pyrophosphate (PP) as indicated in the following reaction:

$$\text{LH}_2 + \text{ATP} + \text{E} \underset{}{\overset{\text{Mg}^{++}}{\rightleftharpoons}} \text{E·LH}_2\text{-AMP} + \text{PP} \tag{4.1}$$

The release of inorganic pyrophosphate and reversibility of the reaction have been demonstrated in several ways (Rhodes and McElroy, 1958a, b).

As will be discussed in detail later, the light reaction can best be described by the following reaction:

$$E \cdot LH_2\text{-}AMP + O_2 \rightarrow (E\underset{\downarrow}{\overset{\overset{\text{O}}{\|}}{-L-}}AMP)^* + H_2O \tag{4.2}$$

$$E\overset{\overset{\text{O}}{\|}}{-L-}AMP + \text{light}$$

The product of the light reaction has many of the properties of dehydro-luciferin (see Fig. 4.2). It now seems most likely that the latter is not produced from the excited intermediate but rather is an oxidation product which does not lead to light emission.

We have not yet been able to isolate the product molecule from which the light quantum is emitted. From the facts obtained so far we know that the fluorescence yield of the intermediate must be practically 100%. It seems most likely that oxygen must add to the enzyme-bound luciferyl adenylate to form an organic hydroperoxide. The exergonic step could then be described as a dehydration process in which one atom of oxygen remains in the intermediate and the other appears in a water molecule. The over-all reaction could be described as follows. We have not included the enzyme or the adenylic acid for simplicity.

$$\text{(a)} \quad LH_2 + O_2 \rightarrow L\overset{\overset{\text{H}}{|}}{-}O \cdot O \cdot H$$

$$\text{(b)} \quad L\overset{\overset{\text{H}}{|}}{-}O \cdot O \cdot H \rightarrow (L{=}O)^* + H_2O \tag{4.3}$$

$$\text{(c)} \quad (L{=}O)^* \rightarrow L{=}O + \text{light}$$

Such a reaction would liberate well over 100 kcal of energy, which is more than enough to account for the energy of the excited state.

When light emission is initiated by the injection of ATP into a reaction mixture containing excess luciferin, one observes a rapid rise in intensity followed by a rapid decrease in the first few seconds, followed by a decay that may last for hours. This decrease in the rate of reaction can be shown to be due to product inhibition. The fact that the light intensity does not fall completely to zero is due in part to the fact that the production of pyrophosphate initially is sufficient to reverse partially the inhibition due to the enzyme-product complex. If inorganic pyrophosphate is destroyed by the addition of pyrophosphatase, rapid and almost complete inhibition of light is observed (see Fig. 4.3). As discussed below, pyrophosphate production from ATP is not inhibited under these conditions. That LH_2-AMP is the active intermediate has been demonstrated by a number of different types of experiments. Synthetic LH_2-AMP will react directly and rapidly with enzyme and molecular oxygen to emit light; neither ATP nor Mg^{++} are required for this reaction. If inorganic pyrophosphate is added to this reaction mixture, a slower rate of light emission is observed, owing to a reversal of the activation step.

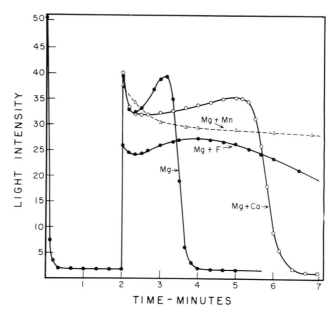

FIG. 4.3.   Effect of inorganic pyrophosphatase and pyrophosphate on light emission. The initial rapid decline of light intensity after the addition of ATP is due to product inhibition and the removal of pyrophosphate which favors the removal of L-AMP from the enzyme. The addition of pyrophosphate at 2 min stimulates light emission until it is removed by hydrolysis. Inorganic pyrosphosphatase is present to hydrolyze the pyrophosphate produced in the activation reaction (McElroy and Seliger, 1961).

It is known that at least one of the oxidation products of luciferin is dehydroluciferin. Dehydroluciferin is also activated by ATP and Mg^{++} in the presence of enzyme as indicated in the following reaction:

$$L + ATP + E \overset{Mg^{++}}{\rightleftharpoons} E\cdot L\text{-}AMP + PP \qquad (4.4)$$

If dehydroluciferin is added to a reaction mixture prior to the addition of luciferin, light production is completely inhibited. Since the fluorescence of dehydroluciferyl adenylic acid when bound on the enzyme is much lower than that of free dehydroluciferin, it is possible to study the kinetics of the reaction using the fluorescence of free dehydroluciferin as an assay method. It was found that the equilibrium constant for the activation step at pH 7.1 was 2.5×10^5. Furthermore the dissociation constant, K, as defined by Eq.(4.5) below was determined to be 5×10^{-10}.

$$K = \frac{(E)_{free} (L\text{-}AMP)_{free}}{(E \cdot L\text{-}AMP)} \tag{4.5}$$

This tight binding of dehydroluciferyl adenylate to the enzyme and the ability of inorganic pyrophosphate to react reversibly with the complex accounts for the over-all kinetic behavior of the light-emitting reaction.

If ATP and luciferin are added to an enzyme preparation under anaerobic conditions, no light emission is observed. However, pyrophosphate liberation proceeds normally, indicating that the activating reaction is not inhibited. Subsequent introduction of oxygen gives a normal light reaction with no evidence of initial inhibition. In fact, the flash height is considerably higher than the normal aerobic flash height because of the accumulation of enzyme-bound luciferyl adenylate.

In an effort to find conditions which would favor continuous light emission at a high rate, we have tried a number of compounds which might conceivably reverse the product inhibition. Although inorganic pyrophosphate will stimulate light emission from a product-inhibited reaction mixture, it will not stimulate total utilization of luciferin because of the removal of the activating step. The only other compound of the many tested which will stimulate light emission from an inhibited reaction is coenzyme A (Airth et al., 1958). The increase in intensity observed is proportional to the amount of CoA added, over a limited range of CoA concentrations. The luminescence will continue at this higher level for a time period that is proportional to the total amount of CoA added. Once the CoA is exhausted the rate of reaction returns to its original low level. Dehydroluciferyl-CoA has been isolated from such a reaction mixture. It has been shown that L-CoA in the presence of the enzyme will react with AMP to form E·L-AMP. If C^{14}-labeled AMP and PP^{32} are added to such a reaction mixture, both labels can be recovered in the ATP. CoA is the only compound that will stimulate AMP exchange. Dephospho-CoA and other derivatives of CoA are completely inactive.

Since CoA and PP will stimulate the normal light reaction it suggests that the product has many of the properties of dehydroluciferin discussed above and below. The following reaction sequence is adequate to explain the secondary stimulation of light emission by CoA or PP when excess luciferin and ATP are present.

(a) $ATP + LH_2 + E \rightarrow E \cdot LH_2 - AMP + PP$

(b) $E \cdot LH_2 - AMP + O_2 \rightarrow E \cdot \overset{\overset{O}{\|}}{L} - AMP$ (inhibitory complex) $+ H_2O + light$

(c) $E \cdot \overset{\overset{O}{\|}}{L} - AMP + PP \rightarrow \overset{\overset{O}{\|}}{L} + E \cdot ATP$

(d) $E \cdot \overset{\overset{O}{\|}}{L} - AMP + CoA \rightarrow E + \overset{\overset{O}{\|}}{L} - CoA + AMP$

(4.6)

Both pyrophosphate and CoA remove the inhibitor from the enzyme allowing reaction (4.6a) to proceed normally. These reactions would also explain the PP- and CoA-dependent AMP exchange reactions.

ATP is the only nucleoside triphosphate that will function in the production of light; deoxy-ATP is completely inactive. Various nucleotide derivatives of luciferin were prepared and tested for light-producing activity. Only the 5'-adenylic acid derivatives of luciferin were active. The earlier observation that the 3'-adenylic acid was active was due to the presence of a 5'-adenylic acid impurity. In crude extracts ADP will function for light emission due to the presence of an active myokinase. In addition, in the presence of ADP, a number of other triphosphates will support luminescence due to the presence of active transphosphorylases. Crude extracts prepared in the cold will often contain enough ADP to give light responses with a number of nucleotide triphosphates.

Luciferase appears to have a dual role in light emission. The first step is concerned with the luciferin-ATP activation reaction to form LH_2-AMP and the second involves the catalytic utilization of oxygen to form the excited state (Hastings et al., 1953; McElroy et al., 1953a,b). Although we are not in a position to describe completely the organic mechanism involved in this oxidation reaction, there are certain facts which are important and which must be considered in any proposed mechanism. We know, for example, that the total light output is directly proportional to the amounts of ATP and luciferin present, i.e., both substrates are used (Seliger and McElroy, 1959). Furthermore, we know from previous studies that one light quantum is emitted for each luciferin molecule used, for very low concentrations of luciferin at alkaline pH. As the pH is decreased below neutrality, the oxidation of luciferin does not always lead to light emission. To study this phenomenon carefully it was necessary to determine the emission spectrum under these conditions.

Figure 4.4 shows the emission spectrum of the *Photinus pyralis* firefly light reaction *in vitro* in glycylglycine buffer at pH 7.6. The peak emission for the bioluminescence is 562 mμ, with the band ranging from 500 to 630 mμ. As an absolute minimum, therefore, the energy requirement for the bioluminescence should exceed 57 kcal/mole. Measurements of the spectrum using dissected *Photinus pyralis* organs stimulated with ATP gave approximately the same emission. The reasons for this are discussed below.

One might expect that some product would occur in the light-emitting step which would have a fluorescence emission spectrum similar to the

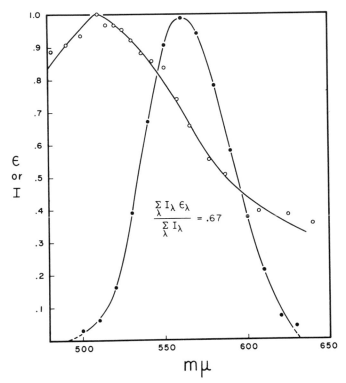

FIG. 4.4. Emission spectrum of firefly light *in vitro*. Bioluminescence (solid circles) in glycylglycine buffer at pH 7.6. The peak emission is at 562 mμ. Superimposed upon this emission spectrum is the normalized photon spectral efficiency of the photocathode of the phototube used in the quantum yield measurements (Seliger and McElroy, 1960b).

bioluminescence emission spectrum and which could possibly be identified as the light-emitting species. Detailed fluorescence studies on luciferin, dehydroluciferin, and the adenylic derivatives have been made. We will not review all these data here. In summary, the evidence clearly indicates that the adenylic derivative of an oxygenated product is the most likely emitter (McElroy and Seliger, 1963b).

As the pH of the firefly *in vitro* solution is lowered it can be observed that the intensity of the yellow-green bioluminescence decreases, leaving a dull brick-orange glow (Seliger and McElroy, 1960a,b). This variation in bioluminescence emission with pH is shown in Fig. 4.5. As can be seen, at neutral (and alkaline) pH, there is a predominant emission band in the yellow-green region. At intermediate pH, a red emission band appears

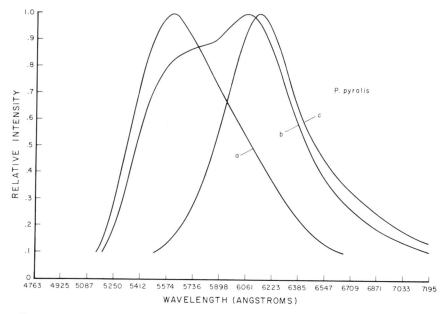

Fig. 4.5.   Effect of pH on the *in vitro* bioluminescence emission of *Photinus pyralis*. The spectra are normalized to 1 at the peak emission. (a) Purified *P. pyralis* luciferase plus *P. pyralis* luciferin, pH 7.6. (b) Purified *P. pyralis* luciferase plus *P. pyralis* luciferin, pH 6.5. (c) Purified *P. pyralis* luciferase plus *P. pyralis* luciferin, pH 6.0. From Seliger and McElroy (1960b).

at 614 mμ and, at pH values below 5.5, the yellow-green emission is completely suppressed and only the red band is evident. At acid pH, the number of light quanta emitted per luciferin molecule oxidized is markedly lower than 1 and indicates a predominantly dark reaction. However, at alkaline pH, although the rate of light emission is reduced to a fraction of the rate at pH 7.6, the quantum yield is essentially unity. The change of yield with pH corresponds rather closely in form to the fluorescence yield of luciferin and oxyluciferin at various pH's except for the fact that the pK has been shifted essentially one pH unit toward the acid range of the bioluminescence quantum yield. This may represent the interaction of the enzyme with the phenolic OH group or possibly the amino group of AMP, altering in effect the fluorescence or chemilumi-nescence properties of the luciferyl-adenylate compound.

Several other factors have been shown to modify the emission spectrum of the bioluminescence reaction *in vitro*. The shift from the 562 mμ peak to the 614 mμ peak, under acid pH, and the sharp decrease in quantum yield in acid pH are additional evidence that the enzyme is important in influencing the excited state. Recently we have attempted to measure the pK of the red light-emitting species by observing only the red light as a function of pH. Although the data are not conclusive, there is a sugges-tion that the pK is near 6.8. It is possible, therefore, that a histidine residue may be involved in the binding of the luciferyl adenylic acid to the enzyme, which in turn may affect the pK of the excited state and consequently the color of the light emitted. Since zinc is known to bind to the imidazole group of histidine, we have recently tried this ion on the light-emitting reaction. At as low a concentration as 5×10^{-4} M ZnCl$_2$ red light appears as a significant part of the emission spectrum and at higher concentrations most of the emitted light is red. This "red shift" has also been observed for Cd^{++} and Hg^{++} ions and as the temperature is increased. The color of the light thus depends upon the nature of the binding of the intermediate to the enzyme and histidine is probably involved (McElroy and Seliger, 1963b).

Macaire (1821) was the first to observe the red color of firefly light. He did this by placing the firefly at higher temperature. We assume that this change in the color of the light is similar to that observed when the pH or salt environment is changed.

It is clear from these observations that small changes in the environ-ment or changes in the configuration of the enzyme, possibly due to slight differences in amino acid composition, can alter the color of light emitted and consequently may be the basis for the small differences in peak emission noted for various fireflies even including the red light emitted by the railroad worm. In Figs. 4.6 and 4.7 are shown *in vivo* emission spectra of Jamaican and American fireflies, respectively. These spectra show the range of color that can be observed, from the green of *pennsylvanica* to the deep yellow of *plagiophthalamus* ventral organ (Seliger and McElroy, 1964; Seliger *et al.*, 1964b).

As described above, L(+)-LH$_2$, appears to have all of the chemical properties of natural luciferin except that no light emission is observed when it is mixed with enzyme, ATP, and Mg^{++} (Seliger *et al.*, 1961b). However, the initial activation by ATP to form the adenylic acid derivative and pyrophosphate proceeds normally. In addition, L(+)-LH$_2$ is a potent competitive inhibitor of luminescence. The data indicate that in the activation step the luciferase makes no distinction between D and L forms of luciferin. As will be discussed later, only D($-$)-luciferyl adenylate is oxidized.

Both L- and D-luciferin can be converted to dehydroluciferin by heating in alkaline solution or by oxidation with ferricyanide. The chemical data establish with certainty that the adenylic acid oxidation products from

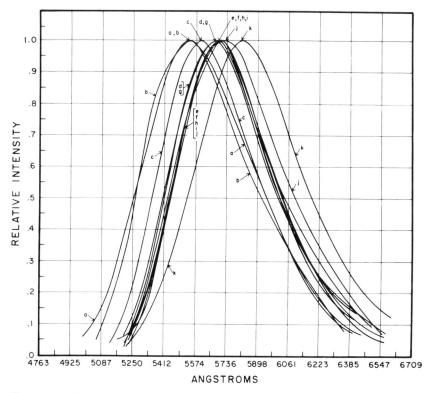

FIG. 4.6. Normalized bioluminescence emission spectra of various Jamaican fireflies. (a) *Pyrophorus plagiophthalamus* (thoracic organ); (b) *Diphotus sp.*; (c) *Photinus pardalis;* (d) *Photinus xanthophotis;* (e) *Photinus leucopyge;* (f) *Photinus melanurus;* (g) *Photinus nothus;* (h) *Photinus evanescens;* (i) *Photinus ceratus or morbosus;* (j) *Photinus gracilobus;* and (k) *Pyrophorus plagiophthalamus* (ventral organ). From Seliger *et al.* (1964b).

the two forms of luciferin are identical. The adenylic acid derivatives of both luciferins will emit light in the presence of strong base in the organic solvent, dimethylsulfoxide (Seliger and McElroy, 1962).

As might be expected from the structure, it is possible to convert synthetic L($+$)-luciferin to D($-$)-luciferin in an alkali-catalyzed isomerization. If L($+$)-luciferin is heated to approximately 80°C in 1 N sodium hydroxide and in the absence of oxygen, one obtains significant amounts of D(−)-luciferin as judged by its ability to produce bioluminescence. One can follow these isomeric changes by the measurement of optical rotation. In dimethylformamide solvent the synthetic D(−)-and L(+)-luciferins have specific rotations for the sodium D doublet of minus and plus approximately 30°, respectively.

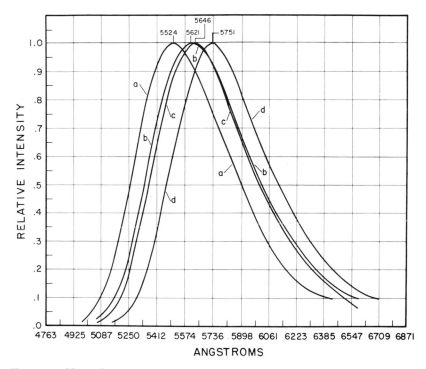

FIG. 4.7. Normalized bioluminescence emission spectra of some American fireflies. (a) *Photuris pennsylvanica;* (b) *Photinus pyralis* ♂ and ♀; (c) *Photinus marginellus;* (d) *Photinus scintillans* ♂ and ♀. From Seliger *et al.* (1964b).

1.1. UTILIZATION OF OXYGEN

We have found that one mole of oxygen is consumed per mole of D(−)-luciferin oxidized in the enzymatic reaction. No oxygen is consumed in the case of L(+)-luciferin. Using an oxygen microelectrode in a closed system, we have measured both the rates of oxygen consumption and the total oxygen consumption for D(−)-luciferin, L(+)-luciferin, and racemic mixture of D(−)L(+)-luciferin. In all cases the total oxygen consumed is directly proportional to the initial amount of D(−)-luciferin present. The presence of L(+)-luciferin affects only the reaction rate. Catalase has no effect on oxygen consumption, indicating that no free hydrogen peroxide is formed during the light reaction. Further, the rate of oxygen consumption follows the same kinetics as the light intensity versus time (McElroy and Seliger, 1963b).

1.2. COMPARISON OF VARIOUS FIREFLIES

In Table 4.1 we present the results of a number of tests on cell-free extracts of various fireflies (McElroy and Harvey, 1951). In these experiments crude extracts of the lanterns were made and, when all the light

TABLE 4.1

THE PRODUCTION OF LIGHT BY FIREFLY LANTERN EXTRACTS IN THE PRESENCE OF ADENOSINE TRIPHOSPHATE AND *Photinus pyralis* LUCIFERIN[a]

Firefly	ATP	LH$_2$	ATP + LH$_2$
Jamaican			
Photinus lobatus lobatus	+	−	+
Photinus pallens	+	−	+
Photinus lobatus morbosus	+	−	+
Photinus commissus	+	−	+
Photinus melanotis	−	−	+
Diphotus montanus	−	−	+
Photuris jamaicensis	−	−	−
Glowworm (unidentified)	−	−	+
Pyrophorus (sp.) from Puerto Rico	+	−	+
American			
Photinus pyralis	+	−	+
Photinus scintillans	+	−	+
Photuris pennsylvanica	+	−	+
Glowworm (*Photinus pyralis*)	−	−	+

[a] Symbols: +, light; −, no light; LH$_2$, luciferin; ATP, adenosine triphosphate.

had disappeared, ATP and *Photinus pyralis* luciferin were added separately and in combination. In most cases, all that was necessary to restore light was ATP, indicating that both luciferin and luciferase were still present in the extract. In some cases where the ATP reaction was negative, light could be restored by adding luciferin and ATP. Additional experiments using *Photinus melanotis* and *Diphotus montanus* indicated that the luciferin and ATP disappear at about the same rate in crude extracts. Some glowworms appear to behave in this manner also. Harvey and Haneda (1952) have also found that ATP will restore light in crude extracts of Japanese fireflies. The negative result obtained with *Photuris jamaicensis* was due to the loss of enzyme in the crude extract. Chromatographic evidence obtained using crude extracts of *Photinus pallens, Photinus pyralis, Photinus scintillans, Photuris pennsylvanica,* and *Pyrophorus plagiophthalamus* indicate that the luciferin is identical from all these different forms.

Buck (1948) has reviewed in detail the various hypotheses concerning the mechanism and control of the firefly flash. The detailed architecture of the light organ seems to preclude the possibility that oxygen availability is rate limiting and could be the basis for control of the firefly

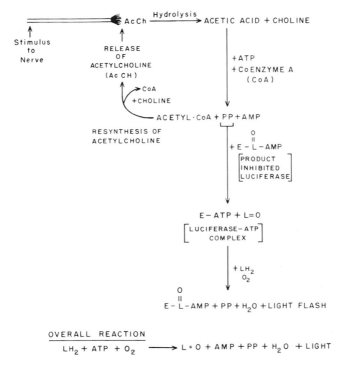

FIG. 4.8. Proposed mechanism for control of firefly flash (McElroy and Seliger, 1962b, reprinted with permission of *Sci. Am.*).

flash (Hastings and Buck, 1956). The scheme presented in Fig. 4.8 is one possible way of controlling the flash in the intact firefly, which is in keeping with our biochemical knowledge. The scheme suggests that light emission is normally prevented by an inhibitory product of the light reaction that is tightly bound to the enzyme. The triggering of light emission by the motor nerve impulse is due to the release of pyrophosphate in the photosite cytoplasm, which, in turn, releases the enzyme so it may react with luciferin. McElroy and Hastings (1955, 1956) and McElroy and Seliger (1961, 1962b) proposed that controlled release of pyrophosphate may take place by an acetylcholine-coenzyme A–ATP cycle. In insects, acetylcholine has not been found at nerve-muscle junctions but is present in high concentrations in the insect nerve cords (Mikalonis and Brown, 1941). The observations of D. S. Smith (1963) concerning the location of the nerve terminal processes in the light organ constitutes a more serious objection to the suggestion that a transmittor passes from the axon directly to the photosites. The tracheolar cell lies between the nerve terminals and the surface of the photosites. From the spatial relations between the endings and the tracheal and photosite cells it may be inferred that the triggering action of the nerve impulse may act in one of two ways. Smith suggests that (1) "either the photosites may be stimulated to produce light by the secretion of a substance from the axon which chemically alters or otherwise reaches the photosites by the intervening cytoplasm; or (2) the role of the nerve terminals in the latter may be analogous to the situation of striated muscle. The arrival of a nerve impulse, initiating a depolarization of the membrane of the tracheolar cell, might then be channeled between the photosites." It is possible that pyrophosphate supply may be critical in triggering light emission. High concentrations should inhibit and low concentrations should stimulate.

It should be remembered, however, in considering this hypothesis that pyrophosphate would not be expected to pass readily across membranes and that in the above scheme two cell membranes and an intracellular space separate end-cell and photosite cytoplasm. On the other hand, according to the scheme shown in Fig. 4.8 only a small amount of pyrophosphate may be required to trigger the light reaction, since the subsequent production of the substance during the luminescent reaction may allow the spread of light emission from the initial foci. As mentioned previously Case and Buck (1963) and Hanson (1962) found that weak stimulation of the nerve supply to the light organ elicits a neurally mediated flash with a latency of about 65 to 75 msec, while very strong stimulation produces a flash with a latency of only 15 msec, possibly representing a response by direct excitation of the photosites. The neurally mediated latency is far longer than that occurring at myoneural junctions, suggesting that the events taking place between the arrival of the impulse at the nerve terminal of the light organ and the activation of the photosites are relatively slow. The observation that continued strong stimulation gradually leads to a low continuous glow in the gland can be explained by the presence of a low concentration of pyrophosphate. This is what is observed in the test tube when the breakdown of pyrophosphate is prevented. Strong direct stimulation of the glowing gland could lead to inhibition by the release of excess pyrophosphate.

There are however many other factors involved such as production of ATP, etc., which must be studied before a definite decision can be made on this matter of flash control.

There is of course the distinct possibility that light is not emitted from the photogenic glands or the granules in the glands but rather from the end cell. Because of the cytological nature of the bodies in the photogenic cells it has been assumed that these are the glands which are concerned with light emission. There is a lot of indirect evidence to suggest this is true but most of it is cytological in nature. It will be of extreme interest and value to attempt a careful fractionation of these various cell types and to determine the location of both luciferin and luciferase. As is well known, in many luminous forms, such as *Cypridina*, luciferin and luciferase seem to be separated in different glandular tissues and may occur in small granules. Often light emission occurs only after the enzyme and luciferin have been excreted and mixed. Thus it may be possible that luciferin made at one site could be actively transported to the luciferase site owing to a nerve impulse. However, there is no direct evidence for such an idea.

Probably the best known luminous click beetles belong to the genus *Pyrophorus* which contains over one hundred luminous species. Almost all occur in subtropical America. Several outstanding examples have been described as occurring in Texas and Florida. *Photophorus* contain two luminous species which are found in the New Hebrides (*P. bakewelli*) and Fiji Islands (*P. jansoni*). *Campyloxenus pyrothorax* is found in Chile.

Dubois (1885) demonstrated that the luminescence of *Pyrophorus* was due to the action of a heat-labile and a heat-stable substance. If the luminous organ was ground up with water, the homogenate would glow for several minutes but the light finally disappeared. If he added to this cold-water homogenate an extract, which he prepared by boiling a fresh lantern in water, light emission was restored. This is a very famous and important experiment for students of physiology and biochemistry, for it opened the way to a clearer understanding of the chemistry of the reaction. However, it was over 60 years before the substance in the hot water extract which restored light was identified as adenosine triphosphate (McElroy, 1947). The "hot water-cold water" reaction is what we presently refer to as the luciferin-luciferase reaction. However, Dubois did not introduce the terms luciferin and luciferase until 1887 when he was working on the chemistry of the light reaction in the luminous clam, *Pholas dactylus*.

Pyrophorus plagiophthalamus has two different kinds of luminous organs. There are two brilliant, greenish luminescent spots on the posteriolateral margin of the prothorax. Because of their appearance the

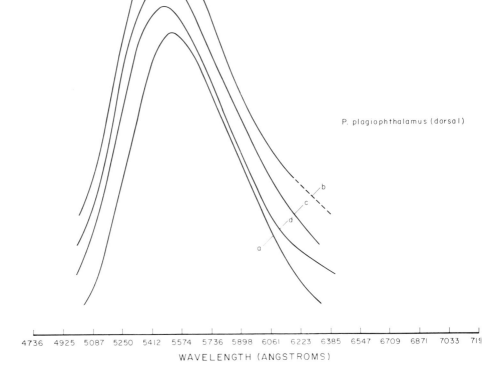

P. plagiophthalamus (dorsal)

WAVELENGTH (ANGSTROMS)

4736   4925   5087   5250   5412   5574   5736   5898   6061   6223   6385   6547   6709   6871   7033   719

FIG. 4.9.   Emission spectra of *Pyrophorus plagiophthalamus* dorsal organ *in vivo* and of purified *P. plagiophthalamus* dorsal organ luciferase with luciferins isolated from different sources. The spectra have been displaced upward so that the relative shapes can be more easily seen. (a) *P. plagiophthalamus* dorsal organ *in vivo*. (b) *In vitro P. plagiophthalamus* dorsal organ luciferase plus native *P. plagiophthalamus* dorsal organ luciferin. (c) *In vitro P. plagiophthalamus* dorsal organ luciferase plus native *P. plagiophthalamus* ventral organ luciferin. (d) *In vitro P. plagiophthalamus* dorsal organ luciferase plus synthetic *Photinus pyralis* luciferin.

insects are often referred to as "automobile bugs." The second type of luminous organ occurs on the first abdominal segment and is visible only when the insects are in flight or when the elytra are extended. The color of the light is yellow and obviously different from the green color that appears from the organs on the prothorax (see Fig. 4.6).

Light emission from *Pyrophorus* appears very slowly (0.5–0.6 sec to reach maximum intensity) but persists for many seconds or minutes before it fades. This is characteristic of the light response one observes in

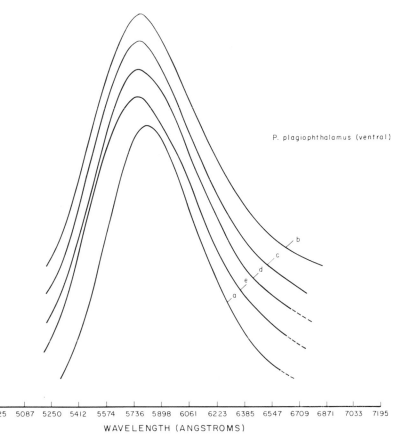

P. plagiophthalamus (ventral)

4763 4925 5087 5250 5412 5574 5736 5898 6061 6223 6385 6547 6709 6871 7033 7195

WAVELENGTH (ANGSTROMS)

FIG. 4.10. Emission spectra of *Pyrophorus plagiophthalamus* ventral organ *in vivo* and of purified *P. plagiophthalamus* ventral organ luciferase with luciferins isolated from different sources. The spectra have been displaced upward so that the relative shapes can be more easily seen. (a) *P. plagiophthalamus* ventral organ *in vivo*. (b) *In vitro P. plagiophthalamus* ventral organ luciferase plus native *P. plagiophthalamus* ventral organ luciferin. (c) *In vitro P. plagiophthalamus* ventral organ luciferase plus native *P. plagiophthalamus* dorsal organ luciferin. (d) *In vitro P. plagiophthalamus* ventral organ luciferase plus native *Photinus pyralis* luciferin. (e) *In vitro P. plagiophthalamus* ventral organ luciferase plus synthetic *Photinus pyralis* luciferin.

luminous glands which have no tracheal end cells. Numerous experiments indicate that the light response is under nervous control and that the air supply is through tracheal trunks whose branches terminate between the photogenic cells. Histological investigations by Dahlgren (1917) and Buck (1948) indicate the absence of tracheal end cells.

Recently a large number of *Pyrophorus plagiophthalamus* from Jamaica has been collected and the luciferin and luciferase from the different

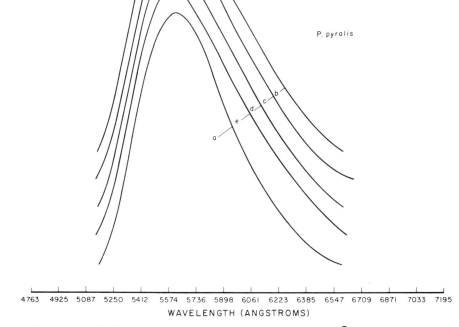

P. pyralis

4763   4925 5087 5250 5412   5574 5736 5898 6061 6223 6385   6547 6709 6871   7033 7195
WAVELENGTH (ANGSTROMS)

Fig. 4.11.   Emission spectra of *Photinus pyralis in vivo* and of purified *P. pyralis* luciferase with luciferins isolated from different sources. The spectra have been displaced upward so that the relative shapes can be more easily seen. (a) *P. pyralis in vivo*. (b) *In vitro P. pyralis* luciferase plus native *P. pyralis* luciferin. (c) *In vitro P. pyralis* luciferase plus synthetic *P. pyralis* luciferin. (d) *In vitro P. pyralis* luciferase plus native *Pyrophorus plagiophthalamus* dorsal organ luciferin. (e) *In vitro P. pyralis* luciferase plus native *Pyrophorus plagiophthalamus* ventral organ luciferin. From Seliger and McElroy (1964).

luminous organs have been partially purified. From these studies it seems certain that the mechanism of light emission in the click beetle is identical to that in the Lampyridae. The luciferins isolated from the two different light organs of *P. plagiophthalamus* are identical to the luciferin of *P. pyralis* as judged by paper chromatography, fluorescence excitation, and emission spectra. Furthermore, the two luciferins will react with *P. pyralis* luciferase in the presence of ATP and Mg++ to give light emission.

   Since the luciferins from two different families of luminous beetles are identical, the question arises as to the mechanism concerned which gives rise to the different colors of light in the various fireflies. By purifying the enzymes from the two different luminous glands of *P. plagiophthalamus* it was possible to study in detail the emission spectrum of the *in vitro* light emission. Even when *P. pyralis* luciferin was used it was easy to demonstrate that the *color of light is determined by the source of enzyme* (Seliger and McElroy, 1964). The *in vivo* and *in vitro* emission spectra are shown in Figs. 4.9–4.11 for *P. plagiophthalamus* dorsal organ, *P. plagiophthalamus* ventral organ, and *P. pyralis*, respectively.

As pointed out previously, it is possible to alter the color of light emitted by *P. pyralis* luciferase *in vitro* by changing the ionic environment. We interpreted this to be due to a change in the tertiary structure of the luciferase molecule, consequently changing the binding of the luminescent intermediate during the course of the light reaction. It appears that genetic changes among species of fireflies may have led to an altered luciferase structure which, in turn, determines the color of light produced. The basic light reaction is not different in the various species and, indeed, seems to be the same for two different families of beetles. Preliminary evidence from experiments on *Phrixothrix*, the beautiful South American railroad worm, suggests that the same mechanism occurs also in the Phenogodidae.

Thus a common luminescent reaction must have evolved early in the Coleoptera.

2. Bacterial Reactions and Mechanism

Numerous workers (see McElroy and Seliger, 1963b, for a review) have established the fact that reduced flavin mononucleotide (FMN), a long-chain aldehyde, oxygen, and bacterial luciferase are the essential components for light emission. Strehler (1953) was the first to demonstrate that reduced pyridine nucleotide would support luminescence in crude extracts of *Achromobacter fischeri*. Following this observation McElroy *et al.* (1953a,b) were able to demonstrate that FMN was required for light emission if a partially purified enzyme was employed as well as an unknown factor which Cormier and Strehler (1953) later identified as a long-chain aldehyde. Further purification of bacterial luciferase by McElroy and Green (1955) and Riley (1963) has not revealed any additional requirements for light emission.

The amount of light emitted in the enzyme reaction is directly proportional to the aldehyde added, indicating the possibility that the oxidation of aldehyde supplies the necessary energy for emission. Kinetic studies suggest the possibility of a requirement for two reduced FMN molecules.

Recently Hastings and Gibson (1963) have shown that $FMNH_2$ reduces the enzyme. This reduced enzyme in turn reacts with oxygen to form a reasonably stable intermediate which can then react with aldehyde *anaerobically* to give off light. Based on inhibition studies with arsenite, they suggest that $FMNH_2$ reduces a disulfide bond on the enzyme to form two SH groups. Presumably one of the SH groups takes up oxygen to form a hydroperoxide which then reacts with the aldehyde to give out light.

Thus in the bacterial system there is evidence of at least three intermediates in the reaction leading to light emission. The separation of the step involving the addition of oxygen and the light-emitting step may be of considerable importance for the interpretation of other luminescent reactions that presumably do not require oxygen. For example, Shimomura *et al.* (1961) have been able to isolate a protein substance from the hydromedusan, *Aequorea aequorea*, which emits light anaerobically when calcium ions are added. It may be that preliminary dark reactions involving oxygen may give some bound intermediate analogous to those proposed by Hastings and Gibson.

The role of the aldehyde in light emission is not understood. The rate of utilization or disappearance of the reduced enzyme-oxygen complex (intermediate 2 of Hastings and Gibson) is not affected by aldehyde. This suggests that there are two pathways for the reaction of the intermediate, one leads to light production in the presence of aldehyde, while the second reaction dissipates the intermediate without appreciable light production. The relative rates of the "dark" and "light" reactions can be altered independently (Spudich and Hastings, 1963). The temperature coefficient for the luminescent pathway is greater. The relative quantum yield is also affected by the amount and the chain length of the aldehyde (Hastings *et al.*, 1963). The luminescent pathway is also favored by conditions of high enzyme or protein concentration. For reasons that are not clear intermediate 2 is stabilized by a high protein concentration, either by the enzyme itself or by added bovine serum albumin.

Hastings and Gibson (1963) question whether the "dark" pathway is in reality "dark" or whether it is the same as the aldehyde-requiring pathway but with a lower quantum yield. The amount of "bound" aldehyde on the enzyme may be important in this respect. Significant light production can be observed without added aldehyde; however, in all cases examined some aldehyde has been shown to be present. It is possible that under some circumstances there could be a repeated utilization of a small amount of aldehyde which favors luminescence. This idea was originally proposed by Strehler. However, subsequent study indicated that total light production was directly proportional to the amount of aldehyde added which suggested aldehyde utilization. The increase in quantum yield with aldehyde and its variation with chain length is assumed to be due to the effective utilization of the aldehyde in the light reaction.

The role or roles of reduced FMN in light emission are still not entirely clear. The reaction of luciferase with reduced FMN is essential for light emission, but as Hastings and Gibson have indicated, this appears to be used, at least in part, for the reduction of the enzyme while the remainder of $FMNH_2$ is rapidly autoxidized by molecular oxygen. McElroy and Green (1955) indicated that DPNH could reduce bacterial luciferase in the absence of FMN and recently we have shown that more light can be obtained from a given amount of $FMNH_2$ if the enzyme is first reduced with glutathione or other reducing agents. These results indicate that reduced FMN is essential for light emission even in the presence of reduced enzyme.

It seems likely that the light-emitting species is an enzyme-FMN complex of some sort. However, the color of bacterial light (ranging from 475 to 505 mμ peak emission) is far from the fluorescence emission peak of FMN which is at 530 mμ. Quantum yield studies indicate that FMN functions as a substrate in light emission. Thus if the isoalloxazine nucleus is the chromophoric group that becomes excited, its electronic structure must in some way be influenced by the enzyme and the aldehyde (Hastings *et al.*, 1963). Recently Terpstra (1962, 1963) has found by filtration of a crude bacterial lysate over Sephadex G-25 that luciferase is partially inactivated. The filtrate contains a fraction which will restore, at least in part, the luciferase activity; however, none of the fractions obtained from the filtrate fluoresces with a peak emission which is comparable to the luminescence maximum. When $FMNH_2$ is added to a bacterial enzyme preparation, some unknown compound is formed which is rapidly converted by UV radiation to a fluorescent substance with a peak emission near 470 mμ. The time during which the first compound is formed is of the same order of magnitude as the time of duration of the light reaction. Terpstra suggests that the compound is a precursor of the light-emitting molecule.

Interesting evidence concerning the importance of the enzyme in determining the color of the light has been obtained recently using luciferase from different strains of luminous bacteria. As observed for the firefly, the source of enzyme and presumably the secondary and tertiary structure of luciferase are important in determining the color of the light emitted. Different strains of luminous bacteria emit slightly different colors of light. As shown in Fig. 4.12, the *in vivo* peak emissions can vary by as much as 200 A. In many cases the *in vitro* emission spectra are the same as the *in vivo* spectra. Thus, using identical mixtures of $FMNH_2$ and aldehyde, it is possible to obtain light emissions whose *in vitro* peak emissions vary as much as 200 A depending upon the sources of luciferase. The binding of $FMNH_2$ and aldehyde to the luciferase appears to be

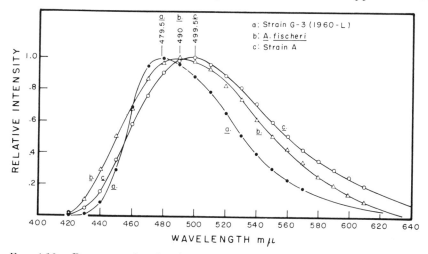

Fig. 4.12.   Representative *in vivo* emission spectra of three species of marine luminous bacteria showing the range of peak bioluminescent emission.

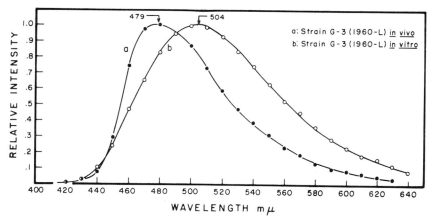

Fig. 4.13. *In vivo* and *in vitro* emission spectra of strain G-3 of a marine luminous bacterium.

the important factor in determining the nature and energy level of the excited state of the emitter. However, in some cases of the luminous bacteria, the *in vitro* emission is quite different from the *in vivo* emission, as shown in Fig. 4.13. We have no explanation as yet for this phenomenon.

3. General Properties of Other Systems

Aequorea aequorea. Shimomura *et al.* (1963) have obtained a partially purified protein from this jellyfish which will emit light upon the addition of calcium ions. The rate of light emission and total light are not altered by first equilibrating all the solutions with pure hydrogen. It seems likely that preliminary dark reactions involving oxygen or other radical-producing oxidants give some bound intermediate analogous to those proposed by Hastings for the bacterial system.

Cypridina. The luminescent system in the marine ostracod crustacean *Cypridina hilgendorfii* has been studied extensively (see Harvey, 1952a, 1953). The luminous material occurs as granules in elongated cells in the submaxillary gland. Observations on the living animal indicate that two varieties of granules pour out from the gland into the surrounding sea water. The larger granules (10μ) contain luciferin and the smaller granules (2μ) apparently contain luciferase. On striking the sea water, both granules dissolve and a blue luminescence is observed. Thus apparently all dark reactions essential for an active luciferin have occurred prior to mixing with enzyme and the general over-all reaction for light appears to be a direct interaction of luciferin, oxygen, and enzyme.

By the use of acetone extraction, ammonium sulfate fractionation, and adsorption and elution from calcium phosphate gels, McElroy and Chase (1951) were able to prepare a highly purified luciferase preparation. On a protein basis the purification was about $150\times$. Recently, Shimomura *et al.* (1961) have used, in addition, DEAE columns to obtain a purified preparation. Unfortunately, however, details concerning recovery on a protein basis are not given, and it is not possible to compare readily their relative purification with other purified samples. Luciferase has an absorption peak typical of that for proteins and with no indication of any bound low molecular weight cofactors. Highly purified luciferase has a slight pink color and there are indications that it may be a metalloprotein. Shimomura *et al.* concluded from sedimentation analysis that the molecular weight of luciferase was about 50,000 and had an isoelectric point of about 4.4. See Chase (1960) for other detailed analysis.

Anderson (1936) was the first to obtain highly purified *Cypridina* luciferin. Subsequent to these studies, important contributions have been made by Harvey (1952a,b, 1955), Harvey and Tsuji (1954), Chase (1949a,b,c, 1960), H. S. Mason (1952), Tsuji (1955) and Tsuji and Harvey (1954). The purification and properties of these various preparations are reviewed by Tsuji *et al.* (1955). Shimomura *et al.* (1957) reported the crystallization of luciferin and subsequent work led Hirata *et al.* (1959) to propose the structure presented in Fig. 3.1. In phosphate buffer at pH 5.6 crystalline luciferin has three major absorption peaks: 270, 312, and 425 mμ. Sie *et al.* (1961) have reported that the excitation at the two longer wavelengths leads to a fluorescence emission at approximately 525 mμ. Excitation at 270 mμ leads to emission at 348 mμ and is most likely due to the excitation band of the indole moiety of the partially degraded luciferin molecule.

Very little is known concerning the mechanism of the light reaction. Chase (1949a,b, 1960) has studied the kinetics of the reaction and has shown clearly that light emission follows simple first-order kinetics which can be adequately described by Michaelis-Menten concepts. The K_m was reported by Chase to be about 6×10^{-7} *M*. Shimomura *et al.* find a similar value using crystalline luciferin and highly purified luciferase.

Recently, Sie *et al.* (1961) have studied the light reaction with respect to fluorescence and bioluminescence emission spectra. They were able to demonstrate that the major fluorescence emission peak at 525 mμ disappeared as luminescence proceeded. Both enzymic and alkali-catalyzed nonenzymic oxidation of luciferin lead to a loss of the two major absorption peaks at 312 and 425 mμ as well as the disappearance of the fluorescence. Tsuji (1955) has described the spectral changes which occur during the luminescent reaction. The corresponding loss of light-emitting properties suggests strongly that the fluorescence observed is due to the luciferin present. If this is so, then there must be extensive alteration of the luciferin molecule in order to excite an intermediate to emit light at 460 mμ. Following emission other rearrangements must occur, for no one has observed, even with highly purified preparations, a fluorescent product with emission properties similar to the bioluminescence emission. Such a mechanism would presumably involve oxygen radicals and would be classified as a direct chemiluminescence, in contrast to a sensitized chemiluminescence. This product formation could also explain the lack of reversibility of the enzyme-catalyzed "oxidation" of luciferin in contrast to the autoxidation which Chase and Lorenz (1945) have studied and in which they have been able to show reversibility. This is true for a number of luciferins from various luminous organisms.

Luminous Fish. Haneda and Johnson (1958) have demonstrated that it is possible to obtain luciferin and luciferase from the glands of two luminous fish, *Apogon* and *Parapriacanthus*. These extracts have many of the properties of the *Cypridina* system and exhibit light emission in cross reactions of substrate and enzyme. Sie *et al.* (1961) studied the spectral composition of the light emitted in the reaction of *Apogon* luciferin and *Cypridina* luciferase and other combinations. In both cases the ranges and peak emissions were identical.

Since relatively large quantities of purified luciferin can be prepared now, it should be possible to study in greater detail the organic mechanism of the light reaction.

Recently a new species of *Cypridina* was found on the Eastern shore off Jamaica. It is very much like *C. hilgendorfii* except that the peak emission is at 4780 A, whereas *C. hilgendorfii* peaks around 4600 A.

Renilla—The Sea Pansy. Cormier (1959, 1960, 1961) has reported the preparation of cell-free extracts of the luminous sea pansy, *Renilla remiformis*. With further purification using ammonium sulfate fractionation and DEAE columns, it was possible to demonstrate that the following components were required for light production: (1) an adenine-containing nucleotide, (2) a crude hot water extract of the organism which contains at least one component that is used during luminescence and has been called *Renilla* luciferin, (3) molecular oxygen, and (4) *Renilla* luciferase.

Recently, Cormier (1962) has made the very interesting observation that the nucleotide required for light emission is neither AMP, ADP, nor ATP as originally reported. He has been able to isolate and identify the trace contaminant in these preparations as 3',5'-diphosphoadenosine (PAP). This isolated cofactor was identified on the basis of its adenine-ribose-phosphate ratio, its chromatographic behavior as compared to authentic PAP, and its identical activity with PAP in the bioluminescent assay system. PAP also functions in the sulfokinase reaction for the transfer of sulfate, and Cormier demonstrated that the unknown factor would replace authentic PAP in this reaction. These results demonstrate once again the extreme sensitivity of light-emitting reactions for detecting small quantities of biologically active materials. The level of contamination of AMP or ADP by PAP may not be significant for most experiments, but the results of Cormier demonstrate the importance of knowing the purity of biological preparations when a presumedly pure compound is required in relatively high concentrations. Obviously this same precaution is also necessary when fluorescence methods are used.

The bioluminescence assay by the method described can detect 10^{-9} M PAP, and it seems likely when both *Renilla* luciferin and luciferase are prepared in pure form that even smaller quantities of PAP can be detected. The reaction seems to be specific for PAP since nucleotides such as 3',5'-diphosphoinosine, 2',5'-diphosphoadenosine, and others tested showed no activity.

Cormier has been able to demonstrate that incubation of PAP with *Renilla* luciferin and luciferase under anaerobic conditions leads to the formation of a relatively heat-stable intermediate. The addition of oxygen to such an anaerobic reaction caused a very rapid emission of light in contrast to the slower rate of the reaction when initiated aerobically with PAP. The reaction is analogous to the firefly reaction in which one can demonstrate the formation of luciferyl-adenylic acid and pyrophosphate under anaerobic conditions when firefly luciferin, luciferase, and ATP are incubated together. The rate-limiting step in the firefly reaction is the initial activation reaction and this also appears to be true for the *Renilla* system. The mechanism of action of PAP in the activation of *Renilla* luciferin is not known. The comparisons between this system and the sulfokinase reaction will be extremely interesting.

Odontosyllis enopla. This luminous annelid appears at the surface of the ocean in Bermuda waters usually with maximum swarming activity 3 days after the full moon and a peak activity at 55–56 min after sunset (Markert *et al.*, 1961). The females appear first, swimming in small circles at the surface and emitting a bright greenish light. The males are apparently attracted to this emission and move, while flashing, to the luminous circle where both eggs and sperm are then ejected into the water. This remarkable lunar periodicity of the "fire worms" of the sea has attracted many travelers (e.g., Columbus), poets, and scientists, including E. N. Harvey (1952a) who first demonstrated a luciferin-luciferase reaction in 1931.

Very little is known of the biochemistry of this interesting reaction (Shimomura *et al.*, 1964). Recently, we have obtained partially purified luciferin and luciferase from frozen and acetone-dried preparations. Like *Cypridina*, the luciferin is readily autoxidized by molecular oxygen but can be reduced to an active form for light emission by cysteine, hydrosulfite, and other reducing agents. The luciferin has a fluorescent emission peak at 510 mμ which corresponds exactly to the bioluminescent emission. Since this fluorescence decreases as the light reaction proceeds, the possibility exists that this may be a sensitized chemiluminescence.

Balanoglossus biminiensis. Dure and Cormier (1961) have reported recently a luciferin-luciferase reaction in extracts from this luminous marine balanoglossid. In addition to luciferin and luciferase, H_2O_2 or an organic peroxide is essential for light emission. This may prove to be a very interesting reaction with regard to an oxygen requirement as discussed above for the comb jellies.

Pholas dactylus. This luminous clam was used by Dubois (1885, 1887a,b) for his classical work on the luciferin-luciferase reaction and the details are reviewed by Harvey (1952a). Recently, Plesner (1959) has studied some of the properties of a crude acetone powder preparation. In such preparations the luciferin has apparently been oxidized, and Plesner reports that DPNH plus FMN will restore light emission. Recently, we have shown that a number of reducing agents will work in such a system and that FMN is not a requirement for light emission. Partial purification of the enzyme by ammonium sulfate fractionation clearly indicates that it is not like the bacterial system. No aldehyde requirement can be demonstrated, and we conclude that in general it has many of the properties of the *Cypridina* system. However, *Cypridina* luciferase will not stimulate light emission with the *Pholas* luciferin. The bioluminescence emission peak for *Pholas* extracts is at approximately 480 mμ. Nicol (1958) observed that the peak emission for light from the intact luminous organ was approximately 490 mμ.

Luminous Fungi. Harvey (1952a) summarizes the extensive literature on the various luminous fungi, and recently a luciferin-luciferase reaction has been described in cell-free preparations which depends upon the presence of reduced pyridine nucleotide (Airth and McElroy, 1959). Airth (1961) has made an extensive analysis of the properties of this system using extracts from *Collybia velutipes* and *Armillaria mellea*. Partially purified luciferase preparations give little light until bovine plasma is added. This and other evidence suggest the presence of an inhibitor of the luciferase reaction. Although reduced DPN or TPN are required, the evidence indicates that this system is entirely different from the bacteria and other systems. Neither flavin nor aldehyde is required. The emission maximum of the luminescence is at 5300 A.

Gonyaulax polyedra. These marine dinoflagellates are both luminescent and photosynthetic. Sweeney and Hastings (1960) have studied under laboratory conditions the interesting diurnal rhythm of luminescence which this and similar forms show under natural conditions (Seliger *et al.*, 1961a). Hastings and Sweeney (1957a,b) have studied the cell-free system in detail and have demonstrated the presence of both a luciferin and luciferase as well as a high salt requirement for maximum light emission. Like *Cypridina* and other systems, the luciferin is labile when exposed to air. Bovine serum albumin is also stimulatory in this system. The light emitted is blue with a major peak at 4700 A.

Since both *Gonyaulax polyedra* and *Pyrodinium bahamence* (McLaughlin and Zahl, 1961) can now be grown in large quantities in the laboratory, they should provide excellent material for a detailed study concerning the organic mechanism which leads to light emission.

There are many other luminous forms that have been studied extensively from an anatomical and physiological point of view and the reader is referred to Harvey (1952a) for a review of this material. A number of studies have been made on emission spectra and most of this research has been summarized in a number of papers by Nicol. *Chaetopterus variopedatus* has an emission whose spectrum extends from 405 to 605 mμ with a maximum at about 465 mμ (Nicol, 1957). The spectral emission curves of four species of polynoid worms, *Harmothoë longisetis*, *Gattyana cirrosa*, *Polynoë scolopendrina*, and *Lagisca extenvata* are similar with a maximum at 515 mμ. *Noctiluca miliaris* has an emission peak at 470 mμ (Nicol, 1958) which is identical to another luminous dinoflagellate, *Gonyaulax polyedra* (Hastings and Sweeney, 1957a,b). Nicol (1958) determined that the peak emission from *Pennatula phosphorea* was at 510 mμ while that of the lantern-fish, *Myctophum punctatum* was at 470 mμ (Nicol, 1960). The latter emission peak is very close to the observed cell-free emission peak of *Apogon* which was determined to be 460 mμ (Sie *et al.*, 1961). Emission peaks for the following organisms have been obtained by Nicol (1958): the siphonophore, *Voglia glabra*, 470 mμ; the scyphomedusa, *Atolla wyvillei*, 470 mμ; the ctenophore, *Beroë ovata*, 510 mμ.

TABLE 4.2

SUMMARY OF CHEMICAL REACTIONS AND PEAK LIGHT EMISSIONS
FOR VARIOUS BIOLUMINESCENT ORGANISMS

Organism	Nature of reactants	Peak light emission(s) (A)
Firefly	$LH_2 + E + ATP + O_2$ $\overset{\displaystyle O}{\underset{\displaystyle \parallel}{}}$	5520–5820
Luminous bacteria	$FMNH_2 + RCH + O_2 + E$	4900–5050
Cypridina hilgendorfii (crustacean)	$LH_2 + O_2 + E$	4600
Cypridina sp. (Jamaica)	$LH_2 + O_2 + E$	4780
Odontosyllis enopla (polychaete worm)	$LH_2 + O_2 + E$	5100
Pholas dactylus (luminous clam)	$LH_2 + O_2 + E$	4800
Omphalia flavida (fungus)	$LH_2 + O_2 + E$	5300
Renilla reniformis (Sea Pansy)	$LH_2 + 3',5'\text{-PAP} + O_2 + E$	Blue
Gonyaulax polyedra (protozoan)	$LH_2 + E + O_2$	4700
Pyrodinium bahamence (protozoan)	$LH_2 + E + O_2$	4770
Apogon (fish)	$LH_2 + E + O_2$	4600

4. Evolutionary Aspects of Bioluminescence

The emission of light by organisms is not restricted to any one group of animals or plants (Harvey, 1952a). The random distribution of a light-emitting system among the bacteria, fungi, numerous invertebrate groups, and even fish suggest that it is an offshoot of a chemical reaction fundamental to all organisms. The widespread occurrence of luminous organisms may be the important key in understanding its origin and evolution.

We think that these light-emitting reactions in organisms were, in fact, detoxifying processes for the removal of oxygen which was necessary for the survival of early anaerobic forms of life. We propose that the use of organic reducing substances to remove oxygen by direct reduction led to the formation of an excited state which could emit light. These reactions were the basis for the origin and evolution of luminescence in organisms.

Most investigators assume that at the time of origin of the primitive forms of life on earth the biosphere was devoid of oxygen. Although traces of oxygen may have been produced by short wavelength UV radiation, it is doubtful that it remained as free oxygen for any length of time. The extensive reducing environment and the large excess of such compounds as Fe^{++} would insure the immediate reduction of any free oxygen.

Although the stepwise oxidative reactions are the primary ones concerned with the efficient supply of energy in modern aerobes, it seems most unlikely that oxygen functioned as the terminal electron acceptor in the primitive forms of life. Free oxygen must have appeared in significant quantities at a much later date as the environment gradually changed from a reducing one to an oxidizing one. Accompanying these changes in evolution was the development of the extensive enzyme systems coupling the primary photochemical reactions with water which acted as the hydrogen donor. Green plant photosynthesis is a consequence of these evolutionary events. Thus it seems certain that the primitive organisms were true anaerobes. The early energy-yielding and energy-coupling processes, however, must have involved dehydrogenation and condensation reactions which were catalyzed first by primitive catalysts, and much later by enzymes. The important point is that life arose and evolved only when such entities were capable of coupling the energy of dehydrogenation to the necessary synthesis of all those compounds which were required for duplication. As we visualize this energy-coupling process today it seems most likely, as suggested by Lipmann, that inorganic phosphate was important in the trapping of energy in the form of a pyrophosphate bond. Thus the first organism must have made use of the rich organic reducing environment to couple dehydrogenation to energy liberation and utilization. The first energy-trapping reaction must have been of the type indicated in the reaction below:

$$AH_2 + B + P \rightarrow A + BH_2 + \sim P \qquad (4.7)$$

An organism could make use of the energy from the oxidation of AH_2 for growth and reproduction as long as AH_2 was present in sufficient quantity in the environment. There are a number of ways in which the oxidized product A could be reduced, but we will defer a discussion of this until later. Needless to say, once AH_2 became exhausted the organism would then be forced to rely upon other dehydrogenation reactions. The successful organisms must have been capable of using less reduced compounds for coupled oxidation-reduction reactions. We visualize the subsequent dehydrogenation and energy coupling as follows:

$$BH_2 + C + P \rightarrow B + CH_2 + \sim P \tag{4.8}$$

During such an evolutionary process it is evident that by using the highly reduced compounds first the anaerobes were gradually creating a more oxidizing environment. At each successive step an electron acceptor was used which was approaching the potential of oxygen itself. Thus, we visualize a stepwise succession of dehydrogenation reactions in which energy liberation was closely linked to synthesis by means of the pyrophosphate bond. If a modest amount of AH_2 could be reformed within the cellular environment by whatever means, it is clear that such a selective evolutionary process could lead to the establishment of a series of electron transport systems which are capable of liberating energy. At some stage in this evolutionary process the reformation of reduced primary substances must have occurred in which sunlight was the source of energy. Life may have originated and evolved to its high level because of the presence of complex organic molecules which were capable of reasonably long-lived excited states. Creation of the excited states by light quanta established for the first time the primitive photochemical event which sometime later in evolution was capable of restoring the highly reducing environment. It seems reasonably certain that the present highly efficient photosynthetic apparatus is much too complicated to have been present in these primitive organisms. A less complicated inorganic metal complex and later an organic metal complex was the most likely system for creating a reducing system driven by light energy. It was the coupling of excitation energy to the electron transport process which created an efficient and inexhaustible supply of reducing energy. A number of organic compounds could have functioned in this manner under anaerobic conditions. For example, in the absence of oxygen, FMN is reduced by light to $FMNH_2$.

In the absence of oxygen in the primitive atmosphere the spectrum of the radiant energy reaching the oceans extended well below the present 2900 A cut-off, so that even the pyridine nucleotides could have been raised to excited states in great abundance. It is a rather significant point that the photochemical production of the leuco form of dyes is favored when oxygen is excluded from the environment. Further, the quenching effect by oxygen of the luminescence of many organic compounds is pictured as being due to an interaction between the ground triplet state of the oxygen molecule and the triplet state of the excited organic molecule. This would leave the latter in a state of high vibrational energy which by interaction with the environment returns to the ground state. The freezing of a solution or adsorption of a substrate onto a lattice such as in a mitochondrion has two effects: there is a reduction in the vibrational and rotational degrees of freedom owing to the bonding, which sharpens the band structure, and there is also a reduction in the vibrational interaction of the excited state with dissolved oxygen. For this reason even organic molecules which show no fluorescence in solution at room temperature will exhibit phosphorescence at low temperatures or when dissolved in a "glass" at room temperature.

The nitrogen heterocyclics of which the adenine molecule is an example are characterized by the presence of nonbonding electrons which give rise to energy levels in the excited molecule lower than the $\pi \rightarrow \pi^*$ levels. It is presumably the triplet $n \rightarrow \pi^*$ level to which the singlet $\pi \rightarrow \pi^*$ level eventually decays, and this triplet level is quenched by oxygen or by interaction with the environment.

In order for a molecule to make efficient use of the available light energy it must have a system of relatively loosely bound π-electrons which can be raised to excited states. These excited states must be reasonably well separated from the vibrational ground states in order that the energy is not dissipated immediately to the environment and in order that the now highly reactive molecule can interact with a suitable donor or acceptor. The molecules involved in the electron transport process are not the cyclic carbon compounds but the nitrogen heterocyclics with their $n \rightarrow \pi^*$ triplet levels.

Porter has shown that in the strict absence of oxygen, triplet-triplet energy transfer is very efficient, and it is therefore conceivable that initially the light excitation was coupled in this way to a suitable reducing substance giving the general reactions:

$$\text{(A)} \quad h\nu + X \rightarrow X^* \rightleftharpoons (X^*)_{\text{triplet } n\pi^*}$$

$$(X^*)_{\text{triplet } n\pi^*} + A \rightarrow A_{red} + X_{ox}$$

$$\text{(B)} \quad h\nu + A \rightarrow A^* \rightleftharpoons (A^*)_{\text{triplet } n\pi^*}$$

$$(A^*)_{\text{triplet } n\pi^*} + \text{Solvent} \rightarrow A_{red} + S_{ox}$$

(4.9)

The advantages of this mechanism over the corresponding singlet-singlet energy transfer in a weakly structured system are the reduction of energy-wasting fluorescence and the longer lifetimes of the triplet states.

Although at the present time, in the well-ordered mitochondrial electron transport apparatus, the reduction of FAD by DPNH and the subsequent reduction of the cytochrome pigments by reduced FAD probably do not involve excited states of these molecules, it is suggestive that the energy-level differences between the blue-fluorescent pyridine nucleotides, the yellow-green-fluorescent flavins, and the red-fluorescent prophyrins are of the order of 10 kcal/mole which is sufficient for the generation of energy-rich phosphate bonds. The nature of the singlet and triplet excited states of these cofactors may have been an important factor in their early selection for electron transport processes. The advent of the lipoprotein mitochondrial matrix for the stepwise reduction involving the terminal oxygen molecule was almost certainly a much more recent development in biological time, relegating the light-energy utilization role to the more efficient aerobic chlorophyll systems.

As a result of the utilization by the anaerobes of the highly reducing substances, and the sensitized photochemical dissociation of water by the action of light on various pigments, it seems quite likely that the amount of oxygen in the environment was gradually increasing, concurrent with the exhaustion of the organic material suitable for anaerobic metabolism. Some authors believe that during this period there developed those chemoautotrophs which were able to oxidize directly ammonia, hydrogen sulfide, or ferrous ion to satisfy their energy requirements. However, if the earlier organisms were strict anaerobes it is evident that growth would be inhibited by the presence of oxygen. Unfortunately, we do not understand why anaerobes fail to grow in the presence of oxygen. There are some organisms that can grow in the presence of a small amount of oxygen, but there are others which cannot grow at all. Obviously, oxygen in some way prevents growth as long as it is present. However, this is an inhibiting rather than a lethal action since removal of oxygen will allow the initiation of growth. The present evidence suggests that oxygen may prevent the growth of strict anaerobes by a number of mechanisms. The formation of hydroperoxides is considered to be one of the primary factors. There are some organisms among the strict anaerobes that can take up oxygen and form hydrogen peroxide. In spite of this oxygen uptake and the accumulation of hydrogen peroxide the anaerobe will still grow if placed in an environment devoid of oxygen. In other words, the organism cannot grow in the presence of oxygen, but the exposure to it does not prevent it from growing later when the oxygen is removed. In addition, there are some anaerobes in which oxygen uptake has not been demonstrated and in which hydrogen peroxide has not been detected. These organisms show the same inability to grow in oxygen. Thus we are forced to the conclusion that oxygen is inhibiting but not lethal, at least over a limited time exposure.

At the time of the appearance of low concentrations of oxygen there was a struggle to survive in a changing environment and it is our contention that the organisms that were successful were those that were able to reduce molecular oxygen directly and quickly. At first this process was an ordinary chemical reduction which in some cases could have been quite slow. However, the most successful organisms were those that acquired a catalyst that accelerated the reduction process. This new enzyme would be an oxidase which catalyzed the oxidation of a reduced organic substrate by molecular oxygen. The oxidase would have the properties of what we presently call luciferase. The direct reaction of the reduced substrate with molecular oxygen is generally a strongly exothermic reaction. We therefore propose that the oxidative reaction left the product or intermediate molecule in a highly excited state. If the specific molecule was capable of fluorescence, the oxidation would have resulted in what we term bioluminescence. Thus the struggle to maintain anaerobic conditions led to the selection of organisms having specific oxidases (luciferase) which catalyzed the rapid removal of oxygen. We propose, therefore, that those organisms that successfully survived the exposure to low oxygen tensions were all potentially luminescent because of the nature of the detoxifying reaction. It seems significant that all luciferase systems that have been carefully studied can catalyze the utilization of reducing substrate (luciferin) at very low oxygen tensions. The firefly, bacteria, and *Cypridina* systems are at least one hundred to one thousand times more sensitive to oxygen than is hemoglobin.

The reduction of oxygen in most cases led to the production of hydrogen peroxide or some organic peroxide radical. Although most organisms can tolerate a small amount of hydrogen peroxide, it is clear that the really successful ones were those that were capable of removing the peroxide by a reducing agent to form water. This must have been the selective pressure which led to the adaptation of a catalyst to use hydrogen peroxide and additional reducing material to form water and an oxidized product. We propose that again the successful organisms were those capable of catalyzing a peroxidase type of reaction. Coupled with the luciferase reaction this led to the elimination of oxygen with the production of water, oxidized product, and light. Undoubtedly during this part of the evolutionary process the development of the iron porphyrin type of catalyst for peroxides occurred. We note, for example, that iron in the Fe^{++} state is a very effective catalyst for the decomposition of hydrogen peroxide. If iron, however, is incorporated into a special structure such as a tetrapyrrole ring then the catalytic activity is increased by several hundred times. Calvin has presented excellent arguments as to why hydrogen peroxide might favor the formation of such iron pyrrole-type compounds.

Simultaneous with the gradual loss of the reducing environment and the appearance of an oxidizing one, a catalase-type enzyme would be favored over the peroxidase system, the former decomposing hydrogen peroxide into water and oxygen. This would lead to the selection of those systems which were capable of reducing oxygen stepwise to form water without *free* hydrogen peroxide being an intermediate. We feel that it was during this period that the iron-porphyrin enzyme systems, possessing the ability to activate oxygen for accepting electrons directly, came into existence. The oxygen could therefore act as the electron acceptor without the intermediate formation of hydrogen peroxide. It was this adaptation which led to the completion of the oxidative phosphorylation system as we see it today. The utilization of oxygen by this primitive aerobic form was then the last major step in the establishment of the electron transport process. By using oxygen as the terminal electron acceptor the

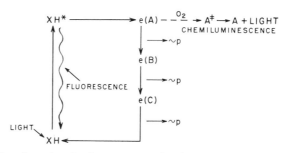

Fig. 4.14.   Coupling of excitation energy to the electron transport process (McElroy and Seliger, 1962a).

entire organic reducing environment was susceptible to oxidation. This must have led to the rapid and final development of the heterotrophic organisms which appeared during this phase of evolution.

Photochemical excitation energy processes must have been highly successful in this environment, using first the electron donors of an inorganic (Fe^{++}) or organic type and finally, with the exhaustion of these, water. Thus we think of the present green plant photosynthesis as being one in which water became the ultimate electron donor leading to the formation of some OH-type compound which was capable of giving rise to molecular oxygen. The emergence of an organism capable of using simple carbon compounds such as CO_2, water, and light as the energy source was the primary evolutionary event which led to the creation of a rich organic environment on earth for a second time (see Fig. 4.14).

With such systems established and with the appearance of the true aerobes, it is evident that the direct reduction of oxygen and the accompanying luminescence were no longer of selective advantage. Therefore, with the appearance of the aerobes luminescence would begin to disappear. However, since light emission was originally intimately connected with the essential energy-liberating electron transport processes it is likely that a number of the light-emitting systems would persist. It is therefore our argument that bioluminescence is a vestigial system in the evolutionary process and that there is, at present, no selective advantage insofar as the primary excitation process is concerned. It is true, however, that during the evolution of various species the luminescent system has been adapted for secondary purposes which have selective advantage. The identification of the female firefly by the male is an excellent example. No such argument involving sexual reproduction advantages can be made for the bacteria. Where luminescence has been put to a use, this use can be important in explaining selective advantage at a secondary level. The primary process which involves an interaction of oxygen with a reducing substance to create an excited state cannot readily be explained except by a more fundamental metabolic system common to all organisms.

The fact that the luminescent systems which have been carefully analyzed have been shown to have entirely different luciferins and emission spectra is an argument in favor of the idea that originally organisms used a variety of complex organic molecules for the reducing system for removal of oxygen and for the creation of an excited state. It seems highly likely that the reducing systems of the potential level of the pyridine nucleotides were the ones used early for energy liberation during dehydrogenation. The establishment of this system probably served as the basis for the evolution of other electron transport steps. This may explain why in photosynthesis we first observe the net reduction of pyridine nucleotides by chlorophyll, even though thermodynamically others might be favored. The flow of electrons propelled by photochemical processes seems always to involve a highly reducing state. If the initial removal of oxygen by the anaerobes made use of reducing systems of this potential level, one might expect to observe the high-energy blue bioluminescence emission most frequently. As the reducing power of the environment decreased other less reducing chromophoric groups could have become involved. It may be significant for our argument, therefore, that most luminescent forms are unicellular (a large number are also photosynthetic) and their emission is in the blue region of the spectrum. A high proportion of all luminous organisms appears in the oceans. The second most frequent forms have their emission in the yellow-green region and, as far as we know, are all multicellular or complex multinuclear structures. Red light emission is quite rare and is known to occur in only one form—the South American railroad worm.

CHAPTER 5

Biological Action of Light

A response to light has been observed in a number of different biological systems including photosynthesis, chlorophyll synthesis, chloroplast constitution, anthocyanin synthesis, seed germination, seedling and vegetative growth, flowering, phototropism, protoplasmic viscosity, photoperiodism, modifications to biological "clocks," chromosome damage, photoreactivation, photoprotection, visual photoreception, chromatophore control, and bactericidal action.

All of the multitudinous effects of light in these systems are initiated by one primary process; light is absorbed by the ground state of a specific molecule providing sufficient free energy either to initiate, or alter the rate of, a chemical reaction. It may provide the activation energy for a *cis-trans* isomerization leading to chemical amplification as in the case of vision. It may provide reducing energy for electron transfer as in photosynthesis. It may act to alter equilibrium concentrations of pigments as in photomorphogenesis. It may modify pigments as in chlorophyll synthesis; it may induce or inhibit enzyme reactions such as presumably occur in phototropism and photoperiodism. In all cases it is important for an understanding of mechanism to separate the primary photochemical act from the subsequent dark chemistry. Since the entire process begins with the absorption event, the most basic and direct line of experimentation is involved with the measurement of the action spectrum for the biological response. In principle, the experimentally observed action spectrum should correspond directly with the absorption spectrum of the involved pigment molecule. This is just a restatement of the Grotthus-Draper law of photochemistry.

Action spectra have appeared in the literature in many forms. In keeping with our discussion in Chapter 1 on the measurement of light, we feel that the proper presentation of an action spectrum should have as the ordinate *the reciprocal number of quanta in the wavelength interval required to obtain any particular constant biological response.* If data are plotted in this way, the action spectrum will be independent of any nonlinear responses with number of quanta absorbed. There is, however, one factor which may be important; some reactions are not linearly dependent upon light intensity. For example, the recombination of active centers may be proportional to the square of the light intensity. It is therefore advisable to work at low light intensities and determine the seriousness of any deviations from a straight line when response is measured at various intensities at a fixed irradiation wavelength. If the rate of product formation is linear with intensity, an action spectrum can be obtained by plotting directly the amount of product formed in any given time interval (normalized to equal incident photon intensities of light) as a function of wavelength. Alternatively, the amount of product formed under conditions of constant exposure can be plotted as a function of wavelength.

1. Organization and Structure of Light Receptor Systems

Control and regulation of cell metabolism depend upon the existence of highly organized structural systems. These include the cell membrane which has the properties of selective permeability, catalytic sites for active transport, phagocytosis, and pinocytosis. The complex membraneous structure of the mitochondrion contains all of the enzymes and cofactors essential for carrying out the process of oxidative phosphorylation. The function of messenger RNA in the synthesis of polypeptides depends upon the presence of the complex polysomal system which is attached to the membrane systems of the endoplasmic reticulum. In the case of photochemical transformations it appears that the pigment concerned with the absorption of light is associated with complex membraneous structures that are capable of carrying out controlled, dark chemical reactions. The most obvious cases are the chloroplast in the photosynthetic apparatus and the rod and cone outer segments in the eye. We will review briefly what is known about the structure of some of these systems in order to bring out the similarities. It seems significant to us that all processes involving several steps in energy transformation are associated with complex membraneous structures which are similar in their architecture.

In discussing these structures it is well to keep in mind some of the limitations of the techniques involved in electron microscopy.

In the electron microscope pictures, contrast in images depends upon the differential scattering and absorption of electrons. In a material containing reasonably homogeneous amounts of carbon, nitrogen, and oxygen atoms there can be little contrast except by the introduction of staining reagents containing high atomic number atoms. In most cases osmium tetroxide is used as both a staining and a fixing agent. It apparently binds together lipids and proteins in pre-existing lipoprotein structures. However, neither the nature of the binding nor the fixation process is well established. Consequently, the interpretation of the structure observed in electron micrographs after osmium staining may be subject to staining artifacts. Potassium permanganate has also been used extensively. It is a rather poor fixative as compared with osmium tetroxide. In examination of potassium permanganate-stained sections of various membranes, additional data regarding the structural organization can be obtained. Whether or not all of these structures represent true structures in the living cell has not been unequivocally established at the present time.

With the exception of the nuclear membrane, cell membranes are double layers of osmiophilic material, with a total thickness of 100 to 200 A. This probably represents two lipoprotein interfaces, with the protein layers on either side of the lipid. The dimensions are compatible with a double lipoid layer, assuming the osmium stain binds at the lipoprotein interface (the nuclear membrane appears to be a continuous closely spaced double membrane with a honeycomb-like inner membrane). Plasma membranes and squid axon membranes have osmiophilic layers about 50 A thick. The double layer lipoprotein grana of the lamellar chloroplast have "fatty" layers about 50 A thick, sufficient to include two phytol tails of chlorophyll, end-to-end. The densely staining flattened sacs in the rod outer segments are double layers, each about 40 A thick. In all of these cases the thickness appears to be sufficient to accommodate an end-to-end double lipid structure with the protein aqueous phase on the outside. Except for certain blue-green algae and photosynthetic bacteria where there are dense granules associated with photosynthetic pigments, the electron microscope indicates a very regular lamellar structure for chloroplasts. In the mitochondrion the membrane consists of two double layers of lipoid molecules sandwiched between two protein layers. The total thickness of these membranes is 170 to 200 A. The five-layered membranes observed in potassium permanganate-fixed specimens are interpreted to be two closely packed membrane elements with an additional protein layer in the middle, giving the typical triple dark-staining membrane elements. Thus many of the complex cellular organelles have a common membraneous structure. We will describe in detail the structure of a few of these systems.

Rod Outer Segment. Each of the discs in the outer segment appears in osmium-fixed sections as a triple-layered membrane element with two densely staining osmiophilic layers separated by a light portion. On the basis of dichroic measurements the lipid molecules are oriented with their long axes perpendicular to the plane of the disc. Upon lipid extraction this positive form birefringence changes to an intrinsic negative birefringence, indicating that the protein component is organized transversely to the rod axis and in the plane of the disc. The model for the unit disc assumes the lipid molecules are located in the light interspaces of the discs, sufficient to accommodate at least one double layer of molecules. Rounded 140 A-thick discs have been isolated from fragmented rod outer segments. These consist of two approximately 30 A-thick membrane elements. When piled on top of each other, the piles represented multiples of 140 A, supporting the picture that the discs are in an aqueous medium and that the lipid material is contained in the disc itself. In potassium permanganate-stained sections (Fig. 5.1) the discs appear with the typical five-layered structure, showing three distinct opaque layers. In mammalian retinal rods there are dense bands 100–250 A thick with less dense layers 200–500 A thick. The dense lipoprotein layer appears to be a double structure. The darkly stained lipoprotein portions are "lobulated bimembraneous flattened sacs." However, in some cone preparations the lamellae appear to be formed by repeated folding of a continuous membrane. In any case the prosthetic groups (retinene in the visual apparatus and the porphyrin ring of chlorophyll a in the photosynthetic apparatus) are at the interface between the lipoprotein layer and the "aqueous" protein layer. This would be consistent with the fact that in the case of the visual pigments there is a physical migration of vitamin A to and from the pigment epithelium, and in the case of the photosynthetic pigments, chlorophyll a must be in the "aqueous" phase in order for fluorescence to be observed. Based on purely geometrical considerations, Wolken (1957) has calculated the maximum cross-sectional area associated with each molecule and finds 225 A^2 per chlorophyll a molecule and 2500 A^2 per retinene$_1$ molecule. This would mean that in the chloroplast the pigment molecules are tightly packed, while in the retinal rod they are loosely packed.

Mitochondrion. In osmium-fixed sections each mitochondrion appears surrounded by a triple-layered surface membrane. The interior mitochondrial structure also shows these triple-layered structures. On the basis of potassium permanganate-stained sections, the membranes appear to be five-layered structures with three opaque layers separated by two light layers (Fig. 5.2). Thus mitochondrial membranes consist of two double layers of lipid molecules sandwiched between two protein layers. In the middle of the membrane the potassium permanganate-stained layer probably corresponds to another thin layer of protein or to the ends of the lipid molecules. So long as mitochondria retain these inner five-layered structures, the oxidative reactions associated with the citric acid cycle and phosphorylation can occur, even though the mitochondria themselves are broken. If these structures are separated from the intact mitochondrion by disruptive procedures, some of the essential water-soluble enzymes and cofactors are lost. Such fragments cannot carry out

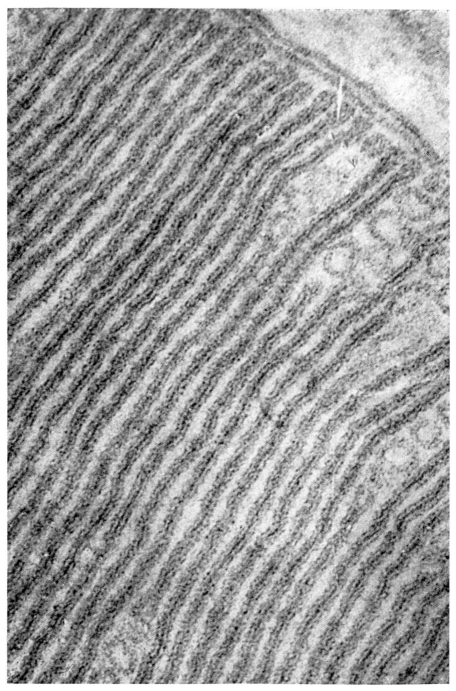

FIG. 5.1. Electron micrograph of the longitudinal section through an outer segment of a guinea pig rod cell fixed in $KMnO_4$ (Sjöstrand, 1960). Each unit disc appears as a five-layered structure with one thicker opaque layer in the middle of the disc. The

many of the oxidative reactions associated with the citric acid cycle; however, they can carry out phosphorylation by the oxidation of succinate or DPNH. If the double layer is now ruptured, the single membrane element can still oxidize succinate and DPNH, but there is no phosphorylation; presumably coupling enzymes have been lost. This single membrane element is called the basic electron-transport particle, ETP (Green,

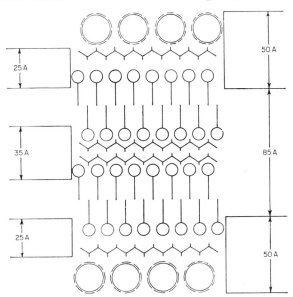

Fig. 5.2. Schematic drawing illustrating a proposed model for the molecular architecture of the compound membranes of the mitochondrion (Sjöstrand, 1960). The molecular pattern is presented in relation to the structure patterns observed in osmium-fixed (right-hand portion) and potassium permanganate-fixed (left-hand portion) material. The large circles represent globular protein molecules. The circles with "tails" are oriented lipid layers. The "backbone" structure indicates protein layers schematically.

1960), and contains 35 lipids, in addition to two flavoproteins, cytochromes a, b, c, c_1, copper, iron, and coenzyme Q. In the ETP there is a lipid core with coenzyme Q located in one segment and cytochrome c in another. Between these is a cytochrome c_1 bridge in the proposed structure shown in Fig. 5.3. Each of the protein components is intimately associated with specific segments of the core. Thus the molecular complexes of the oxidation-reduction components fixed in the lipid structure in combination with nonheme iron and copper form a direct chain of electron transfer from substrate $\rightarrow$ DPN $\rightarrow$ flavins $\rightarrow$ hemes $\rightarrow$ O$_2$. It has been shown on the basis of ESR studies that nonheme iron is reduced

location of this layer corresponds to the middle of the light interspace observed in O_8O_4-fixed material. Magnification $\times 118,400$.

by substrate by DPNH dehydrogenase and that copper is reduced by cytochrome c in cytochrome c oxidase. Analogously, Horio and Yamashita (1962), Chance and San Pietro (1963), and Whatley *et al.*, (1963) have demonstrated that the nonheme iron of ferredoxin (PPNR) is reduced by light in chloroplasts. The lipid matrix appears to be important for all oxidation-reductions in mitochondria.

Chloroplast. In osmium-stained sections discrete regions of the lamellae have thicker (denser) walls. These regions usually occur side by side in several lamellae and form the pigment-containing regions called grana. In some places as many as 10 or 12 discs may be found in a stack, with

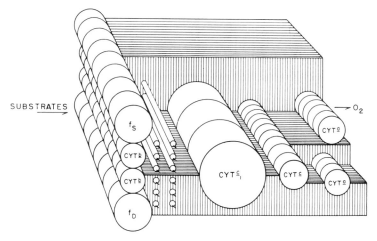

Fig. 5.3. Three-dimensional schematic representation of the structure of the repeating unit of the electron-transport chain. The area with parallel lines represents the lipid core. After Green (1960).

lengths ranging from 0.4 μ to over 2 μ (Fig. 5.4). The lamellar (darkly-staining) portion can be separated from the stroma matrix. These lamellae contain the chlorophyll and, in sonically ruptured spinach chloroplasts, the isolated fragments of lamellae have been shown to be fully active in the photochemical electron transport reactions such as oxygen liberation, the reduction of ferredoxin, and photophosphorylation. Upon further degradation and separation the smallest complete structures (quantasomes) involved in the quantum conversion and oxygen liberation steps have been identified as oblate spheroids of 100–200 A major axis (Park and Pon, 1961, 1963). Aggregates of 5 or 6 quantasomes are essential for the Hill reaction (Park and Pon, 1961). The ratios of chlorophyll, carotenoids, and quinones in isolated quantasomes are identical with those in intact chloroplasts (Lichtenthaler and Calvin, 1964). The minimum

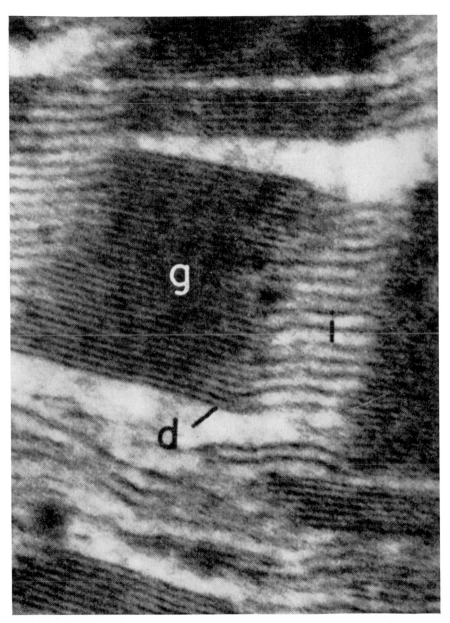

FIG. 5.4. Electron micrograph of a mesophyll plastid of *Zea mays* showing a granum (g) consisting of a stack of discs (d) connected with other discs by interdisc lamellae (i). Starting at the lower end one can follow seven discs clearly from end to end and ascertain that the same two membranes are joined at each end. However, the closed ends cannot be seen clearly in all discs. Magnification ×85,000 (Hodge *et al.*, 1955).

molecular weight of the quantasome is 960,000 and includes 80 chlorophyll *a* molecules, 35 chlorophyll *b* molecules, 24 carotenoids, 23 quinones, lipids, and proteins, the latter contributing 465,000 to the molecular weight. The last figure is based on the assumption that there is 1 mole of manganese per mole of protein (Lichtenthaler and Park, 1963). The stroma material contains the enzymes responsible for carbon dioxide

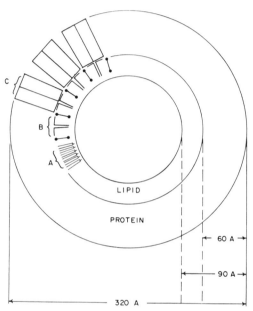

Fig. 5.5.   A proposed structure of the bacterial chromatophore (Bergeron, 1959). The *Chromatium* chromatophore is described as a hollow sphere about 320 A in diameter with a cortex about 90 A thick. The pigment molecules (B) aligned in a monolayer are bounded internally by a phospholipid (A) monolayer and externally by a 60 A-thick protein layer. The "minimal unit" of composition has been used as a structural subunit. The protein has been folded and is related directly to 2 chlorophyll molecules. On the average, the protein is related indirectly to 1 carotenoid molecule and 10 phospholipid molecules.

fixation and other dark reactions. The size of the quantasome and the equal distribution of protein and lipid components are similar to the *Chromatium* chromatophores described by Bergeron (1959) (see Fig. 5.5). Eversole and Wolken (1958) have reported the isolation of "chloroplastin" from *Euglena* using digitonin. The material had a molecular weight of 40,000 and contained 10^{-5} M chlorophyll. The preparation was also able to carry out a light-catalyzed anaerobic conversion of inorganic phosphate into "labile" phosphate. The chloroplast structure, therefore, is not an *a priori* requirement for quantum conversion but an improvement, in that by proper ordering of enzymes and substrates efficient utilization of the products of the primary photochemical act can be achieved.

The chloroplasts isolated from *Euglena* grown in the light exhibit in the electron microscope very regular uniform plates or lamellae. In *Euglena* grown in the dark these structures are disrupted and the chlorophyll disappears. Upon light adaptation the reformation of the lamellae structure appears to be coincident with the reappearance of chlorophyll.

Sager (1959) has reported the following data relative to chlorophyll and carotenoid content in normal and mutant strains of *Chlamydomonas* (Table 5.1).

TABLE 5.1

CHLOROPHYLL AND CAROTENOID CONTENT AND RATIOS IN NORMAL GREEN, PALE GREEN, AND YELLOW STRAINS OF CHLAMYDOMONAS[a]

Sample	Normal green, light-grown	Normal green, dark-grown	Pale green, dark-grown	Yellow, dark-grown
Molecules chlorophyll per cell	2.8×10^9	1.9×10^9	1.6×10^8	Very low
Molecules carotenoid per cell	1.9×10^8	4.7×10^8	8.7×10^5	2.8×10^8
Ratio: chlorophyll/carotenoid	14.8	4.0	178	0
Ratio: chlorophyll/β-carotene	23.6	6.9	241	0

[a] From Sager (1959).

In the *Yellow* dark-grown mutant there is a total absence of lamellae coincident with the absence of chlorophyll, even though the carotenoid content is essentially the same as for the *Normal Green*. In one series of experiments Sager reports that the first detectable lamellae were found from the dark-grown *Yellow* upon illumination when approximately 10% (2×10^8 molecules per cell) of the final concentration of the chlorophyll was formed. Even at a chlorophyll concentration of 8×10^7 molecules per cell no detectable lamellae were found. Other work has shown that the first steps in the morphogenesis of the plastid structure can proceed in complete darkness. These are the formation of vesicles from the inner plastid membrane. The aggregation and fusion into layers of individual discs and the multiplication and parallel arrangement of the lamellar discs are, however, light dependent (Von Wettstein, 1959). Chlorophyll is not essential for the formation of lamellae, but the subsequent stacking and configuration into grana require light. In fact, plastids in the cotyledons of conifers may synthesize chlorophyll enzymatically from protochlorophyll in the dark. In addition, in seedlings of *Picea* there is a complete lamellar system formed in the dark. However, there is no differentiation into grana and stroma regions. It is this subsequent reorganization and full development of the lamellar structure that is the light-dependent process, requiring chlorophyll, and is independent of the synthesis or stabilization of carotenoids. A generalized picture for the development of

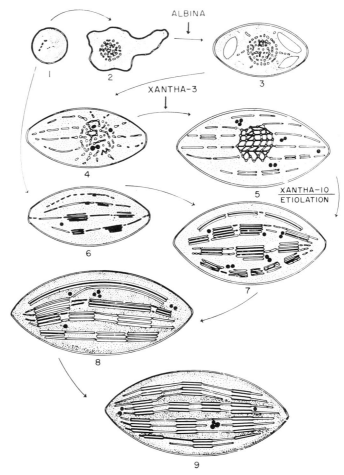

FIG. 5.6. Schematic drawings of the generalized development of chloroplasts (Von Wettstein, 1959).

chloroplasts is given in Fig. 5.6 and the description is adapted from Von Wettstein (1959).

The early proplastids (stage 1) are characterized by a double membrane enclosing a dense stroma, not notably different in structure from the hyaline cytoplasm. In the stroma of the amoeboid proplastids a few vesicles can be distinguished. The vesicles seem to bleb off the inner enclosing plastid membrane. They are thus either formed by this membrane or taken into the proplastid from the surrounding cytoplasm. As the proplastid grows, the number of vesicles increases. The vesicles behave differently depending on the conditions under which the plants are kept.

If germination and seedling growth take place under high light intensity, the vesicles in the proplastids—at least of barley and tomato—will arrange themselves in a few layers as they are synthesized (stage 6). Depending on the size of the proplastid and the number of vesicles formed, three to eight layers are found. At the same time the vesicles fuse to form larger discs. From the very beginning the vesicles are electron optically empty and so are the discs. At independent but restricted regions of these primary layers a multiplication of lamellar discs sets in; packages of two, three, or four discs of relatively uniform diameter arise (stage 6).

If barley or corn plants are kept in diffuse light of low intensity or in darkness, the vesicles will not form layers immediately but aggregate in one or several centers (i.e., prolamellar bodies, "primary grana") as drawn in stage 2. Up to four centers have been found in a single proplastid. As more vesicles are synthesized and accumulated in the center, some fusion within the center to a complex tubular structure takes place (stage 3). Occasionally some reserve starch is deposited in the proplastids. After the proplastid has grown in size, vesicles are protruded radially out of the centers in single rows, and form the primary layers (stage 4). In barley this happens also in complete darkness after 10 days of etiolated growth. If the plants are then illuminated, the development continues with the formation of lamellar discs and their multiplication. The centers sometimes crystallize and are readily incorporated into the lamellar system (stages 5 and 7).

As the lamellar discs multiply, they also grow in surface area and fuse with each other (stage 7). The layers now consist of packages of lamellar discs connected by pieces of double membranes. Eventually a more or less continuous lamellar system is built up (stage 8), and the grana regions are differentiated (stage 9).

Generally it can be noted that the complicated lamellar structure of the chloroplast is developed from structural elements not too different from those found in the hyaline cytoplasm. The basic processes in the morphogenesis of the chloroplast structure are (a) the formation of vesicles from the inner plastid membrane, (b) the aggregation and fusion into layers of individual discs, (c) the multiplication of lamellar discs, (d) the growth and fusion of the lamellar discs to form a continuous lamellar system, (e) the differentiation into grana and stroma regions, and (f) the arrangement of the grana to form single cylindrical columns.

No free ends of lamellae can be observed during the whole development. The fusion of vesicles to discs and from these to larger units constitutes the mechanism for the synthesis of the asymmetrical lamellae of macro-molecular dimensions in the chloroplast.

In Fig. 5.6 the effects of prolonged etiolation and of some genes causing lethality owing to chlorophyll deficiencies are indicated by heavy arrows. Growth of seedlings in the dark for a long period will interfere with the development of the structures after the plastids have reached stage 5. A great number of genes control the series of reactions leading to the

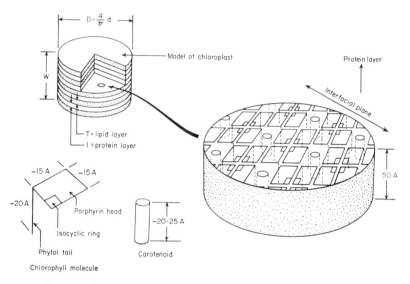

FIG. 5.7. Schematic model of the chloroplast (Wolken, 1959).

lipoprotein lamellae and other structures in the chloroplast. Mutations of these genes are called chlorophyll mutations, although chlorophyll synthesis itself is only rarely blocked in these mutants. According to their seedling color, these mutants are grouped in *albina*, *xantha*, *viridis*, and other classes. Electron microscopy studies in several plant species showed that these mutants are blocked in the differentiation of the plastid struc-tures. The *albina* genes cause blocks in the early stages, impairing the formation of vesicles, the formation of centers, and the aggregation of the vesicles into layers. Although plastids of *albina* mutants may reach the size of normal chloroplasts, their structural differentiation does not go beyond stages 3 and 4. The yellow *xantha* mutants and the yellow-green *viridis* mutants, usually containing carotenoids and some chloro-phyll, have blocks in later stages of the development. *Xantha-3* has a block between stages 4 and 5, and *xantha-10*, between stages 5 and 7. The *viridis* mutants show blocks in the last steps of the structural differentiation, and some of them reach a completely normal structure. Some of these latter mutants are viable, although with a considerable reduction in growth rate.

The chlorophyll appears to be present in monolayers with the porphyrin head at the lipid-protein interface and the phytol tail embedded in the lipid layers (Fig. 5.7). As judged from the depolarization of chlorophyll fluorescence there appears to be little order in the planar axes of the porphyrin heads of the chlorophyll molecules (Arnold and Meek, 1956), making it doubtful that the packing is that characteristic of chlorophyll crystals. Olson and co-workers (1961, 1962), have found that there does indeed exist considerable polarized fluorescence from chlorophyll in *Euglena*. This polarized fluorescence was of longer wavelength (7200 A) than the normal chlorophyll nonpolarized fluorescence and was excited by a pigment absorbing between 6900 to 7100 A. The planes of the emitting chlorophyll molecules have a net component parallel to the lamellar plane with a dichroic ratio of 4 at 6950 A. This contribution could be the result of either of the following structures for point dipoles:

Both of these have a component parallel to the lamellar plane (considered horizontal) and both will result in exciton splitting with a lower energy allowed transition. It was further reported by Butler (1961) that at −196°C there was a transfer of energy from ordinary chlorophyll to the 7050 A absorbing pigment.

The dichroism observed is in agreement with the measurements of Sauer and Calvin (1962) who found for spinach quantasomes a peak in the dichroic ratio at 6950 A, suggesting that essentially 5% of the chlorophyll molecules in a quantasome are oriented parallel or have a component parallel to the lamellar structure. The others apparently do not contribute to dichroism. These molecules presumably have their porphyrin heads in the aqueous phase, forming strongly coupled aggregates. However, there is one problem which arises. If the chlorophyll molecules (P-700) responsible for photoreduction of ferredoxin and photooxidation of cytochrome f indeed show this oriented structure with a lowest allowed singlet transition, they should at the same time exhibit a reasonably high fluorescence yield. This is certainly not in the direction of efficient energy utilization. It may be that the charge transfer complex

$$\text{chl}_{670} \cdots \text{cyt} f \cdots \text{chl}_{700} \cdots \text{X} \cdots \text{Fe-ferredoxin} \xrightarrow{\text{E}} \text{TPNH}$$

effectively quenches the chl$_{700}$ fluorescence *in vivo* and that in the fluorescence depolarization experiments 7000 A light, exciting only chl$_{700}$, could not be utilized subsequent to the cytochrome f oxidation, resulting in the observed fluorescence. It should, therefore, be experimentally verifiable that "non-7000 A" light should markedly reduce the observed 7000 A-excited 7200 A fluorescence in green plant chloroplasts. It should also be possible for the "non-P$_{700}$" chlorophyll molecules to transfer excitation energy to the oriented chlorophyll.

Kholmogorov and associates (1962) have found that amorphous chlorophyll a preparations exhibit no light-induced ESR signals at g = 2.0030. Crystalline chlorophyll even in the presence of *dry oxygen* shows no ESR signals. However, in the presence of oxygen *and* water vapor, a large ESR signal with g = 2.0030 is observed. The peak for excitation was found to be at 7200 A. From infrared spectra of these crystalline chlorophyll a films they observe that dry films exhibit strong C=O and weak O—H absorptions, while the same films exposed to water vapor show weak C=O absorption bands and strong O—H absorption bands. They interpret this as due to the keto and enol tautomeric forms of the cyclopentane ring of the chlorophyll molecule. This process *in pure chlorophyll crystals* is completely reversible. However, they also state that when chlorophyll is crystallized in the presence of yellow leaf pigments (presumably carotenoids), the absorption spectra properties of crystalline chlorophyll are present but are destroyed upon the removal of water vapor. Subsequent exposure to water vapor does not restore the crystalline spectrum.

The authors conclude that the light-induced ESR signal is associated with the transfer of an exciton in the crystal. In the decay of the exciton, O_2, chemisorbed on the crystal surface, traps the electron, effectively forming chl-O_2^-. The freed hole can then migrate through the crystal, giving rise to hole conductivity. The water molecules become part of the structure of the crystal, assuring efficient exciton transfer and hole conductivity. It is quite possible that some of the photoconductivity effects and thermoluminescence effects observed with dried chloroplasts have a similar origin, i.e., O_2, adsorbed on the surfaces, serves as a strong electron trapping center, giving rise to hole conductivity as well as subsequent thermoluminescence. The presence of water may therefore be quite important to our understanding the efficiency of the electron-trapping centers.

While the plasma membrane and the nerve myelin sheath are not normally involved in photochemical reactions, these are included to point up the essential similarity of all membrane structures in biological systems. Into this general type of organized layering of lipids and proteins can be fitted the enzymes, cofactors, pigments, etc., for a specific function. Nerve fibers have been observed to be stimulable by light and the radial nerve fibers in the macula lutea of primates contain xanthophyll, although, so far as we know, this acts only as a blue-absorbing filter, as we shall discuss later.

Plasma Membrane. The plasma membrane appears in potassium permanganate-fixed sections as two opaque layers with an intermediate light layer, all of about equal dimensions, making a total of 75 A thickness. When two cells are in close contact, the boundary consists of a five-layered wall. In different membranes there are sufficient variations in the osmiophilic portions to assume that the lipid to protein ratio of membranes can vary considerably, although this is not necessarily the case. The triple-layered structure, Fig. 5.8, observed on potassium permanganate staining is assumed to represent a double layer of lipid molecules, the opaque layers representing the location of the polar groups of the phospholipids and the light intermediate layer the lipid structure.

Nerve Myelin Sheath. In myelinated nerve fibers the Schwann cells form a multilayered lipoprotein sheath around the axon. This is a highly ordered, layered structure with a periodicity in a radial direction of 120 A. Opaque 25 A-thick layers alternate with lighter 95 A-thick layers, divided in half by an intermediate opaque layer, observed in osmium fixation. Again the double lipid layer shown in Fig. 5.8 is a reasonable one. It has been found that the myelin sheath is formed by invagination of the plasma membrane of the Schwann cell. The axon becomes enclosed by this invagination and completely surrounded by the cytoplasm. Later the invaginated part of the Schwann cell membrane grows considerably and forms a spiral wrapping around the axon. In connection with the packing, the thickness of the osmiophilic layer is reduced from about 60 A to about half, to the 30 A width characteristic of the opaque layer of the main period in the myelin sheath pattern, possibly by loss of a layer of globular protein.

Although the structures described are to a great extent based on interpretation of fixed, dehydrated tissue sections observed in the electron microscope, the preponderance of evidence suggests the basic membrane element to be a layering of protein, lipid, and possibly glycoproteins. The composition of the membrane elements are probably different in only one or a few components when there is functional differentiation. In the case of chloroplast lamellar formation, there is no differentiation into grana and stroma regions except in the light and in the presence of chlorophyll. In the rod outer segments there is a deterioration of the platelet structure in the absence of vitamin A. Thus the function appears to be intimately associated with the final differentiated structure. However, it should be possible in some cases to separate the primary photochemical act from the structure to the extent that one can say that structure is

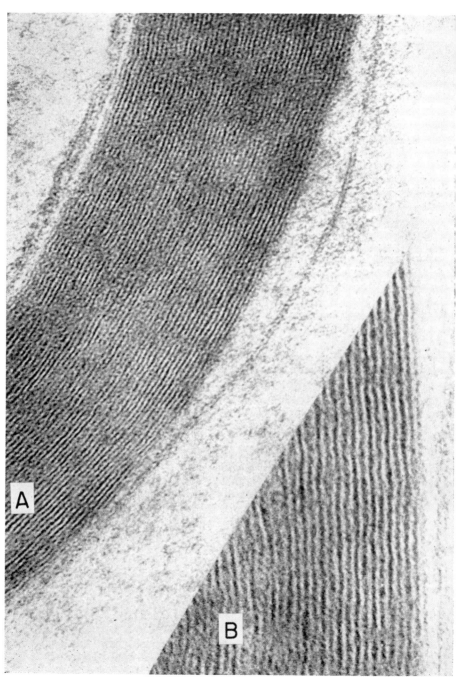

Fig. 5.8.   Electron micrograph of myelin sheath pattern in potassium permanganate-fixed myelinated nerve fiber (Sjöstrand, 1960). The interperiod line is heavily stained. The plasma membrane of the Schwann cell appears triple-layered. Magnification (A) ×136,000; (B) ×240,000.

required for efficient biology but is not a priori required for photochemistry. Chlorophyll can photoreduce other dye molecules in solution, but a lipid charge transfer complex may be the only method of efficiently trapping an electron at ferredoxin by the photoinduced oxidation of cytochrome. If, then, an enzyme is considered to be a catalyst in a chemical reaction, possibly the lipid structure in membranes acts as an assembly line for directing both light and dark chemistry.

2. Photosynthesis

General Considerations. We have discussed a number of photochemical concepts that could be of considerable importance in interpreting biological processes that are effected by light. Of all the photobiological reactions in Nature which have thus far eluded a definitive analysis, photosynthesis by the green plant is probably the most fundamental. In making use of the free energy of sunlight, the green plant is able to fix carbon dioxide and split water to evolve oxygen. Through a series of reactions which are reasonably well known, carbohydrates and other molecules are synthesized. Thus light energy serves indirectly as a source of free energy for all living things.

The immediate end products of metabolism are carbon dioxide and water. Consequently the combination of photosynthetic organisms and heterotrophic organisms gives us a biological machine which continually turns over CO_2 and H_2O on the surface of the earth. It has been calculated that each carbon dioxide molecule in the atmosphere is fixed and incorporated into some plant structure once in every 200 years and that all the oxygen is renewed by plants every 2000 years. During the past 15 years considerable progress has been made in understanding the processes involved in carbon dioxide reduction and fixation in green plants as well as the primary photochemical events. It is not possible for us to review all of the pertinent data concerning the physiological, biophysical, and biochemical investigations on photosynthesis, and consequently the reader is referred to the extensive reviews and monographs: Rabinowitch (1945, 1951, 1956), Bassham (1963), Gaffron (1960), Jagendorf (1962), Whatley and Losada (1964), Clayton (1964) and Blinks (1964).

Photosynthesis in green plants occurs in the chloroplast which consists of a lamellar phase imbedded in a matrix (see Fig. 5.9), surrounded by a membrane. The lamellar structures can be separated from the matrix and recent studies by Park and Pon (1961, 1962), Lichtenthaler and Park (1963) and Park and Biggins (1964) indicate that it is made up of subunits which Calvin (1962) has called *quantasomes*. Aggregates of 6 to 7 quantasomes are capable of performing the light reactions and the associated electron transport processes of photosynthesis. Recently, Lich-

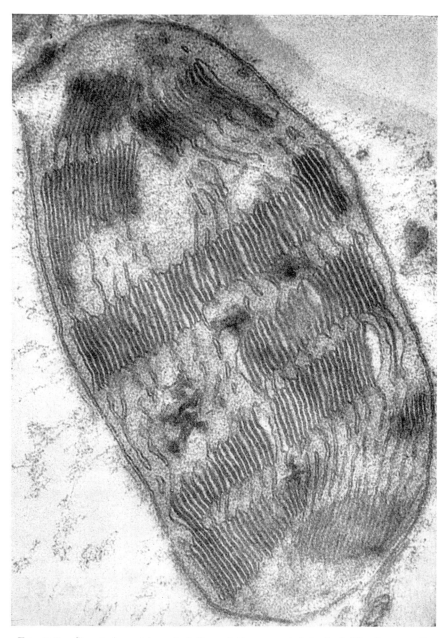

Fig. 5.9. Starch-free chloroplast from fully expanded leaf of *Nicotiana rustica*, fixed in 2.5% $KMnO_4$ (about $\times 7500$). The compartmented structures are grana, and the chlorophyll is thought to be associated with the darkened partitions. A precursor of starch may form in the electron-transport space between the partitions. The grana are embedded in a granular stroma and are connected by anastomosing channels or frets. The whole system, bounded by an envelope, is labile and responds to variations in light intensity, mineral nutrition, herbicides, and the type of fixative that is used to prepare the leaf material for study (T. E. Weier, University of California, Davis; reprinted by permission of the AAAS and the author).

224

TABLE 5.2
PHOTOSYNTHETIC PIGMENTS IN GREEN PLANTS

Pigment[a]	Position of longest wavelength peak *in vivo* (mμ)	Approximate "end" of long wavelength absorption (mμ)	Organism
Chlorophyll *a* (extracted)	665	705	Land plants, algae
Chlorophyll *a*, form 673	673	—	—
Chlorophyll *a*, form 684	684	—	—
Chlorophyll *a*, form 695	695	—	—
Chlorophyll *a*, form 705	705	—	—
Chlorophyll *b*	650	685	Land plants, green algae
Chlorophyll *c*	640	680	Diatoms
β-Carotene	482	520	Land plants, most algae
Fucoxanthin	470	530	Diatoms
Phycoerythrin	566	600	Red algae
Phycocyanin	615	650–670	Red algae, blue-green algae

[a] From Jagendorf (1962).

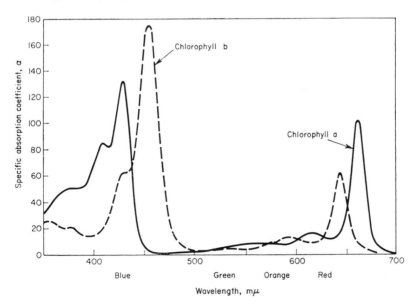

FIG. 5.10. The absorption spectra of chlorophylls *a* and *b* in ether. This curve was determined by Smith and Benetez by measuring the ratio of incident light, I_0, to transmitted light, I, of different wavelengths. Weighed samples of pure chlorophyll dissolved in ether were used. The specific absorption coefficient, a, is defined as: $a = (1/lc) \log_{10} (I_0/I)$, where l is the inside length of the glass cell holding the solution, and c is the chlorophyll concentration in grams per liter. Such reference curves are used to analyze plant extracts for their pigment content (French, 1962).

tenthaler and Calvin (1964) have reported on the chemical composition of the quantasomes, and these results are discussed in a separate section.

The compounds responsible for the absorption of light quanta and their ultimate transformation into chemical energy are the chlorophyll pigments, carotenoids, and certain special accessory pigments. Chlorophyll *a* is found in all green plants. In addition, in most land plants and in most

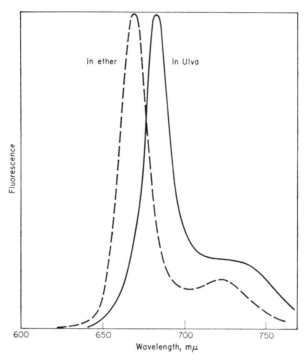

Fig. 5.11.   The fluorescence spectrum of pure chlorophyll *a* in ether compared with that for a sample of living sea lettuce (*Ulva* sp.). The wavelength shift between the two curves shows that chlorophyll as it occurs in plants is different from the extracted material (French, 1962).

algae one finds chlorophyll *b*, *c*, or *d*, while phycobilins are found in red and blue-green algae. The light energy absorbed by these accessory pigments must in some way be transferred to a special chlorphyll *a* molecule before photosynthesis occurs. Some of the spectral properties of these accessory pigments are given in Table 5.2 and in Figs. 5.10–5.12.

Photosynthesis can be summarized by the following equation:

$$6CO_2 + 6H_2O + 672,000 \text{ cal} \rightarrow C_6H_{12}O_6 + 6O_2 \tag{5.1}$$

Green plants are capable of absorbing light energy and of using it to

convert 6 moles of CO_2 and 6 moles of H_2O into 6 moles of oxygen and
1 mole of carbohydrate. This equation for photosynthesis has been recognized for years and merely represents a balance sheet for the process.
The questions that are being asked today relate to the mechanism of this

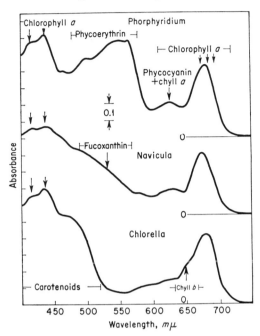

FIG. 5.12.   The absorption spectra of three algae, showing the regions of absorption
due to the different kinds of photosynthetic pigments. The large band at 675–680 mμ
is due to chlorophyll a in all the species. Three arrows show the positions of the separate forms of chlorophyll a, C_a673, C_a683, and C_a695 that overlap to make this
composite absorption band. Chlorophyll a also has a small band around 620 mμ and
two large bands in the blue region, marked with arrows. Nothing is known about the
contribution of the different forms of chlorophyll a to these blue bands. *Chlorella
vulgaris* is a green alga that has much the same pigment system as ordinary leaves.
Chlorophyll b shows as a shoulder in the *Chlorella* spectrum and its blue band also
causes part of the absorption in the carotenoid region at about 480 mμ. *Navicula
minima*, a diatom, shows the absorption of the carotenoid fucoxanthin, which is at
longer wavelengths than most other carotenoids. *Porphyridium cruentum* is a red
salt-water alga with phycoerythrin and a very little phycocyanin (French, 1962).

process. The oxygen released comes from water and the carbon of CO_2
is taken up into carbohydrate. It was soon realized that additional H_2O
was used and that some was re-formed. Thus Van Niel (1931, 1949)
suggested the following general equation:

$$6CO_2 + 12H_2O + energy \rightarrow C_6H_{12}O_6 + 6O_2 + 6H_2O \qquad (5.2)$$

Light energy is used to drive oxidation-reduction reactions or electron transport processes and the over-all effect is the transfer of electrons from water to CO_2 with the resultant formation of organic compounds and oxygen. In the case of the photosynthetic bacteria, electron donors from sources other than water are used, and consequently no oxygen is evolved. Thus a more general equation of photosynthesis which would hold for both bacteria and green plants was proposed by Van Niel:

$$6CO_2 + 12H_2A \rightarrow C_6H_{12}O_6 + 6H_2O + 6A_2 \tag{5.3}$$

where H_2A is any reduced compound which can serve as an electron donor. Carbon dioxide fixation and reduction was assumed to be the same in both types of organisms, while the green plants were the only

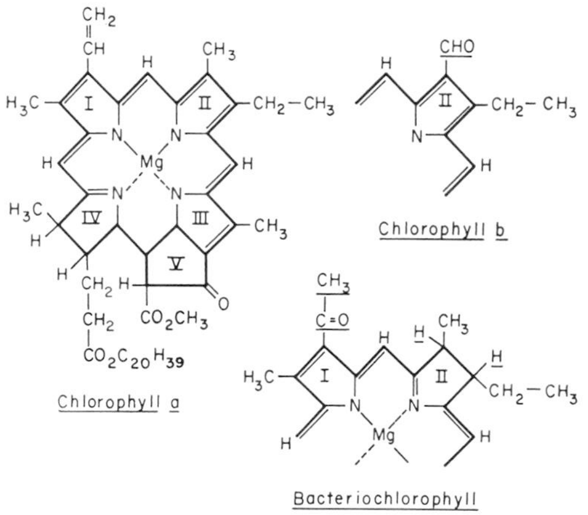

Fig. 5.13. Structures of chlorophyll a, chlorophyll b, and bacterial chlorophyll.

ones capable of using H_2O as the electron donor. The removal of electrons from water ordinarily requires large amounts of energy and as we shall discuss later probably requires a special independent photochemical event. The use by bacteria of reducing power from H_2S or other electron donors is not so demanding in energy.

Light Absorption. The fundamental reaction in photosynthesis is the absorption of radiant energy. Green plants utilize light with wavelengths from 400 mμ to above 700 mμ with variable efficiency for photosynthesis. In the case of certain bacteria, light of wavelengths as long as 950 mμ is effective. The pigments responsible for the longest wavelength absorption are the effective ones for photosynthesis, and light energy absorbed by different pigments at other wavelengths is transferred to these "special" pigments. In the case of green plants the special pigment is chlorophyll *a*, whereas in bacteria it is bacteriochlorophyll (see Figs. 5.13 and 5.14).

If we make use of the general equation $E = h\nu$, where ν is the frequency and h is Planck's constant, then we can calculate the energy available

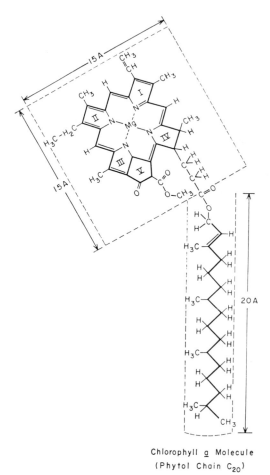

Chlorophyll *a* Molecule
(Phytol Chain C_{20})

FIG. 5.14. Structure of chlorophyll *a* molecule approximately to scale showing porphyrin head approximately 15 A on a side attached to the C_{20} phytol tail. The head can bend relative to the phytol group and it is presumed that the latter is embedded in the lipid matrix.

from each quantum at any wavelength. In biochemistry it is more common to use the units of calories per mole of a substance in describing the energetics of reactions rather than the energy in ergs absorbed by a molecule. We want the photochemical equivalent in molar terms. By using appropriate constants (Appendix I) we obtain the following expression:

$$E = \frac{2.86 \times 10^7 \text{ cal per mole}}{\text{wavelength in millimicrons}} \quad (5.4)$$

Thus a mole equivalent of blue light (an Einstein or 6.025×10^{23} quanta) at a wavelength of 450 mμ is equal to 64 kcal. In the same way light of 680, 700, and 900 mμ is equivalent to 42, 41, and 31.8 kcal/mole, respectively.

In the uptake of CO_2 and H_2O to make organic compounds and to evolve oxygen, the following expression must be satisfied energetically by light:

$$CO_2 + H_2O \rightarrow (CH_2O) + O_2 \quad (\Delta F = +114 \text{ kcal}) \quad (5.5)$$

Thus the minimum *energy* requirement for synthesis by light of wavelength 680 mμ is $114/42 = 2.7$ or in whole numbers, 3 quanta/CO_2 molecule. However, we have already indicated that as a minimum $2H_2O$ must be used for 1 oxygen (O_2) evolved. Thus the absolute minimum quantum requirement must be at least 6. Other photochemical events appear to be essential to provide for additional electron flow. As we shall discuss later, a value of approximately 9.0 quanta/CO_2 molecule fixed as carbohydrate is the minimum quantum yield to be expected under conditions of maximum efficiency for over-all photosynthesis in green plants.

Products of the Light Reaction. The primary products of the light absorption reaction—at least the stable products—are molecular oxygen, ATP, and TPNH. Vishniac and Ochoa (1951), Tolmach (1951a,b), and Arnon (1951) were the first to demonstrate the photoreduction of TPN and DPN by isolated chloroplasts. Frenkel (1954), using particles extracted from photosynthetic bacteria, was able to demonstrate the formation of ATP and ADP and inorganic phosphate, P_i, by light. Arnon *et al.* (1954) demonstrated the light-driven formation of ATP in isolated chloroplasts. These data can be summarized as follows:

$$H_2O + TPN + ADP + P_i \rightarrow \tfrac{1}{2}O_2 + TPNH + ATP \quad (5.6)$$

Thus, light energy is converted into two forms of chemical energy; namely, the reducing energy of TPNH and the chemical bond energy of ATP.

The formation of ATP by light in the green plant is called *photophosphorylation* and involves a light-driven transfer of electrons from chlorophyll to an intermediate capable of reducing TPN with the possible simultaneous formation of ATP. The uptake of CO_2 is assumed to be a subsequent dark reaction utilizing the chemical energy of ATP and TPNH.

Two Pigment Systems in Photosynthesis. Since the work of Emerson

(1956), it has been reasonable to assume that two pigments must be involved in photosynthesis. Efficient photosynthesis does not occur when chlorophyll a is the only pigment receiving the light energy. Emerson observed a decline in the quantum yield for photosynthesis when far-red light ($\lambda > 700$ mμ) was used exclusively, a region of the spectrum where only chlorophyll a absorbs. When the far-red light was supplemented by red light (680 mμ), a synergistic effect of these two wavelengths of light was observed (enhancement effect). Many different hypotheses have been proposed to explain this effect (Franck, 1958). The most likely explanation is that two photochemical reactions are carried out by two separate pigment systems. These suggestions were based primarily upon the analysis of the fluorescence of the photosynthesizing apparatus. In addition, in *Chlorella*, Emerson and Chalmers (1958) demonstrated that in the presence of far-red light the action spectrum for effectiveness of the short wavelength light corresponded to the light absorbed by chlorophyll b or other accessory pigments. In the diatom (*Navicula*) the action spectrum corresponded to the fucoxanthin absorption, in *Anacystis*, a blue-green alga, to phycoerythrin. Thus, efficient photosynthesis seems to require the effective cooperation of two light quanta, each absorbed by a different pigment. Enhancement effects have been observed in a number of different processes representing parts of the photosynthetic process, and a number of different schemes at the electronic and chemical level have been presented to account for the facts. It is not possible for us to review all of the numerous schemes that have been presented. However, the reader is referred to the reviews that have been previously mentioned and, in particular, to the following papers by A. Müller *et al.* (1963b), Arnon *et al.* (1962), Bassham (1962); Duysens and Sweers (1963), Gaffron (1960); Hill and Bendall (1960), Calvin and Adroes (1962), and Franck and Rosenberg (1964) for a discussion of various hypotheses of the mechanism of photosynthesis. Before discussing the evidence supporting the various hypotheses we would like to propose a summary scheme (Fig. 5.15) as a basis for discussion. We feel that this scheme takes into consideration the merits of most of the proposals that have been presented and, we believe, will help the reader to understand the presentation of some of the important experimental work on the photochemical processes involved in photosynthesis.

2.1 A SUMMARY SCHEME FOR PHOTOSYNTHESIS

Outline of the Photochemical Events. The photochemical or photosynthetic unit of green plants, the quantasome, is visualized as consisting of three major subunits. One of these subunits is a lipoprotein complex and contains a special bound form of chlorophyll a (a_1) which Kok (1956a,b,

1957) calls P_{700}. As indicated in Fig. 5.15 the evidence presently available suggests that this part of the photosynthetic apparatus is concerned exclusively with the separation of the electron from P_{700}, creating in a charge-transfer complex a highly reducing component, and oxidized chlorophyll. This charge separation is brought about by a far-red quantum and is the primary or exclusive energy source for the production of a

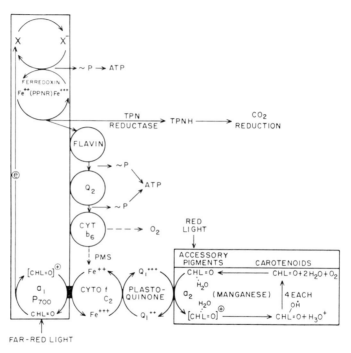

Fig. 5.15.   Proposed scheme for the 2 quantum photochemical reactions in photosynthesis, combining all of the experimental evidence available at the present time.

highly reducing component (X^-) in this lipoprotein complex of the photosynthetic apparatus. The substance X is hypothetical and is placed in the scheme for reasons to be discussed below. We do know that electron transfer must occur and that the reducing substance is capable of transferring its electron to the cofactor, ferredoxin or PPNR (San Pietro and Lang, 1958; Whatley *et al.*, 1963) with the trapping of phosphate bond energy in the form of ATP. The reduced PPNR is capable, in turn, of reducing TPN in the presence of the enzyme TPN reductase, a flavoprotein.

A second subunit of the photosynthetic unit is the "aqueous unit," which is concerned with reduction of the oxidized chlorophyll P_{700} pigment created by the far-red light in the photoreduction step. As indicated, red light absorbed by chlorophyll a_2 or accessory pigments is capable of catalyzing the reduction of the oxidized chlorophyll a_1. Two important cofactors are essential for the coupling process in this reduction. In the case of the green plant, cytochrome f and plastoquinone are essential. In the photosynthetic bacteria, cytochrome c_2 replaces cytochrome f. If this hypothesis is correct, cytochrome f is very closely associated with chlorophyll a_1, while plastoquinone is associated with chlorophyll a_2. The transfer of electrons from plastoquinone to cytochrome f is temperature sensitive, i.e., inhibited at 0°C, and can be affected by a number of other factors. On the other hand, the transfer of electrons from cytochrome f to chlorophyll a_1 is relatively temperature insensitive and, as a matter of fact, can occur at extremely low temperatures (-150°C).

Thus far-red light which creates the general reducing system in photosynthesis also creates an oxidant which is capable of accepting an electron from the reduced cytochrome. The oxidized cytochrome, in turn, is reduced by an electron from plastoquinone which, in turn, is reduced by chlorophyll a_2 which has absorbed a quantum of light in the red region. The net result is the cooperation of two light quanta, one far-red and one red to create a highly reducing atmosphere in the structured lipid phase and an oxidizing atmosphere in the aqueous phase.

In the proposed scheme the oxidized hydrated chlorophyll a_2 molecule, on interaction with the bound water, gives rise to an OH radical. In a series of reactions which are not yet understood, four of these OH radicals apparently interact by (an) enzymatic reaction(s) in which manganese is a specific requirement, to give rise to 1 molecule of oxygen and 2 molecules of water.

The third subunit of the photosynthetic unit is the electron transport system or particle which must be intimately associated with the charge-transfer complex of chlorophyll a_1. The electron derived from the far-red quantum interaction may be transferred from PPNR (ferredoxin) to TPN via a flavoprotein as indicated previously. On the other hand, the observed cyclic electron flow requires the existence of a second flavin component of the chloroplasts or quantasomes which is capable of transferring the electrons from PPNR to quinones through cytochrome b_6, eventually reducing cytochrome f. Thus using only far-red light, it is possible to obtain a *cyclic electron flow* as indicated in Fig. 5.15. Under these conditions red light is not necessary for the photochemical formation of ATP. Tagawa et al. (1963) have been able to demonstrate this in green plants. However, for a net gain of reducing power which is required for CO_2 fixation and reduction and for the ultimate splitting of water in green plants to form oxygen, red quanta are essential.

It still is not clear what the primary role of carotenoids may be in the process of photosynthesis. The only evidence at the present time is that they may be important as protective pigments in preventing a photo-induced destruction of the photochemical apparatus of the cell.

As indicated in the scheme of Fig. 5.15, the electron flow from reduced ferredoxin to flavin through quinones and cytochrome b_6 and ultimately back to cytochrome f may lead to the formation of additional adenosine triphosphate. This cyclic electron flow which is dependent upon ferredoxin and is accompanied by phosphorylation has been called *cyclic phosphorylation* by Arnon (1960) and Arnon *et al.* (1960). This pathway of electron flow apparently does not occur extensively, if at all, in normal photosynthesis, but it does seem likely that some ATP is formed by this pathway. It is possible that in the intact plant some of the TPNH formed could be utilized in the mitochondria, thus forming phosphate bond energy by oxidative phosphorylation. This would eliminate the need for cyclic phosphorylation. It should be emphasized that the generalized scheme of photosynthesis represents a hypothesis which is in agreement with a number of observed facts. There are, however, other interpretations especially of the two-quantum process and the reader is referred to the critical articles of Gaffron (1960) and of Clayton (1964) for a more detailed discussion of these ideas, in addition to those of Franck already discussed.

Quantum Considerations. The generalized hypothetical scheme of photosynthesis which is in keeping with most of the facts indicates that four red quanta are required for the production of 1 oxygen molecule. In order for this cycle to occur, four far-red quanta must also be used to maintain a correct oxidation-reduction balance. The four far-red quanta would lead to the formation of two reduced TPN molecules. The ratio of net TPNH synthesis to oxygen production in the isolated chloroplast has actually been observed to be 2. If this were the total story, however, insufficient ATP would be formed for the total process of photosynthesis to be carried out. According to the CO_2 fixation pathway for the photosynthesis of carbohydrate, three ATP and two TPNH molecules are required for each CO_2 molecule taken up and reduced to carbohydrate (Calvin and Bassham, 1957). Therefore, it may be that some far-red light quanta are used for cyclic photophosphorylation. The evidence presented by Arnon supports the idea that in the far-red light reaction and in the reduction of ferredoxin or TPN, one ATP is formed per TPNH produced. To maintain a correct stoichiometry, the photosynthetic apparatus must therefore be subject to certain regulatory mechanisms. If we initiate photosynthesis with both chlorophylls in the reduced state, then according to the proposed scheme four far-red quanta would lead to the formation of two TPNH molecules and two ATP molecules, while 4 red quanta would be required to restore the reduced state of P_{700} and for the formation of 1 O_2 molecule and 2 molecules of water. In addition, two far-red quanta must be used in cyclic electron flow to generate additional ATP. The absorption of eight additional far-red and eight additional red quanta would yield the total ATP and TPNH indicated in Table 5.3.

TABLE 5.3
CYCLE OF LIGHT AND DARK REACTIONS IN PHOTOSYNTHESIS

Light cycle	TPNH formed	ATP formed	O_2 evolved	H_2O formed
4 "far-red" quanta	2	2	0	0
4 "red" quanta	0	0	1	2
2 "far-red" quanta (cyclic electron flow)	0	3	0	0
8 "far-red" quanta	4	4	0	0
8 "red" quanta	0	0	2	4
Total—26 quanta (14 "far-reds," 12 "red")	6	9	3	6

General Equation for Photosynthesis Minimum-Light Reaction
Light reaction:
$$6TPN + 9ADP + 9P_i + 12H_2O \xrightarrow{\text{light}} 6TPNH + 9ATP + 6H_2O + 3O_2$$
Dark reaction:
$$6TPNH + 9ATP + 3CO_2 \rightarrow (CH_2O)_3 + 9ADP + 9P_i + 6TPN + 3H_2O$$
Summary: (Balance)
$$12H_2O + 3CO_2 \rightarrow (CH_2O)_3 + 3O_2 + 9H_2O$$
Quantum yield:
$$\frac{26}{3} = 8.7$$
$$Photosynthetic\ Quotient = \frac{CO_2}{O_2} = 1$$

This ratio of ATP and TPNH would give a balanced rate of photosynthesis with a quotient of $CO_2/O_2 = 1$ and a quantum yield of 8.7. To maintain this steady state rate of maximum photosynthesis, therefore, 14 far-red quanta must be absorbed for every 12 red quanta absorbed.

It is evident from these considerations that quantum yield measurements for photosynthesis should not be made by measuring only one component of the system. By appropriate uncoupling processes it is possible to run the far-red system by itself. Under these circumstances the quantum yield for TPNH formation would be 2. The yield for ATP formation in cyclic electron flow could be as high as 1.5. Under specialized circumstances when CO_2 uptake is measured *for short periods of time,* a quantum yield of 2 should be possible. Under no circumstances, however, should a quantum yield of less than 4 be observed for oxygen evolution when the measurements are made under steady state conditions, i.e., when oxygen production is maintained at a steady rate.

The requirement for some cyclic electron flow for the generation of necessary ATP when CO_2 is being fixed is an important point. It raises the interesting possibility that CO_2 itself may be important in controlling this cyclic process. It is also important to note that if TPNH is allowed to accumulate then all ATP formation would come from the far-red light system. This would give a TPNH to ATP ratio of 1 which agrees with the experimental findings.

Spectral Changes and the Far-Red Light System. It is now generally agreed that the number of photosensitizing pigment molecules is far in excess of the number of catalytic sites which make use of the immediate photochemical product(s). During the evolution of the photochemical process in green plants, a system has been evolved such that light falling on the system is effectively absorbed by a high pigment concentration. The pigment concentration is so great, as a matter of fact, that even in full sunlight chlorophyll molecules receive a quantum only a few times per second, a rate which is several orders of magnitude lower than the turnover number of most metabolic reactions. Most of these pigments do not participate directly in the photochemical production of oxygen nor in the formation of an active reductant. Rather, the energy absorbed by the pigment molecules is funneled into a few reaction centers. Using rapid flashes of light of short duration, Emerson and Arnold (1932) demonstrated that 1 oxygen molecule was produced per 2000 chlorophyll molecules. If we assume that the quantum requirement for oxygen production is 4, according to the scheme summarized above, this means that one subunit is composed of approximately 500 chlorophyll molecules (2000/4). Numerous investigations have demonstrated that closely packed pigment systems such as are found in the chloroplast can transfer energy from molecule to molecule. In the case of aerobic photosynthetic organisms, this corresponds to the transfer of excitation energy from the accessory pigments to chlorophyll *a* with its red absorption peak at 675 mμ. Kok (1956a,b), Duysens (1956), and Müller *et al.* (1963a,b) have presented numerous experimental data to substantiate the existence of a very small concentration of pigment which is capable of absorbing beyond the 675-mμ wavelength and thus is capable of trapping energy in the far-red. The first evidence for the existence of such a pigment was obtained in fluorescence studies with red algae by Duysens (1952). He observed that the fluorescence of chlorophyll *a* was excited to a larger extent when light was absorbed by accessory pigments (phycobilins) than by chlorophyll *a* itself. This was explained by assuming the presence of two different forms of chlorophyll *a*, chlorophyll a_1 and chlorophyll a_2. One of these, free to fluoresce, was presumably intimately connected with accessory pigments which are capable of activating photosynthesis. The second was presumably connected to another pigment (one which Kok calls P_{700}) which occurs in a very small concentration and inhibits the weak fluorescence of chlorophyll a due to the trapping of the excitation energy.

The widespread occurrence of chlorophyll a in all aerobic photosynthetic organisms, however, indicates that it is the primary pigment in the photochemical process. The accessory pigments such as the carotenoids, chlorophyll b in green plants, and phycobilins in other plants are functional for trapping light quanta but can be effective in photosynthesis only when they transfer the excitation energy to the chlorophyll a active center.

At the present time most workers agree that the rate of photosynthesis sensitized by far-red light may be increased at least a factor of 2 by the addition of supplementary light at a shorter wavelength. As indicated earlier, they strongly suggest the presence of two independent forms of chlorophyll a which are essential for photosynthesis. Duysens and associates have examined the fluorescence of the various chlorophyll molecules and indicate the importance of two different types of chlorophyll a in the transfer of excitation energy from accessory pigments.

Detailed spectroscopic evidence by Kok (1956a,b), Kok and Hoch (1961), and more recently by Müller et al. (1963a,b) have indicated changes in absorption at 703 mμ and at 430 mμ which can be ascribed to the oxidation of a single pigment. According to Kok this is a special form of chlorophyll, P_{700}, with an oxidation-reduction potential of $+0.43$ volts. Witt has made a very careful study of this pigment under various environmental conditions including variations in temperature, pH, and the presence of reducing substances such as reduced phenazine methosulfate (PMS). All the evidence indicates that the substance undergoing spectral changes has two absorption bands, one at 430 mμ and another at 703 mμ which is typical of a chlorophyll a compound. The time required for these spectral changes to occur in light is less than 10^{-5} sec. Furthermore, the magnitude of the change is temperature independent between 0° and 60°C, consistent with the fact that this is a primary light reaction. The decay time for the return to the original state in the dark has a half-time of approximately 10^{-3} sec in *Chlorella*. This decay is also independent of temperature between 0° and 40°C. Using aged chloroplasts, it is possible to add reducing substances in order to change the rate of reduction of the photooxidant. Reduced PMS in the dark reaction shortens the half-time by at least a factor of 100. The initial electron acceptor in this photochemical event is unknown, but this may be a substance which we have called X in the scheme and which could have an oxidation-reduction potential of approximately -0.65 volts. Experimentally, however, the substance which is first reduced and can be observed is ferredoxin or PPNR. The reductant which is formed by the far-red photochemical event can be recognized initially by a change in spectral absorption and subsequently by the appearance of reduced ferredoxin or TPNH.

The evidence that this far-red light system has nothing to do with the oxygen evolution process is provided by the observation that CMU (p-chlorophenyl dimethyl urea), which is a potent inhibitor of the photochemical water-splitting process, has no influence on the absorption changes at 430 and 703 mμ. Witt has observed spectral changes at 430 and 703 mμ at $-150°C$. Under these conditions, however, the reaction is irreversible, i.e., the dark reaction to restore the original state is not observed. Thus, the first photochemical event in photosynthesis involves a far-red pigment system, and light creates an oxidized chlorophyll as the primary product. The pH independence of this photochemical process indicates that direct electron transfer is occurring.

Pyridine Nucleotide Reduction. The far-red system is the one primarily concerned with the initial charge separation and the ultimate reduction of TPN in green plants. Recently Arnon *et al.* (1962), using monochromatic far-red light, have actually separated the photosynthetic reduction of TPN from the oxygen evolution system.

San Pietro and Lang (1958) reported the isolation and purification of a soluble protein which catalyzes the reduction of TPN by illuminated chloroplasts or grana. They suggested that this protein be named photosynthetic pyridine nucleotide reductase (PPNR). A second protein which was partially purified by Keister and San Pietro (1959) was shown to be necessary for the transfer of electrons from PPNR to TPN. This transhydrogenase or TPN reductase was shown to be a flavoprotein which specifically required FAD for activity. Recently, Shin *et al.* (1963a,b) have prepared the enzyme in crystalline form. The first demonstration that a soluble factor can be added back to chloroplasts to reconstitute the electron-transport system was made by Davenport *et al.* (1952) during the course of observations on the reduction of methemoglobin by chloroplasts in the light. They were able to extract a factor which was essential for the light-induced reduction. PPNR and the methemoglobin-reducing factor appear to be identical. Recently, Tagawa and Arnon (1962) have isolated the same protein from spinach and renamed it ferredoxin. The reason for this suggested change in nomenclature is due to the functional similarities between the plant protein and bacterial ferredoxin first isolated by Mortenson *et al.* (1962). The proteins isolated from these two sources are different in their absorption spectrum, iron content, and amino acid composition. Ferredoxin is a very electronegative electron carrier with an $E_0' = -432$ mv at pH 7.55. Tagawa and Arnon (1962) and Shin et al. (1963a,b) have been able to demonstrate that ferredoxin can function as an intermediate in the reduction of TPN by molecular hydrogen provided hydrogenase and TPN reductase are present. The kinetics of TPN reduction by chloroplasts indicated to San Pietro and Lang (1958) that the substrate for TPN reduction was bound in the membrane of the chloroplast. It is for this reason and others that it has been suggested that the highly reducing ferredoxin is located in the lipoprotein structure containing the "dry" chlorophyll (the far-red light system).

The possibility cannot be eliminated that other cofactors capable of trapping the electrons and transferring them to Hill reagents may be present in this lipoprotein structure. However, the evidence suggests that normally TPN operates as the eventual electron trap which then, in turn, can be used for the reduction of CO_2 in the Calvin scheme, or possibly oxidized by the plant mitochondria with the formation of ATP.

Action Spectra for Photosynthesis and Pyridine Nucleotide Reduction. As suggested by the summary scheme presented in Fig. 5.15, the action spectrum for photosynthesis represents the combined action of two major pigment systems and whatever accessory pigments that may be present. The use of very specific background radiation is indicated if the action spectrum of any one system is to be studied. For example, the photooxidation of cytochrome f or c_2 should appear to be much more rapid with red light than with light of shorter wavelength. This is due to the fact that intermediate wavelengths lead to both oxidation and reduction. In the presence of DCMU (dichlorophenyl dimethyl urea) the photooxidation of the cytochrome should be much faster since the reduction of c_2 is inhibited. Amesz and Duysens (1962) have observed these general effects in the blue-green alga *Anacystis nidulans*. However, because of the strong overlapping absorption bands of these pigment systems and the possibility of energy transfer from the red to the far-red system *in vivo*, action spectra are difficult to interpret.

French et al. (1960) have determined action spectra of photosynthesis for *Chlorella* in the presence and absence of background illumination. With a strong 700 mμ light as a background they observed that the action spectra corresponded to that for the fluorescence of chlorophyll a_2. In this case the primary pigment system being activated was the accessory pigment, chlorophyll b, which transfers the excitation energy to chlorophyll a_2. When 650 mμ light was used as background, the action spectrum showed a major peak in the far-red. It is important to note that the two action spectra with background illumination crossed at approximately 683 mμ. As Duysens and Amesz emphasized, the action spectrum without background gave a combined action spectrum whose maximum rate should be 683 mμ with decreasing efficiency on either side of this wavelength. In general, this is what has been observed for the quantum efficiency of *Chlorella* photosynthesis (Emerson and Lewis, 1943).

Oxidation and Reduction of the Cytochromes. Recently Müller *et al.* (1963a,b) have studied the spectroscopic changes which occur in the cytochrome *f* system upon its photochemical oxidation and reduction. By exciting *Chlorella* or chloroplasts with far-red light (720 mμ) it is possible to cause the oxidation of cytochrome without creating any significant reductant from the second pigment system. In addition, upon exciting the chloroplast with far-red light at $-150°C$, oxidation of cytochrome *f* occurs, but reduction is completely stopped. By studying these spectral changes of the cytochrome at 405 mμ it was estimated that the reaction time for oxidation is slow in contrast to the reduction. The oxidation time was estimated to be of the order of 10^{-3} sec. Since the reduction time is much shorter than this, one would expect to find the appearance of the absorption spectrum of oxidized cytochrome under normal conditions. Using various ratios of ferri- and ferrocyanide the oxidation-reduction potential of the cytochrome was estimated to be $+0.4$ volts.

One of the major difficulties in attempting to interpret the experimental work at the present time is the lack of definitive data regarding the relationship between cytochrome b_6 and cytochrome *f*. If oxygen is an electron acceptor in the chloroplast, it may be that cytochrome b_6 is oxidized in much the same way that we think the reaction occurs in mitochondria during oxidative phosphorylation. In the absence of oxygen as an electron acceptor there is evidence that cytochrome b_6 can reduce cytochrome *f*, resulting in cyclic electron flow. The nature of any necessary coupling factors has not been described, however, and the slow rate of photophosphorylation in the absence of added dye such as PMS suggests that the coupling is weak or that a soluble factor has been lost during the isolation of the chloroplasts.

Hill and Bendall (1960) have proposed a scheme in which cytochrome *f* and b_6 are closely coupled. The potential difference of approximately 0.34 volts between the two cytochromes does not prohibit the possibility of a phosphorylation at this site. The observations of Chance and Sager (1957) of a light-induced oxidation of a *b*-type cytochrome in an anaerobic suspension of a pale green mutant of *Chlamydomonas* might be explained as an oxidation due to the photosynthetic oxidant. In the same mutant Duysens (1959) observed a light-stimulated oxidation of reduced pyridine nucleotide. Duysens and Amesz were unable to observe in *Porphyridium* any difference spectrum changes in the blue region which they could attribute to cytochrome *b*. Lundegårdh (1961) has reported the photooxidation of cytochrome c_2 with an accompanying reduction of cytochrome b_6 in leaves of *Agrostis stolonifera*.

Recent experiments of Duysens and Amesz (1962) in the red alga, *Porphyridium cruentum*, confirm the presence of two pigment systems with different photochemical functions. Light in the region where *phycoerythrin* absorbs strongly (560 mμ) is active for oxygen evolution, while longer wavelengths of light (greater than 680 mμ) are more effective for the initial photoreduction of pyridine nucleotide. They report that the excitation of the system by far-red light causes the oxidation of cytochrome f or c_2, whereas excitation of the system by red light leads to the reduction of cytochrome. They propose that the electrons from the oxidation of cytochrome f reduce the pyridine nucleotides which eventually lead to the reduction of CO_2. In their scheme the photochemical reduction of cytochrome f would lead eventually to the oxidation of water and the liberation of oxygen. In support of this latter point is the observation that DCMU, which is a potent inhibitor of oxygen evolution, is also an inhibitor of cytochrome f reduction. Duysens and Amesz propose the following scheme to account for the observations which they have made on these two light systems. The arrows indicate the direction of hydrogen or electron transport.

$$\text{water} \rightarrow \text{system } 2 \rightarrow \text{Q} \rightarrow \text{cytochrome} \rightarrow \text{P} \rightarrow \text{system } 1 \rightarrow \text{XH} \rightarrow \text{PN} \rightarrow CO_2 \quad (5.7)$$

They point out that their scheme is not complete since phosphorylation and other reactions of reduced pyridine nucleotides are not shown. The major concepts in this scheme have been incorporated into the summary diagram presented in Fig. 5.15. The far-red light system is formally identical to system 1 of Duysens and Amesz, while the red light system is identical to their system 2. In their discussion of the evidence it is also evident that the far-red system is the one in which chlorophyll a_1 is imbedded in a lipid structure and corresponds to what Franck (1951) calls "dry" chlorophyll, whereas the red light system contains chlorophyll intimately associated with the aqueous, oxygen-liberating system.

Photoreduction of Plastoquinone. Bishop (1959) has demonstrated that plastoquinone is involved in the electron transport system of photosynthesis. He was able to extract chloroplasts with petroleum ether and eliminate the photolysis of water. However, oxygen evolution and electron transport could be readily re-established by the addition of the petroleum ether extract of plastoquinone.

Duysens and Amesz (1962) place plastoquinone between chlorophyll a_2 and cytochrome f primarily on the basis of fluorescence studies. Duysens and Sweers (1963) observed that the far-red light system quenched the fluorescence of chlorophyll while the red light system raised the fluorescence yield. This is explained on their scheme by the hypothesis that oxidized Q (plastoquinone) which is presumably reduced by the red light system (their system 2) quenches the excited state of chlorophyll a_2 and that QH which is reoxidized by the far-red system cannot quench this excitation energy. It is also possible that plastoquinone can be reduced indirectly by the far-red system and that this, in turn, could create an oxidant (cytochrome f) which could quench the excitation energy from the red (system 2) to lower the fluorescence yield. In other words, red light in the absence of far-red would be expected to give a greater fluorescence intensity since there is no oxidant being created that can quench the excitation energy. However, Witt and his associates (Müller et al., 1963a,b) have presented spectroscopic data which suggest that plastoquinone is the immediate electron acceptor from the chlorophyll a_2 system and that plastoquinone is intimately associated with cytochrome f and serves as its reductant. By measuring the magnitude and the decay time of changes at 475 and 515 mμ, Witt and associates were able to study the effects of temperature, the intensity of the flashing light, pH, duration of the flash, water exhaustion, extraction of the chloroplasts with petroleum ether, and the addition of oxidized and reduced substances on the light effects. The evidence indicates that all these procedures have a strong influence on the absorption spectrum changes at 475 and 515 mμ which indicates that the substance which absorbs at 475 mμ is transformed into a reduced or semireduced form which absorbs at 515 mμ. This substance is, or is in equilibrium with, a plastoquinone.

Plastoquinone is reduced in the light by the red light system and is reoxidized in the dark by cytochrome f in approximately 10^{-2} sec at 20°C. In addition, it is possible to demonstrate that various Hill reagents, such as phenol indophenol and methylene blue, can be reduced directly by the photoreduced plastoquinone. Witt has demonstrated a direct coupling, since by increasing the concentration of these Hill reagents the dark reaction for the oxidation of plastoquinone is shortened from 10^{-2} to 10^{-4} sec. Plastoquinone is continually being oxidized in the chloroplast by the addition of the Hill reagent and under the circumstances an equivalent amount of oxygen production takes place. Thus it is clear that the system which is reducing plastoquinone is the one that is engaged in the splitting of water and the production of oxygen and therefore must be intimately associated with chlorophyll a_2. Further evidence that these changes are intimately associated with the oxygen evolution process is obtained from the observations that CMU, a strong inhibitor of oxygen production, eliminates the spectral absorption changes observed. The removal of water also eliminates the spectral changes which are associated with the reduction of plastoquinone in contrast to the observed changes in chlorophyll itself. Thus it is evident that the spectral absorption changes caused by plastoquinone are closely related to the production of oxygen. The fact that the magnitude of the absorption changes as well as the oxygen production per flash are independent of temperature and light intensities indicates that the reactions are photochemical in nature. Saturation intensities for absorption changes as well as for oxygen production per flash are temperature independent and have the same value. In continuous light, however, the saturation intensity for the absorption changes as well as for oxygen production are strongly temperature dependent, indicating a catalytic transfer of electrons.

Photophosphorylation. As indicated previously, the stable products formed during photoinduced electron transport are molecular oxygen, ATP, and reduced TPN. It is evident from the summary given thus far and indicated in the scheme in Fig. 5.15 that the far-red light system is concerned with excitation of chlorophyll a_1 which leads eventually to the reduction of ferredoxin and ultimately TPNH. The red light system represented by chlorophyll a_2 is concerned with the photolysis of water. The electrons from this excitation process reduce plastoquinone, then cytochrome f which in turn restores the oxidized chlorophyll a_1 to the reduced state. Thus, in the flow of electrons propelled by light quanta the reduction of TPN requires the cooperation of two pigment systems. The ultimate electron donor in the case of the green plant is water which in some way must interact with the manganese-enzyme system of chlorophyll a_2 to produce molecular oxygen.

Arnon (1958) has attempted to determine the stoichiometric relationship between phosphorylation and TPNH formation and Eq. (5.6) expresses this relationship. In the photochemical formation of ATP and TPNH there is a one-to-one relationship, and it is for this reason that a substance X with a higher reducing potential than ferredoxin is indicated in Fig. 5.15. It is proposed that in the oxidation of reduced X by ferredoxin an energy rich phosphate bond is formed. It is also possible that electron flow from the chlorophyll a_2 through the plastoquinone and cytochrome f system to oxidized chlorophyll a_1 could lead to ATP formation. However, there is no direct evidence to support this or the scheme mentioned above involving a more reduced substance formed initially by the far-red light system.

For maximum rates of photophosphorylation in chloroplast preparations it is necessary to add oxidation-reduction dyes such as pyocyanine, FMN, or phenazine methosulfate (Jagendorf and Avron, 1958). Using Swiss chard chloroplast, Avron (1960) has been able to demonstrate a maximum rate of 2500 μmoles of ATP formed per milligram of chlorophyll per hour at 15°C which is much faster than necessary to account for the photosynthetic rate in the intact plant. The electron flow in the presence of PMS has been called *cyclic electron flow* by Whatley *et al.* (1954), since it will occur anaerobically, and the oxidant for the reduced dye is formed photochemically. Recently Tagawa *et al.* (1963) have demonstrated that one can obtain cyclic electron flow without added dyes but that this occurs at a very low rate. However, it is significant that they could demonstrate a ferredoxin requirement for this cyclic phosphorylation. Noncyclic electron flow can occur using oxygen as the electron acceptor or some Hill reagent such as quinone. Jagendorf and Forti (1960) demonstrated that not only is oxygen required for the endogenous phosphorylation but that ATP formation is inhibited by CMU. The data indicate that in chloroplasts, in contrast to bacterial chromatophores, true cyclic flow does not occur to any great extent unless artificial electron carriers are added. Naturally occurring electron carriers which would catalyze rapid cyclic electron flow may exist, but no evidence has been presented to support this idea. As indicated in the scheme, therefore, additional ATP would be formed during cyclic electron flow and this is indicated to take place by the flow of electrons from ferredoxin through flavin, quinones, and cytochrome b_6 and ultimately back to the photo-oxidant, either oxidized chlorophyll a_1 or cytochrome f. If ATP is formed in the oxidation of cytochrome f by oxidized chlorophyll a_1, then it must occur in the presence of CMU which inhibits oxygen evolution. Witt and Arnon have been able to demonstrate cyclic electron flow accompanied by phosphorylation in the presence of this inhibitor. In addition, Tagawa and co-workers (1963) have shown that the phosphorylation which accompanies cyclic electron flow by way of ferredoxin is sensitive to dinitrophenol and antimycin A (similar to mitochondrial phosphorylation), whereas noncyclic photophosphorylation is not sensitive to these inhibitors. Thus if a phosphorylation pathway exists which is noncyclic, it must be a different one than that indicated in Fig. 5.15. The fact that the photooxidation of cytochrome f can take place at $-150°C$ together with the CMU and DNP inhibition data suggests that ATP is not formed in the electron flow between the cytochrome f and chlorophyll a_1. It is for this reason, therefore, that we have proposed a more reducing substance being formed by the far-red system in the initial photochemical event which is capable of forming ATP during its oxidation by ferredoxin.

There is some evidence, however, which suggests that there may not always be a close relationship between ATP formation and TPNH reduction. There is the possibility that phosphorylation by dark reactions may occur following photoinduced charged separation. A number of investigators have suggested that the TPNH to ATP ratio of unity is observed only at saturating light intensities. Black *et al.* (1962) report that the action spectrum for concurrent ATP formation and TPN reduction was different from that of TPN reduction alone. Thus it is possible that ATP formation may be initiated by two different light systems, while TPN reduction depends primarily upon one light system. In addition, there exists the definite possibility that additional ATP formation may occur by cyclic electron flow from ferredoxin rather than from the reduction of TPN. All of these reactions would lead to a lower ratio of TPNH to ATP.

The fact that a stronger reducing agent than ferredoxin may be formed photochemically in the far-red system was suggested by the early work of Gaffron and Rubin (1942) on the production of molecular hydrogen by the green alga *Scenedesmus* and the observations by Gest and Kamen (1949a,b) on the photoproduction of hydrogen by photosynthetic bacteria. Arnon *et al.* (1960) and Losada *et al.* (1961) have reported the photoproduction of molecular hydrogen from *Chromatium* cells using thiosulfate as the electron donor. The most recent evidence would suggest that the stable reducing substance which is formed in the light and is capable of reducing TPN is ferredoxin or PPNR. It has been shown that PPNR can be reduced by illuminated chloroplasts (Horio and Yamashita, 1962; Chance and San Pietro, 1963; Whatley *et al.*, 1963) and by hydrogen and hydrogenase (Tagawa and Arnon, 1962). Recently, Fry *et al.* (1963) have shown that only one-half of the iron in reduced PPNR is in the ferrous state, which is in keeping with the conclusion that ferredoxin or PPNR is a 1 electron carrier and that 1 mole of TPN will oxidize 2 moles of PPNR (Whatley *et al.*, 1963).

Photoinduced Electron Transport—Phosphorylation and Quantum Phenomena. It is of no great significance to calculate the actual energy requirement for the reduction of ferredoxin, TPN, or the unknown X, as long as the light quantum used has an energy sufficient to raise an electron to the oxidation-reduction level of these compounds. Thus even though light quanta with shorter wavelength than the far-red light normally used in the TPN reduction step in photosynthesis have greater energy, the net effect is that 1 quantum is still required for each electron moved. Consequently under all conditions, irrespective of the amount of energy per quantum absorbed by the chlorophyll system, 2 quanta are required to form a reduced TPN molecule. Quantum efficiencies are the only meaningful calculations. The same may be said about the energy available from reduced TPN. Although over 50 kcal of energy are liberated when TPNH is oxidized by molecular oxygen, we know of no biological system that can make more than 3 moles of ATP energy available during the oxidation of 1 mole of reduced pyridine nucleotide. Thus, in considering mechanisms of photosynthesis quantum events are the essential considerations, while thermodynamic calculations are useful only as a measure of the over-all efficiency.

In this connection the question of photoinduced electron flow and phosphorylation must be considered with great care. At the present time there is no good theoretical reason to think that ATP is formed directly upon the photochemical movement of electrons. It must be formed ultimately by electron flow during dark reactions. The results of Nishimura (1962a,b) and Nishimura *et al.* (1962) indicate that ATP formation is due to a dark electron-transport process. Using light flashes of 0.5 msec duration and of high intensity, Nishimura found that the amount of ATP formed in the light was negligible compared with the total delayed phosphorylation. Olson (1962) has reported that one quantum is required per electron for the light-induced oxidation of cytochrome. However, can we obtain an ATP molecule by the transfer of a single electron from cytochrome *f* to oxidized chlorophyll?

The studies of Kok (1961), Gibbs *et al.* (1961), Nakamoto *et al.* (1959), and Vennesland *et al.* (1961) demonstrate also that photophosphorylation can occur with oxygen being the ultimate electron acceptor. Presumably a high-energy reductant is formed by light, and the electrons then flow by means of a number of cofactors to molecular oxygen with coupled oxidative phosphorylation analogous to that occurring in the mitochondria. The demonstration of an oxygen-requiring phosphorylation process in chloroplasts suggests the possibility that this noncyclic type of phosphorylation may be significant in the intact cell. Under certain circumstances only oxygen exchange would be observed and there would be no net oxygen production.

In addition to Krogman's studies (1961), Krall *et al.* (1961) have observed that under conditions where chloroplasts produce ATP with no accumulation of oxygen or net reduction of a Hill reactant there was a vigorous exhange of oxygen. The results suggest that the oxygen-producing system of the chloroplast was operating and that the oxygen produced was immediately reduced by the photoreductant. PMS, which stimulates electron flow and phosphorylation, greatly inhibited the oxygen exchange reaction. Because of these various types of cyclic and noncyclic electron flow with their possibilities of associated phosphorylation it is not yet possible to decide on the sequence of photochemical events and ATP formation with certainty. The best evidence available suggests to us that ATP is formed from the oxidation of a reductant with a more negative potential than that of ferredoxin, and that ATP is not formed during the reduction of TPN.

As indicated in Fig. 5.15 the initial absorption of far-red light leads to charge separation and the ultimate reduction of PPNR. It may appear that the net gain in Gibbs free energy in the initial photochemical event is equal at least to the difference in the potential of the pyridine nucleotide (-0.3 volts) and chlorophyll a_1 ($+0.5$ volts). This ΔE of 0.8 volt would be equivalent to a net gain of 18 kcal/mole Gibbs free energy. Since electron flow is through ferredoxin with oxygen as a possible ultimate electron acceptor, it is possible that the over-all free energy change for the oxidation of the initial reductant is over 29 kcal/mole the ΔE being the difference between ferredoxin potential (-0.46 volts) and oxygen ($+0.8$ volts). The energy per quantum absorbed in the red light system (42 kcal/mole) is more than enough to lead to the reduction of the oxidized chlorophyll a_1.

In bacterial photosynthesis a number of different reducing agents appear to be capable of reducing the photooxidant. The theoretical maximum quantum yield in the bacterial system depends upon the amount of ATP formed during electron flow. Most workers have suggested that ATP formation in bacteria occurs only from cyclic electron flow. However, recently Leadbetter and Whittenbury (1963) have shown that CO_2 assimilation by *Chlorobium thiosulfatophilum* is similar whether H_2 or thiosulfate was used as the electron donor during photosynthesis. Since Larsen *et al.* (1952) have shown that the quantum yield is essentially identical for the two different substrates under similar conditions, this suggests that similar mechanisms are functioning for the generation of reduced pyridine nucleotide and ATP. This argues that for photosynthetic bacteria the chromatophore system is capable of converting radiant energy into chemical potential, making using of noncyclic transport of electrons from a number of different electron donors. The driving force for energy transformation is the light quantum and the potential of the electron donor is relatively unimportant as far as generating assimilatory power is concerned.

Leadbetter and Whittenbury argue from their results that noncyclic electron flow is all that is necessary to account for photosynthesis in bacteria. However, this interpretation is based on the assumption that two quanta are required in the photochemical event to yield an electron suitable for the reduction of pyridine nucleotide. As indicated above, it is possible to interpret this two quanta phenomenon in another way. According to the scheme presented in Fig. 5.15, we would argue that only one quantum is required for the movement of an electron to the potential level of a pyridine nucleotide and consequently to form two TPNH only four quanta are required. The "holes" formed in the photochemical complex can be filled, in the case of bacteria, with H_2, thiosulfate, or other electron donors. In the case of the green plant additional (4) light quanta are required for the transfer of electrons from water in order to fill the "holes." The equivalence of H_2 and thiosulfate in *Chlorobium* does not seem to us, therefore, to argue one way or another for cyclic or noncyclic photophosphorylation. As far as ATP formation is concerned we can make the same arguments as presented previously for the green plants. If ATP formation occurs during the photochemical reduction of pyridine nucleotide then no more than two light quanta in 26 would be needed for cyclic photophosphorylation in order to obtain balanced photosynthesis. On the other hand, if all the ATP comes from cyclic phosphorylation, the quantum yield would depend upon the amount of ATP formed per pair of electrons.

We may obtain as many as three ATP molecules from two electron equivalents by cyclic electron flow or by means of oxidative phosphorylation. Unfortunately, we do not know the answer to this question of ATP formation. If two ATP molecules were formed per electron pair in cyclic flow, then the theoretical minimum quantum yield for bacterial photosynthesis ($2TPNH + 3ATP + CO_2$) would be 7. If only one ATP were formed the quantum yield would be 10.

It should be re-emphasized that the over-all minimum quantum yield for green-plant photosynthesis was calculated to be 9. In the bacteria which lack the O_2-evolving process the hypothetical quantum yield is not too different, which agrees with the experimental facts. It is evident that a great deal of energy of the red light system is used for the splitting of water and for oxygen evolution, a process which evidently is so unique in nature that a second pigment system has evolved in order that this reaction can be carried out efficiently.

3.   Photoperiodism and Photomorphogenesis: The Control of Metabolism by Light

Under natural conditions an organism may show a day-night periodicity for a number of different measurable physiological processes. A large number of animals and plants have been shown to exhibit rhythms, and careful investigations indicate that these internal clocks play an essential role in plant (Hendricks, 1963) and animal behavior and physiology. In the same organism these rhythmic responses are quite accurate, have different periods, and are very reproducible. This leads to the general notion that there exist different types of biological clocks with different periods, all under the control of one master clock. There are numerous examples, some of which we will describe in great detail later. These include phototaxis in *Euglena* (Bruce and Pittendrigh, 1956), zonation of growth in *Neurospora* (Pittendrigh and Bruce, 1959), bioluminescence and cell division in *Gonyaulax* (Hastings and Sweeney, 1957a,b; Sweeney and Hastings, 1958), mating in *Paramecium* (Ehert, 1959), the time sense in the bee (Renner, 1957; Wahl, 1932), running activity in the cockroach (Bünning, 1958), and many others. As Aschoff (1960) has emphasized, there is no need for the clocks that control these rhythmic processes to run continuously. For some purposes the clock could be started at the beginning of the timing cycle and depend exclusively on some environmental influence. However, biological rhythms appear to be controlled by physiological and biochemical processes which are regulated in time by means of a subcellular chemical clock which must have inherent in its mechanism a very accurate method for continuous time-keeping. In addition, the period of most of the rhythmic processes appears to be temperature independent. Thus, one must conclude that there is a chemical mechanism which is temperature compensated.

The term *circadian* which was introduced by Halberg (1953) is meant to describe a biological rhythmic process whose control is endogenous. The average period of the clock is 24 hours and the clock seems to run continuously. The environment has an influence on the circadian rhythm but usually operates only as a synchronizing agent. We will be concerned with a number of rhythmic processes in which light is of great importance as a synchronizing agent. Detailed discussions of many of these phenomena can be found in Withrow (1959).

Other indications of the occurrence of biological clocks are suggested by the way plants and animals are able to recognize the seasons of the year by accurately measuring the changing length of night or day. Examples of this are usually called photoperiodism and are generally familiar to all. For example, the migration of Canadian geese in the autumn is initiated by the change in the length of the nights and is one of the many important and interesting displays of photoperiodic responses in animals. In 1925 Rowan demonstrated that it was possible to stimulate the gonads of the Canadian junco to unseasonable sexual maturity by means of an artificially longer day in autumn. Since that time, the field has expanded tremendously and it is impossible even to summarize briefly all of the studies which have been concerned with the effect of varying day-night lengths on vertebrate and invertebrate physiological functions. The length of the day or the shortness of the night have been demonstrated to have important effects in the sexual cycle of numerous fish, reptiles, and in the reproductive cycle and migration of various birds. Rhythms of various types in man have been studied in great detail and light has been demonstrated to have an important effect on these many activities. The reader is referred to the above two symposia for more detailed information.

The flowering of the poinsettia at the Christmas season is another example of photoperiodism. Other plants grow, flower, and fruit at different times of the year. Thus we come to recognize that some clocks in plants cause flowering in the spring, while other periodic clocks cause flowering in summer and still others, in the autumn. In the fall, trees and woody plants stop growing and go into dormancy in anticipation of the forthcoming cold weather accompanying the progressively shorter days and longer nights.

Many studies have been initiated to unravel the nature of the biological clock that allows the organism to recognize and respond to light. Garner and Allard (1920, 1923) made the initial discovery that the flowering of many plants was controlled by the length of the day. Some plants flowered when the days were short, while others required long days for flowering. This photoperiod response was shown subsequently to be due *not to the length of the day but rather to the length of the night.* This was demonstrated by interrupting the night with a short interval of light. When a small amount of light was given in the middle of a long night, the plants responded as if the night were short. Thus, it was possible to demonstrate that "long day" (short night) plants which fail to flower on "short days" (long nights) could be induced to flower with a brief illumination during the long night. "Short day" plants which would ordinarily flower in long nights were prevented from flowering under similar interruptions of the long night by short light flashes. On the other hand, the interruption of the day with a short period of darkness had no effect on the flowering processes. An exception to this mechanism has been reported by Hillman (1963) for the short day plant *Lemma perpusilla,* in which flowering is affected by a dark period during the "day" period. In addition, Könitz (1959) has reported a marked inhibition of flowering in the short day plant *chenopodium* when the main (13-hour) light period was interrupted by treatments with far-red light for various periods.

Fortunately, the number of quanta required for the night-break light is quite low. Measurements could be made with varying monochromatic light to determine the effective wavelengths for interrupting the flowering processes. The investigators at the Beltsville Laboratories of the Department of Agriculture have shown that a number of very different types of photoperiodic plant responses have about the same action spectrum. For example, the germination of many varieties of seeds is both stimulated and inhibited by light. Action spectra for promotion and inhibition of seed germination are shown in Fig. 5.16. These data are taken from Toole *et al.* (1956). In their paper the ordinate was energy in ergs $\times$ 10^5/cm^2 required to promote or inhibit germination to the 50% level. We have converted their data to reciprocal quanta per square centimeter required to obtain the 50% level so that the ordinate is quantum effectiveness and is therefore a true action spectrum.

Red light is the most effective wavelength for promotion. The formation of anthocyanin pigment in plant tissues as well as the formation of the yellow pigment in the skin of tomatoes are also stimulated by red light and exhibit the same action spectrum. The effect of light on stem elongation and on leaf extension of bean and pea plants also show the same action spectrum. One is led to the rather startling conclusion that all of these responses are controlled by the same red-light absorbing compound.

The system that has been studied in great detail by the Beltsville group is the light-sensitive physiological response that one observes in seed

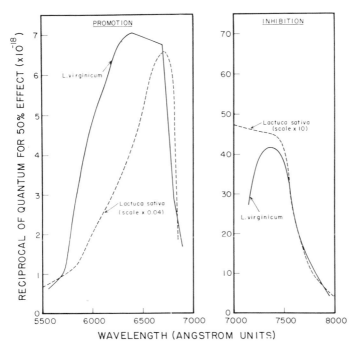

FIG. 5.16. Action spectra for promotion and inhibition of germination of seeds of *Lepidium virginicum* and of *Lactuca sativa* to 50%. Data from Toole *et al.* (1956).

germination. As indicated in Table 5.4 the germination yield of lettuce seed can be increased from a few per cent to almost 100% by red light at about 660 mμ. Also indicated in Table 5.4 and in Fig. 5.16 is the rather interesting result that radiation at longer wavelengths (far-red light), actually inhibits the germination of seed which has been previously stimulated by the red light. The action spectrum for this inhibition indicates that radiation at about 730 mμ is most effective. This red far-red interaction has been investigated in other photoperiodic processes in plants and a similar relationship was found to hold in the case of flowering, growth responses in leaves and stems, and pigment synthesis. When red light of sufficient intensity is used to inhibit flowering completely, it was found that far-red radiation, used immediately after the red, would then initiate flowering. In other words, it is as if the red light irradiation were not given. Thus, the red far-red system is completely reversible and can be demonstrated not only in lettuce seed germination but in the other processes mentioned above. The results suggest that a light-absorbing compound is involved which is converted reversibly from one form to the other. The Beltsville group has called this pigment *phytochrome* and indicates that it can exist in two forms. One, a red-absorbing form (P_{660})

TABLE 5.4

The effect of red light on the germination of lettuce seed[a,b]

Irradiation time (min)	0	$\frac{1}{4}$	$\frac{1}{2}$	1	2	4	8	16
Germination	4	7	8	9	46	61	92	97

The effect of far-red radiation on the germination of lettuce seed which had been irradiated for 10 min with red light[b,c]

Irradiation time (min)	0	$\frac{1}{2}$	1	2	4	8	16	32
Germination	89	96	94	89	76	49	11	8

[a] 580 to 700 mμ, 6000 ergs/cm^2-sec; 2×10^{14} photons/cm^2-sec.
[b] From Butler (1963).
[c] 700 to 800 mμ, 7500 ergs/cm^2-sec; 2.8×10^{14} photons/cm^2-sec.

and the other a far-red absorbing form (P_{730}). P_{660} absorbs red light and is converted into P_{730}. On the other hand, P_{730} absorbs far-red radiation and is converted back to P_{660} as indicated by the equation below:

$$P_{660} \underset{h\nu\ (730m\mu)}{\overset{h\nu\ (660m\mu)}{\rightleftharpoons}} P_{730} \tag{5.9}$$

From the effects of light on lettuce seed germination the phytochrome pigment must be entirely in the P_{660} form when the seeds are kept in darkness. P_{730} is required to trigger the germination process. By irradiating the seed with red light, P_{660} is converted to P_{730}, promoting germination. A subsequent far-red irradiation reconverts P_{730} into the P_{660} form and consequently reinstates dormancy.

Butler (1961, 1963) has used these reversible spectral changes to assay for the pigment and has obtained crude pigment extracts. The reversible photosensitive pigment was extracted with alkaline buffer solutions from corn seedlings kept in the dark. The light-sensitive system could be precipitated by ammonium sulfate and redissolved in buffer without loss of the reversible spectral changes with red and far-red light. The P_{660}–P_{730} difference spectrum of a partially purified solution of this pigment is shown in Fig. 5.17. This difference spectrum was obtained in a double beam spectrophotometer, irradiating the solution in a sample beam with far-red radiation and that in the reference beam with red radiation.

Some photochemical properties of phytochromes have been studied with these partially purified solutions. The photochemical transformations appear to be first order in both directions. The action spectra for the photochemical transformations have been measured by placing the solutions at various wavelengths in the spectrum using a large spectrograph. By measuring at various time intervals and at various wavelengths it was possible to determine the extinction coefficient and the quantum

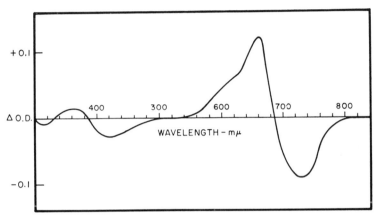

FIG. 5.17.   Difference spectrum of a clear solution of photosensitive pigment from corn seedlings in a commercial double-beam spectrophotometer. Cuvette in sample beam irradiated with far-red radiation. Cuvette in reference beam irradiated with red light (Butler, 1961).

yield for the photochemical process. The results of such a series of measurements are shown in Fig. 5.18. Phytochrome has a very high extinction coefficient and is a very efficient absorber of light as might be expected since it occurs in very low concentrations in the plant. The extinction coefficients of P_{660} and P_{730} are approximately equal at their maxima; the quantum yields for the two forms are different. Red radiation is at least four times more efficient in converting P_{660} to P_{730} than is the far-red radiation in converting P_{730} to P_{660}.

This difference in the quantum efficiency for conversion between the two forms of pigment may be very important for our understanding of the timing mechanism of plants insofar as flowering, seed germination, and other processes are concerned. Borthwick *et al.* (1952) have demonstrated that P_{730} converts spontaneously into P_{660} in the dark and that this dark reaction is the probable process that the plant uses for "measuring" time.

The dark conversion of P_{730} to P_{660} thus provides a timing mechanism and can account for the effect of night-break illumination. When dark-grown corn seedlings are irradiated with red light, all of the P_{660} is converted into P_{730}. If the plants are now placed back in the dark, P_{730} gradually drops to a very low level in the course of several hours. Most of the P_{730} converts to P_{660}. The rate of this dark reversal depends upon the metabolism of the plant and undoubtedly varies from plant to plant. The rate of the dark conversion is much slower at low temperatures and does not occur in the absence of oxygen. The effect of daylight, therefore, on plants is largely that of red light so that at the beginning of the night

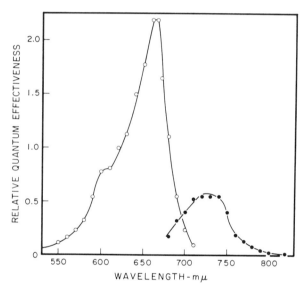

Fig. 5.18.   Action spectra for photochemical transformations of $P_{660} \rightarrow P_{730}$ and $P_{730} \rightarrow P_{660}$.

the phytochrome in the plant is largely P_{730}. As the night progresses, P_{730} slowly converts to P_{660} which is stable in the dark. A long-night effect may therefore depend entirely on P_{660} being the sole form of phytochrome present during the latter half of the night. If the night is interrupted by a brief illumination with red light, the accumulated P_{660} is changed to P_{730}; the dark reaction begins again and the long night has been broken into two short nights. This night-break light can be counteracted, however, by irradiating with far-red light which restores the P_{660}. Under these circumstances it is as if the two short nights were converted back into the long night and the processes continue unbroken.

The fact that sunlight exerts primarily a red light effect must be due to a great extent to the greater quantum yield of the red light reaction and to the fact that plants in darkness store the phytochrome in the P_{660} form. The action of far-red radiation can be observed in those cases where the sunlight is filtered through the chlorophyll of leaves. Red radiation will be absorbed preferentially to the far-red. *Thus seeds on the floor of a dense forest will remain dormant until they are blown or washed into a patch of sunlight, where they will receive the light energy necessary to stimulate and sustain growth.*

Unfortunately, we know very little about the biochemical action of the phytochrome. One form of the pigment, presumably the red-absorbing form, must act either directly or indirectly as a catalyst. The conversion of the far-red to the red pigment form in the dark is temperature sensitive. At the same time the reaction which is being stimulated by the red pigment is also temperature sensitive; both light processes are affected by temperature. The resultant is that the over-all timing process is temperature compensated and is, therefore, essentially temperature insensitive. The change of the pigments and the enzymes they must affect are essential factors in the timing mechanism. Whatever the biochemical basis for the control may be, it seems certain that the far-red absorbing pigment is the controlling factor for flowering as well as other physiological processes in the plant.

The color changes observed in leaves in autumn involve the fading or bleaching of the original pigments and the formation of new pigments. In the ordinary green leaf in summer the effects of the blue-absorbing (and therefore yellow-colored) carotenoids are masked by the blue- and red-absorbing (and therefore green-colored) chlorophylls. In the fall the chlorophylls bleach out and the carotenoids remaining give the leaves their bright-yellow color. The new pigments formed in autumn belong to the anthocyanins and are responsible for the majestic red hues of autumn leaves. Both light and sugars appear to be required for the synthesis of anthocyanin (anthocyanin cyanidin), which is intimately related to the

metabolism of the plant. For example, in the utilization of glucose or glycogen in living cells it must first be converted to glucose 1-phosphate. However, in the synthesis of the red anthocyanin from glucose, phosphate is not required. Thus if phosphate is in limited supply, more glucose is available for anthocyanin production and the leaves turn red. In fact, it has been known for some time that phosphate-deficient soils give rise to leaf-reddening in crop plants. In the fall the nutrient level (including phosphate) in the leaves is much reduced and presumably the photosynthetic pathway of anthocyanin production from leaf sugars, owing to low phosphate concentration, becomes important.

With regard to the mechanism of the red light and the far-red light effects in plants, probably the most extensive work concerning the mechanism has been done on the germination of photosensitive seeds. The conventional method of examining germination in seeds is to observe whether or not the radicle of the embryo has elongated out of its surrounding seed coat. In photosensitive seeds red light promotes the protrusion of the radicle through the coverings, while far-red light inhibits the process. Cutting, pricking, breaking, or removing the seed coverings stimulates the germination of photosensitive seeds even in the dark. Such observations have given rise to a number of hypotheses concerning the role of the seed coats in germination and the relationship of these coats to the photo-controlled process. The recent studies of Ikuma and Thimann (1963) on lettuce seeds are instructive in attempting to discriminate between these various hypotheses. The fruit coat and the endosperm are the main coverings surrounding a mature lettuce embryo. One can disturb both of these coats by a number of processes including pricking, surgical removal, and destructive radiation and obtain full germination in the dark. Such observations gave rise to the idea that the coats prevent the germination process by mechanically restricting the growth of the embryo. Ikuma and Thimann were able to obtain 100% germination of Grand Rapids lettuce seed in complete darkness by removing the endosperm and other embryo coverings. In other words, it appeared that the integrity of the endosperm layer determined the photosensitivity of the seed to red and far-red light. Ikuma and Thimann proposed that the final step in the germination control process is the production of an enzyme whose action enables the tip of the radicle to penetrate through the coat. This makes it possible for the radicle to begin elongating. Seeds injected with cellulolytic, hemicellulolytic, or pectolytic enzyme germinated to nearly the maximum extent in the dark, suggesting the possibility that the red-light effect was to stimulate in some way the formation of an enzyme to weaken the seed coat which is essential for the elongation of the radicle. The suggestion is made, therefore, that the phytochrome system in some way is activated by red light and that in this state it is catalytically active in destroying part of the endoderm and the seed coats to allow the growth and elongation of the radicle. In this connection it is also of interest to consider the effect of gibberellin on the germination process. Gibberellin is known to have the same effect as red light, causing 100% germination of the seeds in the dark. However, it also seems reasonably clear that gibberellin treatment of barley seeds stimulates or activates amylases and possibly other enzymes. Consequently, it is suggested that the germination of the lettuce seed caused by gibberellin is due to its chemically stimulating or activating hydrolytic enzymes which attack the seed coat.

The interpretation of this effect of red light on germination is not entirely supported, however, from work on other systems particularly those related to dormancy. A number of workers and particularly Wareing (1959) have developed the hypothesis that the seed covering contains inhibitors which prevent germination. Wareing has been able to extract from birch seeds, using 80% aqueous methanol, a substance that inhibits germination in normal seeds. This has been partially purified by paper chromatography. By placing birch embryos with the seed coats removed on the damp chromatographic paper they have been able to assay inhibitory and noninhibitory zones. When the embryos were placed on the inhibitory zone, germination occurred in high yield when a long daylight treatment was given. A low germination percentage resulted when a short day treatment was given. On the other hand, isolated embryos plated on filter paper which was moistened only with water germinated equally well in both light and dark. Thus, when the embryos isolated from their seed coats are planted on filter paper containing the inhibitor, their photoperiodic behavior appears to be restored. In addition, they have been able to demonstrate that by washing the seeds for long periods of time with water, it is possible to leach out an inhibitor and under these circumstances the intact seeds do not demonstrate a photoperiodic behavior. Thus, their results are consistent with the hypothesis that the inhibitory effect of the embryo coats is due primarily to the presence of a growth inhibitor. Presumably the red-light effect is to activate the embryo and stimulate enzyme activity which destroys the inhibitor, thus allowing germination to take place. Wareing also concludes that the effect of the inhibitor is due to the interference of oxygen consumption. For example, if the pericarp and endoderm are slit on one side of the seed, one observes a large increase in the germination yield of seeds kept in the dark. When the treated seeds are placed in an oxygen atmosphere, an additional increase in the germination occurs. It should be noted, in addition, that high oxygen tension is completely ineffective if the seeds are left intact. It appears that there may be a number of related mechanisms which can bring about the activation of germination in seeds and it is possible that a combination of all these hypotheses is closer to the truth. Gibberellic acid is also effective in breaking the dormancy of birch seeds just as it is with lettuce seed.

Since the original observations of Flint and Macalaster (1937) who first demonstrated the promotion of germination of seed by red light and its inhibition by far-red, a great volume of literature on the role of light in seed germination has arisen. Reviews are given by Crocker (1936), Evenari (1956), and Toole *et al.* (1956).

From these numerous studies it is clear that a reversible red, far-red photoreaction can profoundly effect many aspects of plant morphogenesis. As indicated above, the effect of red light on the germination rate of Grand Rapids lettuce seed can be mimicked by treatment with either gibberellin or kinetin. In these cases, however, the promotion of germination by these two chemical substances was *not* reversed by far-red light treatment. A. H. Haber and Tolbert (1959) demonstrated that temperature could affect the response of lettuce seed to the stimulation by gibberellic acid and kinetin at room temperatures. Kinetin is also effective at high temperatures in promoting germination of lettuce seed, whereas gibberellic acid loses its effectiveness. Thus, at 35°C gibberellic acid had no effect, but kinetin promoted germination. On the other hand, at 17°C kinetin was ineffective in promoting germination, whereas gibberellic acid retained its promoting ability. It is interesting, however, that kinetin can antagonize the stimulating effects of gibberellic acid. Thus, at 17°C kinetin can cancel the effect of gibberellic acid.

As Hendricks (1959) has pointed out, the features of the photoreaction are determined from measurements of complex physiological responses and these methods are to be contrasted with the usual *in vitro* photochemistry. The important point here is that the functional relation of the response to the degree of pigment interaction in the photoreaction is not immediately known and this does not need to be linear with light energy supplied. Even though the photochemistry that occurs in the plant may be immediately concerned with one pigment transformation, if the latter affects an enzyme, we may obtain tremendous amplification from a few light quanta. However, it is important to realize that responses induced by red light are readily reversible when far-red light is used. As indicated in Table 5.5, there are a number of photochemical processes that are important in higher plants, including the numerous photoreversible reactions. The elucidation of the underlying biochemical mechanisms will do much in identifying the major control mechanisms involved in plant growth and development.

Circadian Rhythms. Photoperiodism in plants is merely one interesting aspect of a biological response in which internal chemical clocks appear to be at work. Many aspects of the behavior of organisms are much better understood when it is realized that both plants and animals frequently have such internal clock mechanisms. Most everyone is familiar with the fact that various activities of plants and animals observed under natural

TABLE 5.5

PRINCIPAL PHOTOCHEMICAL REACTIONS OF HIGHER PLANTS[a]

Photoprocess	Reaction or response	Products	Photoreceptors	Action spectra peaks (mμ)
Energy conversion				
Chlorophyll synthesis	Reduction of protochlorophyll	Chlorophyll a Chlorophyll b	Protochlorophyll	Blue: 445 Red: 640
Photosynthesis	Dissociation of H_2O into 2[H] and $\frac{1}{2}O_2$ and reduction of $[CO_2]$	Reductant [H] Phosphorylated compounds	Chlorophylls Carotenoids	Blue: 435 Red: 675
Regulation of growth				
Blue reactions	(1) Phototropism (2) Protoplasmic viscosity (3) Photoreactivation	Oxidized auxin, auxin systems and/or other components of the cell	(1) Carotenoid and/or flavin (2) Unknown (3) Pyridine nucleotide, riboflavin, etc.	1. Near UV: 370 Blue: 445, 475 2. Uncertain 3. Uncertain
Red, far-red reactions	(1) Seed germination (2) Seedling and vegetative growth (3) Anthocyanin synthesis (4) Chloroplast responses (5) Heterotrophic growth (6) Photoperiodism (7) Chromosome response	Biochemistry completely unknown	Possibly tetrapyrrole	(1–6) Induction by red: 660; reversal by far-red: 710 and 730 (7) Far-red induced, red reversed, spectral details uncertain

[a] From Withrow (1959; reprinted with permission of the AAAS and the authors).

conditions occur at select times of the day. Some animals are most active by day, while some reserve their active movement for the night. Fireflies flash in the early evening hours in the northern hemisphere, while other forms of insects such as mosquitoes can be active either early in the morning or in the evening. Most people familiar with the habits of bats observe that each day these animals begin to fly in the early evening when the insects they feed on are the most abundant. This activity of the bat, however, does not depend upon this particular external stimulus for feeding, for it has been observed that even caged bats will go through controlled activity where there will be a definite pattern of active periods followed by rest periods. This is true even when the animals are kept in complete darkness. However, in this case the rhythm may not be exactly every 24 hours but may be 23 or even $23\frac{1}{2}$ hours depending upon the individual bat. Investigations along these lines indicate that rhythms may be synchronized by an external cycle such as light and dark phases but that the *persistence of the rhythm in the absence of external stimulus* is also present even though it may not be exactly 24 hours. As indicated previously, such rhythms are called circadian rhythms because their period is about (circa) a day (diem). There are, of course, many other rhythms with periods which do not match 24-hour night-day clocks. Many rhythms appear to match the tidal, lunar, or even annual periods of the environment. Thus, there are many different types of rhythms, some of which have timing devices with 24-hour face clocks, whereas others may be 30-day face clocks, and some may even have yearly face clocks. There are, of course, many examples of this, but we shall discuss only one which has been studied in some detail from both a physiological as well as a biochemical viewpoint. This involves the persistent daily rhythms of photosynthesis and luminescence in the dinoflagellate, *Gonyaulax polyedra*.

The beautiful blue and sometimes spectacular luminescence of the sea is most often due to the presence of unicellular organisms, the dinoflagellates. One particular species of dinoflagellate, *Gonyaulax polyedra*, has been isolated and cultured in the laboratory by B. W. Sweeney (1960). These dinoflagellates are brilliantly luminescent and exhibit a diurnal rhythm in their light-emitting capacity. Hastings and Sweeney (1957a,b, 1958) have used these organisms to study many of the aspects of this rhythmic process. Not only have Hastings and associates shown that there persist diurnal rhythms of bioluminescence but in the same organism one can also demonstrate a rhythm of photosynthesis and cell division, each of which continues with a period of approximately 24 hours when the temperature and light intensity are held constant. In this case the phases of the different rhythms are different, i.e., the maximum of each occurs at different times of day. These phase differences are retained when cells are maintained under constant conditions. Figure 5.19 illustrates these rhythms and the phase relationships that are involved and Fig. 5.20 shows the persistence of the luminescent rhythms in cells maintained under constant conditions. Hastings (1964) and Hastings and Sweeney (1959) have been able to demonstrate that the time of day for a maximum in luminescence can be shifted or reset by a light stimulus. A phase shift may be brought about either by a single change in the light intensity or by a single pulse of light, also illustrated in Fig. 5.20. The fact that light perturbations of this sort produce phase shifts indicates

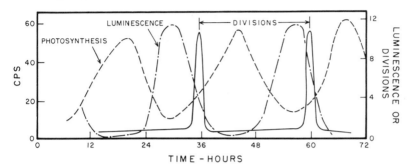

FIG. 5.19. The several rhythms and their phase relationships in *Gonyaulax*. All measurements were made with cells maintained in alternating light and dark periods of 12 hours each, the dark periods being indicated by the black bars on the abscissa. The light intensity during the light period was 960 foot-candles, and temperatures varied from a maximum of 26°C during the light periods to a minimum of 23°C during the dark periods. Cell densities were uniform in all aliquots, about 11,000 cells/ml. Ordinates: on left, counts per second incorporated, corrected for background and controls; on right, luminescence or divisions in arbitrary units. Abscissa: time in hours. For the measurement of rate of incorporation of $C^{14}O_2$, two flasks, each containing 20 ml of cells, were removed and incubated in the light at an intensity of 960 foot-candles for 15 min, in the presence of 12.5 μc of C^{14}. At the same time appropriate controls were incubated in the dark. Luminescence measurements and measurements of cell division were made with aliquots removed at the times indicated (Hastings, 1964).

that the products of the photochemical reaction have a specific effect upon the mechanism responsible for the rhythm. Hastings and Sweeney have measured the action spectrum for the phase shifting in *Gonyaulax* and this is shown in Fig. 5.21. The action spectrum corresponds roughly to the absorption spectrum of whole cells of *Gonyaulax* to which chlorophyll *a*, chlorophyll *c*, and peridinin contribute the major gross features. However, the wavelengths of maximum effectiveness do not precisely correspond with the absorption maxima. In the red region of the spectrum, for example, the maximum absorption of the pigments *in vivo* occurs at 680 mμ, while phase shifting was most effectively accomplished at 650 mμ.

Even larger differences can be seen for the blue region of the spectrum. The pigments *in vivo* show a broad absorption with a maximum of about 440 mμ, whereas the action spectrum shows a relatively sharp peak at 475 mμ. It is of considerable interest in connection with photoperiod responses in other animals that far-red light had no effect on the red light insofar as phase shifting is concerned. Far-red light alone was ineffective in causing a phase shift.

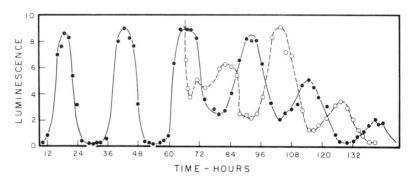

Fig. 5.20.   This figure illustrates shifting of the phase of the rhythm of luminescence by a single exposure of the cells to light. Phase shifting in all the experiments described was carried out in this way. The cells were grown with alternating light and dark periods of 12 hours each (LD), and were pipetted into test tubes about 2 days before irradiation, at some time between 0 and 12 hours on the graph. They were then replaced in LD conditions, and the typical rhythm of luminescence is shown. Cells remaining in the dark after 60 hours continue to show the rhythm with a period of approximately 24 hours, although the amplitude decreases (solid line). Cells treated similarly, except for a 3-hour exposure to light, between 66 and 69 hours, also show the rhythm of luminescence (broken line), but with a shift in the phase, or time at which the maxima occur. The number of hours by which the phase is shifted is dependent upon the intensity and color of the light used for irradiation, as illustrated in the subsequent figures (Hastings and Sweeney, 1960, reprinted by permission of the Rockefeller Institute Press from *J. Gen. Physiol.*).

The concentrations of substrate (luciferin) and the enzyme (luciferase) which are involved in the luminescent reaction of *Gonyaulax* have been found by Hastings to show similar rhythmic fluctuations. Both the enzyme and the substrate are found in greater amounts in cell-free extracts prepared during the dark period as compared to those prepared from cells harvested during the light period. The results definitely establish that the rhythm of luminescence reflects biochemical processes (Karakashian and Hastings, 1963). In biochemical investigation on rhythmic processes Pittendrigh and Bruce (1959) have emphasized correctly that one must be careful in drawing conclusions concerning the control of rhythmic reactions in the cells. It is difficult to find a basis for distinguishing a biochemical system which is the clock from a biochemical system which is controlled by the clock. It would appear in the *Gonyaulax* system that the luminescent reaction is certainly not the clock, and most recent evidence by Karakashian and Hastings (1962) using metabolic inhibitors would suggest that the clock function is dependent upon RNA synthesis. They reach this conclusion on the basis of studies on the luminescent rhythm

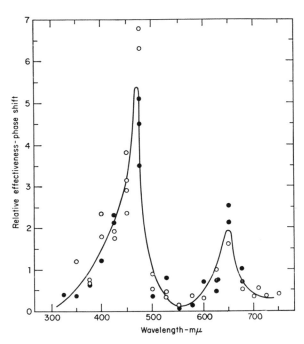

FIG. 5.21. Action spectrum for phase-shifting in *Gonyaulax*. The ordinate is the relative quantum effectiveness at each wavelength (Hastings and Sweeney, 1960, reprinted by permission of the Rockefeller Institute Press from *J. Gen. Physiol*).

and the effect of actinomycin D. The latter compound in very low concentrations influences not only the luminescent rhythm but the photosynthetic rhythm as well (Fig. 5.22). It is interesting that at least one subsequent luminescent cycle was expressed when actinomycin D was added shortly before the onset of the rhythmic increase in luminescence. The results suggest that the actinomycin sensitive step involved in the termination of a given peak of luminescence occurs at some earlier time in the cycle. Subsequent studies suggested that the sensitive period occurs 20–25 hours prior to the maximum of the cycle being effected. Although actinomycin D even at very high concentrations does not immediately decrease the rate of photosynthesis, it does inhibit the rhythm and in

both respects its action is similar to the effect on the rhythm of luminescence. A potent inhibitor of protein synthesis, chloramphenicol, penetrates *Gonyaulax* cells and stops division and growth immediately while
allowing the luminescent rhythm to persist. On the other hand, there is a
very clear inhibition of the rhythm subsequent to the addition of puro-
mycin. This compound promptly inhibits growth and luminescence
including the expression of the cycle. Thus, it would appear that actinomycin D gives a rather specific and selective inhibition of cellular rhythms
without any immediate deleterious effects upon either the cells or upon

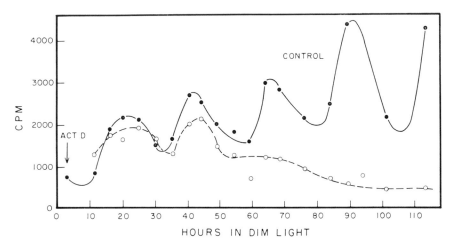

Fig. 5.22.   This illustrates the inhibition of the rhythm of photosynthetic capacity
by actinomycin D. Cells were grown in a cycle of alternating light and dark periods
of 12 hours each and placed in constant dim light and constant temperature at the
end of a light period, zero time on the graph. Actinomycin D was added 4 hours later
(arrow) and the rate of photosynthesis measured as counts of $C^{14}O_2$ incorporated in
20 min at a saturating light intensity (1000 foot-candles). Temperature, 20°C
(Karakashian and Hastings, 1962).

the mechanism responsible for the expression of the rhythm. This suggests that the control and the function of the clock may be related to the
actual functioning of the DNA in the synthesis of specific RNA.

Pittendrigh *et al.* (1958) have presented evidence to indicate that the
biological clock may be analogous or similar to a coupled oscillator system.
On the basis of studies on the nature of transients which occur following
single phase shifting perturbations to rhythmic systems, they propose
that two distinct oscillating systems underlie a particular biological
rhythm. The chemical studies which have just been mentioned by Hastings and associates would suggest the possibility that there may be cyclic
function of specific regions of the genetic material and this, in turn, may

be reflected in the synthesis and degradation of specific informational or messenger RNA molecules. A chemical oscillator could be established if it were possible to demonstrate that the product of the messenger RNA actually repressed the function of DNA. If both the repressor and the messenger RNA had a limited half-life, then one could expect such chemical oscillations in enzymatic and biochemical functions.

It has long been known that these diurnal rhythms of luminescence in marine dinoflagellates occur in the ocean itself. Many of these earlier studies are reviewed in Harvey's monograph (1952a). Recently, investigations by Seliger *et al.* (1962) on the luminescence of the dinoflagellates in Chesapeake Bay and in Falmouth Harbor in Jamaica indicate not only a diurnal rhythm in luminescence but also a diurnal migration of the organisms upward to the surface of the water at night and downward away from the light toward daybreak. Many other marine dinoflagellates have been studied for various other types of rhythms. Rhythms in cell division of many dinoflagellates have been observed and evidence for a daily rhythm in the rate of photosynthesis in phytoplankton samples taken from the sea was observed by Doty and Oguri (1957) and Shimada (1958).

Pohl (1948) described the daily rhythm in the phototaxis of *Euglena gracilis*, the response being maximal in the middle of the day and minimal at night. Studies by Bruce and Pittendrigh (1956–1958) have shown that this rhythm persists for weeks with a period which is close to 24 hours. In addition, they were able to show that the period is essentially temperature independent and that the rhythm may be entrained to light-dark cycles that are different from the normal. Furthermore, as in the case of the luminescent studies, the phase may be changed by a single exposure to light. Thus, it is clear that a *cellular level of organization is sufficient for biological clocks to function* with the essential timing features which have been described for more complex multicellular plants and animals.

The period of a diurnal rhythm appears to be an innate characteristic which need not be learned or impressed upon the organism. In certain cases, however, organisms which are kept in constant darkness or constant illumination fail to show a rhythm or show a rhythm with a period slightly different from 24 hours (Fig. 5.23). In the *Gonyaulax* luminescence, for example, although the rhythm persists in constant dim light, very bright light causes arhythmic responses. This is true for rhythms which have been studied in other organisms such as *Euglena* and *Paramecium*. If *Gonyaulax* cells which have been kept in bright light are transferred to darkness, it can be demonstrated that the diurnal rhythm is again expressed and the phase is related to the time when the cells were placed in the dark. The characteristic natural period is the one always observed, however, even though the cells may not have been exposed to diurnal light or temperature cycles for as long as 3 years. Thus, the setting of the clock requires only one change in light intensity which serves to give the phase information. Probably the most convincing evidence that the period is innate rather than learned is derived from the failure of attempts to change the natural period. By exposing an organism to light-dark cycles with atypical periods, it is possible to "entrain" a rhythm so that

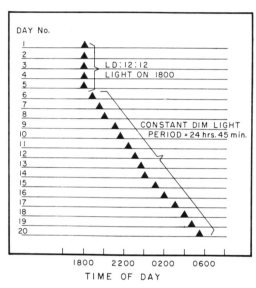

Fig. 5.23. The time of day at which the maximum in the luminescent glow occurred on each of 20 successive days. The values were obtained by measuring the luminescence from this vial approximately every 2 hours. For the first few days the culture was kept on a daily light-dark cycle, and the peak occurred at the same time each day. Subsequently the light remained on continuously, and whereas the rhythm continued, its maximum occurred about 45 min later each day, indicating that the free-running period in this case was 24 hours, 45 min. Temperature, 24°C (Hastings, 1964).

the maximum activity occurs once every cycle that has been artificially established. Thus it is possible to obtain a luminescence maximum in *Gonyaulax* every 16 hours by exposing the organisms to alternating light and dark periods of 8 hours each. However, when the cells are placed back under constant conditions, the rhythm reverts to its natural period (Hastings and Sweeney, 1958). These experiments have been carried out with cells which have been grown for at least 100 cell divisions under an 8-hour light period and an 8-hour dark period. When these cells are placed under constant low light conditions the rhythm reverts to its natural period. Thus, the evidence strongly supports the general conclusion that the mechanism involved is endogenous to the organism. F. A. Brown (1959), on the other hand, has argued that there are a number of other environmental physical factors such as barometric pressure, electric and magnetic fields, or cosmic ray intensity which may fluctuate with a 24-hour frequency and may be responsible for certain biological responses. If the clock system were coupled to these in a manner in which they could be affected by light, the frequency of the biological rhythm should correspond to these external variables. However, since the periods of persistent rhythms are characteristically different from exactly 24 hours, one concludes that such variables are not effecting the frequency of the clock system itself.

The temperature independence of biological clocks is one of its most striking and important features and is often used as an argument in favor of some external pace-making signal. The temperature independence of biological clocks is relative and one does observe some small but important effects (Sweeney and Hastings, 1960). In *Euglena*, for example, Bruce and Pittendrigh (1958) found that the period varied from about 26 hours at 16.7°C to slightly greater than 23 hours at 33°C. There are now well-established biochemical feedback systems with inhibition and stimulation not only of enzyme function but enzyme synthesis which could account for an apparent temperature insensitive enzymatic reaction.

Another important feature of the biological clock which we mentioned previously is the ability to reset the time piece. Diurnal variation in light intensity constitutes one of the principal environmental factors which affect many of the biological clocks which we have mentioned thus far. The rhythms which have been studied in organisms maintained on a 24-hour light-dark cycle always show a 24-hour period rather than its natural period which is usually observed only under carefully controlled laboratory conditions. Any inaccuracies that may occur in the cycle can be reset as one enters a new period. Thus, a correction is made for each cycle each day. It is possible to have the rhythm phase in a way where there is no necessary relationship to local solar day or night. When the organisms are subsequently placed under constant conditions, however, the rhythm persists with its characteristic natural period; the phase being related only to the time of the previous light-dark schedule. Therefore it is evident that the phase of biological rhythms may be fixed or modified by coupling to diurnal light cycles, but when light and temperature are held constant no other periodic or environmental variables have been found to modify the phase. The effect of the light perturbation in bringing about a phase shift does depend upon the cycle in which the light is administered. If a light signal is given to *Gonyaulax* cells, for example, during the time of the cycle corresponding to normal day, little or no phase shift occurs. A similar light signal given during what would be the night phase results in a marked phase shift. The phase is delayed if the light is given at a time corresponding to the late day or early night phase. The phase shift is affected maximally at a time which corresponds roughly to the middle of the night phase. These same generalizations seem to apply not only to lower organisms but to higher organisms as well (Bruce and Pittendrigh, 1958).

Biological clocks as illustrated by circadian rhythms occur in all types of organisms. In addition to the microorganisms which we have discussed in some detail, rhythms are known to occur in many fungi and simple algae and higher plants as well as lower and higher animals. There are only a few major groups in which the occurrence of these rhythms has not been demonstrated. These groups include the bacteria, the blue-green algae, and the bryophytes. This practically universal occurrence of the types of organisms which demonstrate circadian rhythms is equally matched by the diversity of the types of rhythms which they show. In other words, there are many ways in which the presence of an internal clock is made observable. From all of these numerous studies it would appear that much of the physiological, biochemical, and behavioral activity of an organism is more or less controlled or influenced by its internal rhythmic system. In many cases there does not appear to be an obvious adaptive significance to the rhythm which one observes such as the luminescence of *Gonyaulax*. In other cases, however, one can reasonably speculate about the adaptive value and in still others, the adaptive value is obvious.

In summary, biological clocks are widely distributed both with respect to organism and function. They are endogenous and innate and not learned. They can be synchronized to the external environment by a light or temperature cycle. They run at a rate which is very little modified by temperature, i.e., they are temperature compensated. They occur in single-cell microorganisms as well as in higher plants and animals and they serve to satisfy all sorts of adaptive requirements of the organisms. Although very little is known about the underlying mechanisms of these circadian rhythms, it would seem reasonably clear on the basis of recent investigations on the mechanism and control of metabolisms of cells that the synthesis and function of the macromolecules must be inherently close to the spring works of these clocks that have different faces. As more is learned about all of these different phenomena, the air of mysticism with which they have been viewed is dispelled and they are seen to constitute perfectly understandable adaptations of the organisms to the environment.

4. Phototropism and Phototaxis

In this section we distinguish between phototropism, the light-mediated bending of fixed organisms, mainly plants, and phototaxis, the light-mediated locomotion of organisms or cell components. This last condition is included since as will be described below, there is strong evidence that

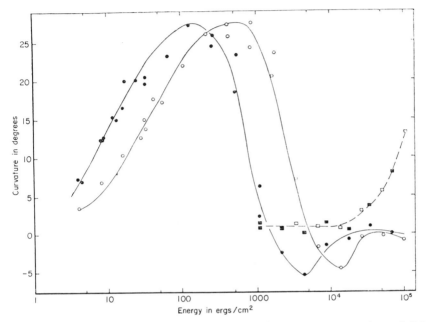

FIG. 5.24. Response curves of *Avena* at two wavelengths as functions of light intensity. Ordinates: curvature after 90 min; abscissa: total light dose. ●, Tip curvature, 436 mμ; ■, base curvature, 436 mμ; ○, tip curvature, 365 mμ; □, base curvature, 365 mμ (Thimann and Curry, 1961).

phototropic curvature results from a phototactic redistribution of auxins. The response to light exhibits the following general characteristics:

(1) There are positive (motion or bending toward light) and negative (motion or bending away from light) effects

(2) Blue-absorbing, yellow, carotenoid or flavenoid pigments are involved.

(3) The magnitude of the response at low light levels increases with light intensity. At higher intensities it appears to saturate and then decrease to zero. Over limited ranges of high intensity light, the response appears to be reversed. This is shown in Fig. 5.24 for the response of *Avena* coleoptiles plotted as a function of light intensity (Thimann and Curry, 1961). The light intensities are plotted in energy units rather than

number of quanta per square centimeter per second. However, since monochromatic light is involved, the conversion is simple. One erg/cm^2 of 436 mμ light is equal to $4360/1987 \times 10^{11} = 2.32 \times 10^{11}$ light quanta/cm^2.

The seedlings of the grasses have been known to be among the most phototropically sensitive of all plants. In the seed the primitive bud or plumule is enclosed in a protective sheath or coleoptile. During germination this coleoptile elongates, pushes its way out of the seed, and grows upward through the soil. Most experimental work on phototropic curvature has been with the coleoptiles of *Avena* (oats), corn, and barley, and the sporangiophores of *Phycomyces*, a fungus. One of the first observations made (by the Darwins) was that the zone of light sensitivity was the top of the coleoptile, while the curving occurred some millimeters below the tip. This led to a large number of studies by many workers for a substance, produced at the tip and transported to lower parts of the coleoptile. It was established that the growth of the coleoptile is controlled by a growth hormone, auxin. The curvature is thus due to an excess of auxin on the dark side for positive phototropism and vice versa. This was demonstrated by Went (1928), who performed the following simple experiment. The auxin from the tip of a decapitated, illuminated *Avena* coleoptile was allowed to diffuse out into two small agar blocks in contact, respectively, with the lighted and shaded sides of the plant. By applying these agar blocks to the sides of other coleoptiles it was found that about twice as much auxin had come from the dark sides as from the light sides of the original coleoptiles. More recently, Briggs *et al.* (1957) have shown that, although the yield of auxin from the lighted side decreases (confirming the old observations), *this does not mean that there is a destruction of auxin by light.* In fact, they found that (1) the yield of auxin from the dark side increases; (2) the total yield of auxin remains constant, independent of illumination; and (3) there is a lateral movement of auxin from the lighted side to the dark side which can be prevented by physically slitting the coleoptile all the way up to the tip; experiments slitting only the base of the coleoptile showed that the lateral migration occurs in the tip. The action spectra for the curvatures of *Avena* coleoptiles (positive) and *Phycomyces* sporangiophores (positive) are shown in Fig. 5.25. In both cases there is a main peak at 445 mμ, secondary peaks at around 430 and 470 mμ, a minimum near 410 mμ, and a broad peak in the near ultraviolet around 370 mμ. Suggestions have been made that the photoreceptor might be a carotenoid or riboflavin. The argument that carotene has no UV absorption was countered by showing that one or more *cis*-configurations, for example, 9,9'-*cis*-β-carotene, could exhibit UV absorption in approximately the same UV region, although not exactly matching the

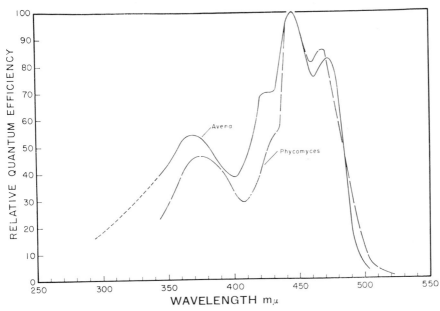

FIG. 5.25.   Action spectra for the positive curvature of *Avena* coleoptiles and *Phycomyces* sporangiophores (Curry and Thimann, 1961).

near UV action spectrum. The absorption spectrum of riboflavin shows general agreement with the action spectrum and in addition can photosensitize the oxidation of indoleacetic acid (IAA) *in vitro*. This will be brought out below.

At the present time even the most detailed action spectra for phototropism do not lead to an unequivocal demonstration of the effective photoreceptor for positive phototropism in coleoptile tips of any of the grasses. The action spectrum for positive phototropism in *Phycomyces* sporangiophores is in striking agreement with that for *Avena*, indicating that the photoreceptors are the same. There is carotene present just below coleoptile tips as well as in a region a few millimeters below the sporangium in *Phycomyces* (Bünning, 1937, 1955; Abbott and Grove, 1959). Work with albino mutants has not been completely conclusive.

To further complicate the picture, there appear to be several types of phototropic response. It has already been stated that at high light dosages the initial positive response may be reduced and even reversed in sign. While low dose curvature begins slowly and shows a light sensitivity centered close to the coleoptile tip, the high dose curvature starts earlier and the sensitive region is distributed along the whole length of the coleoptile. This second type is called "base response" to differentiate it from the "tip response." It is this "base response" which mainly determines the tropism of plants in daylight. According to Thimann and Curry (1961) this positive "base response" is also effected by short wavelength UV light and the action spectrum is shown in Fig. 5.26. This corresponds neither with carotenoid absorption nor with flavin absorption,

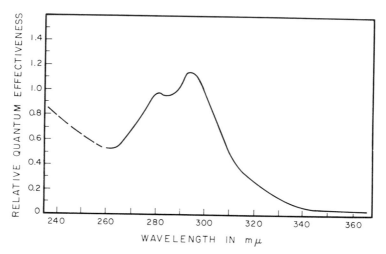

FIG. 5.26. Action spectrum for the *Avena* coleoptile "base" curvature toward UV light (Thimann and Curry, 1961).

and Curry *et al.* (1956) have proposed indoleacetic acid as the light-sensitive pigment, based on qualitative agreement of the IAA absorption and on the fact that IAA can reinstitute lost phototropic activity in decapitated coleoptiles.

Experiments with the sporangiophores of *Phycomyces* have been productive of further information. First of all, the entire sporangiophore is a single cylindrical cell about 0.15 mm in diameter and about 180 mm in height. The growing zone extends for only between 20 and 50 mm below the sporangium. This zone itself is the photoreceptor region and exposure of other portions of the sporangiophore does not cause curvature. The positive phototropism can also be reversed at high light dosages. However, for short wavelength UV light, around 280 mμ, *Phycomyces* exhibits negative phototropism. This differs from *Avena* not only in the sign of the tropism but in that the sensitive region remains the same, while in *Avena* we have already described the shift from "tip response" to "base response." On the basis of the time course of the phototropic reaction the proposal has been made that positive curvature is the simple result of differential growth reactions on both sides due to a cylindrical lens effect of the translucent sporangiophore for visible light. Presumably parallel light would be focussed on the opposite wall. In order to explain the negative phototropic response to short wavelength UV light, the assumption is made that protein absorption is sufficient to render the sporangiophore opaque and that only the illuminated side is stimulated to growth.

There is no evidence for the existence of a growth hormone, although indoleacetic acid is present in sporangiophores. A quantitative argument is presented by Thimann and Curry to show that much more IAA must be applied to the side of a coleoptile to produce a given curvature than that which could be inactivated by photosensitized oxidation by the amount of 4358 A light required to produce the same curvature. However one must be quite careful in this type of comparison in that (1) it is not definite that an *excess* relative to normal amounts of IAA is as effective in producing curvature as an excess relative to decreased amounts as is the actual case in the irradiated organism, and (2) all of the added IAA may not end up in the right place to effect bending, while all of the light-absorbing centers are presumably available to work at high efficiency.

Thus it is not possible to discard IAA as the UV absorber or the photosensitized component. It has also been shown that IAA added to coleoptiles after very high unilateral dosages can stimulate curvature. However, those are extremely high doses of heterogeneous radiation (500,000 ergs/cm² of white light) and the reactions may be quite complex.

Thimann and Curry postulate that the lateral translocation of auxin giving rise to positive phototropism is due to a negative phototaxis of plastids to the lateral wall opposite to the light direction where the contact of wall and plastid favors the movement of auxin through the wall. This they correlate with a positive *geotropic* response where plastids falling to the "bottom" will stimulate growth (and auxin flow). Presumably the negative geotropic response of root tips is an inhibition of auxin transport. In any case, it should be emphasized that most of this is intelligent speculation and much more information will be required before an adequate self-consistent picture evolves. It is known that plastid movements have a peak effectiveness around 450 mμ and that red light, effective in the lamellar chloroplast formation from proplastids, markedly reduces the phototropic sensitivity of *Avena* coleoptiles. Galston (1960) reported that in albino mutants of corn and sunflower with no detectable carotenoids and in *Phycomyces* with less than 1% of normal carotenoids there was full retention of phototropic sensitivity. These presumably had normal quantities of flavins. In summary, therefore, for photo-tropism, the photoreceptor pigment may be flavinoid or carotenoid, most likely a protein-dye complex. Structures are involved as in apparently all other light absorption events in biology. There may be an indole moiety involved, either for direct UV sensitivity or for photosensitized action. The mechanism may be a trigger-type reaction such as in vision, releasing an inhibitor from an enzyme system, permitting the production or transport of auxin. Phototaxis of proplastid structures may be involved in the gross phenomenon so that possibly mechanism studies on phototaxis would be more basic.

The mechanisms of phototaxis are certainly no better understood at the present time than those of phototropism. Reviews have recently been given by Haupt (1959) and Bendix (1960), and only a brief survey of data on *Euglena* and other flagellates will be presented here. While we shall be discussing only phototaxis, it is important to point out that there are many other types of oriented movements in response to stimuli. Very likely the basic dark chemistry has a common origin and it may be only the energy transduction mechanism which varies. These other taxes are:

Anemotaxis—response to currents in air
Chemotaxis—response to chemical stimuli
Geotaxis—response to gravitational field
Menotaxis—light compass reaction; movement at a fixed angle with
   respect to the stimulus
Rheotaxis—response to currents in water
Thigmotaxis—response to tactile stimuli

The action spectrum which is the same for both positive and negative photaxis in the photosynthetic flagellate *Platymonas subcordiformis* (Halldal, 1961) is given in Fig. 5.27. So far as we can see, except for the peaks at around 280 and 470 mμ, the spectra for phototaxis and photo-tropism do not coincide well. The threshold for (+) phototaxis is given as 0.15 erg/cm^2-sec at 405 mμ and that for (−) phototaxis is 0.80 erg/cm^2-sec at 405 mμ. These correspond, respectively, to 0.3 × 10^{11} photons/cm^2-sec and 1.6 × 10^{11} photons/cm^2-sec.

The magnitude *and the sign* of the phototactic response are extremely sensitive to ionic strength, pH, temperature and light intensity, and wave-length. For example, by adjustments in relative ionic concentrations of Ca^{++}, Mg^{++}, and K$^+$, *Platymonas* populations can be made all positive or all negative or any desired combination, such as 50% positive and 50% negative. In 0.5 M NaCl, 0.01 M CaCl$_2$, 0.02 M MgCl$_2$, 0.005 M KCl,

0.001 M KHCO$_3$ at pH 7.5, a *Platymonas* population will show a 90–100%
negative (−) reaction, lasting, in the dark, for at least 12 hours. Within
24 hours a few minutes of white light reversed the light responsivity from
negative (−) to positive (+). Halldal (1960) has measured the action

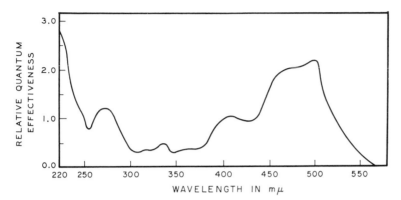

FIG. 5.27.   UV action spectrum of negative phototaxis in *Platymonas* (Halldal,
1961).

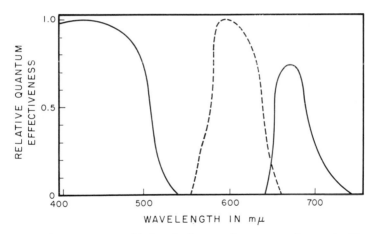

FIG. 5.28.   Action spectra of induced phototactic response changes in *Platymonas*.
Solid line, the negative to positive transformation; dashed line, the positive to negative transformation (Halldal, 1960).

spectra for the (−) → (+) transformation and also for the (+) → (−)
transformation. These are given in Fig. 5.28. There is some uncertainty
about the data in the 640–660 mμ region. Violet through green light
(400–540 mμ) and red light (660–740 mμ) induce a (+) response in a (−)
population. Light between 540 and 650 mμ was ineffective in changing a

(−) population. Conversely violet through green light and red light were ineffective in changing a (+) population and only light between 560 and 630 mμ with a peak at 590 mμ could induce a (−) response in a (+) population. These reversals are themselves reversible by changing back and forth from blue to yellow irradiation, with an induction time of about 5 min. The (+) inducing action spectrum is similar to that of photosynthesis. However, the phototactic action spectrum extends only through the blue region. It must also be remembered that even at 590 and 685 mμ, the peaks of the (−) and (+) induced action spectra, the intensities required were very high, around 21,000 ergs/cm²-sec for approximately 5 min or between 6–7 × 10¹⁵ quanta/cm²-sec, at least 200,000 times higher intensity than the threshold for (+) or (−) phototaxis.

Nicol (1959), for the mysid *Praunus neglectus*, found a threshold response of 29 × 10⁻⁶ μwatts/cm² for light with a spectral range 420–540 mμ and a peak emission at 475 mμ. This corresponds to approximately 7 × 10⁷ photons/cm²-sec of a constant light source. Seliger *et al.* (1962) calculated that the dinoflagellate *Pyridinium bahamence* emits 10¹⁰ photons per flash lasting approximately 0.1 sec. Assuming an average photon intensity of 10¹¹/sec we can calculate the maximum distance at which a dinoflagellate flash can elicit a response. This is

$$\frac{10^{11}}{4\pi r^2} = 7 \times 10^7$$

$$r \simeq 11 \text{ cm}$$

From our own visual observations in Oyster Bay in Jamaica this value places the dinoflagellate within the volume of water that is agitated to luminescence by a darting fish. Does the dinoflagellate therefore light up as a self-offering, or does it signal to a passing fish that it has been disturbed by a delectable copepod? Presumably the 50-μ diameter dinoflagellate, when swallowed along with the copepod, can escape through the gill system.

The action spectrum for *Euglena* phototaxis is the same as that for *Platymonas* over the region 400–600 mμ. No data is available for the UV region. The crustacean *Daphnia* which is normally indifferent to light becomes strongly (+) phototactic in the presence of excess CO_2 in the water. *Parmecia* is (+) geotactic in light but (−) geotactic in darkness.

A series of changing taxes leading to protective adaptation is illustrated by the sand flea. These are usually found in dark places under masses of seaweed on the beach, being negatively phototactic. If they are disturbed, they become strongly menotactic. This directs them at a fixed angle toward the water depending on where they have grown up (this may be effected by a dichroic sensitivity due to the polarization of reflected light from the water surface). Once in the water they again exhibit negative phototaxis; this response directs them to the bottom and into crevices. A large body of information is included in the book by Mast (1911) on the responses of organisms to light.

In both phototropism and phototaxis none of the pigment systems has been positively identified. The reactions, especially those of phototaxis appear to be related analogously to the photoreversible $P_{680} \rightleftharpoons P_{720}$ pigment in photomorphogenesis and the pH, temperature, etc., effects have their components in photoperiodism. Much basic photochemistry and biochemistry remains to be done.

5. Direct Stimulation by Light

Actually the phototaxis exhibited by *Euglena* and *Platymonas* are examples of direct stimulations by light and could be included in this section. However, we have reserved this title for a special type of light-induced reaction; the direct stimulation of muscle contraction without the direct intervention of the sympathetic or the parasympathetic nervous system. In some amphibia, *Rana esculenta* and *R. temporaria*, *Salamandra*, *Bufo*, *Hyla*, in the eel *Anguilla* and in the fish *Esox*, *Perca*, and *Salmo*, it is possible to observe a light-stimulated contraction of the iris sphincter pupillae *in completely excised irises*. So far as is known this type of stimulation in the iris and possibly the stimulation of certain chromatophore cells are the only cases of naturally occurring, reversible, direct muscle stimulation by light. The iridial stimulation was first reported by Brown-Séquard in 1847 and by Steinach in 1892. Weale (1956) reported a partial action spectrum based on threshold sensitivity for *Rana temporaria* using a tungsten lamp and Ilford filters. More recently, Seliger (1962) has published an action spectrum for contraction in excised eel irises, shown in Fig. 5.29. The action spectrum for contraction has a peak which agrees with the eel rhodopsin absorption maximum (Wald, 1961). Inasmuch as the rhodopsin is the rod pigment-opsin complex and the iris sphincter pupillae evolves from the pigment epithelium of the retina, the "muscle pigment" might be the same as the visual pigment.

The sphincter pupillae consists of two or three layers of concentrically disposed, spindle-like, pigment-containing smooth muscle fibers. The intervening tissue and stroma are free of pigment. By fixing and staining these pigmented muscle cells in the dark as well as in the light, Steinach was able to furnish direct anatomical proof of the contraction. A second surprising observation is that after complete atropine mydriasis the dilated iris will continue to exhibit the same maximal contraction as the nondrugged iris.

The mechanism is a real biological phenomenon in the sense that it is operative in the live animal. Weale (1956) found, for example, that in *Rana temporaria* the threshold for pupillary contraction was only 10 times higher than the photopic threshold of seeing, presumably to protect the cones from excessive light. (In the rods there is a light-induced migration of the epithelial pigment granules, even in the excised eye.)

We know nothing at the present time about the nature of the photochemistry involved in the triggering of the smooth muscle fiber by light,

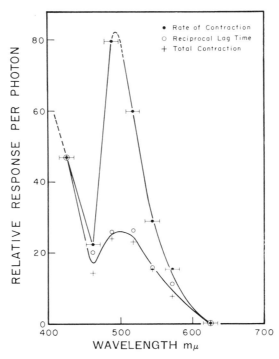

FIG. 5.29. Action spectrum for contraction in eel iris sphincter (Seliger, 1962; reprinted by permission of the Rockefeller Institute Press, from *J. Gen. Physiol.*).

particularly since innervation of nerve endings by cholinesterase does not inhibit the light-induced contraction. The effect is not observed in human irises. In the human eye the contraction of the iris sphincter is under nervous control and is activated only by light incident on the retina. The pupil diameter varies inversely with the square root of the light intensity, so that the pupil area varies inversely with light intensity. However, in the eel, Seliger reports that the *diameter* varies inversely with light intensity, suggesting a more direct reaction and therefore a more primitive origin for this light stimulation effect.

Cells containing pigment that can disperse or concentrate, thereby changing the tint of the skin of the organism in which they lie, are known as chromatophores. The chromatophores of cephalopods consist of a nucleus and a central pigment-containing sac surrounded by many radiating uninucleate smooth muscle fibers. When the fibers are relaxed, the pigment sac is a small dense speck contributing essentially nothing to the coloration of the animal. When the muscle fibers contract, the pigment sac is pulled out to form a thin disc whose diameter is up to sixty times the diameter of the relaxed condition. The total volume of pigment remains the same, except that it becomes a thin sheet instead of a dense sphere.

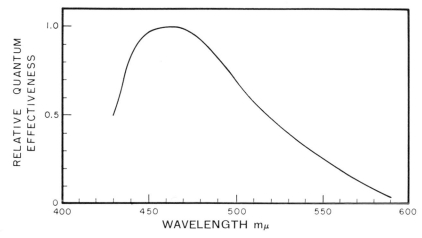

FIG. 5.30. Relative quantum effectiveness of the chromatophores in *Diadema setosum*. Normalized at 468 mμ. (Redrawn from Yoshida, 1956.)

If the pigment is brown or black, the cell is a *melanophore;* yellow chromatophores are *xanthophores* and red chromatophores are *erythrophores*. The coloring agents that have been isolated are melanins, carotenoids, astaxanthin (the pigment giving the boiled lobster its characteristic lobster-red), and guanine. Both nervous and direct light-stimulation effects have been observed. Many organisms exhibit rhythms of pigment dispersion (solar, tidal, or semilunar) as discussed in Section 3. Yoshida (1956) reported for the sea urchin, *Diadema setosum*, the action spectrum for chromatophore spreading shown in Fig. 5.30. The data are redrawn to present relative quantum effectiveness. When a pigment-free area of the chromatophore was illuminated by a minute light spot, the pigment dispersed to cover the illuminated area. Yoshida concludes that photosensitivity resides in the chromatophore itself, probably in the hyaloplasm.

Hogben and Winton (1923) found no evidence to support the possi-

TABLE 5.6
DRUGS AND TREATMENTS FOUND TO HAVE SOME EFFECT ON
CHROMATOPHORE CONCENTRATION OR RELAXATION

Drug or treatment	Darken-ing	Lighten-ing	Species	Reference
ACTH	+			
Acetylcholine[a]		+	*Parasilurus*	Enami (1955)
Acetylcholine	+		*Salmo gairdneri*	Robertson (1951)
α-MSH[b]	+		*Rana temporaria*	Johnsson and Hogberg (1952)
ATP	+		*Rana*	Lerner and Takahashi (1956)
Corticotropin	+		*Rana temporaria*	Johnsson and Hogberg (1952)
Epinephrine	+		*Xanopus laevis*	Burgers (1956)
Epinephrine		+	*Phoxinus*	Abolin (1925)
Hydrocortisone[c]		+	*Rana pipiens*	Wright and Lerner (1960)
Melatonin[d]		+	*Rana pipiens*	Lerner *et al.* (1958)
Norepinephrine	+		*Parasilurus*	Enami (1955)
Potassium chloride	+		*Salmo gairdneri*	Robertson (1951)
Serotonin	+		*Rana pipiens*	Davey (1959)
Thyroxin	+		*Carassius auratus*	J. Müller (1953)
Thyroxin[c]		+	*Rana pipiens*	Wright and Lerner (1960)
Asphyxia	+		*Salmo gairdneri*	Robertson (1951)
High temperature	+		*Salmo gairdneri*	Robertson (1951)
Pressure		+	*Salmo gairdneri*	Robertson (1951)
Hypophysectomy		+	*Rana temporaria*	Hogben and Winton (1923)
Dehydration		+		Fingerman (1963)
Illumination	+			Fingerman (1963)
Constant darkness	24 hour rhythm			Fingerman (1963)

[a] This is unexpected since melanin dispersing fibers are supposed to be cholinergic; can also reverse darkening action of MSH (Wright and Lerner, 1960).

[b] Hormone "intermedin" (melanophore-stimulating hormone), specific requirement for Na^+. Structure: Acetyl-Ser-Tyr-Ser-Met-Glu-His-Phe-Arg-Tyr-Gly-Lys-Pro-Val-NH_2.

[c] Reverses darkening action of MSH.

[d] Prevents darkening by MSH in concentrations of 10^{-12} g/ml presumably secreted by pineal gland.

bility that nerves play a role in controlling the melanophores of *Rana temporaria* and *Rana esculenta*, our friends from the light-induced iris sphincter contraction. The eel, *Anguilla*, also shows chromatophore response except in this case apparently both nerves and hormones are involved. Table 5.6 lists some of the drugs and treatments which have been found to have some effect on chromatophore concentration or relaxation. Again we must conclude that essentially nothing is known of the photochemistry of the direct stimulation by light.

6. Vision

The human eye, including the nervous system for evaluating the visual responses, is by far the most versatile light detector known. It can detect and measure light intensities varying over 13 orders of magnitude. It can see the light from a 6th magnitude star, corresponding to a "candle" at a distance of 13 miles, assuming no intervening absorption in the atmosphere. Based on a statistical analysis of the responses of observers at the threshold of dark adapted vision, a single absorbed photon can stimulate a retinal rod (Hecht *et al.*, 1942), although for "seeing" the minimum number that must be absorbed at the retina in a short flash appears to be about 10 (Pirenne, 1962, p. 131). Because of the tremendous range of light adaptation the eye does not accurately measure absolute intensities. However, at normal visual photopic intensities it can differentiate sources which differ in brightness by as little as 2%. All this is done by an optical system consisting of a light-tight cavity containing a continuously sensitive light detector which extends over the interior surface of the posterior chamber. This retina is an extremely thin partially transparent membrane, at most 200 μ thick. Within this very specialized nerve ending are contained, in a highly organized array, a pigment epithelium, approximately 125 million rods and 6.5 million cones, and approximately 1 million optic nerve fibers. The surface density of rods and cones varies greatly over the retina. For rods the surface density varies linearly with eccentricity* from a value of about 40,000/mm² at 80° eccentricity to a maximum of 160,000/mm² at 20° eccentricity, dropping quickly to zero at 0.5°. For cones the concentration outside of 10° eccentricity is about 5000/mm², rising sharply within 0.5° eccentricity to 147,000/mm². According to Schultze (1866), the distributions of rods and cones in fresh preparations follows a mosaic pattern very closely, much more so than is observed in fixed and stained preparations. The type of mosaic pattern observed by Schultze in the periphery of the human retina is shown in Fig.

---

* In physiological optics it is common to use values of eccentricity in degrees, i.e., the angular distance in the external field measured from a point which forms its image in the center of the fovea. A visual angle of 1° of arc corresponds to a distance on the retina of approximately 4.85 μ. Thus within an eccentricity of 0.5°, corresponding to the rod-free area of the retina, we have the entire color- and sharp-vision region of the human eye, containing essentially all of the cones.

5.31. A schematic cross section of the human retina is also shown in Fig. 5.31.

The eye of the horseshoe crab is an amazingly simple system. It is a compound eye composed of individual ommatidia, similar to those of the insects. However, each ommatidium has its own separate nerve fiber proceeding directly to the brain. In the classic studies of H. K. Hartline and

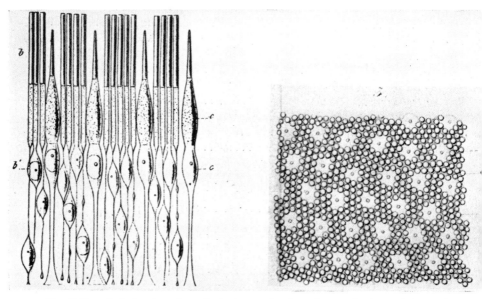

FIG. 5.31. *Left:* Section through the layer of rods and cones in the periphery of the human eye; (b) rods; (c) cones. It is probable that the cones actually have fine tips, not shown in this drawing, which reach up to the level of the ends of the outer segments of the rods. Magnification ×500. From Schultze (1866) (figure reproduced in Pirenne, 1948). *Right:* The mosaic of rods and cones in the periphery of the human retina, as seen under the microscope in a fresh preparation; (b) rods; (c) cones. Magnification ×500. From Schultze (1866) (figure reproduced in Pirenne, 1948). The light entering the eye through the pupil reaches the receptors in the sectional view from below.

his associates the electrical responses due to a nerve fiber from a single ommatidium were observed. It was found that the magnitudes of the nerve impulses from a single nerve fiber were essentially the same. However, the frequency of discharge was dependent upon the light intensity. More specifically, the frequency is dependent upon the number of light quanta absorbed within an integrating time of a fraction of a second. For example, a weak flash lasting 0.1 sec produced the same frequency response as a flash 1000 times as intense with a duration of only 0.0001 sec. By recording, therefore, the reciprocal of the intensity of spectral light required to produce identical responses or frequencies of nerve discharges, one can generate a true action spectrum, which for *limulus*, peaks at approximately 527 mμ. The light sensitive ommatidium can be stimulated by just a few quanta, like the human rods.

The eyes of chickens and pigeons contain a large predominance of cones over rods so that these animals are almost completely night-blind. In the cones of these animals there are differently colored oil globules at the joint between the inner and outer segments. These are thought to act as individual color filters for the incident light and, in a manner similar to the autochrome process of color photography, may provide the basis of color differentiation in vision. The three pigments in the oil globules have been isolated and identified as carotene (greenish-yellow), xanthophyll (golden-yellow), and astaxanthin (red).

Color vision in pigeons has been demonstrated in a remarkable series of experiments on trained pigeons by Guttman and Kalish (1958). The birds can be taught to peck at a disc illuminated with monochromatic light and this conditioned color response is then tested as a function of wavelength of stimulating light. The frequency of pecking plotted as a function of wavelength is then remarkably symmetric about the training wavelength, independent of the intrinsic spectral sensitivity curve for the pigeon. It was shown that the pigeon can be trained to wavelengths differing by only 50 A.

The change from cone vision to rod vision, like that from slow to fast film in the camera, involves a change from the fine-grained cone mosaic in the fovea to the coarser-grained rod mosaic in the remainder of the retina. It is not that the cones are smaller in diameter than the rods. It is apparently due to the fact that each cone connects to its own optic fiber while there are not sufficient nerve fibers to provide individually for the rods. Thus an area involving many rods may affect one nerve fiber, giving, in the photographic analogy, a coarse-grained or poor resolution image. This, however, rather than being a limitation is an advantage, for the eye is one device in dim light and quite another in bright light. At low intensities the eye is primarily a device for detecting light, not pattern.

Hecht (1945) proposed that there is a spontaneous dark noise in the rods of the retina. This is analogous to the dark noise observed in electron multiplier phototubes owing to thermal emission of electrons from the photocathode surface. In the latter, owing to the lowering of the photoelectric work function of the photocathode so that red light is capable of producing a photoelectric effect with high efficiency, some of the electrons from the photocathode itself can escape. In the phototube we can cool the photocathode to increase the signal to noise ratio and, in fact, this is normally done for infrared photon detectors. In the eye this is not as easily accomplished. It is for this reason that the absorption of a single quantum is not sufficient for vision. Its effect can be confused with the spontaneous excitations. This could also serve to explain the observations reported by Helmholtz (1896) that the after images of white light were indeed "blacker" than the field "seen" in absolute darkness. This dark noise

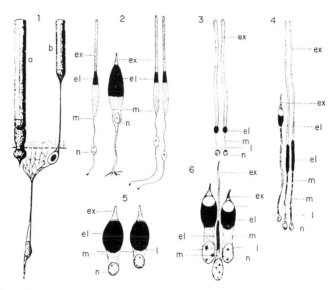

Fig. 5.32.   Drawings of rods and cones from various species. From Verrier (1935). (ex) External or outer segment; (el) ellipsoid or inner segment; (m) myoid; (l) outer limiting membrane; (n) nucleus.

1. Rods of the frog *Rana pipiens* (*a*), common or rhodopsin rod; (*b*) "green rod"; (*c*) outer limiting membrane. 2. *Left:* rod and cone, respectively, from periphery of human retina. *Right:* cones from fovea of human retina. 3. Rods of the conger eel (*Conger vulgaris*). 4. Cone and rods of a bird (*Tyto alba*). 5. Cones of a snake (*Tropidonotus piscator*). 6. Cones and rod of a pig (*Sus scrofa*).

theory postulates that in the living rods there occur spontaneous excitations of rhodopsin molecules, causing the same nervous response as produced by actual light quanta. This spontaneous firing of impulses by ganglion cells has been observed in electrophysiological preparations of animal retinas. Thus perception of light is a *comparison* between the total of excitations that occur in the *presence* of light and those that occur in the *absence* of light.

Visual receptors vary quite markedly in size and shape. In the human retina the foveal cones are very slender and elongated while the cones in the peripheral portions look like cones, from whence comes the name.

Drawings of receptors from various species are shown in Fig. 5.32. A diagrammatic cross section of a mammalian rod based on electron microscopy observations is shown in Fig. 5.33. The inner segment below the cilium (CC in Fig. 5.33) is very densely packed with mitochondria. On exposure to light the retinene formed by the bleaching of the visual pigment is rapidly converted wholly within the retina to vitamin A, which

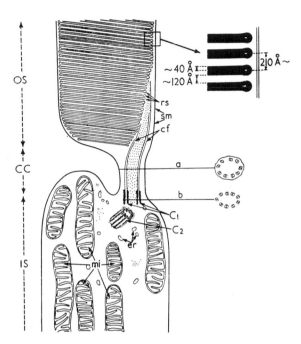

FIG. 5.33. Diagrammatic cross section of a mammalian rod based on electron microscopy observations. The extents of the outer segments (OS), the connecting cilium (CC), and the inner segments (IS) are indicated on the left of the section. C_1 and C_2 are the two centrioles; rs, the rod sacs; cf, the ciliary filaments; sm, the surface membrane; mi, mitochondria; and er, endoplasmic reticulum. To the right are shown cross sections through (a) the connecting cilium, and (b) the centriole C_1 (De Robertis, 1960; reprinted by permission of the Rockefeller Institute Press, from *J. Gen. Physiol.*).

then migrates to the pigment epithelium. During dark adaptation these processes are reversed. In the completely dark adapted retina there is no vitamin A in the pigment epithelium. Going one step further in the resolution of structures, Fig. 5.34 is a schematic model of the rod outer segment (Wolken, 1961) showing one retinene molecule per protein (opsin) molecule and the possible mode of double-layer stacking of the rhodopsin molecules forming the electron microscope-resolvable lipoprotein layers.

The structures and relative sizes of the retinene molecule, and the β-carotene molecule are given in Fig. 5.35. The prosthetic group (retinene) of the visual pigment molecules faces outward into the aqueous phase, making it easy for the subsequently reduced retinene (vitamin A) to migrate to and from the pigment epithelium.

The eye can adapt rapidly to light. It takes a much longer time to adapt to darkness. One of the characteristic features of the dark adaptation (log threshold intensity versus time) curve is the sharp break indicating the transfer of function from cones at high luminance to rods at low

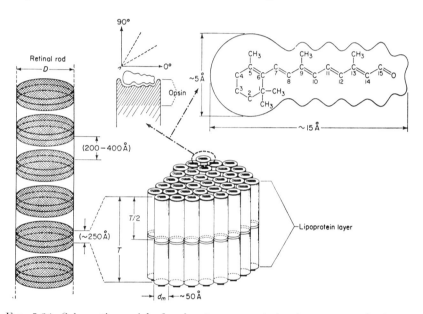

Fig. 5.34. Schematic model of rod outer segment showing one prosthetic group ("retinene") per visual pigment protein molecule ("opsin") and the possible mode of packing of the molecule in the sacs (Wolken, 1961).

luminance. The course of dark adaptation measured in the near periphery of the retina is shown in Fig. 5.36. The abscissa gives the time measured in minutes from the moment (time zero) when the adapting light was switched off. The logarithmic ordinate scale gives the threshold intensity, that is, the physical light intensity just sufficient for the subject to see the test field.

In the experiments of Fig. 5.36 the eye of the subject had been previously light-adapted during 3 min (from time -3 to time zero) to a white field subtending 35° at the eye, the luminance of which was 1550 millilamberts (1 mL = 3.183 candela/m^2). The center of this field was at an eccentricity of 7° on the temporal side of a fixation mark. A red fixation point adjustable in intensity was used for the threshold measurements after the adapting field had been switched off. The subject fixed his gaze on this point while the test field was exposed in flashes of 0.2-sec duration. The diameter of the test field subtended an angle of 3° at the

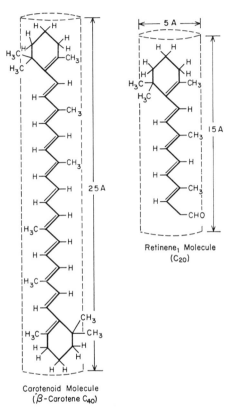

FIG. 5.35. Structures and relative sizes of the retinene₁ molecule and the β-carotene molecule.

eye and its center was placed 7° temporally to the fixation point. The images of adapting field and test field were thus both formed around the same retinal position. Such an arrangement ensures a good control of the conditions of light adaptation—it is difficult to light-adapt evenly the whole area of the retina. The test light consisted of a band of short wavelengths below 460 mμ.

Under these conditions the field appears definitely violet or blue to the subject for the first five points of the curve, that is, up to about 7 min.

For the later measurements it appears almost colorless and its outlines look much more blurred than previously. This is evidence that, at the beginning, the measurements refer to cone function and later involve rod function.

At first the threshold drops rapidly; then, from about 5 to 7 min it levels off nearly or completely to a plateau. This first part of the curve is the cone branch. The threshold value at the end of it is called the cone threshold. After the cone-rod transition time ($t \approx 7$ min in Fig. 5.36), that

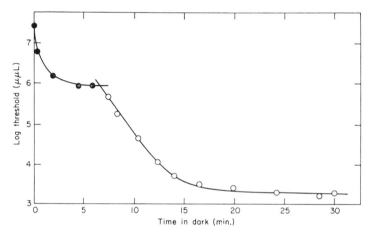

Fig. 5.36.   The time course of human dark adaptation following a 3-min period of adaptation to white light of luminance 1550 millilamberts (1 mL = 3.183 candelas/m²). Test field 3° in diameter placed temporally to fixation point at an eccentricity of 7° and lit in 0.2-sec flashes of light from the violet end of the spectrum. Natural pupil (see text for further details). The experimental points represent single measurements on one subject. For the first five points the light appears blue or violet even at threshold, whereas for the remaining points threshold flashes appear colorless or almost colorless. The threshold luminances are expressed in micro-microlamberts ($1 \mu\mu L = 10^{-12}L$) (Hecht and Shlaer, 1938).

is, when the appearance has changed over from violet to grayish, there follows another less rapid drop; from about 15 min onward the change in threshold is much slower, and after 25 min a further stay in the dark leads to little further change. This second part of the curve is the rod branch. The threshold value at the end of it is called the final rod threshold. Under the present conditions the total threshold change between time $t = 0$ and $t = 30$ min covers about four log units, i.e., the threshold is about 10,000 times smaller at the end than at the beginning of dark adaptation; the sensitivity, defined as 1/(threshold luminance), is thus about 10,000 times higher at the end than at the beginning. The cone threshold, after about 7 min in the dark, is about 2.7 log units, i.e., about 500 times, higher than the final rod threshold. The relative spectral sensitivities of the dark adapted fovea and the dark-adapted peripheral retina of the human eye are shown in Fig. 5.37. These data are taken over

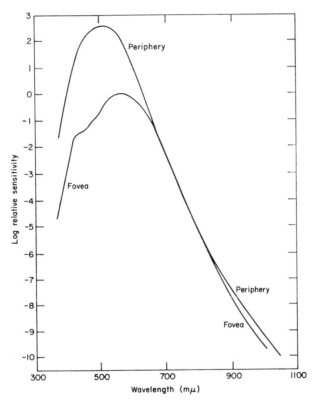

FIG. 5.37.   Relative spectral sensitivities of the dark-adapted fovea and the peripheral retina. The spacing of these curves is appropriate for 1° test fields fixated within the fovea or placed 8° above the fovea, and exposed for 1 sec. Below 750 mμ, the foveal curve is a composite function based on the original data of a number of workers; the peripheral function is from Wald (1945). To these the data for the far-red and infrared have been joined (Griffin *et al.*, 1947).

a tremendously larger range than the ordinary absorption spectrum of rhodopsin, but there is no reason to believe that the shape of the absorption curve of rhodopsin does not follow this curve, even out to 1 μ where the number of quanta required for threshold sensitivity is about 6×10^{12} greater than at 507 mμ. The agreement between the observed rod sensitivity and the absorption curve for extracted human rhodopsin is shown in Fig. 5.38. This is another indication that most of the selective absorption by ocular media occurs in the lens. The large depression and minor inflections in the region 440–520 mμ of the fovea sensitivity curve of Fig. 5.37 are due to absorption by the xanthophyll pigment in the *macula lutea* or yellow spot.

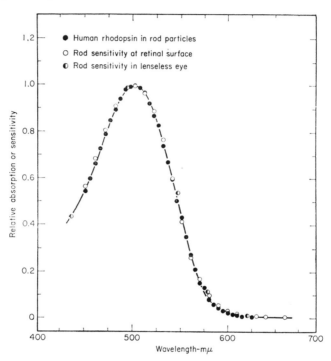

FIG. 5.38.   Absorption spectrum of human rhodopsin and the scotopic spectral sensitivity. The absorption spectrum of rhodopsin, measured in human rod particles, is compared (a) with the scotopic sensitivity function, expressed on a quantum basis and corrected for ocular transmission so as to represent the sensitivity at the retinal surface, and (b) with the scotopic sensitivity, on a quantum basis, of the lensless human eye. In the latter, the principal colored structure of the eye, the yellow lens, having been removed in the operation for cataract, the spectral sensitivity becomes very close to the intrinsic sensitivity of the living retina (Wald and Brown, 1958; reprinted by permission of AAAS and the authors).

Rushton (1961) has pointed out that one must be careful in the photochemical theory of vision to distinguish between increment threshold determinations and visual pigment concentrations and the absolute threshold determinations and visual pigment concentrations; the former is essentially independent of the amount of pigment bleached, following the simple Weber-Fechner relation, while the latter depends directly upon the bleaching. In experiments on a rod monochromat with no measurable foveal pigments Rushton and co-workers were able to demonstrate that rod function is present at any given level at the same time whether there are cones present or not. In other words, the rod portion of Fig. 5.36 continues in the absence of foveal cones. A completely linear relation between pigment-bleached and log-absolute threshold sensitivity was found. Every quantum still produces an effect. However, when as a result of partial light adaptation the threshold rises by a factor of 100, the "noise" of the system has been increased to such an extent that 100 times as many incident quanta are required to see the "signal" above this noise.

Although the histology and neurophysiology of the retina and the optic nerves are important in our understanding of the mechanism of vision and conversely must figure in any proposed mechanism of vision, it will be impossible in this summary to describe these in any detail. The reader is referred to the excellent summaries by Polyak (1957), Granit (1947, 1962), Müller-Limmroth (1959), Brindley (1960), Galifret (1960), and Jung and Kornhuber (1961).

By far the largest number of fibers visible in the optic nerve have diameters below 1–2 μ. There is some evidence that fiber size is a guide to function and that the larger the fiber size the greater the conduction velocity of the impulses. For example, in the electrical stimulation of the optic nerve of the cat, Chang (1956) found anatomical evidence for fiber diameters in the 1–2 μ, 4–5 μ, and 9–10 μ ranges, corresponding to conduction velocities of the volleys of impulses of 17, 30, and 70 m/sec, respectively. Light stimuli produce spikes which are much less synchronous than those set up by electrical shocks, making the situation much more complex in terms of the interaction at the large ganglion cells of the rapidly conducted spikes which set up evoked potentials and the asymmetric (in time) arrival of the more slowly conducted spikes. It has been suggested that the wavelength of the stimulating light may be connected with conduction velocity (Chang, 1952; Lennox, 1958a,b). Lennox found in the cat optic nerve that blue responses were preferentially signaled by slow fibers and red responses by fibers of higher conduction velocities. These results must form a consistent part of any mechanism of color vision and imply a high degree of complexity to the cone structures as well as an integration phenomenon at the ganglia.

One of the major observations in the electrophysiology of vision made in the last century was that an electrical impulse from the retina was produced twice, once at onset of illumination and again at the cessation of illumination. Thus, any theory of vision must make provision for the "off" effect as well as the "on" effect. There appear to be four distinct portions to the electroretinograms (ERG) of rods and cones, the relative sizes of which vary both with the nature of the receptor and the light stimulus. These are shown in Fig. 5.39. At the onset of light there is a small negative *a* wave followed by a larger positive *b* wave ("on" effect). The *b* wave sometimes drops below the base line just before the *c* wave or secondary rise begins. At the cessation of light there is a sharp *d* wave ("off" effect). The *a* and *b* waves are never absent. The *c* waves appear to be associated mainly with rod vision and even when present initially disappear with light adaptation, where presumably rod sensitivity is replaced by cone response. The *d* wave conversely appears to be associated with cone eyes. Both can be affected by drugs. However, there is some question as to whether or not some of the observed electrophysiological effects are not the result of algebraic addition phenomena. The field is exceedingly complex both technically and in the interpretation of the

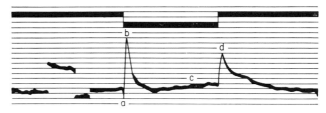

Fig. 5.39.   Mammalian cone retina of the squirrel, *Sciurus carolinensis leucotis*, with marked *a* wave and narrow pointed *b* and *d* waves. Calibration: 0.5 mv; time, 1 sec (Arden and Tansley, 1955).

myriad of observed phenomena relating to mutual inhibition and localization effects, flicker, adaptation, etc.

It is possible to exchange instantaneously two "white" lights *without initiating a new electroretinogram response*, provided the mechanical switching equipment operates with sufficient speed and accuracy. However, no such exchange can be made between a red and, say, a blue stimulus, regardless of their relative intensities in light-adapted turtle, frog, and human eyes. This clearly indicates that there are different types of cones contributing to the ERG. It has been found in some cases, such as the pure cone retina of the grey squirrel, *Sciurus carolinensis leucotis*, that the action spectrum for ERG signals is singularly narrow, with a much smaller half-width (65 mμ) than the broad-band absorption curves observed for the visual pigments (120 mμ) (Arden and Tansley, 1955). These narrow-band action spectra are termed *modulator* curves, and it was found that the messages delivered to the optic nerve fibers were organized into broad-band *dominator* curves and narrow-band *modulator* curves. It is the *dominator* type of curve that is usually reported to show that the ERG scotopic and photopic action spectra coincide with the absorption spectra of the visual pigments involved. The *dominators* are responsible for the Purkinje effect (Pirenne, 1962). The present ideas are that the narrow-band *modulator* action spectra are the result of neural interaction. All this suggests an interaction which, as a consequence of overlap in sensitivity between different photopigments, sharpens up an action spectrum without too much influence on the position of the actual maximum. This would be analogous with auditory tonal representation on the basilar membrane which is sharpened up by neural interaction. Granit (1962) states that "photochemical explanations need not be excluded merely because photochemists so far only have found broad-band absorption curves and like to think that these are the only ones present."

6.1. UNITS AND SENSITIVITY OF THE EYE TO LIGHT

It would be well here to correlate the various units of intensity and illumination that have been used in measurements in vision, since again it would be preferable to specify quanta per second per square centimeter per unit wavelength interval. However, a number of different systems have been used, most of them based on photometry. While they are in one sense simpler to use in the physiology laboratory where "white light" sources are generally available, it is sometimes extremely difficult to correlate the photometric data with the quantum events occurring in photochemistry. Examples of this have been given in Chapter 1. Normally all photometric units of intensity and illumination are in *photopic* units, i.e., the photopic spectral efficiency curve of the Standard Eye is used in the integration over the radiant emission of the blackbody radiation. However, for rod vision the spectral sensitivity curve of the eye is shifted to peak at 5070 A. The scotopic luminous efficiency curve is quite different from the photopic luminous efficiency curve (see Fig. 1.2).

An extended source of *luminance* 0.75×10^{-6} candela/m² or 0.235×10^{-6} millilambert corresponds to the absolute threshold of scotopic or rod vision in white light of color temperature 2400°K (Pirenne, 1962, p. 125). This threshold represents the 50% frequency of seeing. The most sensitive subject found had a threshold of 0.2×10^{-6} candela/m². In the photometric system the brightness of a luminous surface is defined for a perfectly diffusing surface, obeying Lambert's cosine law of emission or reflection. The unit is the *lambert* and is the brightness of a perfectly diffusing surface radiating or reflecting 1 lumen/cm². The brightness in candelas per square centimeter is reduced to lamberts by multiplying by π. Thus, *luminance* as used corresponds to brightness and conversion of candelas per square meter to lumens per square centimeter (see Table 1.2) gives 0.235×10^{-6} millilambert. As a comparison, the darkest night sky on the same scale has a brightness of 4.4×10^{-6} millilambert. The scotopic mechanical equivalent of light is 1745 lumens/watt. This factor already implies an integration over the scotopic luminous efficiency of the eye.

In Chapter 1 we derived the photopic mechanical equivalent of light, 0.00147 watts/lumen. If we substitute in the integration the scotopic efficiency curve we can obtain the scotopic mechanical equivalent of light, 0.000573 watts/scotopic lumen. In reciprocal numbers these are 680 lumens/watt and 1745 scotopic lumens/watt, respectively. The two sets of units are joined by the definition of the blackbody radiator at 2042°K. In each system the brightness of the blackbody cavity is 60 photopic or 60 scotopic candelas/cm². A source of brightness 0.75×10^{-6} candelas/m² of scotopic white light corresponds to

$$0.75 \times 10^{-6} \frac{\text{candela}}{\text{m}^2} \times 10^{-4} \frac{\text{m}^2}{\text{cm}^2} \times \frac{\text{lumen}}{\text{steradian-candela}} \times \frac{\text{watts}}{1745 \text{ lumens}}$$

$$= 0.43 \times 10^{-13} \text{ watt/cm}^2\text{-steradian} \quad (5.10)$$

Now we must go one step further to determine how much of this light actually strikes the retina. The posterior nodal distance of the eye is 1.668 cm. Thus the illumination, D, of the retina by a field of luminance, B, substending a wide angle at the eye is given by

$$D = \frac{1}{(1.668)^2} T_\lambda A B = 0.36 T_\lambda A B \quad (5.11)$$

where T_λ is the fractional transmission of the eye media and A is the area of the pupil. At 5070 A, $T_\lambda \sim 0.5$ and for the dark-adapted eye $A \sim 0.5$ cm². Therefore the illumination produced on the retina by the image of such a field is

$$D = 0.36 T_\lambda \times A \times 0.43 \times 10^{-13} \text{ watt/cm}^2 = 0.0387 \times 10^{-13} \text{ watt/cm}^2 \quad (5.12)$$

The energy flux of 1 photon/sec is given by

$$\text{watts} = \frac{1987}{\lambda[\text{Angstroms}]} \times 10^{-18} \quad (5.13)$$

so that at 5070 A the minimum number of photons per square centimeter per second that is detectable by the eye is

$$\frac{5070}{1987} \times \frac{0.0387 \times 10^{-13}}{10^{-18}} \simeq 10^4 \quad (5.14)$$

The next step is to determine how many of these photons are absorbed by the rods themselves. Rushton (1956) has reported that the maximum optical density of rhodopsin in living human rods is 0.15, corresponding to an absorption of 29% of all light of 5070 A falling on the retina. Therefore the total number of quanta absorbed per second, assuming a retinal area of 1.5 cm² is

$$\text{number of photons absorbed by rods per second} = 4300 \qquad (5.15)$$

From our previous figures on the concentration of rods in the retina, assuming about 13.4×10^6 rods/cm², we see that *on the average* during each second, only *one out of 4700 rods absorbs a quantum of light.*

It would be instructive to compare these threshold values of rod sensitivity with astronomical data on the faintest visible stars. In astronomy the brightness of stars as "seen" by any receptor is expressed as a stellar magnitude, m, related to the *irradiance* in watts per square centimeter received from the star. Visible stars are grouped into magnitudes ranging from 1 to 6. Those of 1st magnitude are the brightest. Those of 6th magnitude are just visible to the naked eye. About 1830 Herschel found that a 1st magnitude star gave 100 times as much light as a 6th magnitude star. As a basis for comparison of threshold visual intensities with stellar magnitudes we can consider the following: the stellar magnitude, m, is defined by

$$m = -2.5 \log_{10} I/I_0 \qquad (5.16)$$

The apparent magnitude of a candle at 1 meter is -14.2 (*Smithsonian Physical Tables*, 1959, Table 827). The illumination at 1 meter from 1 candle is 1 lumen/m². Solving Eq. (5.16) for I_0 in units of illumination [lumen/m²] we obtain

$$m = -2.5 \log_{10} \frac{I}{0.12 \times 10^{-5}} \qquad (5\ 17)$$

Thus for a 6th magnitude star the illumination at the pupil will be

$$I = \frac{0.12 \times 10^{-5}}{0.602 \times 10^3} = 2 \times 10^{-9} \text{ lumen/m}^2 \qquad (5.18)$$

For an effectively point source emitting white light and observed in extrafoveal vision the threshold illumination for seeing at the pupil has been calculated to be 4×10^{-9} lumen/m² and is reported to vary between 0.8×10^{-9} and 11×10^{-9} lumen/m², depending on the subject (Marriott *et al.*, 1959). These data are in excellent agreement with the stellar magnitude observations. At 5070 A 10^{-9} lumen/m² is equivalent to 146 quanta/sec-cm².

In other cases in the literature the intensities of light stimuli are sometimes expressed in terms of energy flux originating from an area of a
diffuse source subtending a solid angle of 1 square degree at the eye
(1 square degree = 3.05×10^{-4} steradian). Sometimes the *retinal illumination* is given in *trolands* (an extended diffuse source of brightness 1
candela/m², as seen through a pupil area of 1 mm² produces a retinal
illumination of 1 troland). In still other cases the *illumination at the pupil*
is given in lumens per square meter or lux. For rod vision the units must
take into account the scotopic efficiency curve. Thus we can have photopic flux (lumens) or scotopic flux (scotopic lumens) and either photopic
or scotopic trolands. Usually if the noun is not preceded by the adjective
scotopic, it is understood to be *photopic*.

It is not immediately obvious as to how these other methods of defining
light stimuli relate to one another or to the number of quanta in any
given wavelength interval. The dark-adapted pupillary area is
approximately 50 mm². Thus a diffuse source of 1 candela/m² will produce through the dark-adapted pupil a retinal illumination of 50 scotopic
trolands. From Eq. (5.11) we see that this also corresponds to a retinal
illumination of $D = 0.36 T_\lambda A B$. Since $T_\lambda \sim 0.5$, $A = 0.5$ cm² and
1 candela/m² is equivalent to $\pi \times 10^{-4}$ scotopic lumens/cm², the retinal
illumination produced by a diffuse source of 1 candela/m² is

$$D = 0.36 \times 0.5 \times 0.5 \times \pi \times 10^{-4} = 0.283 \times 10^{-4} \text{ scotopic lumens/cm}^2$$
$$(5.19)$$

Therefore,

$$1 \text{ scotopic troland} = \frac{0.283 \times 10^{-4}}{50}$$
$$= 0.566 \times 10^{-6} \text{ scotopic lumens/cm}^2 \quad (5.20)$$

Equation (5.20) gives the relation between scotopic trolands and scotopic lumens per square centimeter. The interconversion will be complete
if we can now relate scotopic trolands to energy flux per square degree
and to photons per second-square degree. A brightness of 1 candela/m²
corresponds to the emission of 10^{-4} lumen/cm² per steradian. One square
degree equals 3.05×10^{-4} steradian and 1 scotopic lumen equals $1/1745$
watts of 5070 Å radiation. Therefore, since 1 candela/m² gives 50 scotopic trolands

$$1 \text{ scotopic troland} = \frac{1 \times 10^{-4}}{50} \times 3.05 \times 10^{-4} \times \frac{1}{1745}$$
$$= 3.5 \times 10^{-13} \text{ watts/sq. deg} \quad (5.21)$$

incident at the pupil. In terms of quanta per second of wavelength 5070 Å we have

1 scotopic troland of retinal illumination is produced by:

$$3.5 \times 10^{-13} \times \frac{5070}{1987} \times 10^{18} = 8.9 \times 10^5 \text{ photons/sec-sq. deg} \quad (5.22)$$

incident at the pupil.

For flashes lasting of the order of 0.1 sec and below the threshold for scotopic seeing remains constant and independent of flash duration. As quoted in Pirenne (1962, p. 131), between 2.1 and 5.7×10^{-10} erg of 5100 A light gave a 60% probability of seeing. This corresponds to between 54 and 148 quanta incident on the cornea and agrees quite well with the stellar magnitude calculations. Between 8 and 21 quanta are absorbed by the rods directly.

In the case of foveal vision the limit of perception of a constant intensity point source of light corresponds to an illumination at the cornea of 10^{-7} photopic lumen/m^2 or about 41,000 quanta/sec-cm^2 incident on the cornea. For brief exposures and small field sizes Miller (1959) found a threshold for 5500 A light in foveal vision of 2440 quanta. This is approximately a factor of 20 lower than for a steady source in foveal vision but still a factor of 20 higher than the threshold for extra-foveal or rod vision. In the latter case the steady point source and the flashing point source have approximately the same threshold.

6.2. STRUCTURAL FEATURES

There are two types of defects common to lens systems. One is spherical aberration due to the fact that the lenses are not ideally thin. The marginal portions of the lens focus rays of light from a point source in different planes than the central region, producing a blur circle. Around the middle of the 18th century, J. Dolland described a method of correcting for spherical aberration by the use of converging and diverging lenses, analogous to the correction for chromatic aberration, and this is the technique used in high resolution photography. However, Descartes in the 17th century had shown that spherical aberration could also be corrected for by grinding lens surfaces in other than sections of spheres. The eye follows Descartes' ideas more closely. The cornea has a flatter curvature at its margin than at its center and the lens is denser (has a higher index of refraction) at its center than at its periphery. In both cases these irregularities compensate for the tendency of an ordinary spherical lens to refract more strongly at its margin than at its center.

Since the protein structures of the various transparent membranes and optics in the eye cannot be too different in composition, it would be rather difficult for the eye to correct for chromatic aberration in the same way as is done in glass optics, i.e., by the use of coverging and diverging lenses of different materials and different indices of refraction such as crown and flint glass. The next best approach involves a sharpening of the image at the expense of light intensity. Toward the center of the retina whose area is about 1.5 cm² and which sweeps through a visual angle of 240° there is a small shallow pit whose diameter approaches 5°. In the center of this depression the *fovea* is defined physiologically as the rod-free area with a diameter of roughly 1°. In a fixed human retina the first rods seen in the preparation including the rod-free fovea are at radial distances from the center of the fovea of about 0.13 mm, giving an area of the order of 5×10^{-4} cm². This corresponds very closely with the area of 6.6×10^{-4} cm² calculated from the 1° diameter physiological measurement, using a posterial nodal distance of 1.668 cm. It is with this tiny patch of retinal surface, smaller than the head of a pin, that all of the high resolution, detailed seeing is done. As will be discussed later this is also the color vision center of the eye. Only three-thousandths of 1% of the light-sensitive extension of the brain is specialized for what is such a major fraction of our civilized existence, i.e., reading, measuring, creating in the arts, all depending more or less on the detailed vision of the fovea. Lower mammals lack a rod-free area of physiological fovea. It is best developed in primates and in birds.

The severed optic nerve in the adult salamander can regenerate and re-establish the complex network of nerve-fiber connections between the eye and the brain. This has since been observed for fishes, frogs, and toads, but not for mammals. The question naturally arises as to how this mass of tens of thousands of fibers, winding back into the brain, could establish the orderly pattern of vision. Is it a matter of a "learning" process in which a superabundance of fibers is produced and only those linkages corresponding to "experience" are retained, or is each fiber specific and "coded" to a specific nerve cell in the visual area of the brain? Sperry (1956) has reported the results of some ingenious experiments on newts which indicate that the more plausible "learning" mechanism is not operative but that the regeneration is due to information already contained in the tissues themselves. In newts whose optic nerves were severed the eyeballs were also rotated surgically in their respective sockets by 180°. During the period of nerve regeneration the animals were blind. At the end of 25 to 30 days the critical visual responses to stimuli were systematically reversed as though only the eyeball had been rotated by 180°, i.e., an object above the newt was "seen" below and one to the rear was "seen" in front. This reversed vision remained entirely uncorrected by experience. It is as though each fiber were endowed with some quality which differentiates it from its neighbor and which marks it as having originated from its particular spot on the retinal field. Thus if the eyeball (and consequently the retina) is inverted (rotation through 180°), vision will be inverted. The type of coding, information retrieval, and feedback network, both physiological and biochemical, that must be involved in this operation staggers the imagination. When we realize that we do not yet comprehend even the primary photochemical event in vision (except that it is absorption of light by a pigment system), we can get some feeling for the enormity of our ignorance in this one specialized area.

In the relatively bright light required for highest visual acuity the iris diaphragm can stop down so that only the central portion of the lens system is used. This reduces the chromatic aberration somewhat. If we examine the dispersion curve for the protein of which the lens is constructed we find that in the red and yellow portions of the spectrum it is a slowly varying function, obeying the Drude equation. However, there is a strong absorption band in the near ultraviolet, extending into the blue, so that the lens actually appears yellow. Because of this absorption the dispersion in the visible blue region of the spectrum is high. (We cannot use the argument that the lens itself by absorbing blue light and near ultraviolet light acts to reduce components in the white light spectrum which would produce greater chromatic aberration, since if the lens were colorless the dispersion in the blue would automatically be much smaller.) The absorption by the lens is characteristic of its structure. The retina itself is sensitive to ultraviolet, and aphakic persons equipped with glass lenses can read fine print in UV light, completely invisible to ordinary eyes.

The threshold intensity on changing to cone or foveal vision from rod or peripheral vision must increase by about three orders of magnitude. However, *in the red region the foveal and peripheral visual sensitivities are essentially the same.* This is shown strikingly in Fig. 5.37 where the logarithms of the relative sensitivities of the dark-adapted fovea and peripheral retina are plotted as functions of wavelength. It is as though a strong blue-absorbing filter has been used to reduce the blue sensitivity of the entire system and thus minimize the greater contribution to chromatic aberration of blue light.

In addition, only in monkeys, apes, and man of all the mammals is the foveal depression colored yellow. This patch, called the *macula lutea* or yellow spot, contains a green-leaf carotenoid pigment, xanthophyll (Wald, 1949, 1950, 1959) and acts as a further blue-absorbing filter, absorbing a major fraction of blue light incident on the fovea. This further reduces chromatic aberration. Recapitulating, the visual organ, presumably not being able to make compound lenses too different in indices of refraction,

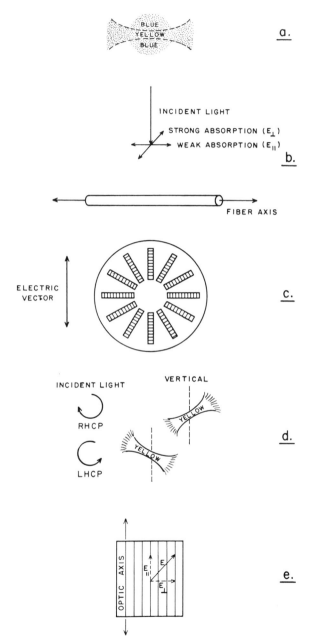

FIG. 5.40. Drawings illustrating various aspects of the Haidinger Brush phenomenon. (a) Appearance of the Haidinger brush when the electric vector of the light incident on the retina is vertical. The region labeled yellow varies in intensity

has settled for (a) a reduction in lens stop, (b) an approximately 1000-fold reduction in blue sensitivity relative to red sensitivity in cones relative to rods, and (c) the macula lutea blue absorption filter. The macula lutea is involved in the Haidinger Brush Phenomenon.

6.3. THE HAIDINGER BRUSH PHENOMENON

This fascinating and easily observable phenomenon, first reported by Haidinger (1844), is a beautiful example of how we can infer certain structural properties of the visual system using the eye itself as a detector. When an observer stares at a clear white field through a polarizer such as a Nicol prism or a sheet of Polaroid film, a faint "Polarisationsbuschel" is seen at the fixation point, consisting of a double-ended, yellowish, brush-like figure with the intervening areas appearing bluish (Fig. 5.40a). If the electric vector of the light is vertical, the yellow brush appears horizontal with the blue areas above and below the brush. The image fades out in a few seconds if there is no motion on the retina, but this is a physiological effect. As the polarizer is rotated, the yellow brush and its blue areas remain visible and rotate directly with the polarizer, so that when the electric vector of the light is horizontal the yellow brush is now vertical and the blue areas are to the left and right of the brush. However, at angles other than the vertical and horizontal the yellow brush appears fainter and in some cases at ± 45° to the vertical it seems to fade out almost completely. This can easily be seen by looking with one eye through a polarizer at a white cloud or the blue sky and rotating the polarizer slowly. In the former case the blue areas appear light blue and in the latter they are a deeper shade, but this is due primarily to color saturation. In fact, with a little practice the Haidinger Brush can be seen without any polarizer. If one looks at the blue sky at 90° from the direct sun, the yellow brush is observed in a horizontal position. In this case there is a maximum of linearly polarized scattered sunlight with the electric vector of the scattered light in a vertical plane. The effect is usually more easily seen with one eye than the other, and some persons

of color with different observers but is usually a straw yellow. The blue area will vary in extent and also in shade. (b) Fiber axis of radial nerve fiber. (c) Schematic of distribution of radial nerve fibers according to Helmholtz, producing maximum blue absorption at 3 o'clock and 9 o'clock and minimum absorption at 6 o'clock and 12 o'clock. The short lines perpendicular to the fibers are presumed to represent the planes of maximum polarizability of the blue-absorbing pigment molecules. (d) Appearance of Haidinger Brushes when right-hand circularly polarized light and left-hand circularly polarized light respectively are incident on the retina. (e) Vector diagram illustrating the resolution of plane polarized light, represented by the E vector, into components perpendicular and parallel to the optic axis of a birefringent structure.

cannot observe these brushes, possibly owing to the absence of yellow pigment in the foveal depression. Helmholtz (1896) states that when Haidinger first reported this phenomenon he, Helmholtz, was unable to see it. However, 12 years later he was able to see it quite clearly. Stokes (1883) reported that the brushes are seen only in the presence of blue light and are not observed for green, yellow, or red light. The explanation given by Helmholtz was based on the assumption that the radial nerve fibers of H. Muller, observed in the macula lutea or yellow spot, were doubly refracting and blue-dichroic. He assumed that the optic axis was along the fiber axis and that strong blue absorption occurred when the electric vector of light passing perpendicular to the fiber axis was itself perpendicular to the fiber axis, and weak blue absorption occurred when the electric vector was parallel to the fiber axis (Fig. 5.40b). It is as though the radial nerve fibers act as a radial dichroic filter directly in front of the cones of the retina. According to this explanation and the drawing in Fig. 5.40c, polarized light with its electric vector vertical passes perpendicularly through the optic axis of the radial fibers. Weakest absorption of blue would take place at 12 o'clock and at 6 o'clock and maximum absorption at 3 o'clock and 9 o'clock. For the other angles intermediate, absorption of blue would occur. In this case white light minus blue (yellow) would fall maximally on the cones in the horizontal plane giving rise to a horizontal yellow brush.

We cannot put forward a similar explanation for the appearance of the blue areas. It is our opinion that the blue areas are "psychological," i.e., the sensation of blue is stimulated in the brain by the yellow sensations coming from the brush area, as though additional blue (complement of yellow) were incident on contiguous areas of the cones. For example, we have looked at the blue sky through a circular yellow filter in such a way that the periphery of the yellow filter covered half of the field of view. At the periphery there is seen a slight purplish halo which is presumably induced by the yellow incident light.

Actually, the assumption of the double refraction of nerve fibers described by Helmholtz leads to difficulty in trying to explain the effects observed for circularly polarized light. This phenomenon, so far as we know, was reported only by Shurcliff (1955). For incident *right-hand circularly polarized light* the brush is seen only at an angle of 45° to the vertical, going from the upper right to the lower left quadrant. Conversely, for incident *left-hand circularly polarized light* the brush is seen only at an angle of −45° to the vertical, going from upper left to lower right (Fig. 5.40d). These are independent of the rotation of the linear polarizer-quarter waveplate combination producing the circularly polarized light. We have verified this effect by repeating Shurcliff's experiments.

In order to explain this latter phenomenon let us first consider how one obtains circularly polarized light. A description can be found in Jenkins and White (1950) and only the pertinent points will be covered here. It is the characteristic of doubly refracting crystals that the index of refraction for light vibrating parallel to the optic axis, $\eta_\parallel$, is different from $\eta_\perp$ (for light vibrating perpendicular to the optic axis). This is a property of the crystal structure and composition and depends upon planes of maximum polarizability in the crystal. From the definition of the index of refraction, we see that the light beams in the crystal will travel at different speeds, depending on whether the electric vector is vibrating $\parallel$ or $\perp$ to the optic axis. By *definition*, a negative crystal is one in which the value of $\eta_\parallel - \eta_\perp$ is negative; that is, light vibrating $\parallel$ to the optic axis (extraordinary ray) travels *faster* in the crystal than light vibrating $\perp$ to the optic axis (ordinary ray). (In the laboratory where retardation plates made of doubly refracting material are used, the designations of "fast axis" and "slow axis" refer to the vibration directions of the faster and slower traveling rays.)

The simplest method for producing circularly polarized light is to use a quarter waveplate, where the faces of the plate are cut parallel to the optic axis. If we have plane polarized light incident on the face of the quarter waveplate with the direction of vibration of the incident light making an angle of $+45°$ with the optic axis, vibrating in the first and third quadrant (*always looking end on, against the oncoming light beam*), Fig. 5.40e, we can resolve the electric vector into two components, one $\mathbf{E}_\parallel$, parallel to the optic axis, and the other, $\mathbf{E}_\perp$, perpendicular to the optic axis. Since $\theta = 45°$, $|\mathbf{E}_\perp| = |\mathbf{E}_\parallel|$. If the quarter waveplate is a positive crystal, i.e., $\eta_\parallel > \eta_\perp$, the component with $\mathbf{E}_\perp$ will travel faster than the component with $\mathbf{E}_\parallel$. The thickness is such that there is a phase difference of exactly $90°$ or one-quarter wavelength between $\mathbf{E}_\parallel$ and $\mathbf{E}_\perp$, with $\mathbf{E}_\parallel$ lagging behind $\mathbf{E}_\perp$. From Fig. 5.40e, $\mathbf{E}_\parallel$ lags $90°$ behind $\mathbf{E}_\perp$. Mathematically we can express this by saying that $\mathbf{E}_\perp$ is given by $|\mathbf{E}_\perp| \cos \omega t$ and $\mathbf{E}_\parallel$ is given by $|\mathbf{E}_\parallel| \cos (\omega t - \pi/2)$. Thus at $\omega t = 0$, the resultant vector is $\mathbf{E}_\perp$ and as time increases, say at $\omega t = \pi/2$, the resultant vector is $\mathbf{E}_\parallel$. From Fig. 5.40e it is as though the resultant vector were rotating counterclockwise with time, giving by definition *left-hand circularly polarized light* (LHCP). If the original linearly polarized light vector had made an angle of $-45°$ with the optic axis, $\mathbf{E}_\perp$ would be given by $-|\mathbf{E}_\perp| \cos \omega t$ and $\mathbf{E}_\parallel$ would be given by $|\mathbf{E}_\parallel| \cos (\omega t - \pi/2)$. Again by following the resultant vibration at $\omega t = 0(\mathbf{E} = -\mathbf{E}_\perp)$ and at $\omega t = \pi/2(\mathbf{E} = \mathbf{E}_\parallel)$, it is as though the resultant vector were rotating *clockwise*, giving *right-hand circularly polarized light* (RHCP).

The inverse process also holds. If right-hand circularly polarized light is incident on the quarter waveplate of Fig. 5.40e, the transmitted light will be plane polarized in the first and third quadrants at $+45°$ to the optic axis. For incident LHCP light passing through the positive quarter waveplate, the transmitted vector would also be plane polarized, but this time in the second and fourth quadrants, making an angle of $-45°$ with the optic axis.

Now let us return to Fig. 5.40c. If each of the radial nerve fibers is assumed to be a birefringent positive crystal, as assumed by Helmholtz, of such thickness as to be a quarter waveplate for blue light, each will produce plane polarized light from circularly polarized light. However, because of the radial symmetry this would result in vibration vectors with a radially symmetric distribution, or effectively unpolarized light. Therefore the double refraction of the radial nerve fibers in the *macula lutea* cannot explain the observed Haidinger brushes in RHCP and LHCP light. In order to explain the Haidinger Brush in linearly polarized light all that is required is that the blue-absorbing and consequently yellow-colored pigment molecules of the macula lutea be arranged in the Schwann cell plasma membranes in the plane of the membranes, and in such a way that when the Schwann cell membrane wraps spirally around the nerve fibril, there will always be pigment molecules with their planes of highest polarizability (absorption) perpendicular to the fiber axis as shown in Fig. 5.40b.

On the basis of the discussion above and Fig. 5.40d, where RHCP light produces a first and third quadrant brush or plane polarized light vibrating in the second and fourth quadrants, it is necessary that there be effectively a uniaxial plate in the optical system of the eye whose retardation is approximately $4500/4$ A or $4500(n + \frac{1}{4})$A, where n is an integer. The cornea is approximately 0.5 mm thick. If the protein molecules of which it is composed are stacked in a horizontal plane, the value $(\eta_E - \eta_0)$ need only be 2×10^{-4}, much smaller than the values of about 100×10^{-4} usually observed for other biological fibers. Conversely, a thin fiber film of normal positive birefringence need be only 11 μ thick. This effective quarter waveplate would also explain the fading of the Haidinger brushes in plane polarized light at 45° to the vertical. The optic axis of the proposed "birefringent plate" remains fixed. Therefore when the incident plane polarized light is at 45° to the vertical (and therefore also to the horizontal), circularly polarized light will be produced with equal components perpendicular to one another. Thus for plane polarized light at 45° to the optic axis of the birefringent plate or at 45° to the vertical no Haidinger brushes should be seen. At other angles the light is elliptically polarized and there is always a net component giving rise to the brush.

6.4. BIOCHEMISTRY OF VISION

Rhodopsin. The photosensitive pigment in the rods of the frog retina was first discovered by Boll (1877). The retina of the dark-adapted frog, excised in the dark or in dim red light has a deep reddish-purple hue which bleaches on exposure to white light. This "Sehrot" or "Sehpurpin" or rhodopsin is a protein-dye complex of opsin and an 11-*cis* form of retinene, whose structure, as well as that of the all-*trans* form of retinene, is shown in Fig. 5.41. When rhodopsin in the retina is exposed to white light, the result is a photoisomerization to the yellow-colored all-*trans* form through several intermediate steps which Wald and his group have named prelumirhodopsin, lumirhodopsin, metarhodopsin I, and metarhodopsin II. The all-*trans*-retinene₁ or vitamin A₁ aldehyde is converted rapidly to vitamin A₁ in the presence of alcohol dehydrogenase and DPNH. The reverse reaction from all-*trans*-vitamin A₁ proceeds through an isomerase

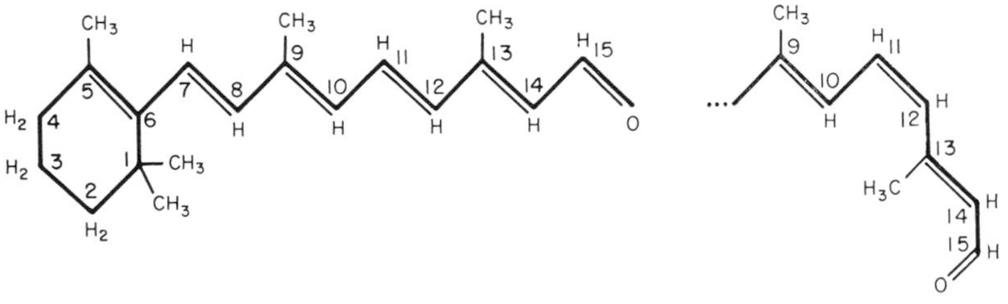

all-<u>trans</u>-Retinene₁ (vitamin A aldehyde)   11-cis isomer

FIG. 5.41.   Structures of all-*trans*-retinene₁ and the hindered 11-*cis* isomer.

to 11-*cis*-vitamin A₁, then through the reverse reaction with alcohol dehydrogenase and DPN to form 11-*cis*-retinene₁, which rapidly combines with the protein opsin to reconstitute the rhodopsin molecule. These steps are shown schematically in Fig. 5.42. The heavy lines in Fig. 5.42 indicate the more energetically favored and thus the more rapid reactions in this cycle. It has been demonstrated that the Q_{10} for light adaptation, which is essentially the same type of photoequilibrium process as discussed in Chapter 2, Section 3.5 is essentially unity, while the Q_{10} for dark adaptation, the enzymatic synthesis of rhodopsin from vitamin A₁, is about 3.85 (Griffith and McKeown, 1929).

One of the earliest symptoms of vitamin A deficiency is a tremendous rise in scotopic visual threshold or "night-blindness." There is even a further participation by vitamin A and presumably rhodopsin. This is concerned with the physical integrity of the rod structure itself. It has been shown that in rats maintained on a vitamin A-deficient diet not only is there a rapid onset of night blindness, but there is irreversible deterioration in the physical structure of the rod segments themselves. This is analogous to the deterioration or lack of formation of the lamellar structures in the grana of chloroplasts in the absence of chlorophyll, discussed in Section 1.

The dashed line in Fig. 5.42 between all-*trans*-retinene and 11-*cis*-retinene indicating an isomerase reaction is placed there on purely physiological evidence. It has been observed that after a short flash exposure of the retina to a high intensity light source, bleaching a large quantity of rhodopsin, the dark adaptation is very rapid, much faster than the dark adaptation after long exposure to light. It is assumed that in the latter case the *trans* retinene has been converted to *trans* vitamin A and that rhodopsin resynthesis must proceed through the steps outlined.

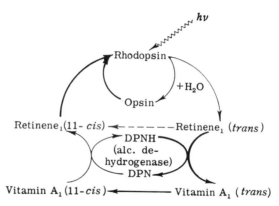

Fig. 5.42.   Cycle of photochemistry and dark chemistry in the formation of rhodopsin in the dark and the dissociation by light.

However, after a bright flash a direct isomerization of the initial product, *trans*-retinene to 11-*cis*-retinene, would be much more rapid, corresponding to the observed physiological response.

The intermediates formed on the exposure of rhodopsin to light have been observed *in vitro* by Wald and co-workers at various temperatures by virtue of their absorption spectra. At liquid nitrogen temperatures irradiation of rhodopsin produces "prelumirhodopsin." Upon warming to below −50°C a different isomer, "lumirhodopsin" is formed. At around −20°C "metarhodopsin," absorbing at 465 mµ, comes into being and above −15°C in the presence of water the "metarhodopsin" hydrolyzes rapidly to a mixture of retinene₁ and opsin. These temperature-dependent intermediate stages are very similar in description to the photochromic changes in the spiropyrans discussed in Chapter 2, Section 3.5, where the temperature and presumably the viscosity of the solution affected the stability of the various photoisomers that were formed upon initial irradiation of the room-temperature stable material. For example, Wald (1963) has found that at pH around 5.3, metarhodopsin is in the form metarhodopsin II, absorbing mainly in the ultraviolet at 380 mμ, while at pH around 7.7, he observes what he calls metarhodopsin I, absorbing at 480 mμ.

It has been pointed out that on the basis of stability of structures the all-*trans* form of retinene$_1$ should be expected to be the prevalent form. By reference to Fig. 5.41, it can be seen that a *cis* configuration could normally occur only adjacent to the methyl groups at positions 9 and 13. A *cis* configuration at the 7 or the 11 position would overlap the methyl

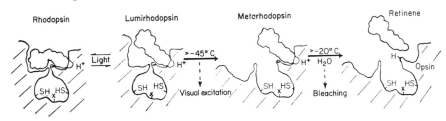

FIG. 5.43. The action of light on rhodopsin. The absorption of light by rhodopsin isomerizes its 11-*cis* chromophore to the all-*trans* configuration, yielding as first product the all-*trans* chromoprotein lumirhodopsin. This labilizes the protein, opsin, which rearranges to a new configuration, yielding a second all-*trans* chromoprotein, metarhodopsin. This second process exposes reactive groups on opsin: two SH groups, and one proton-binding group and may be responsible for triggering visual excitation. Vertebrate metarhodopsins are unstable, and above about −20°C hydrolyze to opsin and all-*trans* retinene, the process that corresponds to bleaching (Hubbard and Kropf, 1959; reproduced by permission of the New York Academy of Sciences).

groups and require therefore not only a bend but a *twist* from planarity of the molecule. However, it appears that it is just for this reason that the bent and twisted 11-*cis* isomer of retinene$_1$ is the visual pigment in biological systems. Wald has used this bent and twisted configuration in a modified "lock and key" enzyme mechanism to point out the possible advantages of this choice. This is shown in Fig. 5.43. He postulates that only the 11-*cis* isomer can "fit" into the enzyme surface to form the protein-dye complex. The only action of light therefore is to photoisomerize the 11-*cis* to the all-*trans* configuration, at which time it no longer "fits" and is hydrolyzed and removed from the enzyme as *trans*-retinene$_1$. Spectrophotometric curves showing stages in the synthesis and bleaching of rhodopsin in solution are shown in Fig. 5.44. The small band at 350 mμ in both sets of curves is the β-band of rhodopsin. Retinene$_1$ has its peak at 380 mμ.

In a polar solvent at room temperature and under continuous irradia-

tion to photoequilibrium it is found that rhodopsin is converted to 25%
11-*cis*, 50% *trans*, and 25% other *cis* isomers such as 9-*cis* or isorhodopsin.
At −60°C, where hydrolysis is inhibited, Wald has shown that after
irradiation of rhodopsin to photoequilibrium, removal of the light, and

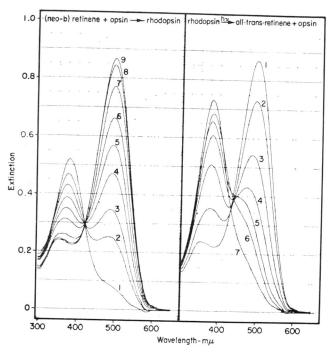

Fig. 5.44.   Synthesis and bleaching of rhodopsin in solution, at 22.5°C, pH 7.
Left: a mixture of 11-*cis*(neo-b)retinene and cattle opsin is incubated in the dark, and
absorption spectra are recorded (1) at 0.3, (2) at 2.5, (3) at 5, (4) at 10, (5) at 18, (6) at
30, (7) at 60, (8) at 120, and (9) at 180 min. The absorption band of 11-*cis*-retinene
(λ_{max} 380 mμ) falls regularly, as that of rhodopsin (λ_{max} 498 mμ) rises. *Right:* the
rhodopsin formed at the left (1) is now exposed to light of wavelengths longer than
550 mμ for various intervals, and the absorption spectra are recorded immediately
afterward. The total irradiations were (2) 5 min, (3) 10, (4) 15, (5) 30, and (6) 120 sec.
The residue was exposed for 45 sec longer to light of wavelengths longer than 440 mμ;
then spectrum (7) was recorded (Wald, 1961).

subsequent warming, 50% of the original rhodopsin remains as 11-*cis*
rhodopsin and 9-*cis* rhodopsin while only 50% has been bleached to
retinene and opsin. He infers from this that light itself can isomerize
trans-lumirhodopsin to either 11-*cis*- or 9-*cis*-retinene. If we assume that
the quantum yield for the photosiomerization of 11-*cis* retinene to *trans*
meta- or lumirhodopsin is 1, the relative quantum yields have been

estimated to be

$$\underset{0.5}{\overset{1}{11\text{-}cis\text{-retinene}}} \rightleftharpoons \underset{0.3}{\overset{0.09}{trans\text{-}\text{retinene}}} \rightleftharpoons 9\text{-}cis\text{-retinene} \tag{5.23}$$

The visual pigments are usually extracted from retinas in aqueous solutions of digitonin, sodium cholate, sodium deoxycholate, or cetyltrimethyl ammonium salts; the most commonly used extractant is digitonin, a glucoside with a molecular weight of 1288. This apparently forms micelles containing the pigment-enzyme complex. These extracts contain, in addition to the visual pigments, other light-absorbing impurities, so that true absorption spectra are usually masked. However, since it is the specific property of visual pigments that they are bleached upon exposure to light, it is often possible to use the technique of measurement of difference spectra to obtain the true absorption spectrum of the visual pigment. An example of this technique is shown in Fig. 5.44 (Dartnall, 1957). The photoproduct absorption spectrum usually varies with pH in the same manner as a pH indicator so that even here one must be careful in comparing *in vitro* pigments with physiological action spectra. Difficulties in interpretation may arise because of partial bleaching or heterogeneous solutions of visual pigments, in which cases the kinetics of bleaching must be investigated. The absorption spectra of chicken rhodopsin and iodopsin compared with the rod and cone sensitivities of various animals are shown in Fig. 5.45.

The first direct evidence for the presence of vitamin A in eye tissue is due to Wald (1933, 1935), who later demonstrated the relationship between visual purple, vitamin A, and retinene extracted from the retinas of the frogs *Rana esculenta*, *R. pipiens*, and *R. catasbeiana*. On exposure of dark-adapted retinas to light the observed color changes from a deep red to orange. This then fades slowly to become colorless, the sequence going from

$$\underset{\lambda_{abs} = 500 \text{ m}\mu}{\text{rhodopsin}} \rightarrow \underset{\lambda_{abs} = 385 \text{ m}\mu}{\text{retinene}_1 + \text{opsin}} \rightarrow \underset{\lambda_{abs} = 328 \text{ m}\mu}{\text{vitamin A}_1 + \text{opsin}} \tag{5.24}$$

In fishes there is a different visual pigment termed porphyropsin (Wald, 1937a,b), giving retinene on exposure to light. In this case the sequence is

$$\underset{\lambda_{abs} = 522 \text{ m}\mu}{\text{porphyropsin}} \rightarrow \underset{\lambda_{abs} = 405 \text{ m}\mu}{\text{retinene}_2 + \text{opsin}} \rightarrow \underset{\lambda_{abs} = 355 \text{ m}\mu}{\text{vitamin A}_2 + \text{opsin}} \tag{5.25}$$

In both cases the solvent was chloroform. According to Wald, true land and marine vertebrates have the rhodopsin system; true fresh water vertebrates have the porphyropsin system; in equivocal systems there may be mixtures or temporal changes in relative content. Every visual pigment system known at the present time is based on either vitamin A_1

or vitamin A₂. The peak positions of all the "rhodopsins" and "porphy-
ropsins" extend from 462 mμ in the shrimp *Euphasia pacifica* and 478 mμ
in the deep sea hatchet fish *Argyropelecus affinis* to 536 mμ in the grudgeon
Gobio gobio and 543 mμ in the rudd *Scardinius erythrophthalmus*. Wald
also found a 575 mμ light-absorbing pigment in chicken retinas which he
named iodopsin. Dartnall (1953) established that, regardless of the wavelength of λ_{max} of the various visual pigments, if the extinctions were

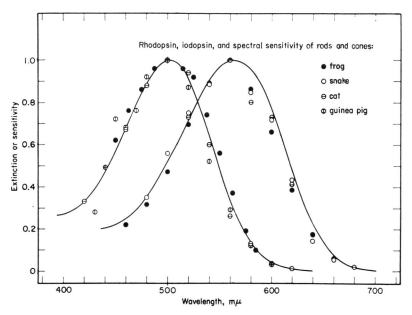

Fig. 5.45. The absorption spectra of chicken rhodopsin (λ_{max} 502 mμ) and iodopsin
(λ_{max} 562 mμ) compared with the rod and cone sensitivities of various animals. The
lines show the absorption spectra of the visual pigments, the points, electrophysiological measurements of spectral sensitivity (quantized). Rod data: frog (Granit,
1947, p. 292), cat (Donner and Granit, 1949), guinea pig (Granit, 1942). Cone measurements: frog (Granit, 1942), snake (Granit, 1943), cat (Granit, 1943) From Wald,
(1961).

plotted as a function of wave number, the relative shapes were identical.
He was able, therefore, to construct a nomogram so that the shape of an
entire spectrum can be inferred from the accurate measurement of λ_{max}.
It is also suggested by Dartnall that all of the taxonomic names and
color names be discarded in favor of the visual pigment and then the
λ_{max}. Thus rhodopsin and porphyropsin would be visual pigment 500
and visual pigment 522, respectively.

Munz (1957) has concluded that the peak wavelengths of the main
visual pigments in fishes are correlated with their habitats; the bathy-

pelagic fishes have λ_{max} around 480 mμ, surface pelagic fishes, around 490 mμ, and rocky shore fishes, around 500 mμ. In clear oceanic water the λ_{max} of transmitted light at a depth of 200 meters is 475 mμ. Nearer the surface λ_{max} of the transmitted light is around 490 mμ.

The prosthetic groups of all visual pigments appear to be based on the 11-*cis* form of either retinene$_1$ or retinene$_2$. If this is the case it is conceivable that the ranges of λ_{max} of the various retinene$_1$ and retinene$_2$ pigments are due to slight differences in the protein to which the pigment is complexed depending on the species. This would be analogous to all of the cases of firefly bioluminescence that we have examined thus far. That is, as described in Chapter 4, the different emission colors of bioluminescence in various firefly species are due to slight differences in the species enzymes. The species enzyme differences are not so great that they will not cross-react with the same substrate. However, it is only the enzyme which determines the color of the light emission.

The absorption spectrum of a visual pigment such as pigment 498 of cattle is composed of three extinction bands. One is in the UV region at 280 mμ and is due to the aromatic amino acids (tyrosine and tryptophan) of the "opsin." Light absorbed at 280 mμ does not contribute to bleaching. At longer wavelengths, in the near UV region there is a subsidiary band or *cis* peak with 20–30% of the height of the main band in the visible. In all visual pigments the separation of the main and *cis* peaks is 9700 $\pm$ 400 cm^{-1}. Finally, there is the visible region or main extinction band. Light absorbed by both the main and *cis* peaks is effective in bleaching and the quantum yield appears to be unity.

It is interesting that the stereoisomerism in visual pigments was found essentially by accident. Hubbard and Wald (1951) originally prepared visual pigment by placing in darkness opsin from cattle, liver alcohol dehydrogenase, DPN, and a fish-liver oil concentrate as a source of vitamin A$_1$. When they later used crystalline vitamin A$_1$, hardly any rhodopsin was found. Since fish-liver oils were known to contain isomers of vitamin A$_1$, it was obvious to them that a particular form of vitamin A$_1$ was required. It is in the reaction of retinene$_1$ with opsin that the particular stereoisomer is required. All-*trans* vitamin A$_1$ is easily oxidized by liver alcohol dehydrogenase and DPN. However, only the *neo-b* or 11-*cis* isomer produces rhodopsin.

As further evidence that the shape of the molecule is important, the *iso-a* or 9-*cis* isomer of vitamin A_1,

can be oxidized by liver alcohol dehydrogenase and DPN and will then combine with cattle opsin to give a pigment 487, or isorhodopsin. Apparently no other isomer will combine with opsin and there is no evidence that isorhodopsin is formed *in vivo*.

Since the shape of the stereoisomer is important in the complexing with the protein opsin and the effect of light is to convert the *cis* form to the *trans* form which can no longer be bound to opsin, Wald's hypothesis that rhodopsin is a zymogen or proenzyme which is activated by light appears to be an inviting one. However, at the present time we do not know the mechanism of amplification in the visual process.

For a complete summary of the visual pigments thus far isolated the reader is referred to Dartnall (1962, appendix to Part II).

6.5. COLOR VISION

Goldsmith (1961) has surveyed the evidence for color vision in insects. The conclusions are based on three lines of experiments. First, insects can be trained to associate food with monochromatic lights when precautions are taken that the response is not due to intensity differences but to wavelength alone. Second, the optomotor response, reaction to movement of vertical stripes in the visual field, can be observed when strips of a given color alternate with gray. A brightness and shade of gray can usually be found such that movement of the stripes does not elicit a response from the insect. The gray and colored stripes are then said to have equal "brightness." However, if instead of the alternating colored-gray sequence, one substitutes a color 1, color 2, color 1, color 2, . . . sequence where previously color 1 and color 2 were determined in separate experiments to have equal "brightness" with the same gray stripe, an optomotor response is usually obtained. Third, electrophysiological observations on compound eyes have indicated the presence of receptors maximally sensitive in different regions of the spectrum.

Again it should be emphasized that in many papers when the relative effectiveness of various wavelengths of light has been measured, the authors have plotted as a function of wavelength the reciprocal of the relative energy flux for a constant effect. This they call the action spectrum (Bertholf, 1932; Wolken, 1957; Cameron, 1938; Heintz, 1959; Goldsmith, 1961). In all of these cases the *reciprocal of the number of incident photons per second* should have been plotted. Since these are relative curves, they can be corrected simply. If we call the peak sensitivity wavelength λ_0, all ordinates at wavelengths, λ, should be multiplied by λ_0/λ. Thus an equal ordinate reciprocal energy plot in the blue and in the red regions means that the receptor is about *twice* as sensitive to blue as to red light. These λ_0/λ corrections should be applied to the ordinates of Fig. 5.46.

In the determination of action spectra and their interpretation as corresponding with the absorption spectra of the particular visual pigments involved in color vision one must be careful that a distinction can be made between "sensitivity" and "preference." For example, at different times in the life cycle of the cabbage butterfly, *Pieris brassicae*, the response action spectra are different (Ilse, 1928, 1937). Feeding cabbage butterflies tend to alight on blue or yellow flowers and can be attracted to blue or yellow paper models. Green, on the other hand, is not visited. Females of the same species when laying eggs, however, seek out green stems and leaves and blue-green and green paper models, ignoring blue, red, and yellow.

Behavioral experiments imply the existence in bees of receptors maximally sensitive in the near UV, the blue, and the yellow-green regions of the spectrum. The dark-adapted compound eye of the worker honeybee has a peak sensitivity around 535 mμ. Figure 5.47 shows the spectral sensitivity of the compound eye of the blowfly *Calliphora* (Goldsmith, 1961). Again the ordinate correction λ_0/λ must be made to obtain quantum sensitivity. Both the ultraviolet and yellow-green have been observed in ERG measurements, and Goldsmith has extracted a pigment 440 from the heads of honeybees which, upon exposure to light, bleaches and liberates retinene.

One question that can be asked is why, if UV light is present in daylight, insects but not vertebrates are sensitive to it. The answer may be that in compound eyes chromatic aberration is not a problem, whereas in single-lens vertebrate eyes chromatic aberration of ultraviolet as compared with yellow would be extreme.

MacNichol and coworkers (1961) have investigated the electrophysiology of color vision in the goldfish *Carrassius auratus*. Slow potential changes in response to flashes of light can be obtained from localized regions in the isolated retinas of fishes by means of ultra-micropipette electrodes (Svaetichin, 1956). Intraretinal potentials are of two types, there is a luminosity or "L" response which appears abruptly when the electrode reaches a particular depth in the retina. There is a sudden nega-tive resting potential of 25–35 millivolts and above this there occur large

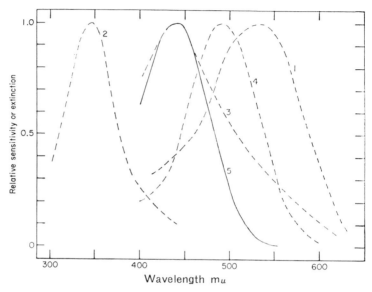

Fɪɢ. 5.46.   The spectral sensitivities of the several receptor systems described for the honeybee (broken curves) and a visual pigment from the bee (solid curve). (1) The spectral sensitivity of the dark-adapted compound eye of the worker bee; maximum, 535 mμ (Goldsmith, 1959). (2) The UV receptor system of the compound eye of the worker bee as revealed during adaptation with a bright yellow-green light. The maximum lies near 345 mμ. A receptor with similar spectral sensitivity is found in the compound eye of the drone and the ocellus of the worker (Goldsmith, 1959). (3) Spectral sensitivity of the compound eyes of drones. The maximum is at about 440 mμ; however, as other preparations may be more sensitive at longer wavelengths, there is evidence that a green-sensitive receptor (curve 1) is also contributing to this curve (Goldsmith, 1958a,b). (4) The 490-mμ receptor of the ocellus of the worker bee. (Goldsmith and Ruck, 1957). (5) The absorption spectrum of the light-sensitive retinene-protein pigment extracted from the heads of honeybees (Goldsmith, 1961).

negative going electrical changes during illumination. The other type is a chromatic or "C" response. In this case at short wavelengths the large negative going responses are indistinguishable from the "L" responses. However, as the incident wavelength is increased, the amplitudes of the negative going pulses decrease until around 575 mμ the amplitudes are zero. At wavelengths longer than 575 the sign of the electrical responses is reversed, becoming positive. This is shown in Fig. 5.48 and indicates a possible mechanism of color vision. Frogs and cats lack the "C" response and have no demonstrable color vision. In the intraretinal recordings made by MacNichol *et al.* the units gave an on-off type of discharge,

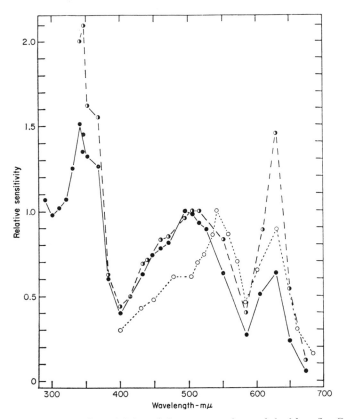

Fig. 5.47.   The spectral sensitivity of the compound eye of the blow-fly, *Calliphora*. Open circles and dotted curve (Autrum and Stumpf, 1952). Half-filled circles and dashed curve (Walther and Dodt, 1957). Filled circles and solid curve (Walther and Dodt, 1959). The height of the 630-mμ peak increases (relative to the maximum in the green) with increasing intensity. The height of the UV maximum varies in different preparations, but it does not vary in a consistent fashion as a function of intensity (Goldsmith, 1961).

that is, they gave a short burst of impulses when the light was turned on and another short burst when the light was turned off. They were also able to show that it the peak of the short wavelength sensitivity, 450–500 mμ, the response, was pure "on," i.e., electrical pulses were observed only during the illumination, while at the peak of the long wavelength sensitivity, 650–700 mμ, the response was pure "off," a pattern of optic nerve discharges that follows the "C" response. Figure 5.49 shows a threshold intensity versus wavelength plot for a typical ganglion cell. In addition, it was found that an "off" discharge appeared to be combined with a strong inhibitory effect on the "on" discharge while the "on" discharge did not affect the "off" discharge sensitivity. For example, after adaptation to deep red light when the sensitivity of the "off" response would be greatly decreased, an "on" threshold could

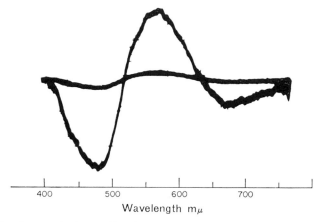

Wavelength mμ

Fig. 5.48.   Intraretinal recording from goldfish. Chromatic (C) type response to a linear change in wavelength as a function of time from 400 to 750 mμ. The time required for this change was 0.75 sec. The two traces are simultaneous high- and low-gain records of the same response; d-c recording was used on both channels. The peak to peak amplitude of the high-gain response was approximately 9 mv. The intensity of the stimulus at 600 mμ was 30 μwatts/cm^2. The stimulus was not equal energy. See Wagner *et al.* (1960) for energies at other wavelengths and other details concerning the technique by which this record was made. The stimulus was circular in outline and 5 mm in diameter, centered over the electrode.

be observed throughout the entire visible spectrum. After dark adaption the "on" response was completely inhibited in the red region.

Other ganglion cells (Fig. 5.50) have been found in which the converse is true. That is the "off" response has a blue-green maximum and the "on" response has a red maximum. Still others have an identical spectral sensitivity for both "on" and "off" responses. When there are separate action spectra for "on" and "off" responses, these peak at 500 and 650 mμ. When the action spectrum is the same for both "on" and "off," the peak is at 600 mμ, intermediate between the former two and just about broad enough to accommodate both curves.

Recently, Marks (1963) and Marks *et al.* (1964) have presented direct microspectrophotometric evidence for the existence of three separate cone pigments in primate retinas and for the first time have demonstrated the presence of a blue-absorbing pigment with an absorption maximum at shorter wavelengths than the reds.

In the Young theory of color vision there are presumed to be in the retina different visual pigments, each with differing absorption spectra.

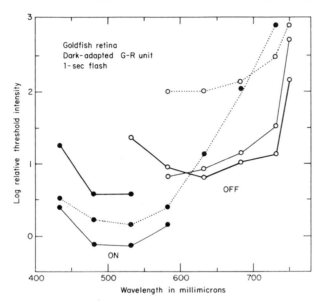

FIG. 5.49. Intensity necessary to elicit "on" and "off" threshold responses from a single ganglion cell at different wavelengths before, during, and after exposure to a red-adapting light. Heavy solid line indicates thresholds before adaptation; dotted lines, thresholds during red-adapting light; thin solid line, thresholds approximately 10 min after extinction of the adapting light. Test stimuli 1.0 sec in duration were repeated at 2-sec intervals. For intensity of test stimulus 0 log units = 2.3×10^{-2} μwatts/cm^2 for all wavelengths. Energy of red-adapting light, 5.2 μwatts/cm^2. Duration of adaptation was 27 min. Adapting light was obtained by filtering incandescent light 2850K color temperature through a Wratten 898 B filter (MacNichol et al., 1961).

The absorption of light by a red-absorbing pigment can then produce a nervous response which is distinguishable from that produced by a blue-absorbing pigment or a green-absorbing pigment. This infers that *in situ* the pigments are sufficiently separate so that the nervous system can sense their excitation separately. A large number of investigators have searched for these separate absorption spectra by the method of partial bleaching in which different spectra are obtained by bleaching away pigments one at a time with light in various spectral regions (Wald, 1937a,b;

Dartnall, 1952; Wald *et al.* 1955; Dartnall *et al.* 1961; Weale, 1955a,b; Crescitelli, 1958; Munz, 1958; Bridges, 1962). The principal difficulty in this technique is that the spectra of pigments present in small amounts may be masked by others. The microspectrophotometric method applied to individual cone outer segments has the additional advantage of determining whether one or more pigments are present in a single photoreceptor. Marks found peak wavelengths at 450, 525, and 630 mμ. From

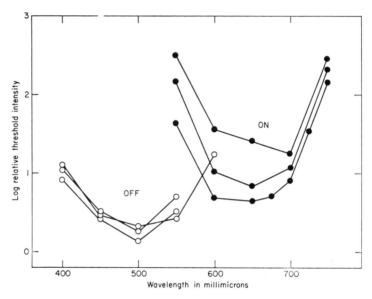

Fig. 5.50. Intensity necessary to elicit various types of threshold responses from a single ganglion cell at different wavelengths. Stimuli 0.5 sec in duration were repeated at 2-sec intervals. Intensity plotted on a log quantum scale. 0 log units = 2.6 $\times$ 10^{11} quanta/sec per cm^2 for all wavelengths. NOTE: During the 30 min necessary to determine the three sets of each curve, the "off" components changed slightly or not at all, while the "on" component became progressively more sensitive with little change in shape, a result confirming earlier evidence of a degree of independence of the two processes (MacNichol *et al.*, 1961).

visual examination it was found that (a) a single cone has a single pigment, (b) about half of the cones in the goldfish retina are members of symmetrical, joined twin cones. Out of 30 pairs examined, 29 were red-green pairs and only 1 was a blue-green pair. (c) Of the remaining half of single cones the ratios of red:green:blue were 2:4:1. (d) Assuming that the extinctions of the cone pigments are similar to those of the rod pigment, the concentrations of pigments in cones and rods are approximately the same.

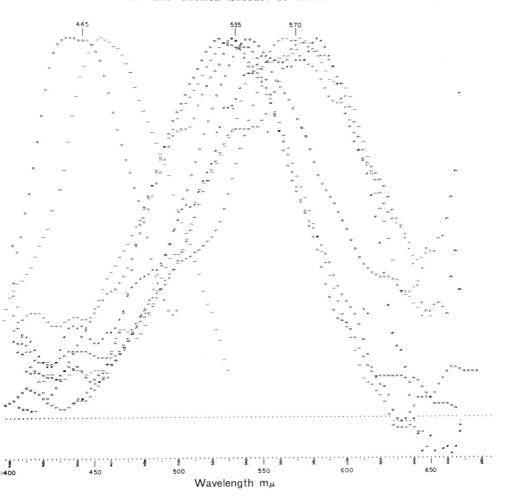

FIG. 5.51. Absorption difference spectra from ten individual primate cones, corrected for bleaching by the measuring beam. Curves recorded from monkey cones (*Macaca neimistrina* and *M. mulata*) are represented by numbers, those from humans by open parentheses. Maximum absorption, 3 to 6%, except the human blue, 0.4% is decreased by light scattered past the receptor. This figure is a photograph of the original record plotted automatically by a digital computer (Marks *et al.*, 1964; reprinted by permission of the AAAS and the authors).

In the living human fundus Rushton (1958), by reflection densitometry, has found two pigments at 540 mμ and at 590 mμ. P. K. Brown and Wald (1963) have reported two pigments in humans at 535 and 655 mμ and in monkeys at 527 and 565 mμ. Marks *et al.* (1964) have conclusively demonstrated that visual cone pigments are segregated in individual cones and that there are three major absorption peaks at around 445, 535, and

570 mμ. So far as we know, this is the first direct determination of single primate cones by coincidental microscopic selection and microspectrophotometry, although Hanaoka and Fujimoto (1957) have reported on the microspectrophotometric examination of single carp cone outer segments. The difference spectra obtained by Marks *et al.* (1964) are shown in Fig. 5.51.

7. Deleterious Effects

The term radiation damage covers a wide variety of deleterious effects in biological systems. These range from high energy ionizing radiation through UV and visible light to infrared and radio frequency energy. In order to discuss these in some sensible manner it has become customary to catalog the radiations by the mechanism and the nature of the effects that they produce. Ionizing radiation is unselective in its effects on molecules of all kinds. Short wavelength ultraviolet is absorbed specifically by nucleic acids (2600 A) and proteins (2800 A). Long wavelength UV, visible, and infrared light is selectively absorbed by pigment systems and produces changes in the electron population of electronic and vibrational levels. Radio frequency radiation interacting with the dipole moments of macromolecules can cause internal heating as well as denaturation by alteration of effective charge equilibria.

We shall be concerned in this section only with UV and visible radiation. However, in order to demark clearly the term UV damage it will be necessary to compare the radiations on either side of the energy spectrum of ultraviolet. There is no sharp distinction between ultraviolet and X-rays since the differentiation between inner-shell and outer-shell electrons becomes meaningless for low atomic number elements. We therefore consider that on the short wavelength side the UV region covers that energy range where excitation rather than ionization is predominant. This would extend the ultraviolet from the vacuum region ($\lambda < 1900$ A) through the region of DNA and protein absorption to the absorption by the aromatic amino acids of proteins ($\lambda \sim 3600$ A). Visible radiation is absorbed by special pigment molecules in the cell, usually in conjunction with an enzyme system or, as in the case of photodynamic damage, with the pigment molecule adsorbed at some specific site in the cell. The light regions effective in photoreactivation and photoprotection exhibit action spectra characteristic of pigment absorptions.

One of the major differentiations in radiation damage mechanisms is the influence of oxygen. For example, in the X-ray killing of *E. coli* B/r, removal of free oxygen (nitrogen atmosphere) reduces the observed killing to 30% of that observed in the presence of oxygen, assuming the effects are independent and therefore additive. In the case of UV killing of *E. coli* B, there is no reduction of killing in the absence of oxygen. In addition, the reactivation from radiation damage and the dependence on light, temperature, and chemical treatment are quite different for damage by ionizing radiations and UV light. A comparison of the killing effects of X-rays on *E. coli* B/r and ultraviolet on *E. coli* B is shown in

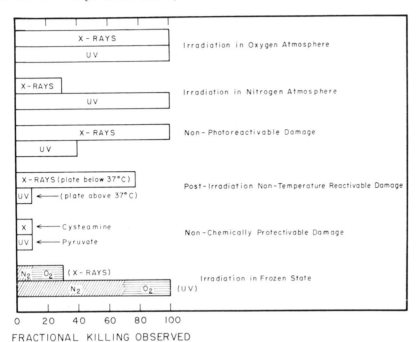

FRACTIONAL KILLING OBSERVED

Fɪɢ. 5.52.   Comparison of data for X-ray killing of *E. coli* B/r and UV killing of *E. coli* B (Jagger and Stapleton, 1957). In the case of chemical protection pyruvate has a smaller protective effect on X-ray killing than on UV killing. The cross-hatched portions of the bottom bar graphs and labeled N₂ indicate the amount of damage remaining if oxygen is removed prior to freezing the bacteria. While the ratio of X-ray damage without oxygen/X-ray damage with oxygen appears to be the same whether frozen or in the liquid state, the data indicate that for UV damage there is a difference between the mechanisms of damage in frozen and liquid states.

Fig. 5.52, based on data reported by Jagger and Stapleton (1957). As can be seen, there are apparently quite different mechanisms at work in effecting the observed damage. It is possible that, because of the high energies involved in ionizing radiations, subsequent degradation can produce excitations identical to those produced by ultraviolet. In addition, the unselective and "bull in a china shop" damage produced by ionizing radiations may have several competing subsequent chemical events, making an analysis of the kinetics and damage very complex. For exam-

ple, the DNA of transforming principle does not show an oxygen dependence for X-ray inactivation.

It is rather difficult to compare the relative biological effectiveness
(RBE) of ionizing radiation and UV radiation, since by virtue of their
different modes of action, penetration, and energy loss characteristics
the RBE referred to equal incident energy can have widely differing
values. As a best approximation McLaren (1957) has presented a comparison of the quantum yields for UV inactivation and the ionic yields for
X-ray inactivation of several enzymes and of tobacco mosaic virus
(TMV). These are shown in Table 5.7. As an example of some of the

TABLE 5.7

COMPARISON OF IONIC YIELDS FOR X-RAY INACTIVATION AND QUANTUM YIELDS
FOR UV INACTIVATION OF ENZYMES AND TOBACCO MOSAIC VIRUS[a]

Substance	UV quantum yield	X-rays ionic yield
Lysozyme	0.024	0.01
Ribonuclease	0.026	0.03
Trypsin	0.01	0.018
Carboxypeptidase	0.001	0.16
Phosphoglyceral dehydrogenase	0.0032	0.93
TMV	0.00004	0.0001

[a] From McLaren (1957).

difficulties in comparing directly the different types of radiation damage
(and in a sense giving more information as to the nature of both the
damage and the enzyme function) consider the case of chymotrypsin, a
digestive enzyme of 25,000 molecular weight. With X-rays the ionic
yield for inactivation of esterase activity is approximately three times
higher than for protease activity. However, for ultraviolet the quantum
yield for inactivation, though dependent on pH, is the same for both
activities.

7.1. BACTERICIDAL ACTION OF UV LIGHT

The presently available data indicate a very low quantum yield for
UV radiation. Wyckoff (1930a,b) showed that for high energy electrons
and for X-rays a single absorption by a colon bacillus was sufficient to
cause death. However, from the slopes of the survival curves for UV
radiation below 3000 A he calculated an effective absorption of only
0.03 (Wyckoff, 1930c). At 2699 A the quanta required per bacterium
killed was calculated to be 4.2×10^6. Coblentz and Fulton (1924)
reported that for *B. coli*, 1 sec at 24×10^{-6} watt/mm^2 in the wavelength

region of 1700–2800 A was required. Assuming a mean cross-sectional area of the bacterium of 8×10^{-7} mm^2 and a mean wavelength of 2200 A, we can arrive at a quanta per bacterium yield of 21×10^6. However, this number is necessarily high since the killing criterion of Coblentz and Fulton was a blank culture plate and they did not plot survival curves. Gates (1930) obtained a value of 88 ergs/mm^2 for one-half survival at 2660 A, which would correspond to a mean lethal dose of 127 ergs/mm^2. At this wavelength 1 quantum has an energy of 0.75×10^{-11} erg. Assuming the same cross-sectional area as before, 8×10^{-7} mm^2, we can calculate a quanta per bacterium lethality yield of 13.5×10^6. Hollaender and Claus (1936) reported for *E. coli* 13.1×10^{-6} ergs per bacterium of 2650 A, giving a quanta per bacterium yield of 1.75×10^6. Lea and Haines (1940) reported for *B. coli* at 2537 A a mean lethal dose of approximately 8000 ergs/cm^2. Again using the conversion factors and bacterial cross section we arrive at a quanta per bacterium lethality yield of 8.2×10^6. These data are summarized in Table 5.8.

TABLE 5.8

NUMBER OF UV QUANTA REQUIRED FOR BACTERIAL KILLING

Investigators	Bacteria	Wavelength range (A)	Number of quanta per bacterium killed
Coblentz and Fulton (1924)	*B. coli*	1700–2800	21×10^6
Gates (1930)	*Staphylococcus aureus*	2600	13.5×10^6
Wyckoff (1930a,b,c)	*B. coli*	2699	4.2×10^6
Hollaender and Claus (1936)	*E. coli*	2650	1.75×10^6
Lea and Haines (1940)	*B. coli*	2537	8.2×10^6

On the basis of these data and weighting the Hollaender and Claus data more heavily since they corrected for the protective action of non-viable organisms, we can arrive at a rough estimate of 3×10^6 UV quanta per bacterium killed (LD) in the peak absorption of nucleic acid. Lea and Haines (1940) also showed that over a range of 500 in irradiation intensity the Mean Lethal Dose remained unchanged (irradiation times 0.1 sec to 1 min). Further, Rentschler and associates (1941) showed that the LD remains unchanged even if the total dose is delivered in 10^{-6} sec. These results together with those of the lethal effects of ionizing radiation (see Lea, 1956) suggest that killing is caused by a single unit action, the absorption of a single UV photon.

Using the data of Lea (1956), the LD for 0.15 A X-rays is 5700 roentgens. The volume of a *B. coli* is approximately 0.4 μ^3 with a density of essentially 1. Assuming 1 roentgen = 93 ergs/g we have $5.7 \times 10^3 \times 93 \times 0.4 \times 10^{-12} = 2.1 \times 10^{-7}$ ergs per bacterium for X-radiation. The energy per bacterium for a LD of UV radiation of 2650 A is then $3 \times 10^6 \times 0.75 \times 10^{-11} = 2.2 \times 10^{-5}$ ergs per bacterium. Thus 100 times more energy and approximately 2×10^6 times as many quanta as UV radiation than as X-radiation must be delivered to a bacterium for the same killing effect. The UV radiation is absorbed at the sensitive site and the ionization from X-rays is nonspecific. The low quantum yield of 3×10^6 UV photons per bacterium inactivated would suggest, according to the single-unit action mechanism proposed by Lea and others, that a very improbable event occurs as a result of UV excitation of the nucleic acid bases. Possibly this improbable event out of 3×10^6 excitations of the nucleic acid structure can result in an interaction of the first excited state with the solvent or with other bases to effect a lethal group transfer. It is presumably much more likely that this type of transfer can occur as a result of ionization. However, the mechanism of UV action may proceed by a different pathway than ionization. Marcovich and Latarjet (1958) report that a photoreversion by visible light of induction in the lysogenic bacteria strains *E. coli* K12(λ) and K12S is observed for induction by ultraviolet but not for induction by X-rays. This can be interpreted as the formation by ultraviolet of reasonably stable intermediates which they term *preduction*, while in the case of the much higher energy X-rays there is irreversible damage to the backbone of the DNA molecule. As will be described in Section 8, most UV damage appears to be reversible by near UV and visible light, whereas X-ray and corpuscular particle damage are not.

7.2. PHOTODYNAMIC ACTION

Although it is generally accepted that Raab (1900) published the first extensive paper showing the killing action of fluorescent dyes on *Paramecium caudatum* in the presence of light, it has been pointed out by Santamaria (1960) that Marcacci (1888) described a light-dark effect on alkaloids added to bacteria, plants, and eggs. These same dyes in the dark were harmless or at most only slightly toxic. The term *photodynamische Erscheinungen* or photodynamic action was given by Tappeiner (1900).

The basic mechanism is the photosensitized oxidation of the absorbate by molecular oxygen. The term photodynamic action is usually reserved for the killing effect on the entire organism or the inactivation of enzymes; a more specific description would be *sensitized photooxidation*, since the sensitizing molecules remain unaltered at the end of the reactions. Almost all of the photodynamic substances known are fluorescent either in solution or adsorbed on a surface. However, not all fluorescent dyes exhibit photodynamic action and substances such as sodium benzoate which is nonfluorescent can sensitize photooxidations.

The characteristics of photodynamic action are:

(a) Oxygen is specifically required and is used stoichiometrically with a quantum yield close to unity.

(b) The reaction is essentially independent of temperature, corresponding to a true primary photochemical reaction.

(c) The reaction is irreversible, i.e., it cannot be reactivated by light, temperature, or chemical additions.

(d) The rate of the reaction is zero order with respect to sensitizer concentration and the total number of adsorbate molecules reacted is directly proportional to the total quanta absorbed. The sensitizer molecules act as catalysts.

Of itself photodynamic action is an example of the occurrence of sensitized photooxidation in living systems and is an important tool for the elucidation of the mechanism of oxidation by molecular oxygen, similar to studies of bioluminescence. However, for a long time there has been observed a correlation between the photodynamic action of dyes and the light-mediated carcinogenetic action of these same dyes. Unna (1896) first noted the possible influence of light on skin diseases which frequently lead to cancer. Hyde (1906), on the basis of a statistical study, implicated the melanin of colored races as a protective (light-absorbing) agent in skin cancer. Mice tarred *and* exposed to long-wave ultraviolet (both are required) developed malignant growths much more rapidly than when exposed to ultraviolet alone. Since the tars absorb in the long-wave ultraviolet, this was the first laboratory demonstration of the *photosensitization* of carcinogenesis. Finally, Bungeler (1937a,b) showed the specific correlation between the photodynamic dyes eosin, hematoporphyrin, and antrasol (fluid tar containing fluorescent compounds) and their photosensitization of carcinogenesis.

On the basis of an extensive series of experiments on the sensitized photooxidation of dyes such as leuco methylene blue by trypaflavine, both adsorbed on substrates such as silica gel, paper, wool, cellophane, etc., Kautsky proposed that the process of sensitized photooxidation proceeds as follows:

$$S \xrightarrow{h\nu} S^* \tag{5.27}$$

$$S^* + O_2 \rightarrow S + O_2^* \tag{5.28}$$

$$O_2^* + AH_2 \rightarrow A_{ox} \tag{5.29}$$

where AH_2 is the reduced form of the substrate molecule.

What is omitted in this mechanism is a chemical balancing. Kautsky in his other papers, and much later Kholmogorov *et al.*, (1962), in experiments on light-induced ESR signals in microcrystalline films of chlorophyll, reported a strict dependence of these effects on the *presence of a minute amount of water vapor as well as oxygen*. It is therefore very likely that reaction (5.29) requires the participation of water, presumably forming a hydrated A molecule or permitting molecule A to form a zwitterion. In this case the formation of the peroxide form A—O—O⁻ would be accelerated and the resultant steps would be similar to the chemiluminescence oxidation steps described in Chapter 3. The net balanced reaction could then be described by

$$O_2^* + AH_2 \xrightarrow[\text{in presence of } H_2O]{} \overset{\text{O}}{\overset{\|}{A}} + H_2O \tag{5.30}$$

where A=O is the oxidized form of AH_2. This is the same mechanism that we have proposed for the bioluminescent oxidation of firefly luciferin on the basis of the same stoichiometric utilization of one O_2 molecule per molecule of luciferin, LH_2, oxidized (McElroy and Seliger, 1963a,b). With this modification to Kautsky's mechanism there is no disagreement with the alternative set of steps that have been proposed by others,

$$S \xrightarrow{h\nu} S^* \tag{5.31}$$

$$S^* + O_2 \rightarrow SO_2 \tag{5.32}$$

$$SO_2 + A \rightarrow A_{ox} + S \tag{5.33}$$

The reactions

$$S \xrightarrow{h\nu} S^* \tag{5.34}$$

$$S^* + A \rightarrow S + A^* \tag{5.35}$$

$$A^* + O_2 \rightarrow A_{ox} \tag{5.36}$$

and

$$S + A \rightarrow (SA) \tag{5.37}$$

$$(SA) \xrightarrow{h\nu} (SA)^* \tag{5.38}$$

$$(SA)^* + O_2 \rightarrow S + A_{ox} \tag{5.39}$$

have also been proposed. The last two sets are difficult to separate from one another. Usually the complex (SA) would have an absorption spectrum different from S alone and, so far as we know, the absorption spectra of the sensitizing dyes do not change upon adsorption. Kautsky went as far as adsorbing dyes S and A on separate silica gel particles, so that aside from the minor contribution by emission and reabsorption phenomena the latter two mechanisms can be ruled out, at least to explain Kautsky's results. Fiala (1949) observed in polarographic studies of the photodynamic effect of fluorescein dyes on blood serum that the sensitizing dye complexed immediately with the substrate. On this basis the latter two sets of reactions must still be included in any discussion of photodynamic action. Gaffron (1927a,b) observed the formation of H_2O_2 following the irradiation of amylamine in the presence of chlorophyll, further supporting the idea of peroxide formation.

In Table 5.9 is given a list of dyes that have been found to exhibit photodynamic action, as well as some that do not exhibit photodynamic action. The last column indicates their reported light-mediated carcinogenetic ability. Since there is no obvious direct relation between the production of tumors in vertebrates and the death of protozoa or enzyme inactivation, one must search much deeper into the mechanism of these effects to uncover a possible common factor. It has been reported by Kihlman (1959a,b) for acridine orange adsorbed onto the nucleic acids in root tip cells that all of the known types of chromosome structural damage such as chromatid breaks, isolocus breaks, and exchanges are produced by irradiation with visible light, similar to that observed with X-ray damage. Porphyrins as a rule show a pronounced photodynamic action. However, in the form of hemes—i.e., when a metal such as Mg or Zn is incorporated into the porphyrin nucleus and they are incorporated into the cell structure—they do not exhibit this photodynamic action. It may be that the direction of the reaction is channeled effectively and that these can now act as efficient electron donating or accepting centers. Cytochrome c in photosynthesis is a good example. It may also be that the carotenoids present in the chloroplast grana serve to protect against photodynamic action in the presence of excess light.

The depolymerization of in vitro DNA solutions and of polymethacrylic acid (PMA) solutions by X-rays, ultraviolet, nitrogen mustards, and HO_2 radicals is also observed as a result of photodynamic action. Since all of the above are mutagenic agents, there is another correlation here. Even in the case of some of the photodynamic substances the kinetics of the reaction appear to be different. In the photooxidation of blood serum in the presence of 3,4-benzpyrene, the lowering of the oxygen tension is directly proportional to the quanta absorbed indicating a zero-order reaction (Santamaria, 1960). However, in the presence of hematoporphyrin the decay is exponential (first order).

Calcutt (1954) found a large group of carcinogens, other than the polycyclic hydrocarbons, that was also photodynamic. These are acetylaminofluorene, p-dimethylamino azobenzene, chloroform, and carbon tetrachloride, which cause liver tumors, the estrogens—estrone and diethylstilbestrol, the arsenicals—methylarsonic acid, diethylarsonic acid and various substitutions, nitrogen mustard, and urethane. Santamaria

TABLE 5.9
PHOTODYNAMIC ACTIVITY AND LIGHT SENSITIZED CARCINOGENICITY
OF VARIOUS SUBSTANCES

Substance	Photodynamic activity	Photocarcinogenetic activity
Acetylaminofluorene	+ +	+ + +
Acridine orange	+ + +	
Acriflavine	+ + +	
Alizarin red	−	−
Allylarsonic acid	+ +	+
Anthracene	−	−
Anthrasol	+ + +	+ + +
1,2-Benzanthracene	+ + +	?
1,2-Benzfluorene	−	−
3,4-Benzpyrene	+ + + +	+ + + +
Carbon tetrachloride	+ +	+ +
Chlorpromazine	+ +	
Chloroform	+ +	+ +
Cholanthyrene	+ + +	+ + +
Cholesteridine	−	−
Cholesterol	−	−
Chrysene	+	−
Cresyl violet	+	
Methyl orange	−	
Crystal violet	+ +	
1,2,5,6-Dibenzanthracene	+ +	+ +
Diethylarsonic acid	+	+
Diethylstilbestrol	+ + + +	+
p-Dimethylaminoazobenzene	+ +	+ + +
9,10-Dimethyl-1,2-dibenzanthracene	+ + +	+ + +
1,2-Dimethylphenanthrene	?	−
1.9-Dimethylphenanthrene	+	−
Eosin	+ + +	
Ergosterol	−	−
Erythrosin	+ + +	
Estrone	+ + +	+
Ethylaminopropylarsonic acid HCl	+	+
Fluorene	−	−
Hematoporphyrin	+ + +	
2'-Methyl-4-aminostilbene	+ + +	+
Methylarsonic acid	+	+
5-Methyl-1,2-benzanthracene	+ +	+ +
2-Methyl-3,4-benzphenanthrene	+	+ +
20-Methylcholanthrene	+ + +	+ + + +
2'-Methyl-4-dimethylaminostilbene	+ + +	+
Methylene blue	+ + +	
Neutral red	+ +	

TABLE 5.9 (*Continued*)

Substance	Photodynamic activity	Photocarcinogenetic activity
Perylene	−	−
Phenanthrene	−	−
Promazine	+ +	
Pyrene	+	−
Riboflavin	+ +	
Rose bengal	+ + +	
Thioflavine TG	+	

(1960) found a complete parallelism between the changes in optical rotation in blood sera denatured by heat or H_2O_2 and that produced by photodynamic action, indicating an unfolding of the protein. In his polarographic studies the photodynamic effect of hematoporphyrin brings about at first an increase in the SH groups and upon continued exposure a decrease in SH groups, corresponding to a molecular unfolding process and a subsequent oxidation of the SH groups. This increase and a subsequent decrease in optical rotation was repeated for crystallized albumin, glycyltyrosine, and chloroacetyl tyrosine in the presence of light and hematoporphyrin, indicating that the oxidation of the aromatic groups may play a role in the unfolding of the polypeptides.

As can be seen there is a large body of evidence linking photodynamic activity with photosensitized carcinogenesis, and with photosensitized mutagenesis both in bacterial DNA and in chromosomal DNA. Photodynamic action is strictly oxygen dependent, while UV damage and a large fraction of X-ray damage can occur in the absence of oxygen. This alone points up a major difference in mechanism of damage. It may be that damage to DNA is the common factor in their correlation or possibly the photodynamic damage to protein can feed back in some way to the reproductive system of the intact cell in addition to killing. It should be pointed out that the long wavelength limit for carcinogenesis by UV light alone is 3200 A (Blum, 1959) but it has not been established whether nucleic acid or protein is the chief absorbing substance.

Mechanism of Photosensitized Oxidation. Oster *et al.* (1959) and Bellin and Oster (1961) have proposed that there is a complex formed between S* in a long-lived metastable state and an oxygen molecule. They discard the idea of the direct excitation of O_2 molecules on energetic grounds. Presumably by this they are referring to the experiments of Gaffron (1936). The arguments used is as follows. If the O_2 molecule is excited to become "active oxygen," the lowest accessible electronic level would be the $^1\Sigma_g^+$ state at 7623 A, 37.5 kcal above the ground state. Therefore, one might expect that substances absorbing light beyond 7623 A would not be able to excite the $^1\Sigma_g^+$ state of oxygen. Gaffron showed that bacteriophaeophytin was an effective photodynamic agent even at 8000 A. However, in solution or weakly bound on an absorbing surface, the levels of the O_2 molecule may very well be shifted in energy so that the data of Gaffron do not conclusively eliminate the production of active oxygen. It must be emphasized that the experiments of Bellin and Oster (1961) and the more recent work of Zwicker and Grossweiner (1963) have concerned photosensitized oxidation in homogeneous aqueous solution; in the former the oxidation of p-toluenediamine by thiazines (methylene blue), xanthines (fluorescein), thiazoles (thioflavin), acridines (proflavin), azines (neutral red), porphyrins (hematoporphyrin), and riboflavin and, in the latter, the oxidation of phenol by eosin. In a previous paper by Grossweiner and Zwicker (1961) fluorescein, erythrosin and rose bengal were also used. The processes occurring in the eosin photosensitized oxidation of phenol are as follows:

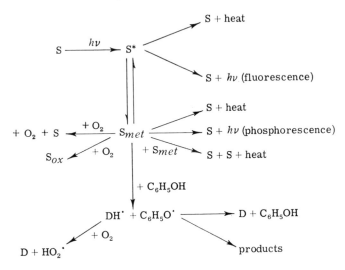

For further reference to these reaction steps the reader is referred to papers by Oster and Adelman (1956), Imamura and Koizumi (1956), and Lindquist (1960). Following Zwicker and Grossweiner, the excited singlet state is not quenched by oxygen. However, based on the absence of slow fluorescence in the presence of oxygen and the strong oxygen retardation of flash bleaching and the acceleration of triplet disappearance in the presence of oxygen, it is inferred that triplet eosin, S_{met}, is deactivated by oxygen. They were able to show by fast flash spectrophotometry that flash irradiation of oxygen-free eosin bleaches the ground state absorption spectrum and produces a transient absorption spectrum which they assign to the triplet state of eosin. The primary photochemical reaction between the phenolate ion and triplet dye results in an electron transfer from the phenolate ion to triplet eosin:

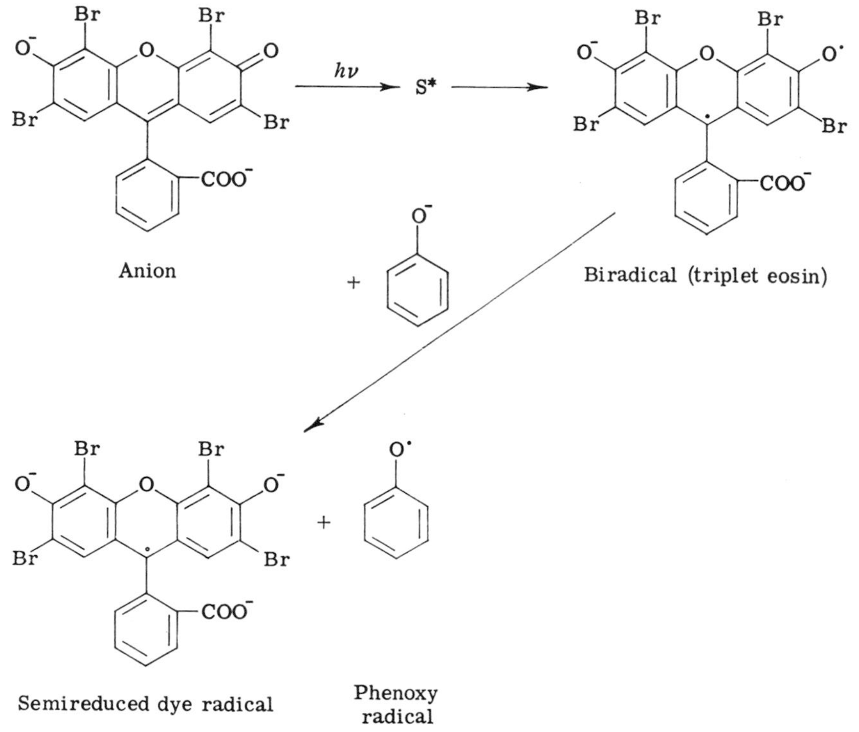

In the case of eosin the first-order triplet lifetime is 2.4×10^{-3} sec. The triplet structure shown is only one of a number of resonant forms.

   This primary photochemical electron-transfer reaction, producing radicals, may be the general mechanism of photosensitized oxidation of aromatic amino acids in the case of proteins. For example, Grossweiner and Zwicker (1963) reported that the transient spectra of UV-irradiated tyrosine and tryptophan have been identified as the p-alanine-phenoxyl radical and the 3-indolyl radical, respectively. Eosin-sensitized photolysis gave the same products in visible light. This was extended to the eosin-sensitized photolysis of ovalbumin for which the same radicals have been tentatively identified. In addition, they have observed evidence of binding of eosin dimers to aromatic amino acid sites on the enzyme. This binding can be responsible for a reduced efficiency of oxygen quenching of triplet eosin and can thus enhance the photodynamic effect in oxygen-containing systems. This would be in agreement with the observations of Oster and Oster (1962) to the effect that there is a lowered oxygen quenching of triphenylmethane dyes adsorbed on polymethacrylic acid. These data emphasize the fact that while solution photochemistry can provide a basis for examining the mechanism of energy and or electron transfer, the photodynamic action of various substances may be qualitatively and quantitatively different when adsorbed on membranes or on protein or nucleic acids. The obvious change in the fluorescence of triphenylmethane dyes when adsorbed on silica gel is a good example. There is still the question of the formation of active oxygen. The experiments of Kautsky implied a diffusion of an oxygen molecule from one place to another, presumably as a gas molecule. In solution the mechanism could very well be the formation of a peroxy radical in the same way in which we have assumed the steps in chemiluminescence to occur.

8. Photoreactivation

Photoreactivation is the light-mediated recovery of a biological system from radiation damage, also known as "photorecovery," "photorestoration," and "photoreversal." The photoreactivating light is of longer wavelength than the damaging radiation. The existence of ultraviolet or "chemical" rays or "actinic" rays has been known since 1801. It was not until 1858 that UV radiation was associated with sunburn and not until 1877 that the bactericidal action of sunlight was discovered. In 1887 it was shown that a glass filter could absorb carbon arc radiation that was passed by quartz, setting the long wavelength limit at 3200 A. This was, in a sense, the first crude action spectrum for UV effects. It was not until 1904 that Bie showed that the radiations from a carbon arc in the 2000–2905 A regions were 10–12 times more effective for bactericidal action as the much more intense radiations above 2950 A. The most reasonable explanation is that some specific substance or substances are absorbing this radiation and therefore the action spectrum should coincide with the absorption spectrum. However, one must be careful that other absorptions do not mask the identification. Gates (1930) demonstrated that the UV absorption by bacteria does not correspond to the bactericidal action spectrum. Rather than the major 2800 A peak protein absorption, the 2600 A peak nuclei acid absorption corresponded with the bactericidal action spectrum. Thus in 1928 Gates suggested that the effects of ultraviolet on nucleic acids underlie the action of ultraviolet on cells.

Hollaender and Emmons (1941) shows that the action spectrum for the induction of mutations resembled the bactericidal action spectrum. Later

Giese (1945) showed that the retardation of cell division also showed a similar action spectrum. However, it is extremely important in any subsequent discussion of the mechanism of the reversal by light of UV damage to point out that not all effects of ultraviolet have an action spectrum like that of bactericidal action. It has been pointed out by Giese (1961) that the immobilization of ciliary activity in protozoa and the UV sensitization of protoplasm to heat have action spectra corresponding to protein absorption. Still other UV action spectra are found for the initiation of parthogenetic development in eggs and the hemolysis of red blood cells.

Hertel (1904) first reported that visible light interfered with the UV-induced inhibition of protoplasmic streaming in *Elodea*. However, the first really quantitative measurements on photoreactivation were made by Hausser and von Oehmcke (1933) and Whitaker (1941) on UV-induced browning of banana skin and inhibition of growth in *Fucus*, respectively. The beautifully descriptive photograph of Hausser and von Oehmcke (Fig. 5.53) shows the reversal of short wavelength UV blackening, by the 3663 A, 4047 A, and 4358 A strong emission lines of a low-pressure mercury discharge lamp. The light bands on the dark background are "photoreversed" regions if the irradiated banana is considered to be a spectrographic plate.

Witkin (1946) discovered a radiation resistant strain of *Escherichia coli* B, namely *E. coli* B/r which appeared among the survivors of strain B subjected to a large UV dose. However, Kelner first demonstrated the nature of photoreactivation in the reversal, by visible light, of the UV-induced inactivation of spores of *Streptomyces griseus* (Kelner, 1949a) and of the UV-induced inactivation of *E. coli* (Kelner, 1949b). Previous to this there had been several sources indicating that postirradiation treatment of UV-irradiated cells could mediate the UV effects. Among these Hollaender and Emmons (1941) found that the survival of UV-irradiated fungus spores was increased by postirradiation storage in liquid menstrua. Latarjet (1943) had found that in some cases the survival of UV-irradiated cells increased if they were kept cold for a period following irradiation. In Kelner's own words (Kelner, 1961), "When we first observed recovery of U.V.-irradiated *Streptomyces griseus* spores in our culture dishes we ascribed it to post irradiation storage of the cells in the cold. When experiment did not support a 'cold-reactivation' we searched for environmental factors which might be causing the overwhelming recovery."

The idea of the recovery of a "target" or the repair of a "hit" was a new one and was further strengthened by the work of Dulbecco (1949, 1950) on the photoreactivation of bacteriophage in *E. coli* and by Novick and Szilard (1949) and Newcombe and Whitehead (1951) on the photo-reversibility of UV-induced mutagenesis.   Since 1949 a wide distribution of photoreactivation has been demonstrated in both the plant and the animal kingdom (Dulbecco, 1955; Jagger, 1958, 1960a,b; Blum *et al.*,

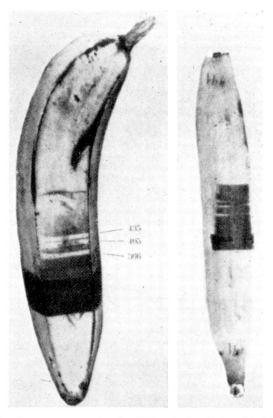

Fig. 5.53.   Two bananas exposed to far-UV radiation (below 3000 A). The bananas were exposed over a masked region indicated by the blackened control banana at the right. Subsequently the left banana was placed at the exit port of a monochromator upon whose entrance slit was focused the light from a low pressure mercury discharge tube. The width of the entrance slit and the linear dispersion of the spectrometer can be estimated from the photoreversed bands at 366, 405, and 436 mμ as indicated in the photograph (Hausser and von Oehmcke, 1933).

1957; Rieck and Carlson, 1955). Apparently, whatever ultra-violet does to the cell is photoreversible, from loss of viability to inhibition of growth, of enzyme synthesis, and of mutation.

The relative action spectra for the photoreactivation of killing in *E. coli* B/r and *Streptomyces griseus* are shown in Fig. 5.54. The relative efficiency per quantum on a logarithmic scale is plotted versus incident wavelength. The curves are shown in their correct relative relationship. It is seen that photoreactivation is much more efficient for *S. griseus* than for *E. coli* B/r and that there is more than one chromophore involved in

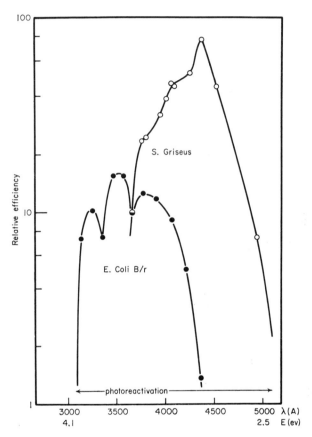

FIG. 5.54. Action spectra for photoreactivation of killing in *E. coli* B/r and in spores of *S. griseus*. The relative efficiency per quantum on a logarithmic scale is plotted versus incident wavelength. The curves are shown in their correct relative efficiencies and illustrate the spectral ranges involved (Jagger and Latarjet, 1956; Kelner, 1951; reprinted by permission of the Rockefeller Institute Press, from *J. Gen. Physiol.*).

photoreactivation. Damage caused by ultraviolet in the range 2180 to 3130 A can be photoreactivated by light in the range 3130 to 5490 A. Photoreactivation is produced right down to those wavelengths that begin to have a killing effect.

In general, it is felt that photoreactivation is the repair or elimination and replacement of a UV-caused lesion to DNA. The evidence from the action spectra for UV killing, the photoreactivation in phage in *E. coli* DNA synthesis and in UV-irradiated transforming principle DNA is quite conclusive. In addition to these obvious DNA repairs, photoreactivation has also been reported for enucleated amebae (Skreb and Errera, 1957), in nerve cell cytoplasm (Pierce and Giese, 1957), in mitochondria in yeasts (Pittman and Pedigo, 1959), in UV inactivation of chlorophyll synthesis and chloroplast development in *Euglena gracilis* (Lyman *et al.*, 1959), and in RNA plant viruses (Bawden and Kleczkowski, 1959). Apparently UV-induced lesions to *self-duplicating structures*, whether nuclear or cytoplasmic, and not necessarily DNA are able to be photoreactivated.

For any given dose of ultraviolet given in the dark and which results in a certain survival fraction, there is a larger dose of ultraviolet which, after maximum photoreactivation, will result in the same survival fraction. The ratio

$$\frac{\text{dark UV dose}}{\text{photoreactivated UV dose}} = \text{dose reduction factor (DRF)} \quad (5.40)$$

If photoreactivation were able to completely reverse the UV damage, the denominator of the above ratio would approach ∞ and DRF $\to$ 0. Thus the photoreactivable sector is defined as (1-DRF). For *E. coli* B/r, DRF = 0.4 and is quite constant. While constancy is not strictly the case for other systems, it is a good approximation. The value of DRF varies from 0.68 for phage T1 in *E. coli* B to 0.20 for phage T4 in the same host.

The time at which photoreactivating light is applied, relative to the UV irradiation, is also quite important. Under normal metabolic conditions, photoreactivability falls off rapidly, not necessarily related to cell division. There can be a rapid loss of photoreactivability in starved bacteria while a considerable retention of photoreactivability after division in *Paramecium*. It has also been retained in bacteria kept at $-10°C$ for 6 days and in spherical plant viruses for as long as 14 days. For killing in *E. coli* B/r it falls exponentially to zero in 2-3 hours, whereas for mutation in the same bacterium it drops exponentially to zero in only 20 min. In some systems such as potato virus X, photoreactivation is not possible until about 30 min after innoculation of the host, corresponding to the time required for the nucleic acid to be freed of its protein coat (Chessin, 1958).

In other systems the surprising result is that there is a *negative time lag* for photoreactivation, which Jagger has termed *photoprotection*, first discovered by Weatherwax (1956) for killing *E. coli* B and by Miki

(1956) for killing in *E. coli* K12. Jagger and Stafford (1962) have made an accurate measurement of the action spectra for photoprotection from UV killing in both *E. coli* B and *Pseudomonas aeruginosa* and have found them to be identical to one another within experimental error. However, the action spectrum for photoprotection in *E. coli* B is quite different in shape from that for photoreactivation in *E. coli* B/r and *E. coli* B, shown in Fig. 5.55. The action spectrum for photoprotection lies entirely in the near ultraviolet, while that for photoreactivation extends out into

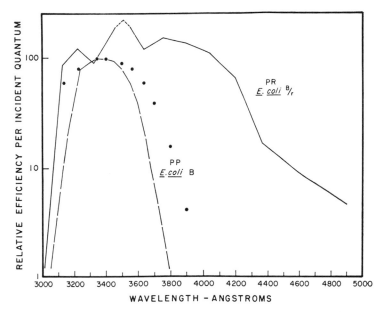

FIG. 5.55. Action spectra (solid and dashed lines) for photoprotection (PP) in *E. coli* B and *P. aeruginosa* and for photoreactivation (PR) in *E. coli* B/r, both normalized to 100 at 3341 A. The dark circles represent the normalized absorption spectrum of the β isomer of DPNH showing the strong correlation between DPNH absorption and photoprotection (Jagger and Stafford, 1962).

the visible regions of the spectrum. At a UV survival level of 2%, photoprotection is about one-third as efficient as photoreactivation in terms of incident quanta. Jagger and Stafford suggest that DPNH is the chromophore involved in photoprotection, although the mechanism appears to be completely unknown. Photoprotection appears to be independent of temperature ($Q_{10} = 1.05$) and of dose rate at low doses (Jagger, 1960a,b).

Although the process of photoreactivation was shown to have a Q_{10} of 2 and an activation energy of 17 kcal/mole, both indicative of an enzyme-controlled reaction, it was not until 1957 that Goodgal and associates demonstrated an increase of activity of UV-inactivated *Hemophilus influenzae* transforming factor with both visible light and an extract of *E. coli* which appears to be heat labile. Rupert (1958, 1960) also found photoreactivation in *H. influenzae* with extracts of baker's yeast which is itself capable of photoreactivation. The yeast agent is nondialyzable and heat labile; it is progressively inactivated by trypsin or chymotrypsin; it can be salted out and recovered quantitatively. In all these respects it can be classified as an enzyme. In the dark it forms a stable complex with UV-irradiated DNA. In this bound condition it is stabilized against heat- and heavy metal-inactivation. The photoreactivating enzyme (PRE)-DNA complex sediments in the ultracentrifuge at the rate characteristic of DNA. In the light the complex dissociates to generate a "repaired" DNA structure and free enzyme. The presence of photoreactivable damage in UV-irradiated nontransforming DNA can be detected by the competitive inhibition of the rate of "repair" of UV-irradiated transforming DNA when the former is added to a solution of the latter. By the criterion of competitive inhibition, photoreactivable damage occurs in all kinds of mammalian, bacterial, and viral DNA tested including single-stranded ϕX-174 DNA but not with irradiated RNA (Rupert and Herriott, 1961). An interesting aspect in terms of mechanism is that photoreactivable damage (as inferred by competitive inhibition of "repair" of UV-irradiated *H. influenzae* transforming principle) develops in heat, acid, and alkali-denatured DNA. However, it does not occur when the *irradiation* of denatured DNA takes place below pH 3.0–3.5, the region in which purine and pyrimidine amino groups become protonated. Melted irradiated DNA inhibits to the same extent as native DNA. The sequence of irradiation and melting or acid or base denaturation is not important. The UV "lesions" are quite stable to acid and heat treatment. UV-irradiated poly deoxy-AT does not inhibit. However, UV-irradiated poly deoxy-GC inhibits slightly. UV-irradiated 35% 5-bromouracil DNA competes for the PRE. However, the "lesions" in the 5-bromouracil DNA are not "repaired" so that the inhibitory power remains.

Mechanisms. The purine and pyrimidine components of nucleic acids are almost entirely responsible for the absorption spectra of DNA. The chromophore for photoreactivation could be a strained nucleic acid or protein structure. Gates (1934) showed that UV inactivation of pepsin caused an increase in absorption in the 3100–4000 A region, and Dulbecco (1950) found that heavily irradiated T2 phage developed an absorption band in the 3000–3800 A region. Along the same lines Wang (1959a,b) found that ultraviolet produced a spontaneous reversible change in thymine and its derivatives. He suggested that the spontaneous reversal may not take place on DNA but that photoreactivation might provide the necessary energy for reversal. The effects of UV irradiation on hetero-cyclic compounds have been studied by Shugar (1960, 1961), Shugar and Baranowska (1960), Shugar and Wierzchowski (1957, 1958), and Wierz-chowski and Shugar (1960). Uracil and cytosine show a partially reversible hydration if irradiated in aqueous solution. Thymine, adenine, and gua-nine were found almost insensitive to ultraviolet. However, as summarized by Beukers and Berends (1961), freezing an aqueous thymine solution brings about a tremendous change in UV sensitivity. Very low doses of ultraviolet bring about 60–80% conversion of thymine into an irradiation product of thymine (IPT) which reverts rapidly to thymine upon irradia-tion *after thawing*. Beukers and Berends (1960, 1961) suggest the following dimer as the IPT:

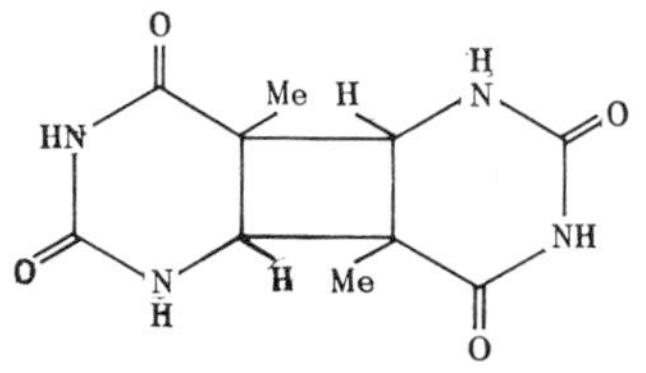

Beukers *et al.* (1960) and Wacker *et al.* (1960) have been able to isolate thymine dimer by hydrolysis of UV-irradiated DNA. This would indicate that the formation of thymine dimer is one of the significant chemical events in UV damage in microorganisms, particularly since UV irradia-tion of dilute aqueous solutions of thymine dimer causes conversion to thymine with a quantum yield between 0.4 and 1. In a further series of experiments Wulff and Rupert (1962) have shown that in samples of UV-irradiated DNA incubated *in vitro* with PRE from bakers' yeast plus visible light, the yield of thymine dimer in formic acid hydrolyzates is only 10% of the thymine dimer extractable from UV-irradiated DNA not subjected to PRE *plus* visible light. Both native PRE *and* visible light are necessary to destroy the UV-formed dimers in the DNA, presumably converting the dimers back to thymine. Even more recently Setlow *et al.* (1963) have reported that in UV-sensitive *E. coli* B_{S-1}, which shows a marked inhibition of DNA synthesis when colony formation is reduced by 50%, the kinetic data fit a model in which polymerization along a "primer" takes place until blocked (by a thymine dimer). They calculate that 2 ergs/mm² of 2650 A radiation (2.7×10^{11} photons/mm²) produces one such block in a DNA strand 350 μ in length—a length equal to one-half of a replicating DNA strand in *E. coli*. In *E. coli* B/r there appears to be an enzyme system that is able to eliminate thymine dimers *in the dark*, based on the recovery of DNA-synthesizing ability in the dark after UV irradiation. However, since thymine dimers are not split in the radiation resistant B/r strain, there is the possibility that polymerization occurs at a slow rate around the block or that the dimer is cut out of the DNA by nucleases and is replaced by two thymines. The low-dose experiments are significant since they permit work with the first effects of ultraviolet which are more likely simpler because they represent single damage events. It may also be that cytosine in DNA and uracil in RNA react under UV irradiation to form dimers. There is also the possibility of forming cytosine-thymine and uracil-cytosine dimers.

It seems quite likely that photoreactivation of UV-irradiated DNA operates primarily by a reversal of UV damage at the reactivable site, presumably a light-driven enzymatic splitting of a dimer on the DNA molecule. There may also be dark "healing" processes as in *E. coli* B/r.

There are differential responses of lethality and mutagenesis to photoreactivation. The addition of the dye acriflavine to postirradiation plating media in UV-irradiated *E. coli* B/r WU36, a tyrosine-requiring substrain of B/r (Witkin, 1963), interferes with the photoreversal of induced phototropy but not with the photoreversal of killing. It reduces the efficiency of photoprotection against induced phototropy while having no such effect on photoprotection against UV killing. The results are interpreted as evidence that not all UV damage occurs in the same pathway, making for a more complicated (at present) picture. Much less is known about photoreactivable systems not specifically involving DNA and almost nothing, about the mechanism of photoprotection.

9. Optical Asymmetry in Biological Systems

It appears to be rather general that all substances produced in the laboratory *without the action of living organisms* are racemic mixtures of antipodes. On the other hand, optical asymmetry appears to be a specific characteristic of living systems. The biologically active amino acids appear to be optically pure and to exhibit an exclusiveness in the asymmetry sign of their rotatory properties. Essentially all of the natural or biological amino acids are of the L configuration. The presence of racemic sugars is extremely rare and it is believed that, as in the case of malic acid, the newly-formed material is optically pure and only later passes into the racemic form.

One of the basic questions to be asked regarding the origin of living systems is what role does optical dissymmetry or asymmetry play in molecular biochemistry? Optical activity is only one of the examples of dissymmetry in nature. The piezoelectric effect, anisotropy of polarizability, preferential crystal packing, can be included. For example, the polymer $(-)$-α-methylbenzyl methacrylate has different properties from the polymer made from a racemate of $(+)$ and $(-)$ antipodes (Beredjik,

1956). It would be expected that in the intra molecular interaction of various portions of a macromolecule and in the intermolecular interactions of packings of molecules, the resultant configuration would be dependent upon the nature of the antipode.

The second general question to be asked is what kind of physical-chemical mechanism would be involved in the production of the original asymmetry? Van't Hoff (1908) originally demonstrated the production of net optical activity from racemic solutions by bringing them into contact with optically active substances in the form of solvents and adsorbents. There are cases of selective crystallization of enantiomorphs from solution, in optically active solvents, by seeding with a crystal of one antipode, or by spontaneous crystallization due to slight differences in solubility (Van't Hoff and Dawson, 1898). Excellent discussions of these techniques and of the general problems of asymmetric synthesis under the action of circularly polarized light are given by Terent'ev and Klabunov-skii (1960) and Klabunovskii (1960).

From a thermodynamic viewpoint the mixing of optically active components into a racemate liberates energy, while their separation requires work. Consequently, the optically active state is not as stable as the racemic state. The question arises, therefore, as to how such a condition of thermodynamic excess free energy can be obtained in enzymatic reactions in biological systems.

There are two general methods of obtaining optically active compounds. One is by the splitting-up of racemates and the other by asymmetric synthesis. If we have the enantiomorphs D-A and L-A in equal concentration then the splitting-up process can be represented as follows:

$$\text{D-A} \underset{k_D'}{\overset{k_D}{\rightleftharpoons}} \text{D-B}$$

$$\text{L-A} \underset{k_L'}{\overset{k_L}{\rightleftharpoons}} \text{L-B}$$

(5.41)

In this case optical asymmetry in the B compounds is obtained when $k_D \neq k_L$.

In asymmetric synthesis a symmetric initial substance A is transformed as follows

(5.42)

Kuhn (1936) has solved the rate equations for these two cases and has shown that the formation of an optically active product in both cases is a transient phenomenon, the final equilibrium being a racemized mixture, even for $k_L \gg k_D$ and $K \gg 1$. However, the ratio of the time required for racemization to the time required for maximum optical activity of an L enantiomorph is given by $k_L/k_D \times K/2$ in the synthesis case and by k_L/k_D in the splitting-up case. Obviously the stability of the optically active state is much greater in asymmetric synthesis than in the separation of a racemic mixture.

In nature there are several methods for achieving and maintaining this optical purity.

(a) In biosynthesis a series of n steps such as Eq. (5.42), where in each case $k_L \gg k_D$ and $K \gg 1$ will give a ratio of essentially

$$\left(\frac{k_L}{k_D} \times \frac{K}{2} \right)^n$$

with a much greater optical purity.

(b) Many enantiomorphs are "stereo-autonomic," i.e., under certain conditions one may be less soluble than its isomer and can be removed by precipitation from the racemizing reaction, thus pushing the reaction toward optical purity.

(c) Specialized enzyme systems exist (racemases, deaminases, oxidases, etc.) for the rapid metabolism of "unnatural" isomers of amino acids, since in catalysis the appearance of small quantities of "unnatural" isomers is inevitable. This more rapid breakdown to inactive keto-acids would be essentially a correction mechanism.

(d) The fundamental physiological processes of synthesis and growth proceed more rapidly with asymmetric than with racemic material, thus contributing by natural selection to precursor systems such as (a) above.

There are two lines of reasoning that have been proposed to explain the optical asymmetry found in nature. One of these assumes that in the formation of racemates, say by crystallization, while the mean optical activity is zero, individual occurrences of slight excesses of one optical form over the other could occur on the basis of statistical fluctuations from the mean. An inequality of a right or left form could have occurred when some systems were in process of synthesis and this inequality could have been preserved by asymmetric catalysis. In fact, Markwald (1904) was able to synthesize optically active valerianic acid from methylethylmalonic acid in the presence of optically active brucine.

This hypothesis would place the presently observed optical activity of biological amino acids as a purely chance initial event which was carried through in all further syntheses.

The second approach assumes that in some manner there could be produced in the primitive "soup" elliptically polarized light which, by virtue of the Cotton Effect of circular dichroism (1896), would be absorbed more strongly by one of the optical isomers of a racemic solution. This increased absorption and therefore the increased activation of one of the isomers could have produced either an increased rate of decomposition or an increased rate of photochemical synthesis, resulting over a long period in either a predominance of the antipode in the decomposition reaction or the more strongly absorbing isomer in the synthesizing reaction, depending on the competition between degradation and synthesis. The former reaction has been demonstrated in the laboratory by Kuhn and Braun (1929) and Kuhn and Knopf (1930).

Asymmetric synthesis, starting from optically inert materials was reported by Karagunis and Drikes (1933a,b) who obtained optically active products by the addition of chlorine to triarylmethyl radicals in circularly polarized light of wavelengths 4300 and 5890 A. Davis and Heggie (1935) reported the addition of bromine to 2,4,6-trinitrostilbene in CCl_4, C_6H_6, glacial acetic acid, and nitrobenzene solution in circularly polarized light of 3600–4500 A to form dextro rotatory trinitrostilbene dibromide. This value upon continuous irradiation in RH circularly polarized light rose from zero to a maximum value of 0.04° and then decreased to zero (Tenney and Heggie, 1935a,b). In addition, the product upon standing either in unpolarized light or in RHCP light loses its optical activity irreversibly. The authors attributed this to the decomposition of the unstable product.

Mechanisms for the Occurrence of Excess LHCP or RHCP Light in Nature. H. Becquerel (1889) found that the plane of polarization of scattered sunlight deviated very slightly from that predicted by Arago, i.e., the plane of the electric vector of the atmosphere-scattered light was not exactly perpendicular to the plane formed by the line from the sun to the scattering volume. Instead, the plane of the electric vector was rotated by a slight angle as a ray traversing a column of air under the influence of a magnetic field. This is the Faraday effect of the rotation of the plane of polarization of a beam of plane polarized light when the magnetic lines of force are parallel to the beam direction. The deviation found by Becquerel for blue skylight was approximately 23 min and there appeared to be a slight daily variation in this value. When a ray of light traverses an optically inactive but *absorbing medium such as the primitive soup of primaeval times, in the magnetic field* of the earth it will become elliptically polarized with the major axis of the ellipse rotated through the same angle.

Such ellipticity is associated with circular dichroism. The rotation and the Verdet constant, V, are determined by the equation

$$\alpha = VHl \qquad (5.43)$$

where H is the magnetic field strength, l is the pathlength, and V is given in angular minutes per centimeter per gauss. The sign is conventionally positive when the rotation takes place in the direction of the electric current which creates the field H. The earth's magnetic field can be approximated by assuming a bar magnet within the core about 17° off the axis of rotation. The two places on the earth's surface where the fields are vertical are called the magnetic poles. By definition, the north pole of a compass needle points to the earth's magnetic pole in the northern hemisphere. Therefore this pole must be in reality a "south" pole. If we use the "right-hand rule" from electromagnetism, i.e., if the right hand is held with the fingers curled with the thumb extended, the direction of the magnetic field will be in the direction of the thumb if the current flow producing the field is in the direction of the fingers. The magnetic field owing to the earth's internal magnet would therefore be produced by a circular current rotating counterclockwise to an observer in the northern hemisphere looking upward in the direction of the sky. Consider now a beam of plane polarized UV light incident on the anaerobic surface of the primitive ocean. Looking upward from below the surface, the plane of polarization of the incident beam in water will appear to rotate in the direction of the "electric current" which creates the field; in this case the rotation of the plane of polarization is counterclockwise. This is equivalent to LHCP light traveling *faster* than RHCP, or $\eta_L < \eta_R$. Since the index of refraction is related to the absorption, it would be expected that close to the absorption band of water in the ultraviolet, around 1850 Å, there would be a greater absorption of RHCP light. Therefore, in the UV region above 1850 Å, in the ocean, or inland water where we expect that the original macromolecules were produced by photochemical polymerization, there was a slight excess of LHCP light, due to circular dichroism associated with the Faraday effect near the UV absorption band of water vapor. This same effect could have been produced by H_2O in the atmosphere so that UV light incident on the primitive ocean also contained a slight excess of LHCP light.

In the southern hemisphere this effect is reversed and there should be a slight excess of RHCP light.

There have been many references in the literature to the existence of a slight excess of one form of circularly polarized light over the other (Terent'ev and Klabunovskii, 1960; Gause, 1941), most of which ultimately go back to Byk (1904), who suggested that because of the rotation of the plane of polarization by terrestrial magnetism there must be a predominance of one of the two forms of circularly polarized light. This predominant form, acting for long periods of time on racemic compounds, could conceivably induce optical activity in photochemical reactions.

It is a well-known fact that plane polarized light, reflected from conductors at other than normal incidence, suffers a phase change between the components ‖ and ⊥ to the plane of incidence, giving rise to elliptically polarized light. This is a general observation for all metals and has been observed for radio waves reflected from sea water where the conductivity is only 3 mhos/meter. However, at optical frequencies in the latter case only the dielectric properties of the medium enter. In the reflection of plane polarized light from an air → water interface there can be a rotation of the plane of polarization of the incident owing to unequal reflectivities for light ‖ and ⊥ to the plane of incidence, but the phase change is either zero or π; the reflected light remains plane polarized.

In the case of plane polarized light which is completely internally reflected, *there is a phase change* between the ‖ and ⊥ components of the light vector in the plane of incidence and the reflected light will be elliptically polarized. This was demonstrated by Fresnel who designed a glass rhomb with two internal reflections, each producing a 45° phase shift between ‖ and ⊥ components upon reflection to produce circularly polarized light from plane polarized light initially incident at 45° to the plane of incidence. Thus it is conceivable that internal reflection of polarized light reflected from the bottom of small pools at a water → air interface can produce elliptically polarized light in the water. However, from symmetry we would expect just as much of RH elliptically polarized light as LH elliptically polarized light.

By postulating a circular dichroism associated with the Faraday rotation of plane polarized scattered sunlight we can provide a physical basis for a slight excess of LHCP light in the northern hemisphere. Since the Faraday rotation is reversed by a reversal of the magnetic field, the same argument would predict an excess of RHCP light in the southern hemisphere. Thus we have gone one step further geographically than those proponents of chance statistical fluctuation, and we infer that if this slight excess of circularly polarized light is indeed responsible for biological asymmetry, life may have originated in either the southern or the northern hemispheres. Presumably if we can find optically active organic molecules in very old sediments of the early Archeozoic era, there may be slight differences in their optical rotations, depending on their hemispheric origin. Recently, evidence has been found of optical activity in Precambrian sediments approximately 1 billion years old (Meinschein et al., 1964; Eglinton et al., 1964). However, Cox et al. (1964) have presented evidence, based on the measurement of remanent magnetization of volcanic rocks, that there have been reversals of the earth's magnetic field. They place the last reversal at 1 million years ago and the next most recent reversal at about 2.5 million years. While the mechanism is not completely understood, if this hypothesis is true it would be impossible to test our hypothesis of different optical activities in 1–2 billion-year-old sediments from the north and south hemispheres.

Physical Constants

A	Avogadro constant	6.025×10^{23} (g-mole)$^{-1}$
k	Boltzmann constant	1.380×10^{-16} erg deg^{-1}
a_0	Bohr radius	0.529 A
e	Electronic charge	4.803×10^{-10} esu
		1.602×10^{-19} coulomb
F	Faraday constant	96,522 coulombs (g-mole)$^{-1}$
R_0	Gas constant	8.317×10^7 erg deg^{-1} mole^{-1}
		1.986 cal deg^{-1} mole^{-1}
T_0	Ice point	273.16°K
O^{16}	Mass of oxygen-16	16.000
m	Mass of electron	9.108×10^{-28} g
p/m	Ratio of proton mass to electron mass	1836.1
H	Mass of hydrogen atom	1.008
D	Mass of deuterium atom	2.015
T	Mass of tritium atom	3.017
V_0	Molar volume	2.2421×10^4 cm^3 mole^{-1}
h	Planck's constant	6.625×10^{-27} erg sec
R_∞	Rydberg constant for infinite mass	109,737.3 cm^{-1}
c	Speed of light in vacuum	2.998×10^{10} cm sec^{-1}
σ	Stefan-Boltzmann constant	5.669×10^{-12} watt cm^{-2} deg^{-4}
$\lambda_{\max} T$	Wien displacement law constant	0.2899 cm deg

Conversion Factors

1 electron volt	1.602×10^{-12} erg
	8066.0 cm^{-1}
	12,397.7 A
	23.08 kcal/mole
	11,605.8°K
1 Angstrom	10^{-8} cm
	10^{8} cm^{-1}
	12397.7 ev
	286,000 kcal/mole
	1.987×10^{-8} erg
1 joule at 5560 A	2.80×10^{18} photons
1 μwatt/cm^2 at 5560 A	2.80×10^{12} photons/sec-cm^2
1 cm^{-1}	2.86 cal/mole
1 electron/sec	1.60×10^{-19} ampere
1 kcal/mole	28.6 μ
Mechanical equivalent of heat	4.186 joules/cal
Energy in one quantum	$\dfrac{1987}{\lambda[A]} \times 10^{-18}$ joule
Energy in one Einstein	$\dfrac{286,000}{\lambda[A]}$ kcal
Mechanical equivalent of light (5560 A; wavelength of maximum visibility)	680 lumens/watt
C—C bond dissociation energy	82 kcal/mole
	3490 A
	28,650 cm^{-1}
O—H bond dissociation energy	118 kcal/mole
	2420 A
	41,300 cm^{-1}

Lasers

The principle of microwave amplification by stimulated emission of radiation or MASER operation was first suggested by J. Weber (1952). An operating MASER was first built by Gordon and co-workers (1954). The feasibility of MASER operation at optical frequencies (*Light Amplification by Stimulated Emission of Radiation*) was proposed by Schawlow and Townes (1958). Maiman (1960) succeeded in operating a pulsed ruby LASER, and Javan and associates (1961) developed a continuous wave gas LASER using helium and neon. This was followed 1 year later by continuous wave (CW) action in $CaWO_4:Nd^{3+}$ (L F. Johnson *et al.*, 1962), ruby (Nelson and Boyle, 1962), and $CaF_2:U^{3+}$ (Boyd *et al.*, 1962). Excellent and extensive reviews of this subject can be found by J. Weber (1959), Bennett (1962), Heavens (1962), and Yariv and Gordon (1963). In particular, *Applied Optics* supplement on "Optical Masers" (1962) and *Proceedings of the IEEE* (1963) are special issues devoted entirely to LASERS and quantum electrodynamics.

The properties of the optical maser are contained in the original treatment by Einstein (1916, 1917), describing the emission and absorption of radiation by quantized systems, in which he derived the Planck Radiation Law. Einstein considered that a large number of similar atoms were in a field of radiation. Assuming Bohr's first postulate, each atom is capable of existing in a discrete series of energy states $Z_1, Z_2, Z_3, \ldots$, each characterized by a discrete energy level, $\epsilon_1, \epsilon_2, \epsilon_3, \ldots$, with Z_1 being the lowest allowed state or ground state. If an atom passes from the state Z_m to state Z_n by the absorption of a quantum of radiation of energy $\epsilon_n - \epsilon_m$, it can revert to state Z_m by the emission of a quantum of energy $\epsilon_n - \epsilon_m$. He then assumes three analogous processes.

Absorption. Of the number N_m of atoms in state Z_m, a certain fraction pass every second to the state Z_n by absorption of radiation. The number of such transitions per second is proportional to the density of radiation $U(\nu)$ of the absorbed frequency, ν. Thus the number of transitions per second will be

$$B_{m \rightarrow n} N_m U(\nu) \tag{III.1}$$

where $B_{m \rightarrow n}$ is *Einstein's coefficient of absorption* and is characteristic of the combination of states Z_m and Z_n.

Induced Emission. Owing to an interaction with the radiation field, a number of atoms in the higher state Z_n are *induced* to jump down to state Z_m. The number of induced transitions per second will be

$$B_{n \to m} N_n U(\nu) \tag{III.2}$$

where $B_{n \to m}$ is *Einstein's coefficient of induced emission* and N_n is the number of atoms in the higher state Z_n.

Spontaneous Emission. Owing to the fact that state Z_n is higher in energy than state Z_m, there is a probability that an atom in state Z_n will drop down "spontaneously" to state Z_m. The number of spontaneous transitions per second will be

$$A_{n \to m} N_n \tag{III.3}$$

where $A_{n \to m}$ is *Einstein's coefficient of spontaneous emission.* The spontaneous transition probability is independent of the radiation field. In a radiation field $U(\nu)$ at equilibrium the number of atoms per second passing from Z_m to Z_n must equal the number of atoms per second passing from Z_n to Z_m. This can be written as

$$N_m B_{m \to n} U(\nu) = N_n [B_{n \to m} U(\nu) + A_{n \to m}] \tag{III.4}$$

The Boltzmann statistical distribution of energies is assumed, i.e., in a system in thermodynamic equilibrium at the temperature T, the numbers of atoms in the different quantum states $Z_1, Z_2, Z_3, \ldots$ with energies $\epsilon_1, \epsilon_2, \epsilon_3, \ldots$ are given by $N_1 = a p_1 e^{-\epsilon_1/kT}$, $N_2 = a p_2 e^{-\epsilon_2/kT}$, $\ldots$ where a is a constant and $p_1, p_2, \ldots$ are the statistical weights of the states $Z_1, Z_2, \ldots$. If we now substitute the Boltzmann expression into Eq. (III.4) and solve for $U(\nu)$ we obtain

$$U(\nu) = \frac{P_n A_{n \to m}}{P_m B_{m \to n} e^{\epsilon_n - \epsilon_m/kT} - P_n B_{n \to m}} \tag{III.5}$$

If we assume that $U(\nu)$ increases as T increases, then the denominator of Eq. (III.5) must approach zero as $T \to \infty$. Therefore

$$P_m B_{m \to n} = P_n B_{n \to m} \tag{III.6}$$

If we assume equal statistical weights for the energy levels, this means that $B_{m \to n} = B_{n \to m}$ or that the Einstein coefficient of absorption is equal to the Einstein coefficient of induced emission. Since the Planck radiation law for the energy density of radiation in a blackbody cavity is given by

$$U(\nu) = \frac{8 \pi h \nu^3}{c^3} \cdot \frac{1}{e^{h\nu/kT} - 1} \tag{III.7}$$

the coefficient of spontaneous emission $A_{n \rightarrow m}$ is related to the coefficient of induced emission $B_{n \rightarrow m}$ by

$$B_{n \rightarrow m} = \frac{c^3}{8\pi h \nu_{nm}^3} A_{n \rightarrow m} \qquad (III.8)$$

Accordingly, in a radiation field of a density $U(\nu)$ the ratio, f, of the number of induced emissions per second to the number of spontaneous emissions per second is given by

$$f = \frac{B_{n \rightarrow m} N_n U(\nu)}{A_{n \rightarrow m} N_n} \qquad (III.9)$$

Assume that $U(\nu)$ is derived from blackbody emission at a high temperature T. The stream of radiation emitted by a blackbody in any direction is the same as the stream of radiation traveling in one direction in a blackbody cavity at the same temperature. The energy density in an enclosure, $U(\nu)$, will be twice the energy density, $\psi(\nu)$, due to emission from a blackbody surface, since the radiation emitted by a blackbody surface is emitted over a hemisphere, whereas the radiation in an enclosure is traveling in all directions. Further, for a blackbody surface radiator, obeying a cosine law the emissive power of the surface, $I(\nu)$, is related to the energy density due to emission from the surface by

$$I(\nu) = \frac{c}{2} \psi(\nu) \qquad (III.10)$$

Therefore the intensity of illumination of frequency ν due to a blackbody source of radiation is related to the energy density in a blackbody cavity from Eq. (III.7) by

$$I(\nu) = \frac{c}{4} U(\nu) \qquad (III.11)$$

This is equivalent to a radiation density in the path of $I(\nu)/c$, and the ratio of *induced* to *spontaneous* emission from Eq. (III.9) becomes

$$\frac{1}{4(e^{h\nu/kT} - 1)} \qquad (III.12)$$

In the range of optical frequencies for transitions and blackbody or thermal source temperatures normally available in the laboratory this fraction is very much less than 1. For example, for a transition at 5000 A and an irradiating source at a temperature of 3000°K, $h\nu/kT$ equals 10 and

$$\frac{1}{4(e^{h\nu/kT} - 1)} \sim 10^{-5}$$

The key to the maser or laser action is contained in Eq. (III.9). By the use of resonant cavities it is possible to build up standing waves of radiation at particular frequencies such that the density of the radiation field $U(\nu)$ in a small frequency interval is far and above those densities that would be obtained from thermal distributions.

The line shape of a transition $g(\nu)$ can be obtained from the normalized absorption curve plotted versus ν. Thus we define $g(\nu)\ d\nu$ as the probability that a given transition will result in an absorption (or induced emission) of a photon with energy between $h\nu$ and $h(\nu + d\nu)$, so that

$$\int_0^\infty g(\nu)\ d\nu = 1 \qquad\qquad (III.13)$$

From the definition of $g(\nu)\ d\nu$ we can see that $(Ag(\nu)\ d\nu)$ is the spontaneous transition rate resulting in the emission of a photon of frequency between ν and $\nu + d\nu$, while $(BU(\nu)g(\nu)\ d\nu)$ is the *induced* emission rate, W_{ind}, of a photon of frequency between ν and $\nu + d\nu$, owing to a radiation field of density $U(\nu)$ per unit frequency interval. In a radiation cavity the intensity of radiation of frequencies between ν and $\nu + d\nu$, $I(\nu)\ d\nu$, is given by

$$I(\nu)\ d\nu = cU(\nu)\ d\nu \qquad\qquad (III.14)$$

so that finally, from Eqs. (III.8) and (III.9)

$$W_{\text{ind}} = \frac{c^2}{8\pi h\nu^3}\, g(\nu)I(\nu) \cdot \frac{1}{\tau_0} \qquad\qquad (III.15)$$

where τ_0 is the lifetime for the spontaneous transition, defined by $A = 1/\tau_0$.

In the case of the MASER, owing to the small number of allowed modes in the resonant cavity, the value of the decay time constant for emission of spontaneous radiation into the cavity can be substantially increased above the free space value τ_0 and thus markedly increasing the ratio of induced to spontaneous emission. In the ordinary LASER where the cavity is constructed of reflecting plates separated by a distance large compared to their diameters this no longer holds.

The decay time constant for radiation losses in the laser cavity, t_{loss}, is given by

$$t_{\text{loss}} = \frac{\eta L}{\alpha c} = \frac{Q}{2\pi\nu} \qquad\qquad (III.16)$$

where L is the distance between reflecting plates, α is the fractional loss per pass, and η is the dielectric constant of the medium. Q is the "quality factor" of the resonant cavity. If we now have two levels Z_1 and Z_2 populated by a radiation field $I(\nu)$ with N_1 and N_2 being the population densities at each level, the rate of increase in the energy flux due to induced transitions from Z_2 to Z_1 will be given by

$$\left[\frac{dI(\nu)}{dt}\right]_{2\to1} = h\nu(N_2 - N_1)cW_{ind} \tag{III.17}$$

where $h\nu$ is the energy of a photon of frequency ν. The rate of loss of intensity due to radiation in the cavity is given by

$$\left[\frac{dI(\nu)}{dt}\right]_{loss} = -\frac{I(\nu)}{t_{loss}} \tag{III.18}$$

In order that sustained oscillation be developed, the rate of increase in the induced emission must just compensate for the losses, or

$$h\nu(N_2 - N_1)cW_{ind} - \frac{I(\nu)}{t_{loss}} \geqslant 0 \tag{III.19}$$

This gives the condition that, at threshold of LASER operation, from Eqs. (III.14) and (III.15)

$$N_2 - N_1 = \frac{8\pi\nu^2}{g(\nu)c^3}\frac{\tau_0}{t_{loss}} \tag{III.20}$$

Thus Eq. (III.20) gives the *critical inversion population density* for LASER operation and for any given material and frequency depends only on t_{loss}, defined by Eq. (III.16).

In the solid state lasers such as the ruby laser, optical excitation is used to obtain the required population density inversions. This is generally effected by broad-band absorption, owing to the matrix or host material, followed by rapid, sharp transitions to metastable levels present because of impurities deliberately placed within the host lattice. The impurities are usually transition metal or rare earth or actimide series elements, characterized by sharp spectral transitions. A 3-level laser (ruby — $Al_2O_3:Cr^{3+}$) is shown in Fig. III.1a. The useful absorbing regions for the $^4A_2 \to {}^4F_1$ and the $^4A_2 \to {}^4F_2$ transitions are between 3200–4400 A and 5000–6000 A. In both cases the energy rapidly decays to the ^2E level which has a high fluorescence yield and a mean life $\tau_0 = 3 \times 10^{-3}$ sec. From the integral of the absorption curve for the $1 \to 3$ transition (see Chapter 2, Section 2) the lifetime for the $3 \to 1$ transition is 3×10^{-6} sec. From a comparison of the fluorescence yields for the $3 \to 1$ transition and the $2 \to 1$ transition the lifetime of the $3 \to 2$ transition is 5×10^{-8} sec. Therefore we can neglect both spontaneous and stimulated emission for $3 \to 1$ transition compared to $3 \to 2$ transitions. Most of the absorption ends up in level 2. At room temperature the fluorescence line width is ~11 cm^{-1}. This is reduced to ~0.1 cm^{-1} at liquid N_2 temperature.

Typical data for a ruby laser would have a reflecting plate separation of 10 cm and a 2% loss per pass. This gives from Eq. (III.16) a loss of lifetime $t_{loss} \sim 3 \times 10^{-8}$ sec. The fluorescence wave number $\bar{\nu}$ is 14,418 cm^{-1} (6943 A) and we again neglect multiplicity considerations. If we assume that the shape of the absorption curve $g(\nu)$ is gaussian, then $g(\nu)$ (center frequency) $= 0.94/\Delta\bar{\nu}$, where $\Delta\bar{\nu} = 11$ cm^{-1}.

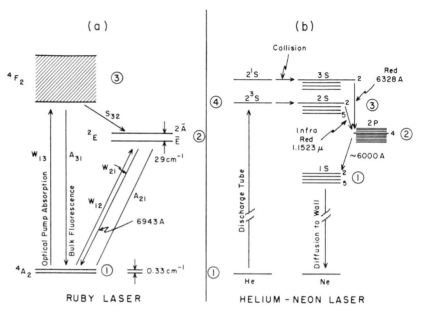

FIG. III.1. Energy level diagrams for a 3-level laser (ruby-Al$_2$O$_3$: Cr^{3+}) and a 4-level laser.

Substituting τ_0, t_{loss}, ν, and $g(\nu)$ into Eq. (III.20) and remembering $\nu = c\bar{\nu}$ we can derive a critical inversion population density of

$$N_2 - N_1 = 26.7\bar{\nu}^2\Delta\bar{\nu}\,\frac{\tau_0}{t_{loss}} \qquad (III.21)$$

$$N_2 - N_1 \sim 10^{16}$$

In ruby the typical atomic concentration is 2×10^{19}/cm^3. There is a slight splitting of the ^2E level which we shall neglect. Thus in order to maintain an inversion density of 10^{16} atoms/cm^3 we must maintain $\sim 10^{19}$ atoms/cm^3 in the excited state. The minimum number of absorptions per second required is therefore

$$\frac{10^{19}}{\tau_0} \sim 3 \times 10^{21}/\text{sec}$$

corresponding to a rate of energy absorption at 5500 A of ~ 1000 joule/

sec. The usual flash duration of a pulsed Xe flash tube is 500 μsec. If we discharge 1000 joules through the tube, the average rate of power dissipation will be 2×10^6 joules/sec. Of this, therefore, $1/2000$ must be of the proper spectral composition and optically coupled to the ruby laser in order to obtain threshold laser action. Of this total energy of the order of 10^{-2} joule appears as laser light. It has been possible with the use of modulated Kerr cells to obtain a more optimum stimulated emission, the so-called "Giant Pulses." These are several orders of magnitude higher in intensity than the ordinary stimulated pulses (McClung and Hellwarth, 1962), since the laser flash is much shorter in duration. While the ordinary laser pulse duration is around 500 μsec, the "Giant Pulse" has a duration of 0.12 μsec. Nelson and Boyle (1962) have obtained continuous power outputs from a ruby laser of 4 mwatts.

A 4-level laser is shown in Fig. III.1b. In this case there is a level ② separated from level ① such that $E_2 - E_1 \gg kT$. Note here that the lifetime of the 2P $\rightarrow$ 1S transition of Ne is much shorter than that of the 2S $\rightarrow$ 2P and 3S $\rightarrow$ 2P transition. With this fact, and since level ② cannot be populated thermally, it can be considered as virtually empty. Here then is the advantage of the 4-level laser. Since level ② is practically empty, N_1, in Eq. (III.20), which now corresponds to the population density of level ②, will be very much less than $N_2 - N_1$ and therefore Eq. (III.21) becomes

$$N_2 \sim 26.7 \bar{\nu}^2 \Delta \bar{\nu} \frac{\tau_0}{t_{\text{loss}}} \tag{III.22}$$

We saw that for an inversion density of 10^{16} atoms/cm^3 in ruby it was necessary to maintain 10^{19} atoms/cm^3 in the excited state. Here in the case of the 4-level laser the population density of excited states is essentially the same number as the critical inversion density. Therefore, much less pump power is required for a 4-level laser, all other factors being equal.

In the case of the He-Ne laser a radio frequency discharge is established in the gas mixture (usually 1.0 mm Hg of He and 0.1 mm Hg of Ne). The He atoms are excited by the electrons in the discharge tube to a variety of excited states. Many of these end up in the metastable 2^3S state of He. In collision with Ne atoms the energy is exchanged and the 2S levels of Ne are populated. It is this 2S level in the He-Ne laser that corresponds to the ^2E level of ruby. In the He-Ne laser τ_0 is 10^{-7} sec. The alignment of the Fabry-Perot plates in order to obtain oscillation is quite critical and fuller discussions of other experimental details are contained in the references.

Very lucid descriptions of the semiconductor laser have been given by Lax (1963) and Schawlow (1963). Basically the requirements for laser

operation are again the inversion of a population of electrons, the co-
herence of the stimulated emission and oscillatory amplification. Since
the description requires a knowledge of solid state theory of filled and
conduction bands, p- and n-type semiconduction, p-n junctions and electrical injection of minority carriers, we shall not attempt to give the
mechanism in this short space. It has been possible to obtain a GaAs
diode laser operating continuously with a photon output of 1.5 watts of
8400 Å light at an over-all efficiency of 30%. By operating the diode laser
in a variable magnetic field the energy levels can be varied and in the
case of InAs the output wavelength can be "tuned." Another method of
continuously adjusting the output wavelength is to vary the operating
temperature of the laser. In other cases (Ga-As-P) the output wavelength
will depend on the ratio of As to P, and can be adjusted from 8400 to
31,000 Å. This, together with accurate temperature control, can produce
a wide wavelength range for high intensity monochromatic radiation.

Tables III.1–III.3 give a summary of the various lasers reported in the
literature, together with their emission wavelengths.

TABLE III.1
IONIC LASERS[a]

Emission wavelength (μ)	Active material	Host material	Operation	Temperature (°C)	Output power[b] (watts)
.3125	Gd^{3+}	Glass	P	−196	
.6100	Eu^{3+}	Y_2O_3	P	20	
.6943	Cr^{3+} (ruby)	Al_2O_3	P and CW	−269 to 350	0.001–0.5
.708	Sm^{++}	CaF_2	P	−196	
1.015	Yb^{3+}	Glass	P	−196	
1.047	Pr^{3+}	$CaWO_4$	P	−196	
1.06	Nd^{3+}	$CaWO_4$	P and CW	−196 to 27	
1.06	Nd^{3+}	Glass	P	27	
1.116 ⎫	Tm^{++}	CaF_2	P	approx. −269 ⎫	
1.189 ⎭	Tm^{++}	CaF_2	P	approx. −269 ⎭	
1.612	Er^{3+}	$CaWO_4$	P	−196	
1.911	$Tn.^{3+}$	$CaWO_4$	P	−196	
2.046 ⎫	Ho^{3+}	$CaWO_4$	P	−196 ⎫	
2.059 ⎭	Ho^{3+}	$CaWO_4$	P	−196 ⎭	
2.223 ⎫	U^{3+}	CaF_2	P and CW	−196 to 27 ⎫	1
2.438 ⎪	U^{3+}	CaF_2	P and CW	−196 to 27 ⎪	
2.511 ⎪	U^{3+}	CaF_2	P and CW	−196 to 27 ⎪	
2.613 ⎭	U^{3+}	CaF_2	P and CW	−196 to 27 ⎭	
2.36	Dy^{++}	CaF_2	P and CW	−183	0.3

[a] Optically pumped.
[b] Efficiency less than 1% of input energy.

TABLE III.2

Gas Lasers[a]

Major emission lines (μ)	Gas	Pumping method	Output power[b] (watts)
.6328⎫ 1.1523⎬ 3.3913⎭	He-Ne	Gas discharge	0.0015
.8446	Ne-O$_2$	Gas discharge	
.8446	A-O$_2$	Gas discharge	
1.1523⎫ 2.102 ⎭	Ne	Gas discharge	
1.6941⎫ 2.0616⎭	A	Gas discharge	
2.0261⎫ 5.5738⎬ 35 ⎭	Xe Xe	Gas discharge	
2.0261⎫ 2.6511⎪ 3.3667⎪ 3.5070⎬ 3.6849⎪ 3.9955⎪ 5.5738⎭	He-Xe	Gas discharge	
2.0603	He	Gas discharge	
2.1902⎫ 2.5234⎭	Kr	Gas discharge	
3.2 ⎫ 7.1821⎭	Cs	He discharge lamp (optical)	

[a] All continuous.
[b] Efficiency less than 1% of input energy.

TABLE III.3

Semiconductor Injection Lasers

Emission wavelength (μ)	Material	Operation	Temperature (°C)	Output power[a] (watts)
.65–.84	GaAsP	P	−175	
.84	GaAs	P and CW	−196 to 20	1.5 (eff $\simeq$ 30%)
.91	InP	P and CW	−253 to −153	
3.1	InAs	P and CW	−269 to −196	

[a] Output power approximately 1 watt (1 photon per electron injected) ($\sim$100% efficiency).

Filter Data for Isolation or Absorption

Type of filter	Properties
Ultraviolet transmitting	
Calcium fluoride, CaF_2	Transmits from below 1800 A to 10 μ in infrared
Corning C.S. 9-54	Transmits from 2100 A to 4.5 μ in infrared
Corning C.S. 7-54	Band in ultraviolet from 2300 to 4200 A; also band in infrared, peak at 7200 A
Carbon tetrachloride, 0.5 cm	Transmits from 2550; useful for eliminating the 2537 A line of Hg
Potassium acid phthalate, 1.0 cm, 5 g/liter (0.0245 M)	Isolates 3130 A group of Hg lines from those transmitted around 3000 A
Naphthalene, 1.0 cm, 12.8 g/liter in isooctane	Isolates 3340 A Hg line. Removes all radiation of wavelength shorter than 3160 A. This is very slightly fluorescent in the 3200–3900 A region
Visible transmitting	
Corning C.S. 3-75 + 1-69	Absorbs in ultraviolet and infrared
Dense flint glass	Absorbs in ultraviolet; transmits to 2.7 μ in infrared
10% w/v solution $CuSO_4$	Absorbs in ultraviolet below 2900 A; absorbs in infrared
Saturated $NaNO_2$ solution	Absorbs in ultraviolet; transmits infrared
Corning C.S. 5-60 + 1-69	Blue transmission
Corning C.S. 4-96 + 1-69	Blue-green transmission
Corning C.S. 4-64 + 1-69	Green transmission
Concentrated $K_2Cr_2O_7$ solution	Far-red transmission
Infrared transmitting	
Corning C.S. 7-57	Transmits .66–4.5 μ; absorbs ultraviolet, infrared
Concentrated $K_2Cr_2O_7$ solution	Transmits above 7000 A
I_2: saturated solution in CS_2	Absorbs ultraviolet, visible
Germanium	Becomes transparent above 1 μ
Silicon	Becomes transparent above 1 μ
PbS	Becomes transparent above 2.8 μ
PbSe	Becomes transparent above 4 μ

The data in this section is taken in part from Bowen (1949) and Kasha (1948).

MEDIUM BAND-PASS TRANSMISSION FILTERS, 10% FULL WIDTH, HALF MAXIMUM

Component	Substance	Solution	Optical path
1	$CuSO_4 \cdot 5H_2O$	100 g/liter	5 cm
2	$NiSO_4 \cdot 6H_2O$	100 g/liter	5 cm
3	$NiSO_4 \cdot 6H_2O$	200 g/liter	5 cm
4	$NiSO_4 \cdot 6H_2O$	300 g/liter	5 cm
5	$NiSO_4 \cdot 6H_2O$ plus	240 g/liter	5 cm
	$CoSO_4 \cdot 7H_2O$	45 g/liter	
6	Corning Filter 7-37	—	5 mm
7	Corning Filter 7-51	—	4 mm
8	Corning Filter 7-54	—	3 mm
9	2,7-Dimethyl-3,6-diaza cyclohepta-1,6-diene iodide	20 mg/100 ml	1 cm
10	Naphthalene	12.8 g/liter isooctane	1 cm
11	K_2CrO_4	0.2 g/liter	1 cm
12	$KHC_8H_4O_4$	5.0 g/liter	1 cm
13	CCl_4	Pure liquid	0.5 cm
14	1,4-Diphenylbutadiene	4.24 mg/100 ml ether	1 cm

Peak wavelength	Peak transmission (%)	Spectral line (A)	Description of filter[a]
2560	32	Hg 2537	Component 5 + 14 + Corning 7-54
3120	40	Hg 3130	Component 3 + 8 + 11
3300	33	Hg 3341	Component 2 + 7 + 10
3620	30	Hg $\begin{cases} 3650 \\ 3663 \end{cases}$	Component 1 + 6 + 9
3250	90	Hg 2537 Hg $\begin{cases} 3663 \\ 3650 \end{cases}$	Corning[b] C.S. 7-54 (transmits from 2300 to 4200 A; > 40% transmission at 2537 A; > 80% transmission at 3663 A)
3650	55	Hg $\begin{cases} 3650 \\ 3663 \end{cases}$	Corning C.S. 7-39 (15% FWHM)
3650	16	Hg $\begin{cases} 3663 \\ 3650 \end{cases}$	Corning C.S. 7-83
3950	13	Hg 4047	Corning C.S. 5-62
4250	15	Hg 4358	Corning C.S. 5-74
4600	8	Xe 4671	Corning C.S. 5-75
4850	6	Xe $\begin{cases} 4807 \\ 4843 \end{cases}$	Corning C.S. 4-104
5150	11	—	Corning C.S. 4-105
5550	12	Hg 5461	Corning C.S. 4-102
5800	4	Hg $\begin{cases} 5770 \\ 5791 \end{cases}$ Ne $\begin{cases} 5852 \\ 5882 \end{cases}$	Corning C.S. 3-110
6050	12	Ne strong lines	Corning C.S. 2-77
6400	17	Ne strong lines	Corning C.S. 2-78

MEDIUM BAND-PASS TRANSMISSION FILTERS, 10% FULL WIDTH, HALF
MAXIMUM (*Continued*)

Peak wavelength	Peak transmission (%)	Spectral line (A)	Description of filter[a]
7200	10	Xe $\begin{cases} 7120 \\ 7284 \end{cases}$ Ne $\begin{cases} 7173 \\ 7245 \end{cases}$	Corning C.S. 2-106
7500	17	Xe 7394	Corning C.S. 7-84
9500	25	—	Corning C.S. 7-85
2.1 μ	45	—	Corning C.S. 7-86 (45% FWHM)
2.4 μ	22	—	Corning C.S. 7-87 (16% FWHM)

[a] From Kasha (1948).
[b] Corning Glass Works, Corning, New York.

NARROW BAND-PASS TRANSMISSION FILTERS, 0.1–1.5% FWHM

Peak wavelength (A)	Peak transmission (%)	FWHM	Description of filter
2100–2300	5–10	120 A	All dielectric interference[a]
2300–3400	5–10	100 A	All dielectric interference
4000–7000	About 40	100 A	Fabry-Perot interference[b,c]
4000–7000	About 50	1.5%	All dielectric interference
4800–7000	About 30	0.1%	All dielectric interference

[a] Baird-Atomic, Inc., 33 University Road, Cambridge, Massachusetts.
[b] Photovolt Corp., 95 Madison Avenue, New York, New York.
[c] Farrand Optical Co., Bronx Blvd. and E. 238th St., New York 70, New York.

Thermopile Method for Determination of Light Intensities

Equipment Recommended

Eppley linear thermopile coated with lamp black or Parson's black. (The calibrations furnished with these thermopiles have agreed with our own to within $\pm 0.5\%$. However, over long time intervals we would recommend recalibration to avoid systematic error.)

Keithley Model 150A microvoltmeter

National Bureau of Standards Standard Lamp of Radiant Energy

a-c Ammeter, 1% full scale accuracy, 0–0.5 ampere

Autotransformer, rheostat, to adjust lamp current

Procedure

The radiant flux at 2 meters from the standard lamp operated at a current of 0.35 ampere is specified in the calibration sheet furnished with the lamp and is around 80 μwatts/cm^2. The nominal responsivity of the Eppley thermopile is 0.04 μvolts per μwatt/cm^2 incident on the exposed thermopile junctions. It is necessary to measure, therefore, about 3 μvolts. With the Keithley instrument, in a room free from air currents and large thermal gradients, using a rotating shutter (1 min on, 1 min off), and recording the microvoltmeter output on chart paper, we have been able to obtain a precision of 0.1% over several hours. For most purposes this is not necessary and a precision of 3% is easily obtained. Assume that K is the calibration constant of the thermopile with the units (μv/μw/cm^2). Then for monochromatic light or the center of a pass-band, the number of photons per second per square centimeter incident on the thermopile face is given by

$$I \ (\text{photons/sec-cm}^2) = \frac{R}{K} \times \frac{\lambda}{1987} \times 10^{12}$$

where R is the net reading in microvolts observed when the thermopile is placed in the light beam to be calibrated and λ is the wavelength of the incident light in Angstrom units.

Notes

1. *Use of Quartz Window Thermopiles*

(a) Fused quartz has average index of refraction of 1.45 over emission region of NBS standard lamp. For normal incidence Fresnel's Law gives a ratio of reflected light to total incident light of 0.034. We have found that for a quartz window of 1-mm thickness the total transmission of energy from an NBS standard lamp will be reduced by a factor of 0.86.

Example. Assume the NBS specifications give 80 μwatts/cm^2 at 2 meters from the lamp. If we have a 1-mm quartz-window thermopile there will be incident *directly on the thermopile element* only

$$80 \times 0.86 = 68.8 \ \mu\text{watts/cm}^2$$

Assume now that we are using the thermopile to calibrate in the ultra-violet at the 3663 A Hg line where $\eta_{\text{quartz}} = 1.474$. In this region quartz is essentially transparent and therefore the fraction of the UV energy incident on the thermopile element will be

$$T_{\text{UV}} = \left[1 - \left(\frac{1.474 - 1}{1.474 + 1} \right)^2 \right]^2 = 0.928$$

Notice that in the determination of the calibration constant K we use not 80 μwatts/cm^2, the specified value, but 68.8 μwatts/cm^2—that incident *directly on the thermopile element*. In a complementary manner when we take the reading R in the UV beam we are measuring only 92.8% of the energy. Therefore, the true UV photon intensity is given by

$$\frac{R}{K} \times \frac{\lambda}{1987} \times 10^{12} \times \frac{1}{.928}$$

(b) When using high sensitivity thermopiles with quartz windows *never* clean the quartz window by rubbing with a soft tissue or even brushing with a soft brush. This will produce a large static charge which can peel the gold foil absorbing element from its mounting on the thermocouple junction. Use an alpha-particle source "static-free" brush or wash gently with a detergent solution, rinse with alcohol, and let dry.

2. *Vacuum Thermopiles*

The responsivity of evacuated thermopiles is of the order of 20–100 times better than air thermopiles. However, frequent recalibration is required to ensure that loss of vacuum or desorption of gases have not changed the K value.

3. General Techniques

(a) A thermopile does not have uniform sensitivity over its area. Therefore the thermopile area should be smaller than the area of the beam being calibrated so that only the average sensitivity is involved. In this way the beam can be scanned by the thermopile to test for uniformity.

(b) If the thermopile is used to measure tiny beams, it is necessary to know the exact area of the thermopile surface, to scan the thermopile area with the beam, and to reposition the beam on the same portion of the thermopile each time a measurement is made.

(c) If a thermopile surface "looks" black it is probably nonselective. However, if it looks grayish or brownish there will be selectivity and the possibility of systematic error.

Ferrioxalate Chemical Actinometer for Determination of Light Intensities

The data in this section are taken in part from Hatchard and Parker (1956).

Actinometry Solution

Mix 3 volumes of 1.5 M $K_2C_2O_4$ with 1 volume of 1.5 M $FeCl_3$ and stir vigorously.

Recrystallize precipitated $K_3Fe(C_2O_4)_3 \cdot 3H_2O$ three times from warm water and dry.

Dissolve 2.947 g in 800 ml H_2O; add 100 ml 1.0 N H_2SO_4; dilute to 1 liter and mix. This gives a 0.006 M solution. Carry out all reactions in dim red light. (For 0.15 M solution dissolve 73.68 g of dried crystals.)

Calibration Solutions

(a) 0.4×10^{-6} mole/ml of Fe^{++} in 0.1 N H_2SO_4 freshly prepared by dilution from standardized 0.1 M $FeSO_4$ in 0.1 N H_2SO_4.

(b) 0.1% w/v 1:10 phenanthrolene monohydrate in H_2O.

(c) 600 ml 1 N $NaC_2H_3O_2$ plus 360 ml 1 N H_2SO_4 diluted to 1 liter.

To a series of 20-ml flasks add varying volumes of (a). Add [10 − (a)] ml of 0.1 N H_2SO_4, 2 ml of (b), 5 ml of (c). Add H_2O to make total of 20 ml. Mix and let stand for 30 min. Run calibration curve of optical density (OD) at 510 mμ as function of Fe^{++} moles added. This curve should be linear with a slope of 1.81×10^{-6} mole Fe^{++}/OD 1.0/cm.

Irradiation

Irradiate actinometer solution for a period sufficient to produce between 0.005×10^{-6} and 3×10^{-6} mole Fe^{++}/ml. Run blank simultaneously.

Determination of Fe^{++} Produced

Pipette 2 ml of actinometer solution into a 20-ml flask. Add 2 ml of (b) and mix. Add 1 ml of (c) and mix [volume of (c) is always one-half

364

that of photolyte]. Add 15 ml H_2O to make a volume of 20 ml and mix. Repeat for blank and let both stand for 30 min.

Notes

Amount of actinometer solution pipetted will depend on Fe^{++} concentration. Adjust so that OD measured is between 0.1 and 0.8; OD of blank should be ≤ 0.01/cm. Do not decompose more than 20% of the ferrioxalate. For very low light intensities use 3 ml of actinometer solution in 10-mm path length absorption cell. Irradiate. Add 0.5 ml of solution (1.8 N in $NaC_2H_3O_2$ and 1.08 N in H_2SO_4 and containing 0.1% w/v of 1:10 phenanthroline monohydrate). Mix. Repeat for blank; let both stand for 30 min. Use blank as reference cell and measure Δ OD. For longer wavelengths use 0.15 M $K_3Fe(C_2O_4)_3$ and multiply values of Fig. 1.18 by 0.952.

Example: Assume we use a 10-mm quartz curvette to calibrate a beam of ultraviolet at 3663 A. Use an aperture so that a known area of solution is irradiated. Assume the aperture has an area of 2 cm^2, the time of irradiation is 1 hour, and the final OD measured is 0.6. At this wavelength all of the incident quanta will be absorbed, even in 0.006 M solution. The total Fe^{++} produced will be

$$0.6 \times 1.81 \times 10^{-6} \text{ mole} = 1.086 \times 10^{-6} \text{ mole } Fe^{++}$$

This corresponds to the absorption of

$$\frac{1.086}{1.26} \times 10^{-6} = 0.862 \times 10^{-6} \text{ Einsteins of 3663 A light}$$

Since the area is 2 cm^2 and the irradiation time was 3600 sec the average light beam intensity *absorbed* is

$$\frac{0.862 \times 10^{-6} \times 6.02 \times 10^{23}}{2 \times 3600} = 0.72 \times 10^{14} h\nu/\text{sec-cm}^2$$

However, there was reflection from the air-quartz and the quartz-water interfaces, so that this effect was due to (.965) (.998) of the light incident on the face of the quartz curvette. Therefore,

$$I_0 = \frac{0.72}{0.963} \times 10^{14} = 0.748 \times 10^{14} h\nu/\text{sec-cm}^2$$

ABSORPTION OF INCIDENT BEAM BY FERRIOXALATE SOLUTION (10 mm)[a]

Wavelength (A)	0.006 M solution (%)	0.15 M solution (%)
Below 3800	100	100
3900	99	100
4000	95	100
4100	87	100
4200	73	100
4300	56	100
4400	40	100
4500	26	98
4600	17	89
4700	—	65
4800	—	43
4900	—	25
5000	—	14

[a] Reflection from air-quartz interface = 0.035 (see example). Reflection from quartz-water interface = 0.002 (see example).

Assay of Adenosine Triphosphate Using Firefly Luminescence

This method of ATP analysis is based on the linear luminescence response of firefly extracts to added ATP. When the light intensity is measured by an appropriate photomultiplier, solutions with ATP concentrations as low as 10^{-9} M can be measured with considerable accuracy. Because of the high sensitivity of this light-emitting system to ATP it is possible to measure samples which have been highly diluted, thus eliminating interference from other compounds.

Preparation of Crude Firefly Extract

Five grams of dried firefly lanterns are ground thoroughly in a mortar and then extracted with 25 ml of cold water; 1 N sodium hydroxide is added to the crude suspension in order to bring the pH to approximately 7.5. After additional grinding for 5 min in the cold the crude suspension is centrifuged for approximately 10 min at 3,000 rpm in a refrigerated centrifuge. The residue is re-extracted with 25 ml of cold water and after centrifugation the supernatants are combined. This crude extract can be used for the assay of ATP. However, the crude preparation has myokinase present, consequently ADP will be converted into ATP and light emission will be observed. Often with high ATP concentrations the enzyme preparation can be diluted to such an extent that the myokinase activity is very low. Because of the photocell sensitivity very dilute luciferase preparations can be used without appreciable loss of sensitivity. Bovine serum albumin should be added (0.5%) if very high dilutions are to be used. This prevents the rapid denaturation of luciferase in dilute solutions.

In addition to the ADP response, very crude luciferase preparations often emit light with other nucleotide triphosphates. This is due to the fact that there are transphosphorylases and ADP present. Dialysis or ammonium sulfate precipitation, however, will remove the ADP and eliminate this response to other triphosphates. Using partially purified enzyme, light emission is initiated only with ATP.

A partially purified luciferase can be obtained by ammonium sulfate fractionation. The details of this preparation have been described by McElroy and Coulombre (1952). A very active preparation which is low in myokinase can be obtained by fractionation at pH 7.3 in the following manner. Add to the crude extract solid ammonium sulfate to 30% saturation, adjust pH, and centrifuge in the cold. Discard precipitate and fractionate the supernatant as follows: 30–40, 40–50, and 50–60% saturation $(NH_4)_2SO_4$. The fractions are collected as described above and dissolved in 0.025 M glycylglycine buffer. Most of the luciferase will be in the 50–60 fraction. The pH of the supernatant is adjusted to pH 3.5 and extracted twice with an equal volume of redistilled ethyl acetate. This will remove all of the luciferin. Add a small volume of water to the combined ethyl acetate extracts and add dropwise 1 N NaOH with shaking until the aqueous layer turns a light yellow. Remove the aqueous layer and repeat the extraction of the ethyl acetate. Concentrate the aqueous extract under vacuum distillation. If 5 g of firefly lanterns were used initially, concentrate the luciferin to 5 ml. The luciferin should be added to the luciferase preparation as needed.

For extreme sensitivity and for high specificity the crystalline enzyme should be used and its preparation has been described by Green and McElroy (1956). Using the partially purified luciferase, the luciferin remains bound to the enzyme in adequate quantities to give a good light response. However, additional luciferin can be added to the enzyme to increase the sensitivity. In addition to the enzyme and luciferin, the only reagents required are 0.1 M $MgSO_4$ and 0.025 M glycylglycine buffer, pH 7.5. ATP can be determined by using the following reaction mixture:

2.1 ml glycylglycine buffer, 0.025 M, pH 7.5
0.1 ml $MgSO_4$, 0.1 M
0.01 ml enzyme—10 mg protein per ml
0.2 ml sample or standard

It is convenient to inject the sample of ATP into the reaction mixture, using a 0.25-ml syringe. The flash height is directly proportional to the ATP concentration. Delay in reaching the maximum is indicative of the presence of ADP and myokinase. With pure ATP 0.5 μg can be determined with an error of less than 1%. In order to eliminate interfering substances, the ATP sample to be assayed may be diluted until one obtains a concentration below 1 μg/ml. Standards of the same or lower concentration should be made fresh in buffer solution and should not be kept longer than 30 min even in ice, since highly dilute samples of ATP are very labile.

Specificity

The assay is specific for ATP, but precursors to ATP will also produce light provided the appropriate enzymes for ATP formation are present. The crystalline enzyme will respond only to ATP. However, as indicated above, crude enzymes are present which are capable of phosphorylating ADP, provided other organic triphosphates are present.

Preparation of Extracts for Assay

In some cases crude boiled tissue extracts can be used for the accurate determination of ATP. However, in most cases there is considerable loss of ATP even with very rapid boiling and cooling. Usually it is best to obtain an extract of the tissue using cold 0.5 M perchloric acid. The concentration and volume of perchloric acid which is most effective for extraction must be determined for each tissue. After extraction the excess acid should be removed by neutralizing with KOH. All extracts should be adjusted to approximately pH 7.5 and then used directly for ATP assay in the light systems described above. With this crude enzyme we have been able to detect 2×10^{-13} mole ATP (10^{-10} g) in 0.2 ml injected into 2 ml of a glycylglycine—Mg^{++}—crude enzyme solution.

Precautions

The light-emitting reaction is extremely sensitive to high salt concentrations, particularly phosphate which is a very potent inhibitor of the initial flash. Consequently, in systems where high salts may be encountered, the samples should be diluted in order to avoid salt concentrations which are greater than 10^{-4} M. Other sources of error include inhibitory effects of substances in the extract, turbidity, precipitation of buffer or activators, anaerobic conditions, and other sources of luminescence. Crude enzyme may be stored in a frozen condition for long periods of time (6 months to a year) without appreciable loss of activity. However, small aliquots of the enzyme should be distributed in reaction vessels, stored in the frozen state, and thawed only at the time of using. Repeated thawing and freezing does lead to some inactivation. The crystalline enzyme must be maintained in an ammonium sulfate solution at refrigerator temperatures. It cannot be frozen and thawed without considerable loss of activity.

Source of Fireflies

Dried firefly lanterns are now available from several commercial sources. Also large numbers can be collected readily during the summer months in most temperate or tropical regions of the world. In fact, a few hours' collection will be adequate for several thousand assays. The fireflies can be dried in a vacuum desiccator over calcium chloride and maintained indefinitely in a deep freeze. The dried lanterns should be removed from the firefly bodies and extracted three or four times with cold acetone in order to prepare an acetone powder. Both the luciferase and the luciferin remain in the powder after acetone extraction.

References

Abbott, M. T. J., and J. F. Grove. *Exptl. Cell Res.*, **17**: 95 (1959).

Abolin, L. *Arch. Mikrobiol. Anat.*, **104**: 667 (1925).

Adelman, A. H., and C. M. Verber, *J. Chem. Phys.* **39**: 931 (1963).

Airth, R. L. In *Light and Life* (W. D. McElroy and B. Glass, Eds.), p. 262. The Johns Hopkins Press, Baltimore, Maryland, 1961.

Airth, R. L., and W. D. McElroy. *J. Bacteriol.*, **77**: 249 (1959).

Airth, R. L., W. C. Rhodes, and W. D. McElroy. *Biochim. Biophys. Acta*, **27**: 519 (1958).

Albrecht, H. O. *Z. Physik. Chem.*, **136**: 321 (1928).

American Institute of Physics Handbook, McGraw-Hill Book Company, New York, 1957.

Amesz, J., and L. N. M. Duysens. *Biochim. Biophys. Acta*, **64**: 261 (1962).

Anderson, R. S. *J. Cellular Comp. Physiol.*, **8**: 261 (1936).

Applied Optics, Suppl. 1: Optical Masers (1962).

Arden, G. B., and K. Tansley. *J. Physiol.*, **127**: 592 (1955).

Arnold, S. J., E. A. Ogryzlo, and H. Witzke. *J. Chem. Phys.*, **40**: 1769 (1964).

Arnold, W., and R. K. Clayton. *Proc. Natl. Acad. Sci. U.S.*, **46**: 769 (1960).

Arnold, W., and E. S. Meek. *Arch. Biochem. Biophys.*, **60**: 82 (1956).

Arnold, W., and H. K. Sherwood. *Proc. Natl. Acad. Sci. U.S.*, **43**: 105 (1957).

Arnold, W., and H. K. Sherwood. *J. Phys. Chem.*, **63**: 2 (1959).

Arnold, W., and J. Thompson. *J. Gen. Physiol.*, **39**: 311 (1956).

Arnon, D. I. *Nature*, **167**: 1008 (1951).

Arnon, D. I. *Brookhaven Symp. Biol.*, **11**: 181 (1958).

Arnon, D. I. In *Light and Life* (W. D. McElroy and B. Glass, Eds.), p. 489. The Johns Hopkins Press, Baltimore, Maryland, 1960.

Arnon, D. I., F. R. Whatley, and M. B. Allen. *J. Am. Chem. Soc.*, **76**: 6324 (1954).

Arnon, D. I., M. Losada, M. Nozaki, and K. Tagawa. *Biochem. J.*, **77**: 23P (1960).

Arnon, D. I., F. R. Whatley, and A. A. Horton. *Federation Proc.*, **21**: 91 (1962).

Arthur, W. E., and B. L. Strehler. *Arch. Biochem. Biophys.*, **70**: 507 (1957).

Aschoff, J. *Cold Spring Harbor Symp. Quant. Biol.*, **25**: 11 (1960).

Audubert, R. *Trans. Faraday Soc.*, **35**: 197 (1939).

Autrum, H., and H. Stumpf. *Z. Vergleich. Physiol.*, **35**: 71 (1952).

Avakian, P., E. Abramson, R. G. Kepler, and J. C. Caris. *J. Chem. Phys.*, **39**: 1127 (1963).

Avron, M. *Biochim. Biophys. Acta*, **40**: 257 (1960).

Baly, E. C. C. *Spectroscopy*, Vol. I. Longmans, Green Company, London, 1924.

Baly, E. C. C. *Spectroscopy*, Vols. II and III. Longmans, Green Company, London, 1927.

Barber, H. S., and F. A. McDermott. *Smithsonian Inst. Misc. Collections*, **117** (1): 1 (1951).

Barbrow, L. E. *J. Opt. Soc. Am.*, **49**: 1122 (1959).

Bassham, J. A. *Sci. Am.*, **206**: 89 (1962).

Bassham, J. A. *Advan. Enzymol.*, **25**: 39 (1963).

371

Baur, E. *Helv. Chim. Acta*, **8**: 403 (1925).
Bawden, F. C., and A. Kleczkowski. *Nature*, **183**: 503 (1959).
Becker, R. S., and M. Kasha. In *Luminescence of Biological Systems* (F. H. Johnson, Ed.), p. 25. American Association for the Advancement of Science, Washington, D.C., 1955.
Becquerel, E. *Compt. Rend.*, **79**: 185 (1874).
Becquerel, H. *Compt. Rend. Acad. Sci. Paris*, **108**: 997 (1889).
Bellin, J. S., and G. Oster. In *Progress in Photobiology* (B. C. Christensen and B. Buchmann, Eds.), p. 254. Elsevier Publishing Co, Amsterdam, 1961.
Bendix, S. W. *Botan. Rev.*, **26** (2): 145 (1960).
Bennett, W. R., Jr. *Appl. Opt.* (Suppl. 1), p. 24, (1962).
Beredjik, N. *J. Am. Chem. Soc.*, **78**: 2646 (1956).
Bergeron, J. A. *Brookhaven Symp. Biol.* **BNL (C28)**, 118 (1959).
Bersohn, R., and I. Isenberg. *Biochem. Biophys. Res. Comm.*, **13**: 205 (1963).
Bertholf, L. M. *Z. Vergleich. Physiol.*, **18**: 32 (1932).
Beukers, R., and W. Berends. *Biochim. Biophys. Acta*, **41**: 550 (1960).
Beukers, R., and W. Berends. *Biochim. Biophys. Acta*, **49**: 181 (1961).
Beukers, R., J. Ijlstra, and W. Berends. *Rec. Trav. Chim.*, **79**: 101 (1960).
Beutler, H., and M. Polanyi. *Naturwissenschaften*, **13**: 711 (1925).
Beutler, H., and M. Polanyi. *Z. Physik.*, **47**: 379 (1928).
Bie, V. *Mitt. Finsens Med. Lysinst.*, **7**: 65 (1904).
Bishop, N. I. *Proc. Natl. Acad. Sci. U.S.*, **45**: 1696 (1959).
Bitler, B., and W. D. McElroy. *Arch. Biochem. Biophys.*, **72**: 358 (1957).
Black, C. C., J. F. Turner, M. Gibbs, D. W. Krogman, and S. A. Gordon. *J. Biol. Chem.*, **237**: 580 (1962).
Blinks, L. R. In *Photophysiology* (A. C. Giese, Ed.), Vol. 1, p. 199. Academic Press, Inc., New York, 1964.
Blum, H. F. *Carcinogenesis by Ultraviolet Light*. Princeton University Press, Princeton, New Jersey, 1959.
Blum, H. F., E. G. Butler, J. J. Chang, R. C. Mawe, and S. E. Schmidt. *J. Cellular Comp. Physiol.*, **49**: 153 (1957).
Boll, F. *Arch. Anat. Physiol., Physiol. Abt.*, **4**: (1877).
Bonhoeffer, K. F. *Z. Physik. Chem.*, **116**: 394 (1925).
Borthwick, H. A., S. B. Hendricks, and M. W. Parker. *Proc. Natl. Acad. Sci. U.S.*, **38**: 929 (1952).
Boudin, S. *J. Chim. Phys.*, **27**: 284 (1930).
Bowen, E. J. *The Chemical Aspects of Light*, p. 278. Oxford University Press, London and New York, 1949.
Bowen, E. J., and B. Brocklehurst. *Trans. Faraday Soc.*, **51**: 774 (1955).
Bowen, E. J., and J. W. Sawtell. *Trans. Faraday Soc.*, **33**: 1425 (1937).
Bowen, E. J., and D. Seaman. In *Luminescence of Organic and Inorganic Materials* (H. P. Kallmann, Ed.), p. 153. John Wiley and Sons, Inc., New York, 1962.
Bowen, E. J., and A. H. Williams. *Trans. Faraday Soc.*, **35**: 765 (1939).
Boyd, G. D., R. J. Collins, S. P. S. Porto, A. Yariv, and W. A. Hargreves. *Phys. Rev. Letters*, **8**: 269 (1962).
Brackett, F. P., and G. S. Forbes. *J. Am. Chem. Soc.*, **55**: 4459 (1933).
Bradley, D. F., and M. Calvin. *Proc. Natl. Acad. Sci. U.S.*, **41**: 563 (1955).
Bremer, T. *Bull. Soc. Chim. Belges*, **62**: 569 (1953).
Bremer, T., and H. Friedmann. *Bull. Soc. Chim. Belges*, **63**: 415 (1954).
Bridges, C. D. B. *Vision Res.* **2**: 125 (1962).

Briggs, W. R., R. D. Tocher, and J. F. Wilson. *Science*, **126**: 210 (1957).
Brindley, G. S. *Physiology of the Retina and Visual Pathway*. Edward Arnold, London, 1960.
Brown, F. A. *Science*, **130**: 1535 (1959).
Brown, H. In *The Atmospheres of the Earth and Planets* (G. P. Kuiper, Ed.), p. 260. University of Chicago Press, Chicago, Illinois, 1947.
Brown, P. K., and G. Wald. *Nature*, **200**: 37 (1963).
Brown-Séquard, E. *Compt. Rend. Acad. Sci. Paris*, **25**: 482, 508 (1847).
Bruce, V. G., and C. S. Pittendrigh. *Proc. Natl. Acad. Sci. U.S.*, **42**: 676 (1956).
Bruce, V. G., and C. S. Pittendrigh. *Am. Naturalist*, **91**: 179 (1957).
Bruce, V. G., and C. S. Pittendrigh. *Am. Naturalist*, **92**: 295 (1958).
Buck, J. B. *Quart. Rev. Biol.*, **13**: 301 (1938).
Buck, J. B. *Ann. N.Y. Acad. Sci.*, **49**: 397 (1948).
Bünning, E. *Planta*, **27**: 148 (1937).
Bünning, E. *Z. Botan.*, **43**: 167 (1955).
Bünning, E. *Periplaneta Americana*, **77**: 141 (1958).
Bungeler, W. *Klin. Wochschr.*, **16**: 1012 (1937a).
Bungeler, W. *Z. Krebsforsch.*, **46**: 130 (1937b).
Burgers, A. C. J. *Investigations into the Action of Certain Hormones and Other Substances on the Melanophores of the South African Clawed Toad, Xenopus laevis*. G. W. van der Weil and Company, Arnheim, Netherlands, 1956.
Butler, W. L. *Arch. Biochem. Biophys.*, **93**: 413 (1961).
Butler, W. L. *Bull. Phil. Soc. Wash.*, **16** (2): 225 (1963).
Byk, A. *Z. Physik. Chem.*, **49**: 641 (1904).
Byk, A. *Z. Physik. Chem.*, **62**: 454 (1908a).
Byk, A. *Z. Elektrochem.*, **14**: 460 (1908b).
Calcutt, G. *Brit. J. Cancer*, **8**: 177 (1954).
Calvin, M. *J. Theoret. Biol.*, **1**: 258 (1961).
Calvin, M. *Science*, **135**: 879 (1962).
Calvin, M., and G. M. Androes. *Science*, **138**: 867 (1962).
Calvin, M., and J. A. Bassham. *The Path of Carbon in Photosynthesis*. Prentice-Hall, Englewood Cliffs, New Jersey, 1957.
Calvin, M., and G. D. Dorough. *J. Am. Chem. Soc.*, **70**: 699 (1948).
Cameron, J. W. M. *Can. J. Res., Sect. D*, **16**, (11): 307 (1938).
Cario, G., and J. Franck. *Z. Physik.*, **10**: 185 (1922).
Cario, G., and J. Franck. *Z. Physik.*, **17**: 202 (1923).
Carrelli, A., and P. Pringsheim. *Z. Physik.*, **17**: 287 (1923).
Case, J. F., and J. B. Buck. *Biol. Bull.*, **125**: 234 (1963).
Cathey, L. *IRE (Inst. Radio Engrs.) Trans. Nucl. Sci.*, [N.S.] **5**: 3, p. 109 (1958).
Chance, B., and R. Sager. *Plant Physiol.*, **32**: 548 (1957).
Chance, B., and A. San Pietro. *Proc. Natl. Acad. Sci. U.S.*, **49**: 633 (1963).
Chandross, E. A. *Tetrahedron Letters*, **12**: 701 (1963).
Chang, H. T. *Res. Publ., Assoc. Res. Nervous Mental Disease*, **30**: 430 (1952).
Chang, H. T. *J. Neurophysiol.*, **19**: 224 (1956).
Charters, P. E., and J. C. Polanyi. *Can. J. Chem.*, **38**: 1742 (1960).
Chase, A. M. *J. Cellular Comp. Physiol.*, **33**: 113 (1949a).
Chase, A. M. *Arch. Biochem.*, **23**: 385 (1949b).
Chase, A. M. *Federation Proc.*, **8**: 24 (1949c).
Chase, A. M. *Methods Biochem. Analy.*, **8**: 61 (1960).
Chase, A. M., and P. B. Lorenz. *J. Cellular Comp. Physiol.*, **25**: 53 (1945).
Chessin, M. *Ann. Appl. Biol.*, **46**: 388 (1958).
Clark, G. L. (Ed.), *The Encyclopedia of Spectroscopy*. Reinhold Publishing Corporation, New York, 1960.

Clayton, R. K. In *Photophysiology* (A. C. Giese, Ed.), Vol. 1, p. 155. Academic Press, Inc., New York, 1964.

Coblentz, W. W., and H. R. Fulton. *Bur. Std. Sci. Papers*, 19: 641 (1924).

Coehn, A., and G. Jung. *Chem. Ber.*, 56: 696 (1923).

Coehn, A., and G. Jung. *Z. Physik. Chem.*, 110: 705 (1924).

Colange, M. G. *J. Phys. Radium*, 8: 254 (1927).

Commoner, B., J. J. Heise, and J. Townsend. *Proc. Natl. Acad. Sci. U.S.*, 42: 710 (1956).

Commoner, B., J. Townsend, and G. E. Pake. *Nature*, 174: 689 (1954).

Compton, D. M. J., and T. C. Waddington. *J. Chem. Phys.*, 25: 1075 (1956).

Condon, E. U. *Phys. Rev.*, 28: 1182 (1926).

Condon, E. U. *Phys. Rev.*, 32: 858 (1928).

Cormier, M. J. *J. Am. Chem. Soc.*, 81: 2592 (1959).

Cormier, M. J. *Biochim. Biophys. Acta*, 42: 333 (1960).

Cormier, M. J. In *Light and Life* (W. D. McElroy and B. Glass, Eds.), p. 274. The Johns Hopkins Press, Baltimore, Maryland, 1961.

Cormier, M. J. *Biol. Chem.*, 237: 2032 (1962).

Cormier, M. J. and B. L. Strehler. *J. Am. Chem. Soc.*, 75: 4864 (1953).

Cotton, A. *Ann. Chim. Phys.*, 8: 347 (1896).

Cox, A., R. R. Doell, and G. B. Dalrymple. *Science*, 144: 1537 (1964).

Crawford, B. H. *Nat. Phys. Lab. G. Brit.*, *Notes Appl. Sci.*, 29 (1962).

Crescitelli, F. *Ann. N.Y. Acad. Sci.*, 74: 230 (1958).

Crocker, W. *Biological Effects on Radiation* (B. M. Duggar, Ed.), p. 791. McGraw-Hill Book Company, New York, 1936.

Curry, G. M., and K. V. Thimann. In *Progress in Photobiology* (B. C. Christensen and B. Buchmann, Eds.), p. 127. Elsevier Publishing Company, Amsterdam, 1961.

Curry, G. M., K. V. Thimann, and P. M. Ray. *Physiol. Plantarum*, 9: 429 (1956).

Dahlgren, U. *J. Franklin Inst.*, 183: 79, 211, 323, 593 (1917).

Daniels, F. (Ed.). *Photochemistry in the Liquid and Solid States*. John Wiley and Sons, Inc., New York, 1960.

Dartnall, H. J. A. *J. Physiol.*, 116: 257 (1952).

Dartnall, H. J. A. *Brit. Med. Bull.*, 9: 24 (1953).

Dartnall, H. J. A. *The Visual Pigments*. Methuen and Company, Ltd., London, 1957.

Dartnall, H. J. A., M. R. Lander, and F. W. Munz. In *Progress in Photobiology* (C. B. Christensen and B. Buchmann, Eds.), p. 203. Elsevier Publishing Company, Amsterdam, 1961.

Dartnall, H. J. A. In *The Eye* (H. Darson, Ed), Vol. 2, p. 523. Academic Press, New York, 1962.

Davenport, H. E., R. Hill, and F. R. Whatley. *Proc. Roy. Soc. (London)*, B139: 346 (1952).

Davey, K. G. *Nature*, 183: 1271 (1959).

Davis, T. L., and R. Heggie. *J. Am. Chem. Soc.*, 57: 377 (1935).

Davydov, A. S. *Theory of Molecular Excitons* (translated by M. Kasha and M. Oppenheimer). McGraw-Hill Book Company, New York, 1962.

De Angelis, W. J. and J. R. Totter. *J. Biol. Chem.*, 239: 1012 (1964).

De Robertis, E. *J. Gen. Physiol.*, 43 (Suppl. 1): 13 (1960).

Dekker, A. J. *Solid State Physics*, p. 372. Prentice Hall, Inc., Englewood Cliffs, New Jersey, 1958.

Donat, K. *Z. Physik.*, 29: 345 (1924).

Donner, K. O., and R. Granit. *Acta Physiol. Scand.*, 17: 161 (1949).

Doty, M. S., and O. Oguri. *Limnol. Oceanog.*, **2**: 37 (1957).

Dufford, R. T., S. Calvert, and D. Nightingale. *J. Am. Chem. Soc.*, **45**: 2058 (1923).

Dufford, R. T., D. Nightingale, and S. Calvert. *J. Am. Chem. Soc.*, **47**: 95 (1925).

Dubois, R. *Compt. Rend. Soc. Biol.*, **37**: 559 (1885).

Dubois, R. *Compt. Rend. Acad. Sci. Paris*, **105**: 690 (1887a).

Dubois, R. *Compt. Rend. Soc. Biol.*, **39**: 564 (1887b).

Dulbecco, R. *Nature*, **163**: 949 (1949).

Dulbecco, R. *J. Bacteriol.*, **59**: 329 (1950).

Dulbecco, R. In *Radiation Biology* (A. Hollaender, Ed., p. 455. McGraw-Hill Book Company, New York, 1955.

Dure, L. S., and M. J. Cormier. *J. Biol. Chem.*, **236**: PC48 (1961).

Dutheil, J., and M. Dutheil. *J. Phys. Radium*, **7**: 414 (1926).

Duysens, L. N. M. Ph.D. Thesis, University of Utrecht, pp. 1–96, 1952.

Duysens, L. N. M. *Ann. Rev. Plant Physiol.*, **7**: 25 (1956).

Duysens, L. N. M. *Plant Physiol.*, **34**: 407 (1959).

Duysens, L. N. M. *Proc. Intern. Congr. Biochem.*, *4th, Vienna, 1958*, **13**: 303 (1959).

Duysens, L. N. M., and J. Amesz. *Biochim. Biophys. Acta*, **64**: 243 (1962).

Duysens, L. N. M., and H. E. Sweers. Microalgae and Photosynthetic Bacteria (special issue) *Plant Cell Physiol. (Tokyo)*, p. 353 (1963).

Eder, J. M. *Phot. Mitt.*, **24**: 74 (1887).

Eglinton, G., P. M. Scott, T. Belsky, A. L. Burlingame, and M. Calvin. *Science*, **145**: 263 (1964).

Ehert, C. P. *Publ. Am. Assoc. Advan. Sci.*, **55**: 541 (1959).

Ehrenberg, A. In *Free Radicals in Biological Systems* (M. S. Blois *et al.* Eds.), p. 337. Academic Press, Inc., New York, 1961.

Einstein, A. *Ann. Physik.*, **37**: 832 (1912a).

Einstein, A. *Ann. Physik.*, **38**: 881 (1912b).

Einstein, A. *Ber. Deut. Phys. Ges.*, **18**: 318 (1916).

Einstein, A. *Physik. Z.* **18**: 121 (1917).

Eley, D. D., G. D. Parfitt, M. J. Perry, and D. H. Toysum. *Trans. Faraday Soc.*, **49**: 79 (1953).

Emerson, R. *Science*, **123**: 673 (1956).

Emerson, R., and W. Arnold. *J. Gen. Physiol.*, **16**: 191 (1932).

Emerson, R., and R. V. Chalmers. *Phycol. Soc. Am. News Bull.*, **11**: 51 (1958).

Emerson, R., and C. M. Lewis. *Am. J. Botany*, **30**: 165 (1943).

Enami, M. *Science*, **121**: 36 (1955).

Engstrom, R. W. *J. Opt. Soc. Am.*, **37**: 420 (1947).

Evenari, M. In *Radiation Biology* (A. Hollaender, Ed.), Vol. III, p. 519. McGraw-Hill Book Company, New York, 1956.

Eversole, R., and J. J. Wolken. *Science*, **127**: 1287 (1958).

Eyring, H., J. Walter, and G. E. Kimball. *Quantum Chemistry*, p. 110. John Wiley and Sons, Inc., New York, 1944.

Fabry, Ch., and H. Buisson. *J. Phys. Radium*, **3**: 196 (1913).

Feofilov, P. O. *The Physical Basis of Polarized Emission*. Consultants Bureau, New York, 1961.

Fiala, S. *Biochem. Z.*, **320**: 10 (1949).

Fingerman, M. *The Control of Chromatophores*. The Macmillan Company, New York, 1963.

Fischer, E., and M. Kaganowich. *Bull. Res. Council Israel, Sect. A*, **10**: 138 (1961).

Flint, L. H., and E. D. Macalaster. *Smithsonian Inst. Misc. Collections*, **96**: 1 (1937).

Förster, Th. *Naturwissenschaften*, **33**: 166 (1946).

Förster, Th. *Ann. Phys.*, **2**: 55 (1948).

Förster, Th. *Disc. Faraday Soc.*, **27**: 7 (1959).

Förster, Th. In *Comparative Effects of Radiation* (M. Burton, J. S. Kirby-Smith, and J. L. Magee, Eds.), p. 300. John Wiley and Sons, Inc., New York, 1960.

Fox, D. In *Electrical Conductivity in Organic Solids* (H. Kallmann and M. Silver, Eds.), p. 239. John Wiley and Sons, Inc. (Interscience), New York, 1961.

Franck, J. *Trans. Faraday Soc.*, **21**: 536 (1926).

Franck, J. *Ann. Rev. Plant Physiol.*, **2**: 53 (1951).

Franck, J. *Proc. Natl. Acad. Sci. U.S.*, **44**: 941 (1958).

Franck, J., and J. L. Rosenberg. *J. Theoret. Biol.*, **7**: 276 (1964).

Franck, J., and E. Teller. *J. Chem. Phys.*, **6**: 861 (1938).

Franck, J., J. L. Rosenberg, and C. J. Weiss. In *Luminescence of Organic and Inorganic Materials* (H. P. Kallmann and G. M. Spruch, Eds.), p. 11. John Wiley and Sons, Inc., New York, 1962.

French, C. S. In *This is Life* (W. H. Johnson and W. C. Steere, Eds.), p. 3. Holt, Rinehart, and Winston, New York, 1962.

French, C. S., J. Myers, and C. G. McLeod. *Symp. Comp. Biol. Kaiser Found. Res. Inst.*, **1**: 361 (1960).

Frenkel, A. W. *J. Am. Chem. Soc.*, **76**: 5568 (1954).

Frenkel, J. I. *Phys. Rev.*, **37**: 17; 1276 (1931).

Fry, K. T., R. A. Lazzarini, and A. San Pietro. *Proc. Natl. Acad. Sci. U.S.*, **50**: 652 (1963).

Gaffron, H. *Chem. Ber.*, **60**: 755 (1927a).

Gaffron, H. *Chem. Ber.*, **60**: 2229 (1927b).

Gaffron, H. *Biochem. Z.*, **287**: 130 (1936).

Gaffron, H. *Res. Photosynthesis, Gatlinburg Conf., 1955*, p. 430 (1957).

Gaffron, H. *Plant Physiol.*, **1B**: 3 (1960).

Gaffron, H. In *Horizons in Biochemistry* (M. Kasha and B. Pullman, Eds.), p. 59. Academic Press, Inc., New York, 1962.

Gaffron, H., and J. Rubin. *J. Gen. Physiol.*, **26**: 219 (1942).

Galifret, Y. In *Mechanisms of Colour Discrimination* (Y. Galifret, Ed.), Pergamon Press, London, 1960.

Galston, A. W. In *Light and Life* (W. D. McElroy and B. Glass, Eds.), p. 670. The Johns Hopkins Press, Baltimore, Maryland, 1960.

Garner, W. W., and H. A. Allard. *J. Agr. Res.*, **18**: 553 (1920).

Garner, W. W., and H. A. Allard. *J. Agr. Res.*, **23**: 871 (1923).

Gates, F. L. *Science*, **68**: 479 (1928).

Gates, F. L. *J. Gen. Physiol.*, **14**: 31 (1930).

Gates, F. L. *J. Gen. Physiol.*, **18**: 265 (1934).

Gattow, G., and A. Schneider. *Naturwissenschaften*, **41**: 116 (1954).

Gause, G. F. *Optical Activity and Living Matter*, p. 52 Biodynamica, Normandy, Missouri, 1941.

Gaviola, E. *Z. Physik.*, **42**: 862 (1927).

Gaviola, E., and P. Pringsheim. *Z. Physik.*, **24**: 24 (1924).

Geiling, L. *Ann. Telecomm.*, **5**: 12; 417 (1950).

Gest, H., and M. D. Kamen. *J. Bacteriol.*, **58**: 239 (1949a).

Gest, H., and M. D. Kamen. *Science*, **109**: 558 (1949b).

Ghosh, J. C. *Z. Physik. Chem.*, **B3**: 419 (1929).

Gibbs, M., C. C. Black, and B. Kok. *Biochim. Biophys. Acta*, **52**: 474 (1961).

Giese, A. C. *Physiol. Zool.*, **18**: 223 (1945).

Giese, A. C. In *Progress in Photobiology* (B. C. Christensen and B. Buchmann, Eds.), p. 273. Elsevier Publishing Company, Amsterdam, 1961.

Gleu, K., and W. Petsch. *Angew. Chem.*, **48**: 57 (1935).

Goldsmith, T. H. *Proc. Natl. Acad. Sci. U.S.*, **44**: 123 (1958a).

Goldsmith, T. H. *Ann. N.Y. Acad. Sci.*, **74**: 223 (1958b).

Goldsmith, T. H. *J. Gen. Physiol.*, **43**: 775 (1959).

Goldsmith, T. H. In *Light and Life* (W. D. McElroy and B. Glass, Eds.), p. 771. The Johns Hopkins Press, Baltimore, Maryland, 1961.

Goldsmith, T. H., and P. R. Ruck. *J. Gen. Physiol.*, **41**: 1171 (1957).

Goodgall, S. H., C. S. Rupert, and R. M. Herriott. In *The Chemical Basis of Heredity* (W. D. McElroy and B. Glass, Eds.), p. 341. The Johns Hopkins Press, Baltimore, Maryland, 1957.

Gordon, J. P., H. L. Zeiger, and C. H. Townes. *Phys. Rev.*, **95**: 282 (1954).

Granit, R. *Acta Physiol. Scand.*, **3**: 137, 318 (1942).

Granit, R. *Acta Physiol. Scand.*, **5**: 108, 219 (1943).

Granit, R. *Sensory Mechanism of the Retina.* Oxford University Press, London and New York, 1947.

Granit, R. In *The Eye*, (H. Davson, Ed.), Vol. 2, p. 537. Academic Press, Inc., New York, 1962.

Green, A. A., and W. D. McElroy. *Biochim. Biophys. Acta*, **20**: 170 (1956).

Green, D. E. *Radiation Res.*, No. 2, p. 504 (1960).

Greenlee, L., L. Fridovich, and P. Handler. *Biochemistry*, **1**: 779 (1962).

Griffin, D. R., R. Hubbard, and G. Wald. *J. Opt. Soc. Am.*, **37**: 546 (1947).

Griffith, R. O., and A. McKeown. *Photoprocess in Gaseous and Liquid Systems*, p. 393. Longmans, Green & Co., London, 1929.

Grossweiner, L. I., and E. F. Zwicker. *J. Chem. Phys.*, **31**: 1141 (1959).

Grossweiner, L. I., and E. F. Zwicker. *J. Chem. Phys.*, **34**: 1411 (1961).

Grossweiner, L. I., and E. F. Zwicker. *J. Chem. Phys.*, **39**: 2774 (1963).

Gurney, R. W., and N. F. Mott. *Proc. Roy. Soc. London*, **A164**: 151 (1938).

Guttman, N., and H. I. Kalish. *Sci. Am.*, **198**: 1, 77 (1958).

Haber, A. H., and N. E. Tolbert. *Publ. Am. Assoc. Advan. Sci*, **55**: 197 (1959).

Haber, F., and W. Zisch. *Z. Physik.*, **9**: 302 (1922).

Haidinger, W. *Ann. Physik*, **63**: 29 (1844).

Halberg, F. J. *Lancet*, **73**: 20 (1953).

Hall, R. T., and G. C. Pimentel. *J. Chem. Phys.*, **38**: 1889 (1963).

Halldal, P. *Physiol. Plantarum*, **13**: 726 (1960).

Halldal, P. *Physiol. Plantarum*, **14**: 133 (1961).

Hanaoka, T., and K. Fujimoto. *Japan. J. Physiol.*, **7**: 276 (1957).

Handler, P., and H. Beinert. Private communication, 1963.

Haneda, Y., and F. H. Johnson. *Proc. Natl. Acad. Sci. U.S.*, **44**: 127 (1958).

Hanson, F. E. *J. Insect Physiol.*, **8**: 105 (1962).

Harvey, E. N. *Bioluminescence*, p. 649. Academic Press, Inc., New York, 1952a.

Harvey, E. N. *Am. Scientist*, **40**(3): 468 (1952b).

Harvey, E. N. *Federation Proc.*, **12**: 597 (1953).

Harvey, E. N. In *The Luminescence of Biological Systems* (F. H. Johnson, Ed.) p. 1. American Association for the Advancement of Science, Washington, D.C., 1955.

Harvey, E. N., and Y. Haneda. *Arch. Biochem. Biophys.*, **35**: 470 (1952).

Harvey, E. N., and F. I. Tsuji. *J. Cellular Comp. Physiol.*, **44**: 63 (1954).

Hastings, J. W. In *Photophysiology*, (A. C. Giese, Ed.), Vol. 1, p. 333. Academic Press, Inc., New York, 1964.
Hastings, J. W., and J. Buck. *Biol. Bull.*, 111: 101 (1956).
Hastings, J. W., and Q. H. Gibson. *J. Biol. Chem.*, 238: 2537 (1963).
Hastings, J. W., and B. M. Sweeney. *J. Cellular Comp. Physiol.*, 49: 209 (1957a).
Hastings, J. W., and B. M. Sweeney. *Proc. Natl. Acad. Sci. U.S.*, 43(9): 804 (1957b).
Hastings, J. W., and B. M. Sweeney. *Biol. Bull.*, 115: 440 (1958).
Hastings, J. W., and B. M. Sweeney. In *Photoperiodism and Related Phenomena in Plants and Animals* (R. B. Withrow, Ed.), p. 567. American Association for the Advancement of Science, Washington, D.C., 1959.)
Hastings, J. W., and B. M. Sweeney. *J. Gen. Physiol.*, 43: 697 (1960).
Hastings, J. W., W. D. McElroy, and J. Coulombre. *J. Cellular Comp. Physiol.*, 42: 137 (1953).
Hastings, J. W., J. Spudich, and G. Malnic. *J. Biol. Chem.*, 238: 3100 (1963).
Hatchard, C. G., and C. A. Parker. *Proc. Roy. Soc. (London)*, A235: 518 (1956).
Haupt, W. In *Encyclopedia of Plant Physiology* (W. Ruhland, Ed.), Vol. 17, Part 1, p. 318. Springer Verlag, Berlin, 1959.
Hausser, K. W., and H. von Oehmcke. *Strahlentherapie*, 48: 223 (1933).
Heavens, O. S. *Appl. Opt.* (Suppl. 1), p. 1 (1962).
Hecht, S. *Sci. Progr. (New Haven) Ser. IV*, 75 pp. (1945).
Hecht, S., and S. Shlaer. *J. Opt. Soc. Am.*, 28: 269 (1938).
Hecht, S., S. Shlaer, and M. H. Pirenne. *J. Gen. Physiol.*, 25: 819 (1942).
Heiligman-Rim, R., Y. Hirshberg, and E. Fischer. *J. Chem. Soc.*, p. 156 (1961).
Heintz, E. *Insectes Sociaux*, 6: 223 (1959).
Helberger, J. H. *Naturwissenschaften*, 26: 316 (1938).
Helmholtz, H. *Handbuch der Physiologischen Optik*, Vol II, p. 570. Leopold Voss, Hamburg, 1896.
Hendricks, S. B. *Publ. Am. Assoc. Advan. Sci.*, 55: 423 (1959).
Hendricks, S. B. *Science*, 141: 21 (1963).
Hertel, E. *Z. Allgem. Physiol.*, 4: 1 (1904).
Herzberg, G. *Atomic Spectra and Atomic Structure* (translated by J. W. T. Spinks). Dover Publications, New York, 1944.
Hill, R., and F. Bendall. *Nature*, 186: 136 (1960).
Hillman, W. S. *Science*, 140: 1397 (1963).
Hirata, Y., O. Shimomura, and S. Eguchi. *Tetrahedron Letters*, 5: 4 (1959).
Hirshberg, Y., and E. Fischer. *J. Chem. Soc.*, p. 297 (1954a).
Hirshberg, Y., and E. Fischer. *J. Chem. Soc.*, p. 3129 (1954b).
Hochstrasser, R. M., and M. Kasha. *Photochemistry and Photobiology* (in press). (1964).
Hodge, A. J., J. D. McLean, and F. V. Mercer, *J. Biophys. Biochem. Cytol.*, 1: 605 (1955).
Hogben, L. T., and F. R. Winton. *Proc. Roy. Soc. (London)*, B95: 15 (1923).
Hollaender, A., and W. D. Claus. *J. Gen. Physiol.*, 19: 753 (1936).
Hollaender, A., and C. W. Emmons. *Cold Spring Harbor Symp. Quant. Biol.*, 9, 179 (1941).
Horio, T., and T. Yamashita. *Biochem. Biophys. Res. Comm.*, 9: 142 (1962).
Hubbard, R., and A. Kropf. *Ann. N.Y. Acad. Sci.*, 81: 388 (1959).
Hubbard, R., and G. Wald. *Proc. Natl. Acad. Sci. Wash.*, 37: 69 (1951).
Hyde, J. N. *Am. J. Med. Sci.*, 131: 1 (1906).
Ikuma, H., and K. V. Thimann. *Nature*, 197: 1313 (1963).
Ilse, D. *Z. Vergleich. Physiol.*, 8: 658 (1928).
Ilse, D. *Nature*, 140: 544 (1937).
Imamura, M., and M. Koizumi. *Bull. Chem. Soc. Japan*, 29: 899 (1956).

International Critical Tables, Vol. V, p. 268. McGraw-Hill Book Company, New York, 1929.

Jagendorf, A. T. *Surv. Biol. Progr.*, **4**: 181 (1962).

Jagendorf, A. T., and M. Avron. *J. Biol. Chem.*, **231**: 277 (1958).

Jagendorf, A. T., and G. Forti. In *Light and Life* (W. D. McElroy and B. Glass, Eds.), p. 576. The Johns Hopkins Press, Baltimore, Maryland, 1960.

Jagger, J. *Bacteriol. Rev.*, **22**: 99 (1958).

Jagger, J. In *Radiation Protection and Recovery*, p. 352. Pergamon Press, London, 1960a.

Jagger, J. *Radiation Res.*, **13**: 521 (1960b).

Jagger, J., and R. Latarjet. *Ann. Inst. Pasteur*, **91**: 858 (1956).

Jagger, J., and R. S. Stafford. *Photochem. Photobiol.*, **1**: 245 (1962).

Jagger, J., and G. E. Stapleton. *Atti Congr. Intern. Fotobiologia, Q, Torino, Italy*, p. 134. 1957.

Javan, A., W. R. Bennett, Jr., and D. R. Herriott. *Phys. Rev. Letters*, **6**: 106 (1961).

Jenkins, F. A., and H. E. White. *Fundamentals of Optics*. McGraw-Hill Book Company, New York, 1950.

Johnson, F. H., H. Eyring, and M. J. Polissar. *The Kinetic Basis of Molecular Biology*, John Wiley and Sons, Inc., New York, 1954.

Johnson, L. F., G. D. Boyd, K. Nassau, and R. R. Soden. *Proc. IRE* (Correspondence) **50**: 213 (1962).

Johnsson, S., and B. Hogberg. *Nature*, **169**: 286 (1952).

Jung, R., and H. Kornhuber, Eds. *The Visual System: Neurophysiology and Psychological Physics*. Springer Verlag, Berlin, 1961.

Kallmann, H., and H. Fränz. *Naturwissenschaften*, **13**: 441 (1925).

Karagunis, G., and G. Drikes. *Naturwissenschaften*, **21**: 697 (1933a).

Karagunis, G., and G. Drikes. *Nature*, **132**, 354 (1933b).

Karakashian, M. W., and J. W. Hastings. *Proc. Natl. Acad. Sci. U.S.*, **48**: 2130 (1962).

Karakashian, M. W., and J. W. Hastings. *J. Gen. Physiol.*, **47**: 1 (1963).

Karreman, G., R. H. Steele, and A. Szent-Györgyi. *Proc. Natl. Acad. Sci. U.S.*, **44**: 140 (1958).

Kasha, M. *J. Opt. Soc. Am.*, **38**: 929 (1948).

Kasha, M. *J. Chem. Phys.*, **20**: 71 (1952).

Kasha, M. *Rev. Mod. Phys.*, **31**: 162 (1959).

Kasha, M. *Radiation Res.*, Suppl. 2: 243 (1960).

Kasha, M. In *Light and Life* (W. D. McElroy and B. Glass, Eds.), p. 31. The Johns Hopkins Press, Baltimore, Maryland, 1961.

Kasha, M. *Radiation Res.*, **20**: 55 (1963).

Kautsky, H. *Trans. Faraday Soc.*, **35**: 216 (1939).

Kautsky, H., and H. Thiele. *Z. Anorg. Allegem. Chem.*, **144**: 197 (1925).

Kautsky, H., and H. Zochar. *Z. Physik.*, **9**: 267 (1922).

Kearney, M. B. *Phil. Mag.*, **47**: 48 (1924).

Keister, D. L., and A. San Pietro. *Biochem. Biophys. Res. Comm.*, **1**: 10 (1959).

Kelner, A. *J. Bacteriol.*, **58**: 511 (1949a).

Kelner, A. *Proc. Natl. Acad. Sci. U.S.*, **35**: 73 (1949b).

Kelner, A. *J. Gen. Physiol.*, **34**: 835 (1951).

Kelner, A. In *Progress in Photobiology* (B. C. Christensen and B. Buchmann, Eds.), p. 276. Elsevier Publishing Company, Amsterdam, 1961.

Kepler, R. G., J. C. Caris, P. Avakian, and A. Abramson. *Phys. Rev. Letters*, **10**: 400 (1963).

Khan, A. U., and M. Kasha. *J. Chem. Phys.*, **39**: 2105 (1963).

Kholmogorov, V. Ye., A. I. Sidorov, and A. N. Terenin. *Dokl. Akad. Nauk SSSR*, 147: 4, 954 (1962).

Khorana, H. G. *J. Am. Chem. Soc.*, 76: 3517 (1954).

Khvostikov, J. A. *Compt. Resid. Acad. Sci. Leningrad*, 4: 14 (1934).

Khvostikov, J. A. *Phys. Z. Sowjet*, 9: 210 (1936).

Kihlman, B. A. *Exptl. Cell Res.*, 17: 590 (1959a).

Kihlman, B. A. *Nature*, 183: 976 (1959b).

Klabunovskii, E. I. In *Aspects of the Origin of Life* (M. Florkin, Ed.), p. 105. Pergamon Press, New York, 1960.

Kleinerman, M., L. Azarraga, and S. P. McGlynn. In *Luminescence of Organic and Inorganic Materials* (H. P. Kallmann and G. M. Spruch, Eds.), p. 196. John Wiley and Sons, Inc., New York, 1962.

Könitz, W. *Planta*, 51: 1 (1959).

Kok, B. *Enzymologia*, 13: 1 (1948).

Kok, B. *Biochim. Biophys. Acta*, 21: 245 (1956a).

Kok, B. *Biochim. Biophys. Acta*, 22: 399 (1956b).

Kok, B. *Nature*, 179: 583 (1957).

Kok, B. *Biochim. Biophys. Acta*, 48: 527 (1961).

Kok, B., and G. Hoch. In *Light and Life* (W. D. McElroy and B. Glass, Eds.), p. 397. The Johns Hopkins Press, Baltimore, Maryland, 1961.

Kondratjew, V. Z. *Physik.*, 48: 310 (1928).

Kornfeld, G. Z. *Physik. Chem.*, 131: 97 (1927).

Koshland, D. E., Jr. *Science*, 142: 1533 (1963).

Krall, A. R., N. E. Good, and B. C. Mayne. *Plant Physiol.*, 36: 44 (1961).

Krogman, D. In *Light and Life* (W. D. McElroy and B. Glass, Eds.), p. 615. The Johns Hopkins Press, Baltimore, Maryland, 1961.

Kuhn, W. *Angew. Chem.*, 49: 215 (1936).

Kuhn, W., and E. Braun. *Naturwissenschaften*, 17: 227 (1929).

Kuhn, W., and E. Knopf. *Naturwissenschaften*, 18: 183 (1930).

Ladenburg, E., and E. Lehmann. *Ann. Physik.*, 21: 305 (1906).

Larsen, H., E. C. Yocum, and C. B. von Niel. *J. Gen. Physiol.*, 36: 161 (1952).

Latarjet, R. *Compt. Rend. Acad. Sci.*, 217: 186 (1943).

Lax, B. *Science*, 141: 1247 (1963).

Lea, D. E. *Actions of Radiations on Living Cells.* Cambridge University Press, London and New York, 1956.

Lea, D. E., and R. B. Haines. *J. Hyg.*, 40: 162 (1940).

Leadbetter, E. R., and R. Whittenbury. *Proc. Natl. Acad. Sci. U.S.*, 50: 1128 (1963).

Lee, J., and H. H. Seliger. *J. Chem. Phys.*, 40: 519 (1964).

Leighton, W. G., and G. S. Forbes. *J. Am. Chem. Soc.*, 52: 3139 (1930).

Lennox, M. A. *J. Neurophysiol.*, 21: 62 (1958a).

Lennox, M. A. *J. Neurophysiol.*, 21: 70 (1958b).

Lerner, A. B., and Y. Takahashi. *Recent Progr. Hormone Res.*, 12: 303 (1956).

Lerner, A. B., J. D. Case, Y. Takahashi, T. H. Lee, and W. Mori. *J. Am. Chem. Soc.*, 80: 2587 (1958).

Levshin, V. L. Z. *Physik.*, 32: 307 (1925).

Lewis, G. N., and M. Kasha. *J. Am. Chem. Soc.*, 67: 994 (1945).

Liang, C. Y., and E. G. Scalco. *J. Chem. Phys.*, 40: 919 (1964).

Lichtenthaler, H. K., and M. Calvin. *Biochim. Biophys. Acta*, 79: 30 (1964).

Lichtenthaler, H. K., and R. B. Park. *Nature*, 198: 1070 (1963).

Lifschitz, J., and O. E. Kalberer. Z. *Physik. Chem.*, 102: 393 (1922).

Lindquist, L. *Arkiv. Kemi*, **16**: 79 (1960).

Linschitz, H. In *Light and Life* (W. D. McElroy and B. Glass, Eds.), p. 173. The Johns Hopkins Press, Baltimore, Maryland, 1961.

Linschitz, H., and E. W. Abramson. *Nature*, **172**: 909 (1953).

Livingston, R., W. F. Watson, and J. McArdle. *J. Am. Chem. Soc.*, **71**: 1542 (1949).

Losada, M., F. R. Whatley, and D. I. Arnon. *Nature*, **190**: 14 (1961).

Lundegårdh, H. *Nature*, **192**: 243 (1961).

Lyman, H., H. T. Epstein, and J. Schiff. *J. Protozool.*, **6**: 264 (1959).

Macaire, J. *J. Phys.*, **93**: 46 (1821).

McClung, F. J., and R. W. Hellwarth. *J. Appl. Phys.*, **33**: 828 (1962).

McClure, D. S. *Solid State Phys.*, **8**: 1 (1959).

McElroy, W. D. *Proc. Natl. Acad. Sci. U.S.*, **33**: 342 (1947).

McElroy, W. D. *Federation Proc.*, **19**: 941 (1960).

McElroy, W. D., and A. M. Chase. *J. Cellular Comp. Physiol.*, **38**: 401 (1951).

McElroy, W. D., and J. Coulombre. *J. Cellular Comp. Physiol.*, **39**: 475 (1952).

McElroy, W. D., and A. A. Green. *Arch. Biochem. Biophys.*, **56**: 240 (1955).

McElroy, W. D., and E. N. Harvey. *J. Cellular Comp. Physiol.*, **37**: 83 (1951).

McElroy, W. D., and J. W. Hastings. In *The Luminescence of Biological Systems* (F. H. Johnson, Ed.), p. 161. American Association for the Advancement of Science, Washington, D.C., 1955.

McElroy, W. D., and J. W. Hastings. *Physiological Triggers*, American Association for the Advancement of Science, Washington, D.C., 1956.

McElroy, W. D., and H. H. Seliger. In *Light and Life* (W. D. McElroy and B. Glass, Eds.), p. 219. The Johns Hopkins Press, Baltimore, Maryland, 1961.

McElroy, W. D., and H. H. Seliger. In *Horizons in Biochemistry* (M. Kasha and B. Pullman, Eds.), p. 91. Academic Press, Inc., New York, 1962a.

McElroy, W. D., and H. H. Seliger. *Sci. Am.* **207**: 76 (1962b).

McElroy, W. D., and H. H. Seliger. *Federation Proc.*, **21**: 1006 (1963a).

McElroy, W. D., and H. H. Seliger. *Advan. Enzymol.*, **25**: 119 (1963b).

McElroy, W. D., J. W. Hastings, J. Coulombre, and V. Sonnenfeld. *Arch. Biochem. Biophys.*, **46**: 399 (1953a).

McElroy, W. D., J. W. Hastings, V. Sonnenfeld, and J. Coulombre. *Science*, **118**: 385 (1953b).

McGlynn, S. P. In *Luminescence of Organic and Inorganic Materials* (H. P. Kallmann and G. M. Spruch, Eds.), p. 282. John Wiley and Sons, Inc., New York, 1962.

McLaren, A. D. *Atti Congr. Intern. Fotobiologia*, *2*, Torino, Italy, p. 123. 1957

McLaughlin, J. J., and P. A. Zahl. *Science*, **134**: 1878 (1961).

McMurry, H. *J. Chem. Phys.*, **9**: 231, 241 (1941).

MacNichol, E. F., Jr., M. L. Wolbarsht, and H. G. Wagner. In *Light and Life* (W. D. McElroy and B. Class, Eds.), p. 795. The Johns Hopkins Press, Baltimore, Maryland, 1961.

McRae, E. G., and M. Kasha. *J. Chem. Phys.*, **28**: 721 (1958).

McRae, E. G., and M. Kasha. In *Symposium on Fundamental Processes in Radiation Biology*, (L. G. Augenstein, R. Mason, and B. Rosenberg, Eds.), Academic Press, Inc., New York, 1964.

Maiman, T. H. *Nature*, **187**: 493 (1960).

Mallet, L. *Compt. Rend.*, **185**: 352 (1927).

Marcacci, A. *Arch. Ital. Biol.*, **9**: 2 (1888).

Marcovich, H., and R. Latarjet. *Advan. Biol. Med. Phys.*, **6**: 75 (1958).

Markert, R. E., B. J. Markert, and N. J. Vertrees. *Ecology*, **42**: 414 (1961).

Marks, W. B. Difference Spectra of the Visual Pigments in Single Goldfish Cones, Ph.D. Thesis, Johns Hopkins University, Baltimore, Maryland, 1963.

Marks, W. B., W. H. Dobelle, and E. F. MacNichol, Jr. *Science*, **143**: 1181 (1964).

Markwald, W. *Chem. Ber.*, **37**: 349 (1904).

Marriott, F. H. C., V. B. Morris, and M. H. Pirenne. *J. Physiol.*, **145**: 369 (1959).

Mason, B. *Principles of Geochemistry*, John Wiley and Sons, Inc., New York, 1960.

Mason, H. S. *Arch. Biochem. Biophys.*, **35**: 472 (1952).

Mast, S. O. *Light and the Behavior of Organisms*, John Wiley and Sons, Inc., New York, 1911.

Mattoon, R. W. *J. Chem. Phys.*, **12**: 268 (1944).

Mees, C. E. K. *The Theory of the Photographic Process*, p. 987. Macmillan Company, New York, 1946.

Meinschein, W. G., E. S. Barghoorn, and J. W. Schopf. *Science*, **145**: 262 (1964).

Melhuish, W. H. *New Zealand J. Sci. Technol.*, **B37**, 142 (1955).

Melhuish, W. H. *J. Phys. Chem.*, **65**: 229 (1961).

Melloni, M. *Poggendorff's*, **28**: 371 (1833).

Mikalonis, S. J., and R. H. Brown. *J. Cellular Comp. Physiol.*, **18**: 401 (1941).

Miki, K. *Japan J. Bacteriol.*, **11**: 803 (1956).

Miller, N. D. *Spring Meeting Opt. Soc. Am.*, *April, 1959*, Paper TD 73, 10 (1959).

Mortenson, L. E., R. C. Valentine, and J. E. Carnahan. *Biochem. Biophys. Res. Comm.*, **7**: 448 (1962).

Mott, N. F. *Phot. J.*, **88B**: 119 (1948).

Müller, A., D. C. Fork, and H. T. Witt. *Z. Naturforsch.*, **18b**: 142 (1963a).

Müller, A., B. Rumberg, and H. T. Witt. *Proc. Roy. Soc. London*, **B157**: 313 (1963b).

Müller, J. *Z. Vergleich. Physiol.*, **35**: 1 (1953).

Müller-Limmroth, H. W. *Elektrophysiologiedes Gesichtssiuns. Theorie und Praxisder Elektroretinographie*. Springer Verlag, Berlin, 1959.

Mulliken, R. S. *J. Phys. Chem.*, **56**: 801 (1952).

Mulliken, R. S., and H. McMurry. *Proc. Natl. Acad. Sci. U.S.*, **26**: 312 (1940).

Munz, F. W. The Photosensitive Retinal Pigments of Marine and Earyhaline Teleost Fishes. Ph.D. Thesis, University of California, Los Angeles, California, 1957.

Munz, F. W. *J. Gen. Physiol.*, **42**: 445 (1958).

Nakamoto, T., D. W. Krogman, and B. Vennesland. *J. Biol. Chem.*, **234**: 2783 (1959).

Nelson, D. F., and W. S. Boyle. *Appl. Opt.*, **1**: 181 (1962).

Newcombe, H. B., and H. A. Whitehead. *J. Bacteriol.*, **61**: 243 (1951).

Nicol, J. A. C. *J. Marine Biol. Assoc. U.K.*, **36**: 629 (1957).

Nicol, J. A. C. *J. Marine Biol. Assoc. U.K.*, **37**: 535 (1958).

Nicol, J. A. C. *J. Marine Biol. Assoc., U.K.*, **38**: 477 (1959).

Nicol, J. A. C. *J. Marine Biol. Assoc. U.K.*, **39**: 27 (1960).

Nishimura, M. *Biochim. Biophys. Acta*, **59**: 183 (1962a).

Nishimura, M. *Biochim. Biophys. Acta*, **57**: 88, 96 (1962b).

Nishimura, M., L. Takeru, and B. Chance. *Biochim. Biophys. Acta*, **59**: 177 (1962).

Novick, A., and L. Szilard. *Proc. Natl. Acad. Sci. U.S.*, **35**: 591 (1949).

Olson, J. M. *Science*, **135**: 101 (1962).

Olson, R. A., W. L. Butler, and W. H. Jennings. *Biochim. Biophys. Acta*, **54**: 615 (1961).

Olson, R. A., W. L. Butler, and W. H. Jennings. *Biochim. Biophys. Acta*, **58**: 144 (1962).

Oster, G., and A. H. Adelman. *J. Am. Chem. Soc.*, **78**: 913 (1956).

Oster, G., and G. K. Oster. In *Luminescence of Organic and Inorganic Materials*

(H. P. Kallmann and G. M. Spurch, Eds.), p. 186. John Wiley and Sons, Inc., New York, 1962.

Oster, G., J. S. Bellin, R. W. Kimball, and M. E. Schrader. *J. Am. Chem. Soc.*, **81**: 5095 (1959).

Padhye, M. R., S. P. McGlynn, and M. Kasha. *J. Chem. Phys.*, **24**: 588 (1956).

Park, R. B., and J. Biggins. *Science*, **144**: 1009 (1964).

Park, R. B., and N. G. Pon. *J. Mol. Biol.*, **3**: 1 (1961).

Park, R. B., and N. G. Pon. *Biochim. Biophys. Acta*, **57**: 520 (1962).

Park, R. B., and N. G. Pon. *J. Mol. Biol.*, **6**: 105 (1963).

Parker, C. A. *Proc. Roy. Soc. (London)*, **A220**: 104 (1953).

Parker, C. A. *Trans. Faraday Soc.*, **50**: 1213 (1954).

Pauling, L., and E. B. Wilson. *Introduction to Quantum Mechanics*, McGraw-Hill Book Company, New York, 1935.

Perrin, F. *Ann. Chim. (Phys.)*, **17**: 283 (1932).

Perrin, J., and F. Choucroun. *Compt. Rend. Acad. Sci.*, **189**: 1213 (1929).

Peticolas, W. L., J. P. Goldsborough, and K. E. Kieckoff. *Phys. Rev. Letters*, **10**: 43 (1963).

Petrikaln, A. *Z. Physik.*, **22**: 119 (1924).

Petrikaln, A. *Naturwissenschaften*, **12**: 205 (1928).

Pierce, S., and A. C. Giese. *J. Cellular Comp. Physiol.*, **49**: 303 (1957).

Pirenne, M. H. *Vision and The Eye*. Chapman and Hall, London, 1948.

Pirenne, M. H. *Ann. N.Y. Acad. Sci.*, **74**: 377 (1958).

Pirenne, M. H. In *The Eye* (H. Davson, Ed.), The Visual Process, Vol. 2, pp. 3–217. Academic Press, Inc., New York, 1962.

Pittendrigh, C. S., and V. Bruce. *Publ. Am. Assoc. Advan. Sci.*, **55**: 475 (1959).

Pittendrigh, C. S., V. Bruce, and P. Kaus. *Proc. Natl. Acad. Sci. U.S.*, **44**: 965 (1958).

Pittman, D., and P. R. Pedigo. *Exptl. Cell. Res.*, **17**: 359 (1959).

Platt, J. R. *J. Chem. Phys.*, **17**: 470 (1949).

Platt, J. R. In *Radiation Biology* (A. Hollaender, Ed.), p. 71. McGraw-Hill Book Company, New York, 1956.

Plesner, P. E. *Publ. Staz. Zool. Napoli*, **31**: 154 (1959).

Pohl, R. *Naturforsch*, **3b**: 367 (1948).

Polanyi, M., and G. Schay. *Z. Physik. Chem.*, **B1**: 21 (1928).

Polyak, A. *The Vertebrate Visual System*, University of Chicago Press, Chicago, Illinois, 1957.

Porter, G., and M. W. Windsor, *J. Chem. Phys.*, **21**: 2088 (1953).

Porter, G., and M. R. Wright. *Discussions Faraday Soc.*, **27**: 18 (1959).

Porter, K., and D. H. Volman. *J. Am. Chem. Soc.*, **84**: 2011 (1962).

Preston, J. S. Photometric Standards and the Unity of Light, *Natl. Phys. Lab. G. Brit. Notes Appl. Sci.*, **24** (1961).

Pringsheim, P. *Fluorescence and Phosphorescence*. John Wiley and Sons, Inc. (Interscience), New York, 1949.

Proceedings of the IEEE, special issue on quantum electronics, **51**: (1) (1963).

Raab, O. *Z. Biol.*, **39**: 524 (1900).

Rabinowitch, E. I. *J. Chem. Phys.*, **8**: 551 (1940).

Rabinowitch, E. I. *Photosynthesis and Related Processes*, Vol. I. John Wiley and Sons, Inc. (Interscience), New York, 1945.

Rabinowitch, E. I. *Photosynthesis and Related Processes*, Vol. II, Part I. John Wiley and Sons, Inc. (Interscience), New York, 1951.

Rabinowitch, E. I. *Photosynthesis and Related Processes*, Vol. II, Part 2. John Wiley and Sons, Inc. (Interscience), New York, 1956.

Rabinowitch, E. I., and L. F. Epstein. *J. Am. Chem. Soc.*, **69**: 63 (1941).
Radio Corporation of America, *Photosensitive Devices and Cathode-Ray Tubes*, Catalog CRP00105B, p. 19. Harrison, New Jersey, 1960.
Radziszewski, B. *Chem. Ber.*, **10**: 70 (1877a).
Radziszewski, B. *Chem. Ber.*, **10**: 321 (1877b).
Radziszewski, B. *Liebigs Ann.*, **203**: 305 (1880).
Randall, J. T., and M. H. F. Wilkins. *Proc. Roy. Soc. (London)*, **A184**: 366 (1945).
Rauhut, M. M., R. C. Hirt *et al. Chemiluminescent Materials* Technical Report No. 1. American Cyanamid Company, Stamford, Conn., August (1963a).
Rauhut M. M., R. C. Hirt *et al. Chemiluminescent Materials* Technical Report No. 2. American Cyanamid Company, Stamford, Conn., November (1963b).
Rauhut, M. M., R. C. Hirt *et al. Chemiluminescent Materials* Technical Report No. 3. American Cyanamid Company, Stamford, Conn., February (1964a).
Rauhut, M. M., R. C. Hirt *et al. Chemiluminescent Materials* Technical Report No. 4. American Cyanamid Company, Stamford, Conn., May (1964b).
Reid, C. *Excited States in Chemistry and Biology.* Academic Press, Inc., New York, 1957.
Renner, M. *Z. Vergleich. Physiol.*, **40**: 85 (1957).
Rentschler, H. C., R. Nagy, and G. Mouromseff. *J. Bacteriol.*, **41**: 745 (1941).
Rhodes, W. C., and W. D. McElroy. *Science*, **128**: 253 (1958a).
Rhodes, W. C., and W. D. McElroy. *J. Biol. Chem.*, **233**: 1528 (1958b).
Rieck, A. F., and S. Carlson. *J. Cellular Comp. Physiol.*, **46**: 301 (1955).
Riley, W. H., Jr. Ph.D. Dissertation, University of Michigan, Ann Arbor, Michigan, 1963.
Robertson, O. H. *Physiol. Zool.*, **24**: 309 (1951).
Robinson, G. W. *Proc. Natl. Acad. Sci. U.S.*, **49**: 521 (1963).
Rose, A. In *Photoconductivity Conference* (R. G. Breckenridge, B. R. Russel, and E. E. Hahn, Eds.), p. 3. John Wiley and Sons, Inc., New York, 1956.
Rosenberg, J. L., and D. J. Shombert. *J. Am. Chem. Soc.*, **82**: 3252 (1960a).
Rosenberg, J. L., and D. J. Shombert. *J. Am. Chem. Soc.*, **82**: 3257 (1960b).
Rothemund, P. *J. Am. Chem. Soc.*, **60**: 2005 (1938).
Rowan, W. *Nature*, **115**: 494 (1925).
Rupert, C. S. *Federation Proc.*, **17**: 301 (1958).
Rupert, C. S. *J. Gen. Physiol.*, **43**: 573 (1960).
Rupert, C. S., and R. M. Herriott. In *Progress in Photobiology*, (B. C. Christensen and B. Buchmann, Eds.), p. 311. Elsevier Publishing Company, Amsterdam, **1961.**
Rushton, W. A. H. *J. Physiol.*, **134**: 11 (1956).
Rushton, W. A. H. *Ann. N.Y. Acad. Sci.*, **74**: 291 (1958).
Rushton, W. A. H. In *Light and Life* (W. D. McElroy and B. Glass, Eds.), p. 721. The Johns Hopkins Press, Baltimore, Maryland, 1961.
Sager, R. *Brookhaven Symp. Biol.*, **11(BNL (C28))**: 101 (1959).
San Pietro, A., and H. M. Lang. *J. Biol. Chem.*, **231**: 211 (1958).
Santamaria, L. In *Recent Contributions to Cancer Research in Italy* (P. Bucalossi and U. Veronesi, Eds.), Vol. 1, p. 167, 1960.
Sauer, K., and M. Calvin. *J. Mol. Biol.*, **4**: 451 (1962).
Sawyer, R. A. *Experimental Spectroscopy.* Prentice Hall, New York, 1944.
Schawlow, A. L. *Sci. Am.*, **209**: 34 (1963).
Schawlow, A. L., and C. H. Townes. *Phys. Rev.*, **112**: 1940 (1958).
Scheibe, G. *Naturwissenschaften*, **26**: 412 (1938).

Scheibe, G. *Z. Angew. Chem.*, **52**: 631 (1939).

Schneider, W. G., and T. C. Waddington. *J. Chem. Phys.*, **25**: 358 (1956).

Schultze, M. *Arch. Mikrobiol. Anat.*, **2**: 175 (1866).

Seliger, H. H. *Anal. Biochem.*, **1**: 60 (1960).

Seliger, H. H. In *Light and Life* (W. D. McElroy and B. Glass, Eds.), p. 200. The Johns Hopkins Press, Baltimore, Maryland, 1961.

Seliger, H. H. *J. Gen. Physiol.*, **46**: 333 (1962).

Seliger, H. H. *J. Chem. Phys.*, **40**: 3133 (1964).

Seliger, H. H., and W. D. McElroy. *Biochem. Biophys. Res. Comm.*, **1**: 21 (1959).

Seliger, H. H., and W. D. McElroy. *Radiation Res.* Suppl. 2: 528 (1960a).

Seliger, H. H., and W. D. McElroy. *Arch. Biochem. Biophys.*, **88**: 136 (1960b).

Seliger, H. H., and W. D. McElroy. *Science*, **138**: 683 (1962).

Seliger, H. H., and W. D. McElroy. *Proc. Natl. Acad. Sci. U.S.*, **52**: 75 (1964).

Seliger, H. H., W. G. Fastie, and W. D. McElroy. *Science*, **133**: 699 (1961a).

Seliger, H. H., W. D. McElroy, E. H. White, and G. F. Field, *Proc. Natl. Acad. Sci. U.S.*, **47**: 1129 (1961b).

Seliger, H. H., W. G. Fastie, W. R. Taylor, and W. D. McElroy. *J. Gen. Physiol.*, **45**: 1003 (1962).

Seliger, H. H., J. B. Buck, W. G. Fastie, and W. D. McElroy. *Biol. Bull.*, **127**: 159 (1964a).

Seliger, H. H., J. B. Buck, W. G. Fastie, and W. D. McElroy. *J. Gen. Physiol.*, **48**: 95 (1964b).

Serono, C., and A. Ciuto. *Gazzetta*, **58**: 402 (1928).

Setlow, R. B., P. A. Swenson, and W. L. Carrier. *Science*, **142**: 1464 (1963).

Shimada, B. M. *Limnol. Oceanog.*, **3**: 336 (1958).

Shimomura, O., T. Goto, and Y. Hirata. *Bull. Chem. Soc. Japan*, **30**: 929 (1957).

Shimomura, O., F. H. Johnson, and Y. Saiga. *J. Cellular Comp. Physiol.*, **58**: 113 (1961).

Shimomura, O., F. H. Johnson, and Y. Saiga. *J. Cellular Comp. Physiol.*, **61**: 275 (1963).

Shimomura, O., J. R. Beers, and F. H. Johnson. *J. Cellular Comp. Physiol.*, **64**: 15 (1964).

Shin, M., K. Tagawa, and D. I. Arnon. *Federation Proc.*, **22**: 100; 589 (1963a).

Shin, M., K. Tagawa, and D. I. Arnon. *Biochem. Z.*, **338**: 84 (1963b).

Shugar, D. In *The Nucleic Acids* (E. Chargaff and J. N. Davidson, Eds.), Vol. 3, p. 39. Academic Press, Inc., 1960.

Shugar, D. In *Progress in Photobiology* (B. C. Christensen and B. Buchmann, Eds.), p. 315. Elsevier Publishing Company, Amsterdam, 1961.

Shugar, D., and J. Baranowska. *Nature*, **185**: 33 (1960).

Shugar, D., and K. L. Wierzchowski. *Biochim. Biophys. Acta*, **23**: 657 (1957).

Shugar, D., and K. L. Wierzchowski. *Postepy Biochem.*, **4** (Suppl.): 243 (1958).

Shurcliff, W. A. *J. Opt. Soc. Am.*, **45**: 399 (1955).

Sie, E. H. C., W. D. McElroy, F. H. Johnson, and Y. Haneda. *Arch. Biochem. Biophys.*, **93**: 286 (1961).

Simpson, O., G. B. B. M. Sutherland, and D. E. Blackwell. *Nature*, **161**: 281 (1948).

Singh, S., and B. P. Stoicheff. *J. Chem. Phys.*, **38**: 2032 (1963).

Sjöstrand, F. S. *Radiation Res.* Suppl. 2: 349 (1960).

Skreb, Y., and M. Errera. *Exptl. Cell Res.*, **12**: 649 (1957).

Smith, D. S. *J. Cell. Biol.*, **19**: 115 (1963).

Smith, R. A., F. E. Jones, and R. P. Chasmar. *The Detection and Measurement of Infra-red Radiation*. Oxford University Press, London and New York, 1960.

Smithsonian Physical Tables, 9th rev. ed. Smithsonian Institution, Washington, D.C., 1959.

Sogo, P. B., N. G. Pon, and M. Calvin. *Proc. Natl. Acad. Sci. U.S.*, **43**: 387 (1957).

Sommer, A. H. *IRE (Inst. Radio. Engrs.) Trans. Nucl. Sci.* **3**: 8 (1956).

Sperry, R. W. *Sci. Am.*, **194**: 48 (1956).

Spudich, J., and J. W. Hastings. *J. Biol. Chem.*, **238**: 3106 (1963).

Stark, J. *Z. Physik.*, **9**: 889, 894 (1908).

Steinach, E. *Arch. Ges. Physiol.*, **52**: 495, 498 (1892).

Stokes, G. G. In *Math and Physical Papers*, Vol. 2, p. 233. Cambridge University Press, London and New York, 1883.

Strehler, B. L. *Arch. Biochem. Biophys.*, **34**: 239 (1951).

Strehler, B. L. *Arch. Biochem. Biophys.*, **43**: 67 (1953).

Strehler, B. L., and W. Arnold. *J. Gen. Physiol.*, **34**: 809 (1951).

Strehler, B. L., and V. Lynch. *Arch. Biochem. Biophys.*, **70**: 527 (1957).

Strehler, B. L., and C. S. Shoup. *Arch. Biochem. Biophys.*, **47**: 815 (1953).

Strickler, S. J., and R. A. Berg. *J. Chem. Phys.*, **37**: 814 (1962).

Strutt, R. J. *Proc. Roy. Soc. (London)*, **A88**: 547 (1913).

Stuart, H. A. *Molekulstruktur*. Julius Springer, Berlin, 1934.

Svaetichin, G. *Acta Physiol. Scand.*, **39** (Suppl. 134): 17 (1956).

Sweeney, B. W. *Cold Spring Harbor Symp. Quant. Biol.*, **25**: 145 (1960).

Sweeney, B. W., and J. W. Hastings. *J. Protozool.*, **5**: 217 (1958).

Sweeney, B. W., and J. W. Hastings. *Cold Spring Harbor Symp. Quant. Biol.*, **25**: 87 (1960).

Szent-Gyorgyi, A. *Science*, **93**: 609 (1941).

Tagawa, K., and D. I. Arnon. *Nature*, **195**: 537 (1962).

Tagawa, K., H. Y. Tsujimoto, and D. I. Arnon. *Proc. Natl. Acad. Sci. U.S.*, **49**: 567 (1963).

Tappeiner, H. v. *Arch. Exptl. Pathol. Pharmakol.*, **51**: 383 (1900).

Tenney, L., and R. Heggie. *J. Am. Chem. Soc.*, **57**: 377 (1935a).

Tenney, L., and R. Heggie. *J. Am. Chem. Soc.*, **57**: 1622 (1935b).

Terenin, A., E. Putzeiko, and I. Akimov. *Discussions Faraday Soc.*, **27**: 83 (1959).

Terent'ev, A. P., and E. I. Klabunovskii. In *Aspects of the Origin of Life* (M. Florkin, Ed.), p. 74. Pergamon Press, New York, 1960.

Terpstra, W. *Biochim. Biophys. Acta*, **60**: 580 (1962).

Terpstra, W. *Biochim. Biophys. Acta*, **75**: 355 (1963).

Thimann, K. V., and G. M. Curry. In *Light and Life* (W. D. McElroy and B. Glass, Eds.), p. 646. The Johns Hopkins Press, Baltimore, Maryland, 1961.

Tollin, G. In *Radiation Res.* Suppl. 2: 387 (1960).

Tollin, G., and M. Calvin. *Proc. Natl. Acad. Sci. U.S.*, **43**: 895 (1957).

Tollin, G., D. R. Kearns, and M. Calvin. *J. Chem. Phys.*, **32**: 1013 (1960).

Tolmach, L. J. *Arch. Biochem. Biophys.*, **33**: 120 (1951a).

Tolmach, L. J. *Nature*, **167**: 946 (1951b).

Toole, E. H., S. B. Hendricks, N. A. Brothwick, and V. K. Toole. *Ann. Rev. Plant Physiol.*, **7**: 299 (1956).

Totter, J. R. *Photochem. Photobiol.* **3**: 231 (1964).

Totter, J. R., E. E. Dugros, and C. J. Rivera. *J. Biol. Chem.*, **235**: 1839 (1960a).

Totter, J. R., V. S. Medina, and J. L. Scoseria. *J. Biol. Chem.*, **235**: 238 (1960b).

Totter, J. R., W. Stevenson, and G. E. Philbrook. *J. Phys. Chem.*, **68**: 752 (1964).

Trautz, M., and P. Schorigen. *Z. Wiss. Phot.*, **3**: 80 (1905).

Tsuji, F. I. *Arch. Biochem. Biophys.*, **59**: 452 (1955).

Tsuji, F. I., and E. N. Harvey. *Arch. Biochem. Biophys.*, **52**: 285 (1954).
Tsuji, F. I., A. M. Chase, and E. N. Harvey. In *The Luminescence of Biological Systems* (F. H. Johnson, Ed.), p. 127. American Association for the Advancement of Science, Washington, D.C., 1955.
Udenfriend, S. *Fluorescence Assay in Biology and Medicine*. Academic Press, Inc., New York, 1962.
Unna, P. G. *The Histopathology of the Diseases of the Skin*, p. 724. Macmillan Co.. New York, 1896.
Valentinei, S., and M. Roessiger. *Chem. Ber.*, **16**: 210 (1924).
Valentinei, S., and M. Roessiger. *Z. Physik.*, **32**: 239 (1925).
Valentinei, S., and M. Roessiger. *Z. Physik.*, **36**: 81 (1926).
Van Niel, C. B. *Arch. Microbiol.*, **3**: 1 (1931).
Van Niel, C. B. In *Photosynthesis in Plants* (J. Franck and W. E. Loomis, Eds.), p. 437. Iowa State Press, Ames, Iowa, 1949.
Van't Hoff, J. H. *Die Lagerung der Atome in Ramme*, 3 Anfl. Braunschweig, 1908.
Van't Hoff, J. H., and H. M. Dawson. *Chem. Ber.*, **31**: 528 (1898).
Vartanyan, A. T., and I. A. Karpovich. *Dokl. Akad. Nauk SSSR*, **111**: 561 (1956).
Vavilov, S. I. *Phil. Mag.*, **43**: 307 (1922a).
Vavilov, S. I. *Z. Physik.*, **22**: 266 (1922b).
Vavilov, S. I *Z. Physik.*, **32**: 236 (1925).
Vavilov, S. I. *Z. Physik.*, **42**: 311 (1927).
Vennesland, B., T. Nakamoto, and B. Stern. In *Light and Life* (W. D. McElroy and B. Glass, Eds.), p. 609. The Johns Hopkins Press, Baltimore, Maryland, 1961.
Verrier, M. L. *Rescherches sur l'Histophysiologie de la Rétine des Vertebres*. Les Presses Universitaires, Paris, 1935.
Vishniac, W., and S. Ochoa. *Nature*, **167**: 768 (1951).
Vogel, H. *Chem. Ber.*, **6**: 1302 (1873).
Von Wettstein, D. *Brookhaven Symp. Biol.*, **11**(BNL(C28)): 138 (1959).
Wacker, A., H. Dellweg, and D. Weinblum. *Naturwissenschaften*, **47**: 477 (1960).
Wagner, H. G., E. F. MacNichol, Jr., and M. L. Wolbarsht. *J. Gen. Physiol.*, **43**: 45 (1960).
Wahl, O. *Z. Vergleich. Physiol.*, **16**: 529 (1932).
Wald, G. *Nature*, **132**: 316 (1933).
Wald, G. *J. Gen. Physiol.*, **18**: 905 (1935).
Wald, G. *Nature*, **139**: 1017 (1937a).
Wald, G. *Nature*, **140**: 545 (1937b).
Wald, G. *Science*, **101**: 653 (1945).
Wald, G. In *Documenta Ophthalmologica* (F. P. Fischer, A. J. Schaeffer, and A. Sorsby, Eds.), Vol. III, p. 94 Uitgeveri, Dr. W. Jank, 'S-Gravenhage, Netherlands, 1949.
Wald, G. *Sci. Amer.*, **183**: 32 (1950).
Wald, G. *Sci. Amer.*, **201**: 92 (1959).
Wald, G. In *Light and Life* (W. D. McElroy and B. Glass, Eds.), p. 742. The Johns Hopkins Press, Baltimore, Maryland, 1961.
Wald, G. Lecture presented at Woods Hole Marine Biological Laboratory, Massachusetts, July, 1963.
Wald, G. and P. K. Brown. *Science*, **127**: 222 (1958).
Wald, G., P. K. Brown, and P. H. Smith. *J. Gen. Physiol.*, **38**: 628 (1955).
Walther, J. B., and E. Dodt. *Experientia*, **13**: 333 (1957).
Walther, J. B., and E. Dodt. *Z. Naturforsch.*, **14b**: 273 (1959).
Wang, S. Y. *Nature*, **184** (Suppl. 4): 184 (1959a).

Wang, S. Y. *Nature*, **184**: 59 (1959b).
Warburg, E. *Sitzber.*, *Preuss. Akad.*, p. 314 (1916).
Warburg, E. *Sitzber*, *Preuss. Akad.*, p. 300 (1918).
Warburg, E. *Sitzber.*, *Preuss. Akad.*, p. 960 (1919).
Wareing, P. F. *Publ. Am. Assoc. Advan. Sci.*, **55**: 73 (1959).
Watts, B. N. *Proc. Phys. Soc. (London)*, **A62**: 456 (1949).
Weale, R. A. *J. Physiol.*, **127**: 572 (1955a).
Weale, R. A. *J. Physiol.*, **127**: 587 (1955b).
Weale, R. A. *J. Physiol.*, **132**: 257 (1956).
Weatherwax, R. S. *J. Bacteriol.*, **72**: 124 (1956).
Weber, G., and F. W. J. Teale. *Trans. Faraday Soc.*, **53**: 646 (1957).
Weber, G., and F. W. J. Teale, *Discussions Faraday Soc.*, **27**: 134 (1959).
Weber, J. *IRE (Inst. Radio Engrs.) Trans. Electron Devices* (1952).
Weber, J. *Rev. Mod. Phys.*, **31**: 681 (1959).
Weigert, F. *Z. Physik. Chem.*, **63**: 458 (1908).
Weigert, F. *Nernst Festschrift*, p. 464 (1912).
Weigert, F., and G. Käppler. *Z. Physik.*, **25**: 99 (1924).
Went, F. W. *Rec. Trav. Botan. Neerl.*, **25**: 1 (1928).
West, W. *J. Chim. Phys.*, p. 672 (1958).
West, W. *J. Phys. Chem.*, **66**: 2398 (1962a).
West, W. *Phot. Sci. Eng.*, **6**: 92 (1962b).
West, W., and B. H. Carroll. *J. Chem. Phys.*, **19**: 417 (1951).
Whatley, F. R., and M. Losada. In *Photophysiology* (A. C. Giese, Ed.), Vol. 1, p. 111. Academic Press, Inc., New York, 1964.
Whatley, F. R., M. B. Allen, and D. I. Arnon. *J. Am. Chem. Soc.*, **76**: 6324 (1954).
Whatley, F. R., K. Tagawa, and D. I. Arnon. *Proc. Natl. Acad. Sci. U.S.*, **49**: 266 (1963).
Whitaker, D. M. *J. Gen. Physiol.*, **24**: 263 (1941).
White, E. H. In *Light and Life* (W. D. McElroy and B. Glass, Eds.) p. 183. The Johns Hopkins Press, Baltimore, Maryland, 1961.
White, E. H., and M. M. Bursey. *J. Am. Chem. Soc.*, **86**: 941 (1964).
White, E. H., F. McCapra, G. F. Field, and W. D. McElroy. *J. Am. Chem. Soc.*, **83**: 2402 (1961).
White, E. H., O. Zafiriou, H. H. Kägi, and J. H. M. Hill. *J. Am. Chem. Soc.*, **86**: 940 (1964).
Wierzchowski, K. L., and D. Shugar. *Acta Biochim. Polon.*, **7**: 63 (1960).
Withrow, R. B. (Ed.) *Publ. Am. Assoc. Advan. Sci.* **55**: 439 (1959).
Witkin, E. M. *Proc. Natl. Acad. Sci. U.S.*, **32**: 59 (1946).
Witkin, E. M. *Proc. Natl. Acad. Sci. U.S.*, **50**: 425 (1963).
Wolken, J. J. *Trans. N.Y. Acad. Sci.*, **19**: 315 (1957).
Wolken, J. J. *Brookhaven Symp. Biol.* 11 (**BNL 512 (C-28)**): 87 (1959).
Wolken, J. J. In *The Structure of the Eye* (G. K. Smelser, Ed.) p. 173. Academic Press, Inc., New York, 1961.
Wright, M. R., and A. B. Lerner. *Nature*, **185**: 169 (1960).
Wulff, D. L., and C. S. Rupert. *Biochem. Biophys. Res. Comm.*, **7**: 237 (1962).
Wyckoff, R. W. G. *J. Exptl. Med.*, **52**: 435 (1930a).
Wyckoff, R. W. G. *J. Exptl. Med.*, **52**: 769 (1930b).
Wyckoff, R. W. G. *J. Exptl. Med.*, **51**: 921 (1930c).
Yariv, A., and J. P. Gordon. *Proc. IEEE*, **51**: 4 (1963).
Yoshida, M. *J. Exptl. Biol.*, **33**: 119 (1956).
Zwicker, E. F., and L. I. Grossweiner. *J. Phys. Chem.*, **67**: 549 (1963).

AUTHOR INDEX

Numbers in italics refer to pages on which the complete references are listed.

A

Abbott, M. T. J., 271, *371*
Abolin, L., 280, *371*
Abramson, A., 117, *379*
Abramson, E., 117, 118, *371*
Abramson, E. W., 156, *381*
Adelman, A. H., 103, 117, 330, *371*, *382*
Airth, R. L., 175, 197, *371*
Akimov, I., 141, *386*
Albrecht, H. O., 156, *371*
Allard, H. A., 250, *376*
Allen, M. B., 230, 244, *371*, *388*
Amesz, J., 239, 241, *371*, *375*
Anderson, R. S., 193, *371*
Androes, G. M., 142, 231, *373*
Arden, G. B., 292, *371*
Arnold, S. J., 166, *371*
Arnold, W., 140, 141, 143, 219, 236, *371*, *375*, *386*
Arnon, D. I., 212, 230–234, 238, 239, 243–245, *371*, *381*, *385*, *386*, *388*
Arthur, W. E., 141, *371*
Aschoff, J., 249, *371*
Audubert, R., 150, *371*
Autrum, H., 315, *371*
Avakian, P., 117, 118, *371*, *379*
Avron, M., 243, *371*, *379*
Azarraga, L., 142, *380*

B

Baly, E. C. C., 27, *371*
Baranowska, J., 339, *385*
Barber, H. S., 169, *371*
Barbrow, L. E., 13, 14, *371*
Barghoorn, E. S., 345, *382*
Bassham, J. A., 223, 231, 234, *371*, *373*
Baur, E., 127, *372*
Bawden, F. C., 336, *372*
Becker, R. S., 90, *372*
Becquerel, E., 131, *372*
Becquerel, H., 343, *372*

Beers, J. R., 196, *385*
Beinert, H., 161, *377*
Bellin, J. S., 329, 330, *372*, *383*
Belsky, T., 345, *375*
Bendall, F., 231, 240, *378*
Bendix, S., 274, *372*
Bennett, W. R., Jr., 349, *372*, *379*
Beredjik, N., 340, 341, *372*
Berends, W., 339, *372*
Berg, R. A., 81, *386*
Bergeron, J. A., 214, *372*
Bersohn, R., 112, *372*
Bertholf, L. M., 313, *372*
Beukers, R., 339, *372*
Beutler, H., 152, 153, *372*
Bie, V., *372*
Biggins, J., 223, *383*
Bishop, N. I., 241, *372*
Bitler, B., 172, *372*
Black, C. C., 245, 246, *372*, *376*
Blackwell, D. E., 36, *385*
Blinks, L. R., 223, *372*
Blum, H. F., 329, 334, *372*
Boll, F., 305, *372*
Bonhoeffer, K. F., 150, *372*
Borthwick, H. A., 253, *372*
Boudin, S., 103, *372*
Bowen, E. J., 96–98, 106, 358, *372*
Boyd, G. D., 349, *372*, *379*
Boyle, W. S., 349, 355, *382*
Brackett, F. P., 40, *372*
Bradley, D. F., 142, *372*
Braun, E., 343, *380*
Bremer, T., 154, 157, *372*
Bridges, C. D. B., 318, *372*
Briggs, W. R., 270, *373*
Brindley, G. S., 291, *373*
Brocklehurst, B., 106, *372*
Brown, F. A., 267, *373*
Brown, H., 123, *373*
Brown, P. K., 290, 318, 319, *373*, *387*
Brown, R. H., 184, *382*
Brown-Séquard, E., 277, *373*

389

Eguchi, S., 156, 193, *378*
Ehert, C. P., 248, *375*
Ehrenberg, A., 127, *375*
Einstein, A., 82, 349, *375*
Eley, D. D., 140, *375*
Emerson, R., 231, 236, 240, *375*
Emmons, C. W., 332, 333, *378*
Enami, M., 280, *375*
Engstrom, R. W., 37, *375*
Epstein, H. T., 336, *381*
Epstein, L. F., 100, *384*
Errera, M., 336, *385*
Evenari, M., 258, *375*
Eversole, R., 214, *375*
Eyring, H., 70, 161, *375*, *379*

F

Fabry, Ch., 43, *375*
Fastie, W. G., 169, 170, 179–181, 197, 265, 276, *385*
Feofilov, P. O., 108, *375*
Fiala, S., 326, *375*
Field, G. F., 156, 179, *385*, *388*
Fingerman, M., 280, *375*
Fischer, E., 145, 147, 148, *375*, *378*
Flint, L. H., 257, *375*
Förster, Th., 110, 112, 113, *376*
Forbes, G. S., 40, *372*, *380*
Fork, D. C., 236, 237, 240, 242, *382*
Forti, G., 244, *379*
Fox, D., 142, *376*
Fränz, H., 151, *379*
Franck, J., 83, 91, 105, 127, 138, 143, 231, 241, *373*, *376*
French, C. S., 225–227, 239, *376*
Frenkel, A. W., 230, *376*
Frenkel, J. I., 113, *376*
Fridovich, L., 160, *377*
Friedmann, H., 154, *372*
Fujimoto, K., 320, *377*
Fulton, H. R., 45, 322, 323, *374*
Fry, K. T., 245, *376*

G

Gaffron, H., 46, 223, 231, 234, 245, 327, 329, *376*
Galifret, Y., 291, *376*
Galston, A. W., 273, *376*

Garner, W. W., 250, *376*
Gates, F. L., 45, 46, 323, 332, 338, *376*
Gattow, G., 166, *376*
Gause, G. F., 344, *376*
Gaviola, E., 93, 109, *376*
Geiling, L., 34, *376*
Gest, H., 245, *376*
Ghosh, J. C., 121, *376*
Gibbs, M., 245, 246, *372*, *376*
Gibson, Q. H., 190, *378*
Giese, A. C., 333, 336, *377*, *383*
Gleu, K., 156, *377*
Goldsborough, J. P., 117, *383*
Goldsmith, T. H., 312–315, *377*
Good, N. E., 246, *380*
Goodgall, S. H., *377*
Gordon, J. P., 349, *377*, *388*
Gordon, S. A., 245, *372*
Goto, T., 193, *385*
Granit, R., 291, 293, 310, *374*, *377*
Green, A. A., 189, 191, 368, *377*, *381*
Green, D. E., 211, 212, *377*
Greenlee, L., 160, *377*
Griffin, D. R., 289, *377*
Griffith, R. O., 125, 305, *377*
Grossweiner, L. I., 127, 330, 331, *377*, *388*
Grove, J. F., 271, *371*
Gurney, R. W., 137, *377*
Guttman, N., 283, *377*

H

Haber, A. H., 258, *377*
Haber, F., 152, *377*
Haidinger, W., 301, *377*
Haines, R. B., 323, *380*
Halberg, F. J., 249, *377*
Hall, R. T., 85, 148, *377*
Halldal, P., 28, 274, 275, *377*
Hanaoka, T., 320, *377*
Handler, P., 160, 161, *377*
Haneda, Y., 182, 194, 198, *377*, *385*
Hanson, F. E., 184, *377*
Hargreaves, W. A., 349, *372*
Harvey, E. N., 156, 168, 182, 193, 196–198, 265, *377*, *381*, *387*
Hastings, J. W., 160, 176, 183, 189–191, 197, 198, 248, 260–264, 266, 267, *378*, *379*, *381*, *386*

SUBJECT INDEX

A

Absorption
dipole oscillator, 109
Absorption spectrum
1,2-naphthoquinone-2-diphenyl-
hydrazone, 147
shift upon adsorption, 136
Accessory pigments, 237
Achromobacter fischeri, 189
Acridine dyes, 74
Acridine orange, 74, 131
chromosome structural damage, 327
Acridone, fluorescence quantum yield, 95
Acriflavine, 74
amino groups, 78
fluorescence quantum yield, 95
Actinomycin D
effect of, 263
inhibition of rhythm of photosynthetic
capacity, 264
Action spectrum
Avena base curvature, 269
Avena coleoptiles, 271
Chlorella luminescence, 140, 141
Chlorella photosynthesis, 239
contraction in eel iris sphincter, 277,
278
effectiveness of short wavelength light,
231
Euglena phototaxis, 276
induction of mutations, 332
limulus nerve discharges, 283
phase shifting in *Gonyaulax*, 261
photoprotection, in *E. coli B*, 337
in *P. aeruginosa*, 337
photoreactivation of killing, in *E. coli
B/r*, 335
in *S. Griseus*, 335
phototactic response changes in *Platymonas*, 275
phototaxis in *Platymonas*, 275
Phycomyces sporangiophores, 271
promotion and inhibition of seed
germination, 250, 251

proper presentation of, 206
pyridine nucleotide reduction, 239
requirement of monochromaticity, 25
Sciurus carolinensis leucotis ERG
signals, 292
Activation energy, 97, 98
Activation energy for quenching, formula
for, 99
Active intermediate, in firefly reaction,
174
Active oxygen, 164, 166, 329, 332
Addition of light quanta, 117
Adenosine diphosphate
in crude firefly extracts, 176
formation of by light, 230
Adenosine triphosphate
assay of, 367
formation of by light, 230
requirement in firefly reaction, 182
Adenosine triphosphate formation
dark electron-transport process, 246
inhibition by CMU, 244
Aequorea aequorea, 193
requirement for calcium ions, 190
Aesculetin, 156
AgBr, energy levels, 137
Agrostis stolonifera, 241
Aldehyde, requirement for, 189
Algae, absorption spectra, 227
Alizarin, 74
Allowed transitions, characterization of,
80
α phosphorescence, 86, 87
Amebae, photoreactivation, 336
9-Aminoacridine, fluorescence quantum
yield, 95
Aminophthalic acid, 157
pH dependence of fluorescence
quantum yield, 158
Anacystis, 231, 239
Anemotaxis, 274
Anguilla, 280
Aniline dyes, 76
Anthocyanin, 255
stimulated by red light, 251

397

I

J

K

L

Labile level, 55
Lactuca sativa, action spectrum for seed germination, 251
Lagisca extenuata, 197
Lambert, 10, 293
Lambert law, 11, 12
Lamellar discs
  formation of, 217
Lamellae, fragments, 212
Lampyridae, 169
Lasers, 349–357
  four level, 354
  three level, 354
Latent image, 128
  formation, quantum yield, 135
Leaf color, effect of phosphate, 255
Leaf extension, stimulated by red light, 251
Leaf-reddening, phosphate-deficient soil, 256
Leaves, color changes in, 255
Left-hand circularly polarized light, 343
  definition, 303
  excess of due to circular dichroism, 344
  northern hemisphere, 345
Lemma perpusilla, flowering, 250
Lens
  color, 299
  selective absorption by, 290
  structural features, 297
Lepidium virginicum, action spectrum for seed germination, 251
Lettuce seed, germination, 251
Light
  biological action of, 206–346
  direct stimulation by, 277
  historical development, 2–5
  least mechanical equivalent of, 9, 12
Light absorption reaction, primary products of, 230
Light emission
  organisms, requirements for, 168
  sensitive assay system, 168
Light from the sun, biological implications, 41
Light intensities
  determination of, 361, 364
  ferrioxalate chemical actinometer, 364

  thermopile, 361
Light quanta, addition of, 117
Light receptor systems, organization and structure, 207
Light sensitized carcinogenicity of various substances, 328
Light sources
  brightness, 18
  efficiency, 18
  energy distributions and intensities, 17–22
Light-stimulated contraction, iris sphincter, 277
Limulus, action spectrum, 283
Line spectra
  mercury, 24
  neon, 24
  xenon, 24
Lithium
  electronic configuration of, 53
  L-luciferin, *see also* Luciferin
    oxidation with ferricyanide, 180
  L(+)-luciferin, *see also* Luciferin
    specific rotation, 181
Lock and key enzyme mechanism, 307
Lone-pair electrons, 63
Long day, 250
Lophine, 156
  chemiluminescence, 161
Luciferase, 193
  definition of, 168
  dual role in light emission, 176
  Renilla, 195
  rhythmic fluctuations, 262
Luciferin, 164, *see also* D-luciferin
  common to all fireflies, 188
  definition of, 168
  enantiomorphs, 172
  reaction with ATP, 172
  Renilla, 195
  rhythmic fluctuations, 262
  structure, 172
Luciferyl adenylate, 172
  chemiluminescence, 180
Lucigenin, 162
  chemiluminescence, 161
Lumen, 10
Lumichrome, 75
Lumiflavin, 75
Luminance, 293

DATE DUE